Micro-organisms

function, form and environment

Micro-organisms
function, form and environment

Second Edition

edited by

Lilian E. Hawker
M.Sc., Ph.D., D.Sc., D.I.C.
Emeritus Professor of Mycology, University of Bristol

and

Alan H. Linton
M.Sc., Ph.D., D.Sc., F.R.C.Path.
Reader in Veterinary Bacteriology,
University of Bristol

Edward Arnold

© Edward Arnold (Publishers) Ltd, 1979

First published 1971
by Edward Arnold (Publishers) Limited
41 Bedford Square, London WC1B 3DQ
Reprinted 1972, 1974
Second edition 1979
Reprinted 1981, 1985

Paper edition ISBN: 07131 2702 3

Filmset by Keyspools Ltd, Golborne, Lancashire
Printed in Great Britain by
Butler & Tanner Ltd,
Frome and London

Foreword

Micro-organisms are living things differing widely in form and life-cycle but resembling one another in their relatively small size, ranging from *ca.* 30 nm for particles of the smaller viruses and *ca.* 1.0 μm diameter for small bacteria to *ca.* 30×10 μm for certain unicellular algae. Since both the problems caused by small size and the methods used to study such small organisms are similar and since unrelated micro-organisms frequently occupy the same habitat and thus influence each other, it is convenient to study them within the same discipline i.e. *Microbiology*. This book is concerned with how micro-organisms live and multiply, their form and the roles they play in their varied habitats. Its aim is to survey the whole field of microbiology in a manner which we hope will be useful to students and specialist workers. It is an up-to-date account replacing two earlier books and while it contains some passages from these much is new.

We thank all those authors, editors and publishers who have provided illustrations. These are acknowledged in the figure captions. Also we are indebted to Helen Parker for preparing the index by computer.

L. E. Hawker
A. H. Linton

Bristol 1978

Contents

Contributors

S. A. Archer, B.Sc., Ph.D. *Lecturer in Plant Pathology, Department of Botany, Imperial College, London.*

A. Beckett, B.Sc., Ph.D. *Lecturer in Botany, Department of Botany, University of Bristol.*

F. W. Beech, B.Sc., Ph.D., D.Sc., F.R.I.C., F.I.Biol. *Reader in Microbiology and Deputy Director, Long Ashton Research Station, University of Bristol.*

R. C. W. Berkeley, B.Sc., Ph.D. *Lecturer in Bacteriology, Department of Bacteriology, University of Bristol.*

R. J. W. Byrde, B.Sc., Ph.D. *Reader in Plant Pathology, Long Ashton Research Station, University of Bristol.*

R. Campbell, B.Sc., M.S., Ph.D. *Lecturer in Microbiology within the Department of Botany, University of Bristol.*

J. G. Carr, B.Sc., Ph.D., D.Sc., F.I.Biol. *Reader in Microbiology, Long Ashton Research Station, University of Bristol.*

L. C. Frost, M.A., Ph.D. *Lecturer in Genetics within the Department of Botany, University of Bristol.*

C. J. Grant, M.A., D.Phil. *Senior Lecturer in Botany, Department of Botany, University of Bristol.*

L. W. Greenham, B.Sc., B.V.Sc., Ph.D., M.R.C.V.S. *Senior Lecturer in Veterinary Bacteriology and Virology, Department of Bacteriology, University of Bristol.*

Lilian E. Hawker, M.Sc., Ph.D., D.Sc. D.I.C., *Emeritus Professor of Mycology, Department of Botany, University of Bristol.*

A. J. Hedges, B.Sc., Ph.D. *Senior Lecturer in Bacteriology, Department of Bacteriology, University of Bristol.*

T. J. Hill, B.V.Sc., Ph.D., M.R.C.V.S. *Lecturer in Bacteriology, Department of Bacteriology, University of Bristol.*

A. H. Linton, M.Sc., Ph.D., D.Sc., F.R.C.Path. *Reader in Veterinary Bacteriology, Department of Bacteriology, University of Bristol.*

K. B. Linton, B.Sc., Ph.D., M.R.C.Path. *Senior Lecturer in Medical Bacteriology, Department of Bacteriology, University of Bristol.*

M. F. Madelin, B.Sc., Ph.D. *Reader in Mycology, Department of Botany, University of Bristol.*

M. H. Richmond, M.A., Ph.D., Sc.D. *Vice-Chancellor, University of Manchester.*

F. E. Round, B.Sc., Ph.D., D.Sc. *Professor of Phycology, Department of Botany, University of Bristol.*

M. A. Sleigh, B.Sc., Ph.D. *Professor of Zoology, University of Southampton.*

D. C. Smith, M.A., D.Phil., F.R.S. *Sibthorpian Professor of Rural Economy, Department of Agricultural and Forest Sciences, University of Oxford.*

G. Turner, B.Sc., Ph.D. *Lecturer in Microbiology, Department of Bacteriology, University of Bristol.*

G. C. Ware, M.A., Ph.D. *Lecturer in Bacteriology, Department of Bacteriology, University of Bristol.*

Part I

Biochemistry and physiology of micro-organisms

Chapter 1

Macromolecules in micro-organisms

Introduction

It is customary to think of bacterial macromolecules solely in terms of single and readily identifiable high molecular weight polysaccharides, lipids, nucleic acids, and proteins, but many more complex situations exist in microbial cells. There are intricate molecular arrangements in which two or more macromolecules are held together in specific ways by complementary ionic and hydrogen bonds. Much of the cell seems to be made up of chemical structures that are more than just an intimate mixture of two types of macromolecule joined together by stable covalent linkages. A good example of this situation is the mucopeptide (sometimes referred to as *peptidoglycan*: the terms are synonymous) that forms the rigid structural matrix of the cell wall of many bacterial species. In this molecule there is close cross-linking between polysaccharide and polypeptide chains in such a way that it is no longer possible to define the limits of the molecule accurately. Thus, with molecules of this type, such concepts as molecular weight and size cease to have much meaning. As far as mucopeptide is concerned, the entire bacterial cell seems to be enclosed by a single sack-like molecule that might not even have a similar repeating structure over the whole of its area.

Although such complex macromolecules do pose problems as far as isolation and identification are concerned, it is possible to consider simple polysaccharides, lipids, proteins and nucleic acids as single molecular entities that can be defined with some precision. In the following sections, therefore, the structures of some examples of each of these four groups will be considered before attention is turned to bacterial mucopeptide as an example of one of the more complex and less well-defined types of microbial macromolecule.

Polysaccharides

All simple polysaccharides have a repeating structure made up of either a single type or alternating types of monosaccharide. One of the simplest is amylose, a polysaccharide which is found in many types of micro-organism and in which the molecule consists of glucose units joined together to form a long chain (Fig. 1.1). The bonds linking the structure join the C_1 atom of one glucose residue to the C_4 atom of the next by an oxygen bridge, often known as a *glycoside bond*. The use of these bonds with each sugar residue leads to a molecule in which all the monosaccharide units have an identical linkage save the first and the last. In the first, the hydroxyl group on C_4 is unsubstituted, while on the last the —OH group on C_1 is in a similar state (Fig. 1.1).

The amylose molecule (Fig. 1.1) is one of the simplest types of polysaccharide, since it is made up solely from a single repeating glucose unit and all the bonds linking the monosaccharides are identical. In many other polysaccharides, however, two

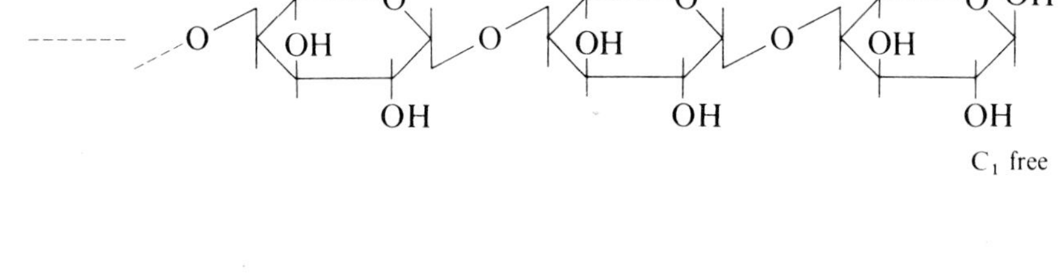

Fig. 1.1 Overall structure of amylose (poly-α-1:4-glucose). The free C_1 end of the molecule is responsible for the reducing properties.

different types of monosaccharide are involved in the chain. An example of such a molecule is the polysaccharide found in the Type III capsular material from certain strains of pneumococci. In this case the structure consists of chains of alternating glucose and glucuronic acid residues arranged as shown in Fig. 1.2

A further, and more complex, class of polysaccharides are those that are branched. Branching occurs where a monosaccharide residue in the chain is attached to three other monosaccharide units

rather than to two, as in any straight-chain polysaccharide. An example of this type of molecule is the dextran molecule synthesized by *Leuconostoc mesenteroides*. This molecule, as in all dextrans (see Table 1.1), is made up entirely from glucose units which, in this case, are joined by a glycoside bond between C_1 of one glucose residue and C_6 of the next, rather than between C_1 and C_4 as with amylose. The branching points in the *Leuconostoc* dextran are introduced where occasional glycosidic bonds are formed between C_1 and C_4 or between C_1

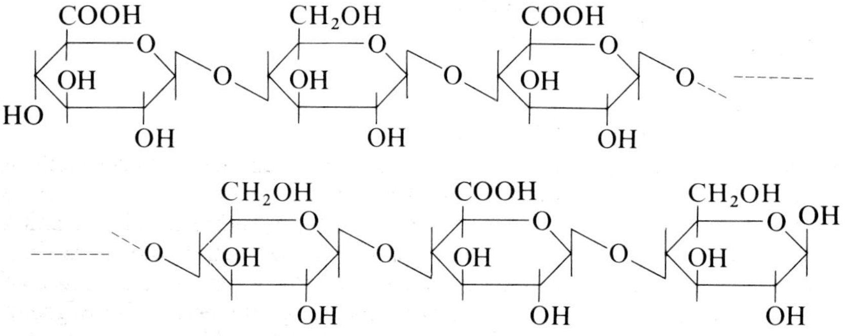

Fig. 1.2 The overall structure of the polysaccharide from the Type III capsular material found in some pneumococci.

Table 1.1 Trivial names and chemical constitution of various microbial polysaccharides

Trivial name	Constitution	Source
Glucan (general term)	Poly-glucose	Many yeasts and bacteria
Dextran	Poly-1:6-glucose	*Leuconostoc* and many other microbial species
Mannan (general term)	Poly-mannose	Yeasts
Amylose	Poly-α1:4-glucose	Many bacteria
Chitin	Poly-β1:4-*N*-acetyl-glucosamine	Fungi
Cellulose	Poly-β1:4-glucose	*Acetobacter xylinum*

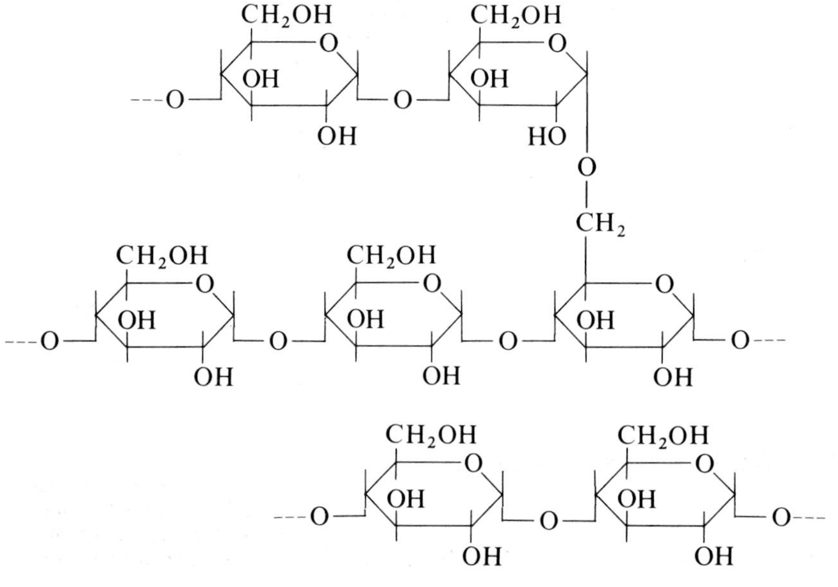

Fig. 1.3 A part of the structure of a branched polysaccharide to show the molecular nature of the branching points.

and C_3 of two glucose residues rather than between C_1 and C_6 (see Fig. 1.3).

Mucopolysaccharides

The term mucopolysaccharide is often used for those polysaccharides containing residues of *N*-acetyl-amino sugars such as *N*-acetyl-glucosamine or *N*-acetyl-galactosamine (Fig. 1.4).

One of the simplest mucopolysaccharides, and undoubtedly one of the most important in many moulds since it forms a great part of their cell walls, is *chitin*. This molecule consists of unbroken chains of *N*-acetyl-glucosamine linked from the C_1 of one residue to the C_4 of the next. In point of fact this polymer is usually present in an organism in association with protein, but whether there are a small number of covalent bonds between the two or whether the mixture is stabilized solely by ionic and hydrogen bonds, is unclear at the moment.

Not all mucopolysaccharides contain amino-sugars unmixed with non-nitrogenous monosaccharides. As an example, the mucopolysaccharide component of the Type XIV capsular material from pneumococci contains *N*-acetylglucosamine, glucose and galactose in the overall molecular arrangement shown in Fig. 1.5. The various general names used to describe simple polysaccharides usually give some idea of their composition and a list of the most common of these names is given in Table 1.1.

Lipids

All lipids share a common property of being soluble in fat solvents, such as chloroform or ether, and almost insoluble in water. A wide range of compounds with different structures fall within this classification and it is convenient to consider lipids under the following groups:
1 Fatty acids
2 Triglycerides
3 Phospholipids, phosphatidic acids and glyco-lipids
4 Steroids
5 Carotenoids

Fatty acids

Fatty acids are long-chain monocarboxylic acids of the general formula R.COOH. Usually the R-group has a long chain, commonly unbranched, consisting of an even number of between 8 and 24 carbon atoms. The fatty acid may be described as *saturated* or *unsaturated* depending on whether the R-group contains any double bonds; acids with one double bond are known as mono-enoic, those with two as di-enoic, and so on. The structure of a number of fatty acids and their trivial names is given in Table 1.2. Little more need be said about their structure at this stage since their greatest importance, apart from the role of the lower examples as metabolic

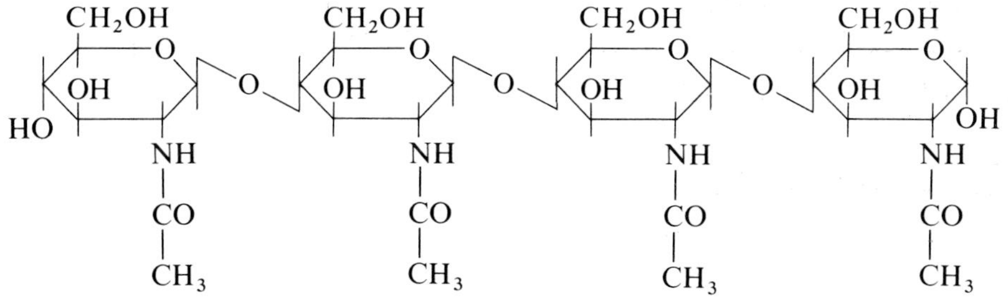

Fig. 1.4 The overall structure of chitin.

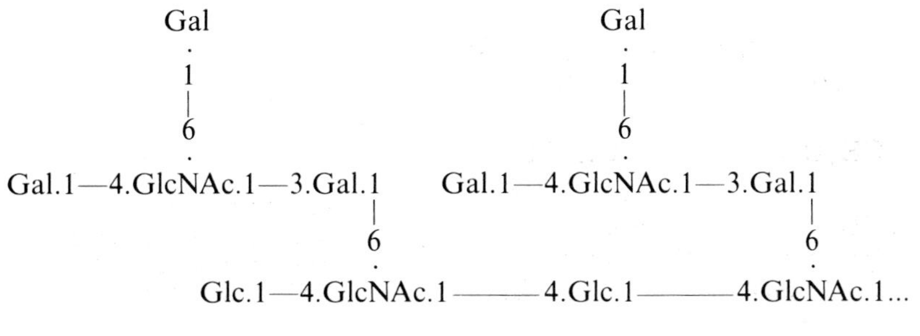

Fig. 1.5 The overall structure of the polysaccharide from the Type XIV capsular material found in some pneumococci.

Table 1.2 Trivial names of some of the various fatty acids to be found in microbial cells

Molecular formula	Common name	Systematic name
Saturated acids:		
$C_4H_8O_2$	Butyric	n-Butanoic
$C_6H_{12}O_2$	Caproic	n-Hexanoic
$C_{10}H_{20}O_2$	Capric	Decanoic
$C_{12}H_{24}O_2$	Lauric	Dodecanoic
$C_{14}H_{28}O_2$	Myristic	Tetradecanoic
$C_{16}H_{32}O_2$	Palmitic	Hexadecanoic
$C_{18}H_{36}O_2$	Stearic	Octadecanoic
$C_{20}H_{40}O_2$	Arachidic	Eicosadecanoic
$C_{24}H_{48}O_2$	Lignoceric	Tetracosanoic
Unsaturated acids:		
$C_{16}H_{30}O_2$	Palmitoleic	Hexadec-9-enoic
$C_{18}H_{34}O_2$	Oleic	Octadec-9-enoic
Doubly unsaturated acids:		
$C_{18}H_{32}O_2$	Linoleic	Octadeca-9,12-dienoic
$C_{18}H_{30}O_2$	Linolenic	Octadeca-9,12,15-trienoic
$C_{20}H_{32}O_2$	Arachidonic	Eicosa-5,8,11,14-tetraenoic

intermediates, is to form part of the higher molecular weight lipids described in the next two sections.

Triglycerides

Triglycerides consist of molecular complexes of glycerol with fatty acids and the generalized structure of all such molecules is:

$$H_2C.O.OC.R_1$$
$$R_2.CO.O.CH$$
$$H_2C.O.OC.R_3$$

In view of the absence of charged groups in the molecule, these structures are sometimes known as neutral fats to distinguish them from the phosphatidic acids described below. In practice, some triglycerides have the same R-group in all positions, but in others two or three different R-groups may be represented in a single molecule.

It should be emphasized that fats as isolated from bacterial cells are unlikely to consist of single molecular entities, largely because of the wide variety of different fatty acid residues involved and the consequent difficulty of separating molecules of such close molecular similarity.

Phospholipids

These compounds, like triglycerides, are also

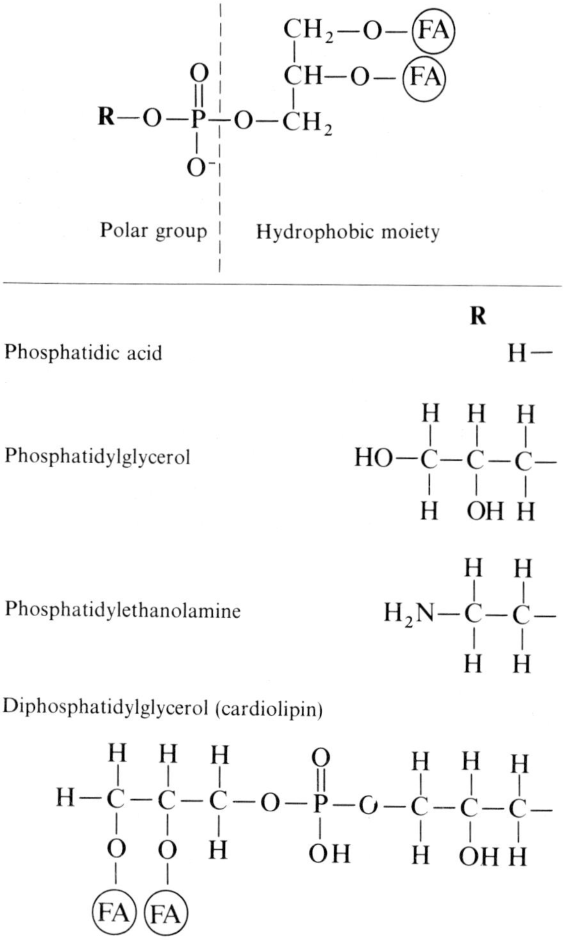

Fig. 1.6 Structures of some phospholipids from bacterial membranes. FA = fatty acid.

substituted glycerols, but all contain phosphorus; many also contain nitrogen. Most of the examples found in bacteria fall into the group known as phosphatidic acids and have the generalized structure:

$$H_2C.O.OC.R_1$$
$$R_2.CO.O.CH$$
$$H_2C.O.P.O\text{-substituent}$$

where R_1 and R_2 are fatty acids and the 'substituent' may be choline, ethanolamine, serine, or inositol (Fig. 1.6 for examples). Where the substituent is choline, the phosphatidic acids are known as *lecithins*.

Steroids

Steroids are unknown in bacteria but certain derivatives are found in fungi where they may be essential for spore formation. One such compound is ergosterol (Fig. 1.7) and this is of particular industrial importance since chemical modification of this compound in the laboratory allows the synthesis of many medicinally important steroids on a commercial scale (p. 366).

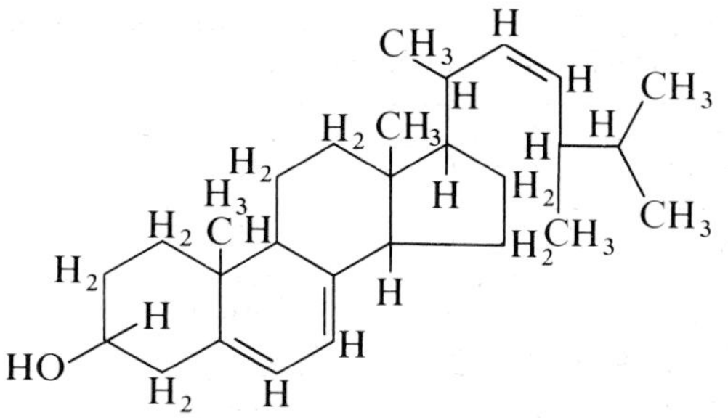

Carotenoids

Carotenoids are a group of lipid substances which have a characteristic orange or red colour. They are compounds responsible for the pigmentation of carrots (hence their name), and also of many bacterial species, particularly the staphylococci and micrococci. All are molecular variants of β-carotene whose structure is as follows:

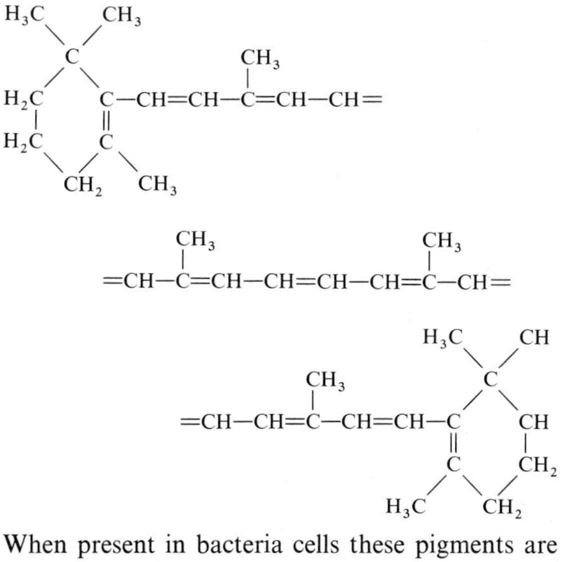

When present in bacteria cells these pigments are usually found in the cell membrane.

Proteins

Proteins may be classified into two types on the basis of their composition. *Simple* proteins consist solely of amino acids while *conjugated* proteins are structures of the same type but with one or more additional molecules attached. This additional molecule is usually known as the *prosthetic group* and is often involved in the physiological function of the protein, particularly if it is an enzyme, and in the immunological specificity of an antigen (p. 330). For example, electron transport is mediated by flavine prosthetic groups in the dehydrogenases and

Fig. 1.7 Ergosterol—a steroid with important medicinal uses that is produced by some fungi.

pyridoxal phosphate is present as a prosthetic group in many transaminases and amino acid racemases.

In both simple and conjugated proteins, the amino acid part of the structure consists of one or more unbranched polypeptide chains formed from amino acid residues joined together by peptide bonds (Fig. 1.8). Although twenty different types of amino acid may be found in such peptide chains, the common basic structure of all (namely $H_2N.CHR.COOH$, where R is one of 20 different chemical residues—see Table 1.3) ensures that the whole polypeptide chain can be formed by joining the amino acids together with a single type of bond. Examination of a number of proteins has shown that not all amino acids are present in the same molecular proportions in each; indeed their distribution may vary quite widely and some may even be missing completely. Although it is impossible to generalize, proteins often contain relatively small quantities of tryptophan, methionine and cysteine but large amounts of the non-polar amino acids valine, leucine and iso-leucine.

Many proteins consist of only one polypeptide chain (which may vary in length from 20 to about 300 residues in different proteins) but many others contain more than one chain, and some as many as six or eight. In such molecules the individual chains

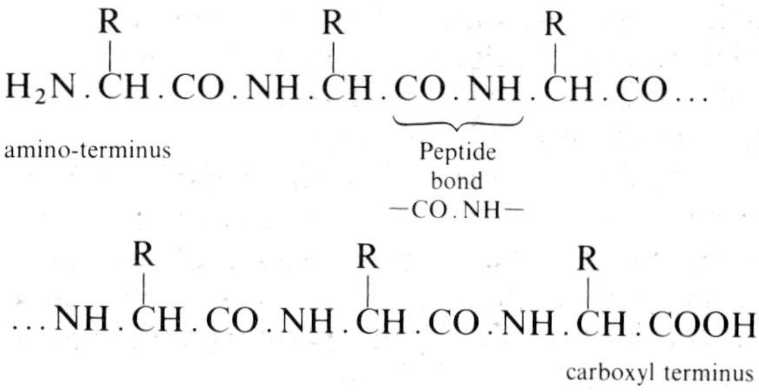

Fig. 1.8 The overall structure of a simple polypeptide chain as found in many proteins. The free —NH₂ residue at the left hand end is the *amino terminus*, and the free —COOH at the other end is the *carboxyl terminus*.

Table 1.3 The sidechain substituents (R-groups) of the amino acids commonly found in microbial proteins

General structure: $H_2N.CHR.COOH$

Amino acid	(abbreviation)	R-group
Glycine	(gly)	H—
Alanine	(ala)	CH_3—
Valine	(val)	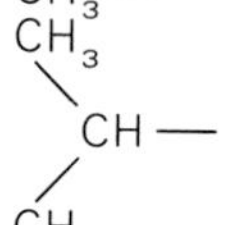 (CH(CH_3)(CH_3))—
Leucine	(leu)	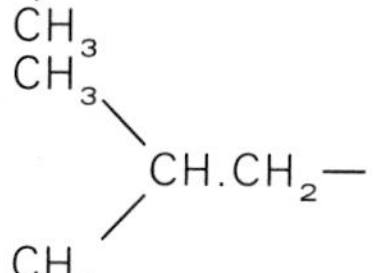 (CH(CH_3)(CH_3)).CH_2—
Isoleucine	(ileu)	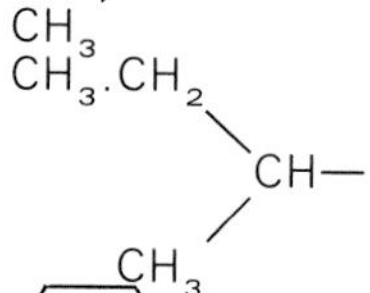 (CH(CH_3.CH_2)(CH_3))—
Phenylalanine	(phe)	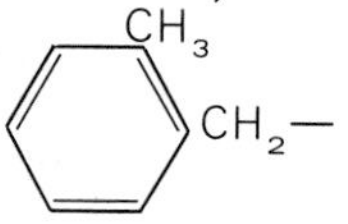 (C₆H₅).CH_2—
Tyrosine	(tyr)	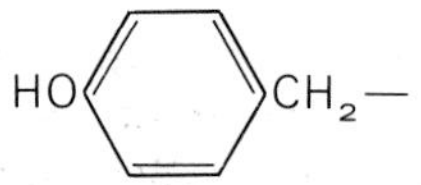 HO.(C₆H₄).CH_2—
Threonine	(thr)	$CH_3.CH(OH)$—
Serine	(ser)	$HO.CH_2$—
Cysteine	(cys)	$HS.CH_2$—
Methionine	(met)	$CH_3.S.CH_2.CH_2$—
Tryptophan	(try)	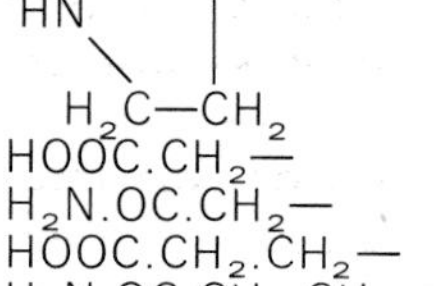 (indole ring).CH_2—
Proline	(pro)	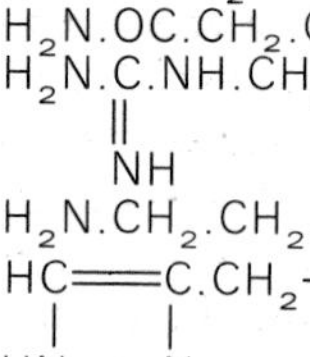 (ring: COOH—HC—CH_2—H_2C—CH_2—HN)
Aspartic acid	(asp)	$HOOC.CH_2$—
Asparagine	(asn)	$H_2N.OC.CH_2$—
Glutamic acid	(glu)	$HOOC.CH_2.CH_2$—
Glutamine	(gln)	$H_2N.OC.CH_2.CH_2$—
Arginine	(arg)	$H_2N.C(\,{=}NH).NH.CH_2.CH_2$—
Lysine	(lys)	$H_2N.CH_2.CH_2.CH_2.CH_2$—
Histidine	(his)	(HC=C.CH_2—, HN N, CH ring)

Note: The proline molecule, being an imino acid, does not fall into the general structural classification of amino acids; the structure shown here is the whole molecule and not just the R-group.

are held together by molecular bridges formed between the thiol groups of two cysteine residues, one in each chain (see Fig. 1.9). Such cross connexions are often referred to as —S.S— (disulphide) bridges and usually occur to the extent of about one to two bridges to every 100 residues, although considerable variation in this value is found.

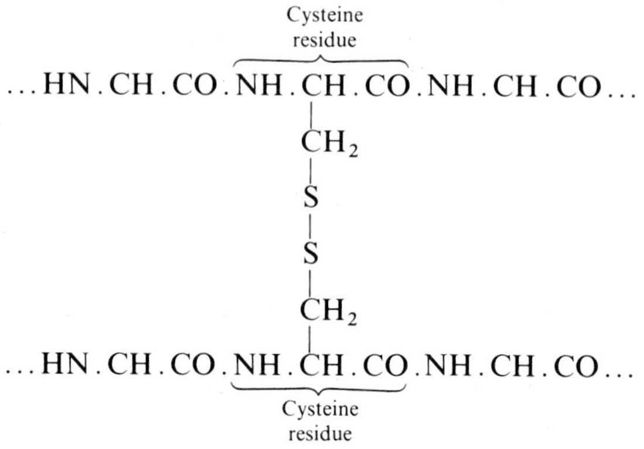

Cysteine residue

$...HN.CH.CO.NH.CH.CO.NH.CH.CO...$
CH_2
S
S
CH_2
$...HN.CH.CO.NH.CH.CO.NH.CH.CO...$

Cysteine residue

Fig. 1.9 The molecular nature of the —S.S— cross bridges that link different polypeptide chains (and sometimes two points on the same chain) in proteins.

Although all proteins are similar in that they consist of polypeptide chains that may be cross-linked with —S.S— bridges, this basic structure admits an enormous variety of detailed structure and, consequently, of detailed properties. This variety depends to a certain extent on the number and length of the polypeptide chains, but chiefly on the type and distribution of the individual amino acid residues along the chains themselves. The chemical nature of the R-groups found in the twenty different types of amino acid (see Table 1.3) show a wide variation in ionic properties, in polarity and in molecular shape (to mention only a few characteristics) and this ensures that each polypeptide chain has a correspondingly wide range of detailed chemical properties depending on the exact distribution of different types of amino acids along its length.

Although the order and nature of the amino acids along the polypeptide chains together with the position of the cross-bridges between the chains (the so-called *primary* and *secondary* structure of the protein) are important factors in the nature of the protein, the crucial element that determines the properties of the molecule is the way in which the structure folds up and this seems to be determined by the amino acid sequence of the polypeptide chains themselves. As an example, Fig. 1.10 shows

Amino terminus:

1

Lys . Val . Phe . Gly . Arg . Cys . Glu . Leu . Ala . Ala . Ala ...

... Met . Lys . Arg . His . Gly . Leu . Asp . Asn . Tyr . Arg . Gly . Tyr ...

2

... Ser . Leu . Gly . Asn . Try . Val . Cys . Ala . Lys . Phe . Glu . Ser ...

... Asn . Phe . Asn . Thr . Gln . Ala . Thr . Asn . Arg . Asn . Thr . Asp ...

... Gly . Ser . Thr . Asp . Try . Gly . Ileu . Leu . Gln . Ileu . Asn . Ser ...

3

... Arg . Try . Try . Cys . Asp . Asn . Gly . Arg . Thr . Pro . Gly . Ser ...

4   3

... Arg . Asn . Leu . Cys . Asn . Ileu . Pro . Cys . Ser . Ala . Leu . Leu ...

4

... Ser . Ser . Asp . Ileu . Thr . Ala . Ser . Val . Asn . Cys . Ala . Lys ...

... Lys . Ileu . Val . Ser . Asp . Gly . Met . Asn . Ala . Try . Val . Ala ...

2

... Try . Arg . Asn . Arg . Cys . Lys . Gly . Thr . Asp . Val . Gln . Ala ...

1

... Tyr . Ileu . Arg . Gly . Cys . Arg . Glu .

Carboxyl terminus.

Fig. 1.10 The primary and secondary sequence of egg-white lysozyme. Note that in this molecule there is only one polypeptide chain which is bridged at intervals by —S.S— bridges. There are four such bridges linking the molecule at intervals; they are formed between the pairs of cysteine residues labelled 1, 2, 3 and 4 respectively.

the primary and secondary structure of egg-white lysozyme, while Fig. 1.11 shows how this structure folds up to form the final, or *tertiary*, structure of

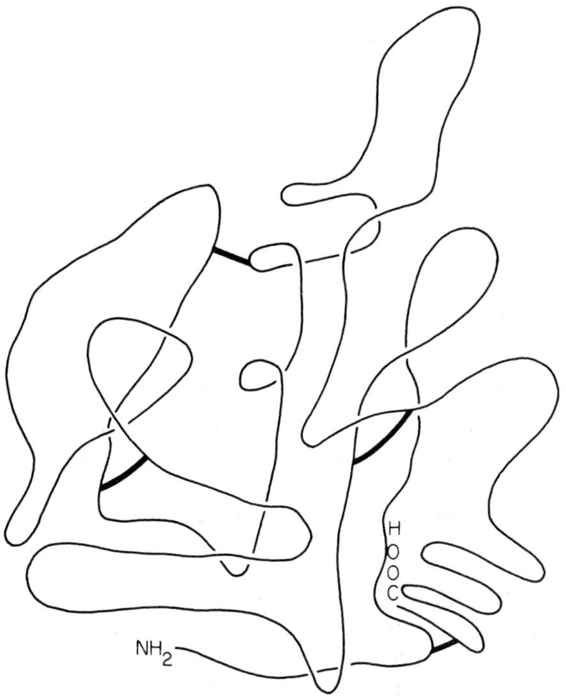

Fig. 1.11 The tertiary folding of the 129 residues of egg-white lysozyme shown in Fig. 1.10. The —S.S— cross bridges are shown as solid bars linking the chain.

the molecule. The nature of this folding of the chains is vitally important since the overall properties of the molecule are determined by the nature of the amino acid chains that are brought into structural juxtaposition with one another and with the external environment in the folded structure and not by the sequence of the amino acids in any one chain.

Protein subunits

Although some proteins consist of one or more distinct polypeptide chains joined together by —S.S— bridges, others may be formed by the interaction of two or more subunits. In some cases the subunits have an identical structure but in others two, or even three, different subunits are grouped together to form a molecule. One type of protein that usually consists of subunits is any enzyme that is subject to feed-back inhibition (see p. 18). In this case one subunit is the enzyme *sensu stricto* and is responsible for recognizing and metabolizing the substrate. The other subunit recognizes the inhibitor, which is usually the product of the pathway. The inhibitory action is mediated by interaction between the subunits in such a way that reaction of one subunit with the inhibitor reduces the action of the other against the normal substrate. Such an effect transmitted between subunits is known as an *allosteric* effect.

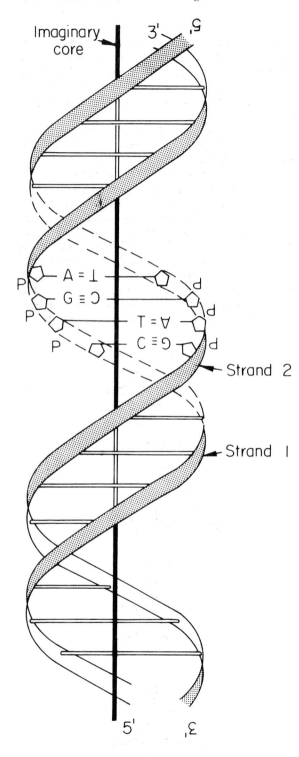

Fig. 1.12 The overall arrangement of residues in the anti-parallel double strand of a DNA molecule. A=adenine, G=guanine, C=cytosine, T=thymine, and P=phosphate.

Nucleic acids

Deoxyribonucleic acid (DNA)

DNA is an extremely long chemical thread made up of two strands, but unlike most threads, the strands are not twisted round one another but wound together round an imaginary core to form a double spiral or, more accurately, 'double helix' (Fig. 1.12). Each thread of the molecule consists of a chain of deoxyribose (Fig. 1.13) and phosphate residues arranged alternately as shown in Fig. 1.14. This arrangement ensures that each individual sugar residue in the chain is joined to two phosphate groups, one substituted on the 3-hydroxyl of the sugar and the other on the 5'-OH group, and that all the phosphate groups in the strand are doubly substituted except the first and the last. Thus one end of a DNA strand carries a single substituted phosphate on C-3 of a deoxyribose residue and the other end has a similar phosphate on the C-5 of the sugar. Thus any chemical change, such as an enzyme reaction, that passes along one of the DNA strands can be thought of as moving in a direction that is either $3 \to 5$ or $5 \to 3$. This concept of direction in a DNA strand is important when one comes to examine the overall structure of the molecule, for the two strands are arranged to lie in the double helical thread so that one lies in the $5 \to 3$ direction and always faced by a strand running $3 \to 5$ (see Fig. 1.11). For this reason, therefore, DNA is said to exist in the form of an *antiparallel double helix*.

In addition to the substitution at the 3 and the 5 positions, each sugar residue in a DNA strand is also substituted at C-1 by one of a number of purine or pyrimidine bases. The bases commonly found in

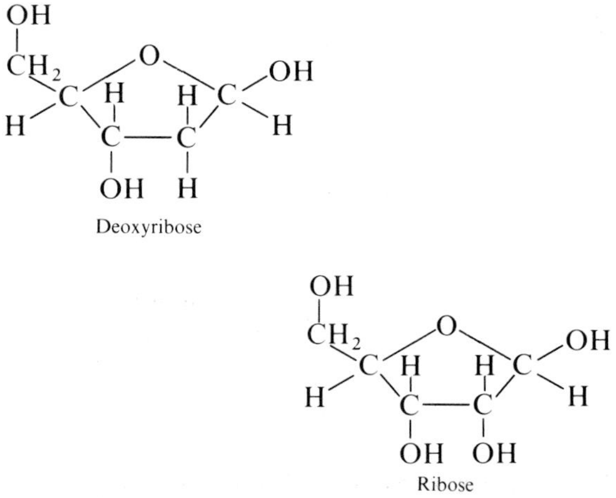

Deoxyribose

Ribose

Fig. 1.13 The structure of deoxyribose and ribose sugars.

DNA are the purines, adenine (A) and guanine (G), and the pyrimidines, cytosine (C) and thymine (T) (Fig. 1.15). Examination of the overall structure of DNA shows that the double-stranded nature of the molecule is maintained by hydrogen bonding between the purine and pyrimidine bases in such a way that an adenine residue is always faced by a thymine while guanine is faced by cytosine (Fig. 1.16). Thus adenine and guanine are said to be a *complementary base pair*, and share two hydrogen bonds, while guanine and cytosine are a similar complementary pair and share three hydrogen bonds.

The fact that there is always an adenine opposite a thymine and a guanine opposite a cytosine in the molecule means that there is an equal amount of adenine and thymine in DNA and the same is true of guanine and cytosine. However, the ratio of each pair of bases (that is $[A+T]/[G+C]$) varies widely from one bacterial species to another. Analysis of the composition of DNA from a number of bacteria shows that the $[A+T]/[G+C]$ ratio may vary from as much as 70/30 in *Clostridium perfringens (welchii)* at one extreme to 28/72 in *Micrococcus lysodeikticus* and 26/74 in *Actinomyces bovis* (Table 1.4). This wide variation in DNA base composition is confined to bacteria and similar relatively

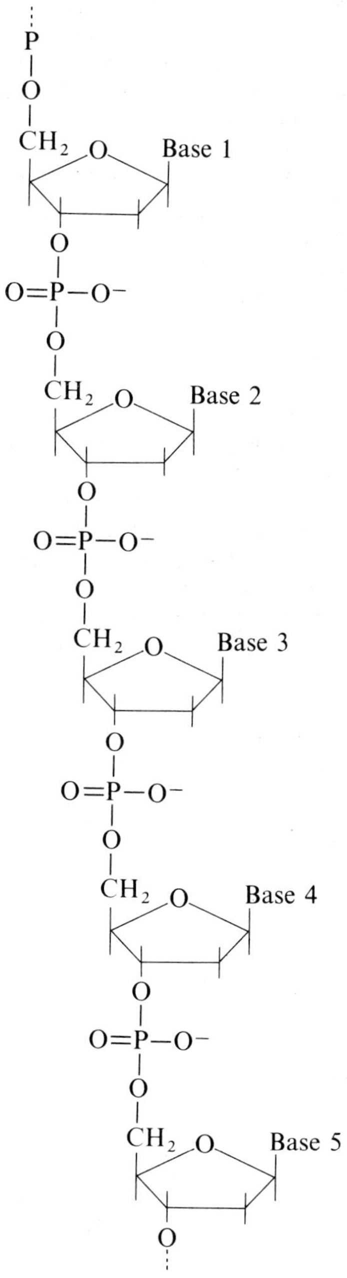

Fig. 1.14 A typical polynucleotide sequence as found in RNA and DNA. In RNA the sugar is *ribose*, while in DNA it is *deoxyribose*.

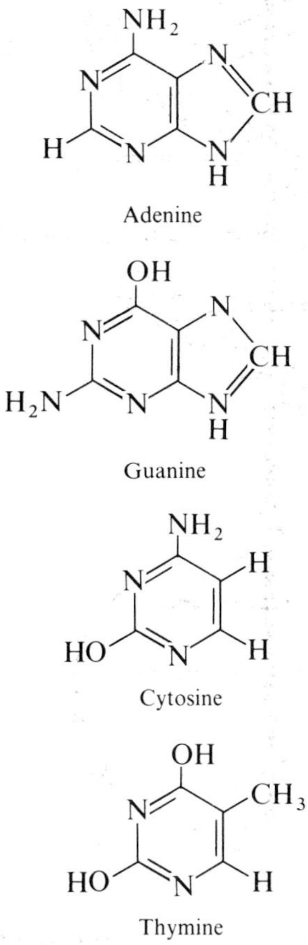

Fig. 1.15 The structure of adenine, guanine, cytosine, and thymine.

primitive forms of life. All vertebrates have a $[A+T]/[G+C]$ ratio of about 60/40 and this is usually taken to indicate a relatively close evolutionary relationship among the vertebrates when compared with bacteria.

It must be stressed that much of what has been

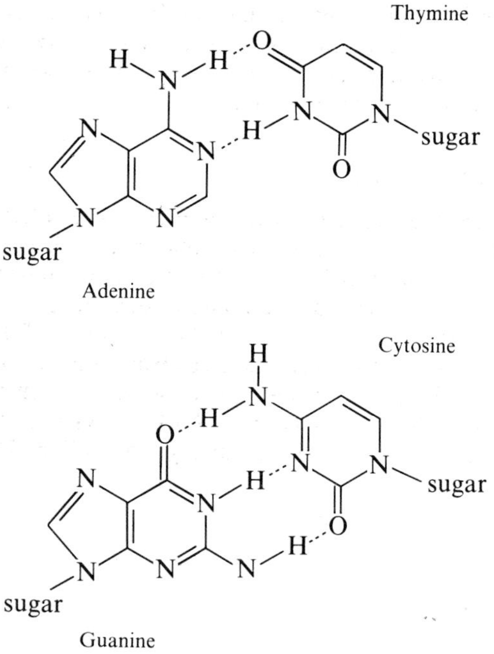

Fig. 1.16 The hydrogen bonding between adenine and thymine and between guanine and cytosine to form the base pairs that constitute part of the DNA double helical structure.

Table 1.4 Percentage of (guanine +cytosine) base pairs in the DNA of some bacteria and rickettsias

Organisms	%G +C	Organisms	%G +C
Clostridium perfringens (welchii)	30	Escherichia coli	52
		Salmonella typhi	54
Rickettsia prowazekii	32	Aerobacter aerogenes	56
Staphylococcus aureus	34	Brucella abortus	58
			60
Proteus vulgaris	36	Pseudomonas aeruginosa	62
Diplococcus pneumoniae	38		64
	40		66
Bacillus subtilis	42	Mycobacterium tuberculosis	68
Rickettsia burnetii	44		70
	46	Micrococcus lysodeikticus	72
Corynebacterium acne	48	Actinomyces bovis	74
Neisseria gonorrhoeae	50		

described in relation to DNA structure has been determined by X-ray crystallography. In the cell the chromosome must be extensively super-coiled. The molecular weight of the DNA of the bacterial chromosome is about 2×10^9, and this amount of DNA, if stretched out, would be about 1 millimetre long, that is about 500 times as long as the cell that contains it.

SINGLE-STRANDED DNA

Although the great majority of the DNA in a bacterial cell is double-stranded, and this is true of most bacteriophages (p. 126) as well, a class of phage containing single-stranded DNA does exist. These are usually very small in size and constitute some of the most simple organisms yet detected.

Ribonucleic acids (RNA)

Superficially RNA has considerable structural similarity to DNA. The molecule has a long chain of alternating sugar and phosphate residues, although the monosaccharide residue is ribose (Fig. 1.13) and not deoxyribose as in DNA. As in DNA, the sugar residues are double substituted, one bond being to the 3'-position of the molecule and the other to the 5'. Once again, the ends of an RNA strand can be distinguished, one by having a free 5' phosphate and the other by a free 3' phosphate.

In RNA the C-1 position of the sugar is substituted with purine and pyrimidine bases, with DNA, but unlike DNA the bases are adenine, guanine, cytosine and *uracil* rather than adenine, guanine, cytosine and *thymine*. In general the incidence of complementary base pairing is a great deal less in RNA than in DNA, and where it occurs adenine pairs with uracil rather than with thymine (see Figs. 1.16 and 1.17).

Despite a basically similar overall chemical structure, three distinct types of RNA can be

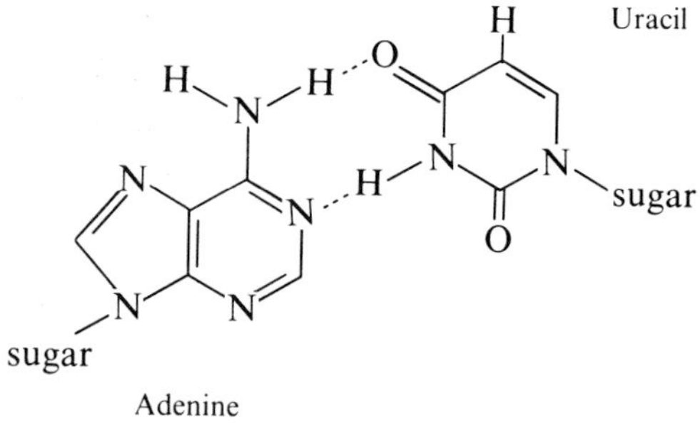

Fig. 1.17 Base pairing of adenine and uracil.

distinguished on the basis of structure and function: *messenger* RNA (or m-RNA), *ribosomal* RNA, and *transfer* RNA (or t-RNA).

MESSENGER RNA (M-RNA)

Messenger RNA is invariably single-stranded and contains only the single purine and pyrimidine bases adenine, guanine, cytosine, and uracil: no base is further substituted (see below). A very large number of different m-RNAs exist in each cell and the variety in structure arises from the fact that the order of bases along the RNA strand is a complementary copy of a stretch of DNA (see Chapter 2). In all cases, the molecules differ solely in their length (which is usually between about 300 and 7000 to 10 000 residues) and in the order of the purine and pyrimidine bases in the strand.

RIBOSOMAL RNA

About 40 per cent of the total mass of each ribosome consists of RNA. Chemical studies show that some, at least, of the material is double-stranded and that some contains substituted purine and pyrimidine bases in place of the normal residues on the C-1 of the ribose residues (see discussion of the structure of t-RNA in Chapter 2). The exact structure of one component of ribosomal RNA, the so-called 5s component, has recently been determined but as yet no function for this molecular species of ribosomal RNA has been discovered, nor is the structure of any other part of the ribosomal RNA known in great detail.

TRANSFER RNA (t-RNA)

There are probably a hundred or more different types of transfer RNA molecule in each bacterial cell and all probably represent relatively minor but important variants of a common structure. All are very small as RNA molecules go (about 70–80 nucleotide residues seems to be an average size) and share the polyribosephosphate structure common to all RNAs. This type of RNA shows a certain limited amount of base pairing, but unlike DNA

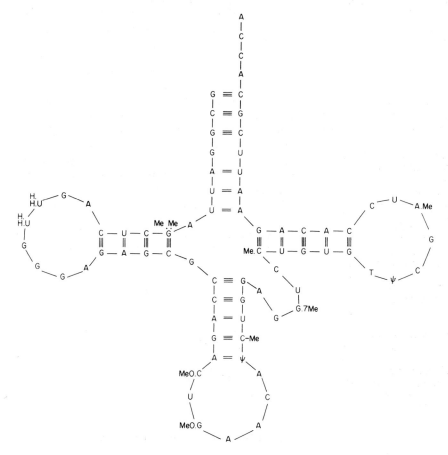

Fig. 1.18 The polynucleotide sequence of phenylalanine t-RNA from yeasts. Abbreviations: A=adenine, C=cytosine, G=guanine, T=thymine, U=uracil, 2Me.G=2-methyl-guanine, diH.U=dihydrouracil, Me.O.G.=O-methyl-guanine, Me.O.C.=O-methyl-cytosine, ψ=pseudouridine, Me.C=methyl cytosine, G.7Me=7-methyl-guanine and A.Me=methyl-adenine.

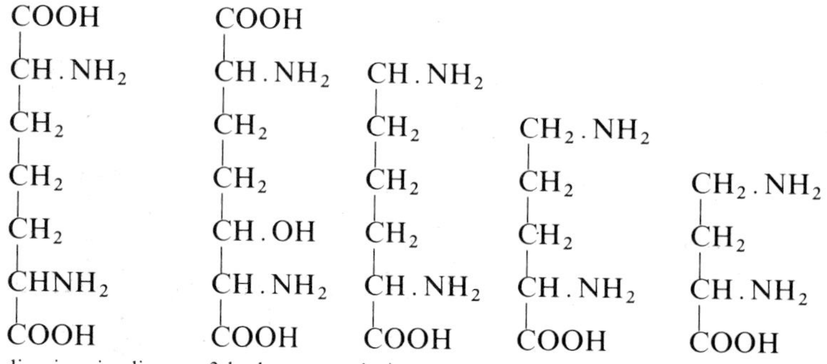

diaminopimelic acid (DAP) | 3-hydroxy-diamino-pimelic acid | lysine | ornithine | 2:4 diamino butyric acid

Fig. 1.19 The structure of certain amino acid components of some bacterial mucopeptides.

(where the pairing always occurs between two distinct DNA strands, p. 10) in transfer RNA a single strand turns back on itself to allow internal base pairing to occur. However, this base pairing does not extend over the whole t-RNA molecule, nor is it uninterrupted, and the result is a key-shaped molecule of the type shown in Fig. 1.18. The detailed structure of t-RNA molecules is described in greater detail in Chapter 2 where the function of these molecules in protein synthesis is also discussed.

Mucopeptide

Whereas the macromolecules that have been described so far are all structures with well-defined limits, this is not so in the case of the mucopeptides (peptidoglycans:mureins) that form the rigid part of the bacterial cell wall, particularly in Gram-positive bacteria. As mentioned previously these compounds consist of relatively enormous sack-shaped molecular nets with a fairly well-defined repeating composition but without any clearly defined limits.

Mucopeptide consists of a linear chain of alternating β1–4 linked *N*-acetyl-glucosamine and *N*-acetylmuramic acid residues. On some or all of the *O*-lactyl groups of the latter are substituted tetrapeptide 'tails' in which there is an alternation of L- and D-amino acids (Fig. 1.19). In the third position where a diamino acid usually occurs, the L-centre of this acid is involved in the peptide linkage. Some of the variations known to occur in the peptide are illustrated in Fig. 1.20. Variation also occurs in the polysaccharide structure 6-*O*-phosphorylated, *N*-glycolylamide-*N*-acetylated *N*-acetyl muramic acid residues, for example, have been identified in some mucopeptides. The number of disaccharide units linked together is of the order of 100, thus the length of polysaccharide backbone is about 100 nm, which precludes a radial arrangement of the chain in the wall but does not permit a conclusion as to whether a circumferential or longitudinal arrangement exists in rods. Whichever is the actual configuration, mucopeptide forms an enveloping macromolecular network which is not, however, static but is constantly being modified by the controlled action of autolytic enzymes to allow

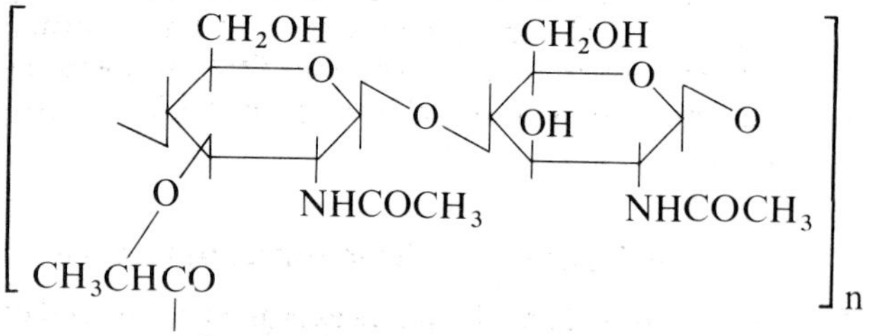

L-ala (gly, L-ser)
D-glu (3-hyg, gly, $glyNH_2$, D-$alaNH_2$)
M-DAP (L-lys, M-DAP, LL-DAP, L-orn, L-hser, etc.)
D-ala

Fig. 1.20 Some variations known to occur in the peptide moiety of the disaccharide–peptide repeating unit of bacterial cell wall mucopeptide. The amino acids in brackets may replace the corresponding 'tail' amino acids or be substituted on to it (After Schleifer and Kandler, 1972)

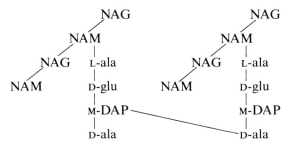

Fig. 1.21 Mucopeptide chemotype I in which direct linkage between peptide 'tails' occur, e.g., in *E. coli*.

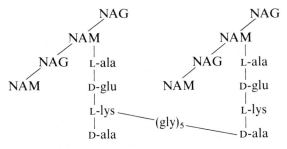

Fig. 1.22 Mucopeptide chemotype II in which linkage is *via* a short peptide, e.g., as in *Staph. aureus*.

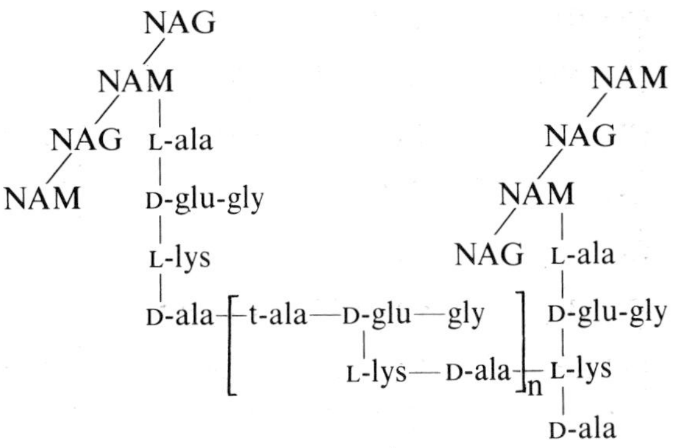

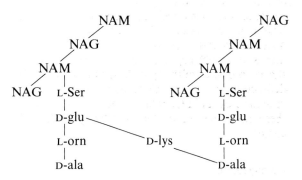

Fig. 1.23 Mucopeptide chemotype III in which the bridge consists of one or several peptides of the same composition as the peptide 'tail', e.g., as in *M. lysodeikticus*.

Fig. 1.24 Mucopeptide chemotype IV in which a diamino acid forms a bridge between free carboxyl groups on D-glutamate and the D-alanyl residues; no diamino acid occurs in the 'tail', e.g., *Corynebacterium poinsettiae*.

growth and cell division.

The biggest variation in gross structure and rigidity of the mucopeptide depends on the degree of substitution of its *N*-acetylmuramic acid residues by tetrapeptide tails and the way in which these are cross-linked. Four different chemotypes representing different cross-linking arrangements, may be recognized (Figs. 1.21, 22, 33 and 24).

Conclusion

In summary, it should be stressed that the preceding account is a rather superficial description of some of the macromolecules that go to make up bacterial cells. But although the molecules described here, or some representatives of them, may account for up to 80 per cent of the dry weight of the cell it does not follow that relatively minor macromolecular components are unimportant to cell life and survival. Were this so, natural selection would probably have seen to it that the molecules had ceased to be produced a long time ago. Typical examples of macromolecules that have been studied a great deal,

but which are not described here for lack of space, are the teichoic and teichuronic acids of bacterial cell walls (p. 151), the poly-D-glutamic acid from the capsules of many *Bacillus* spp. (p. 146), and the complex mucopolysaccharides that are found in the capsular material of various pneumococcal strains (p. 146). For details of these molecules, and many others, it will be necessary to consult a formal textbook of biochemistry or one of the monographs dealing with various classes of bacterial macromolecules.

Books and articles for reference and further study

An integrated list of references is given at the end of Chapter 2, p. 39.

Chapter 2

Biosynthesis and metabolism in Micro-organisms

Introduction

When a bacterial cell grows and divides, the final outcome is that there are now two cells where there was one before. For this process to be completed, therefore, every atom and molecule in the parent cell has to be duplicated and then inserted into its correct place in the developing structure that will eventually become the mature daughter cell.

As described in Chapter 1, bacterial cells are made up largely of macromolecules (proteins, nucleic acids, lipids, polysaccharides, etc.) but these components are always accompanied by a proportion of low molecular weight compounds that are either destined themselves to become part of the new generation of macromolecules, or to be used catalytically in macromolecular biosynthesis, or to take part in the energy metabolism of the cell.

In practice all the atoms that end up in a daughter cell have to be taken up from the environment surrounding the growing parental cell, but there is great variation in the complexity of the molecules that can provide these atoms. In the extreme case the organisms assimilate all the atoms they need in the form of simple inorganic molecules (CO_2, NH_4^+ ions, etc.). Such organisms are said to be *autotrophic*, and their properties are dealt with later in this chapter. In other cases more complex molecules, such as amino acids, sugars, fatty acids and nucleic acid precursors are taken up preformed from outside the cell. Bacteria that behave in this way are called *heterotrophs*. Only rarely do bacteria take up preformed macromolecules from the surrounding environment.

The uptake of preformed molecules by bacteria

The bacterial cell membrane (see Chapter 8) is a semi-permeable structure, and as such it provides a selective barrier to the uptake of molecules from outside the cell. In many cases molecules are excluded completely and cannot therefore act as a source of material for cell growth. With other compounds, on the other hand, entry is by simple diffusion, the internal concentration being de-termined partly by the external concentration of the compound, and partly by the rate of utilization within the cell. A third class of molecules are actively pumped into the interior of the cell using specific energy-coupled uptake mechanisms. The catalytic proteins which carry out this function are known as *permeases*, and are located in the bacterial membrane. These proteins are invariably specific to single compounds, for example, a single sugar or amino acid.

Once inside the cell the molecules that have been taken up in this way may be used for metabolism, and their concentration within the cell at any one time is determined by the balance between the rate of uptake and the rate of utilization. Normally this internal concentration of free small molecules within the bacterium is kept to a minimum. Nevertheless, it is detectable and is known as the free 'pool' of molecules. Thus there is an 'amino acid pool', a 'nucleotide pool', and so on.

Biosynthesis of small molecules

Any molecules that cannot be taken up preformed from outside the bacterial cell have to be synthesized within it, and the processes involved can conveniently be considered under the following headings:
1 Biosynthesis of the low molecular weight organic molecules
2 Biosynthesis of the macromolecules themselves
3 Provision of energy in a convenient molecular form to achieve the necessary biosynthesis.
We will now consider each of these topics in turn.

Biosynthesis of low molecular weight organic molecules

The various types of macromolecules that go to make up a bacterial cell commonly consist of repeating units (see Chapter 1). Sometimes identical units are involved—as in amylose—but at the other end of the scale of complexity proteins consist of chains of 20 different amino acids arranged in a specific order yet joined by a common type of bond.

Repeating structures of this kind have many advantages from the point of view of biosynthesis. In the case of amylose, the biosynthesis of the glucose, or its uptake from outside the cell, need only be followed by a single repetitive reaction to build up the straight-chain poly-glucose molecule we know as amylose. With protein synthesis the process, although more complex, still only requires the provision of the individual amino acids followed by a standard reaction which couples them together through peptide linkages. What sets protein (and nucleic acid) synthesis apart from the biosynthesis of other macromolecules is the need to get the individual units in the right order in the chain. The first requirement for the biosynthesis of all macro-molecules, therefore, is a supply of the individual units that go to make up the molecule in question.

When one comes to examine the overall patterns of the biosynthetic pathways in a micro-organism, it is clear that they have evolved to produce as economical an arrangement as possible. Normally the chemical modification introduced at each step is small. Thus in the pathway leading to the aromatic amino acids, for example, reaction 4 (Fig. 2.1) involves the removal of 2 H atoms, reaction 5 the addition of a phosphate residue, reaction 6 a molecular rearrangement, and so on. All are simple changes.

Two further examples of economy are shown in Figs. 2.1 and 2.2. In Fig. 2.1, the pathway branches; that is the early steps are concerned in the production of three different amino acids, all needed to synthesize proteins. In Fig. 2.2 the early enzymes handle two distinct but related molecules, thus effectively doing the job that might be expected to require two different enzymes.

Perhaps the greatest economy can be seen in the pathway shown in Fig. 2.3. Here the biochemical interactions involved in the synthesis of a number of protein amino acids from glucose is outlined, and much of the pathway is common to all the amino acids concerned. Furthermore this pathway is an example of a 'cycle', which is itself an economy, since it feeds back to re-use those parts of the precursors not needed in the product itself. These are only a few of the many biosynthetic pathways to be found in bacterial cells. Further details should be sought in a textbook of biochemistry.

The phenotypic expression of biosynthetic pathways

The amount of material that needs to be provided by each pathway varies greatly from organism to organism and from situation to situation. At one

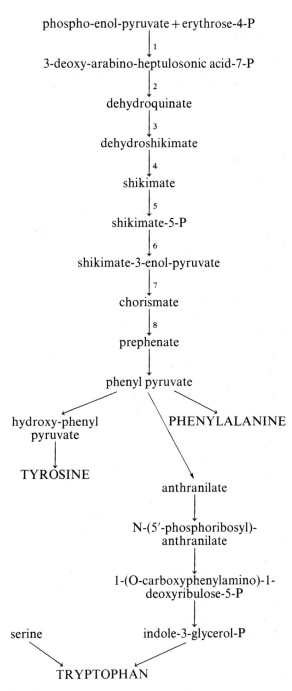

Fig. 2.1 The biosynthetic pathway leading to the aromatic amino acids, phenylalanine, tyrosine and tryptophan.

extreme a pathway may be completely missing, or totally inactive, when its product is available preformed from outside the cell. But even when endogenous biosyntheses is needed, the amount of product required varies greatly from one case to

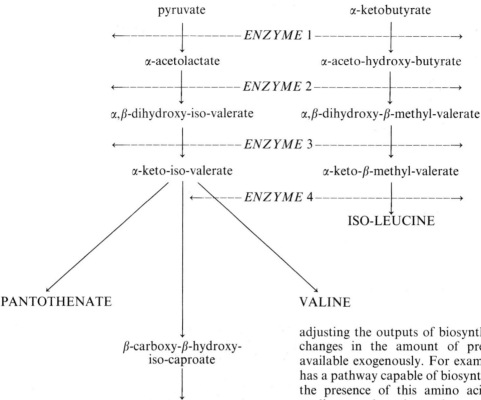

Fig. 2.2 Biosynthesis
of leucine, iso-leucine,
valine, and
pantothenate.

another. For example, a duplicating cell requires
several thousand times as much of a protein amino
acid as it does of a co-factor molecule—such as
biotin, thiamine or pantothenate—where very few
molecules/cell will be needed.

As a consequence, the output of the various
biosynthetic pathways must be adjusted to meet the
cell's requirements; and since there is only one copy
of a gene that specifies the formation of each protein
involved in a biosynthetic pathway, some method
has to be found to 'set the pace' of gene expression.
Possible mechanisms for this process are discussed
later.

On top of the built-in and fairly inflexible 'pace-
setting' of gene expression, however, many cells
have sensitive and highly flexible methods of
adjusting the outputs of biosynthetic pathways to
changes in the amount of preformed material
available exogenously. For example, even if a cell
has a pathway capable of biosynthesizing histidine,
the presence of this amino acid in the growth
medium renders the pathway redundant, temporarily at least, and it is a great advantage to the
economy of the cell to be able to shut down a
pathway temporarily until exogenous material is
exhausted.

In practice, the regulation of a biosynthetic
pathway is achieved in two fundamentally distinct
ways: either (1) the *rate of function* of the pathway is
lowered, or (2) the *relative concentration* in the cell
of the enzymes of the pathway is reduced by
blocking further synthesis. The first of these two
mechanisms is called *inhibition*—and often *feedback
inhibition* because of the details of the mechanism
involved—and the second is *repression*.

FEEDBACK INHIBITION

In any concatenated series of biochemical reactions,
such as a biosynthetic pathway, the throughput of
material passing along the pathway can be influenced by blocking any one of the steps. Normally
feedback inhibition of biosynthetic pathways is
exercised by the *product of the last enzyme* of the
pathway on the *function of the first enzyme*—hence
the name of the phenomenon (Fig. 2.4).

The effect of this type of regulation, therefore, is
to ensure that the throughput of any biosynthetic

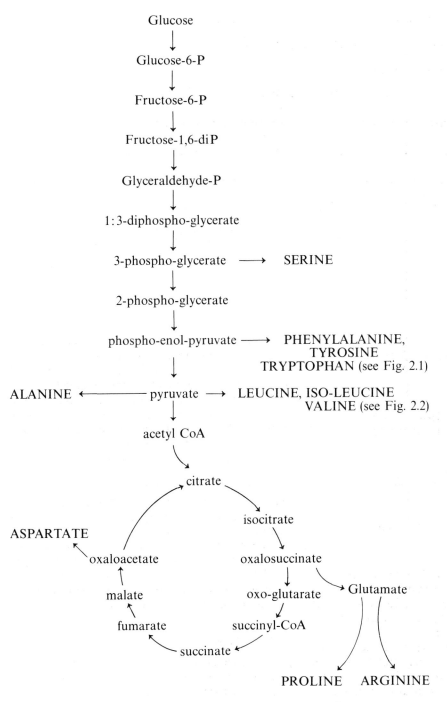

Glucose

↓

Glucose-6-P

↓

Fructose-6-P

↓

Fructose-1,6-diP

↓

Glyceraldehyde-P

↓

1:3-diphospho-glycerate

↓

3-phospho-glycerate ⟶ SERINE

↓

2-phospho-glycerate

↓

phospho-enol-pyruvate ⟶ PHENYLALANINE, TYROSINE TRYPTOPHAN (see Fig. 2.1)

↓

ALANINE ⟵ pyruvate ⟶ LEUCINE, ISO-LEUCINE VALINE (see Fig. 2.2)

↓

acetyl CoA

citrate

isocitrate

ASPARTATE

oxaloacetate   oxalosuccinate

malate   oxo-glutarate   Glutamate

fumarate   succinyl-CoA

succinate

PROLINE   ARGININE

Fig. 2.3 Biosynthetic routes from glucose to various protein amino acids.

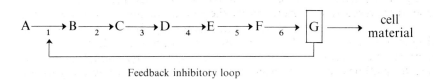

A ⟶ B ⟶ C ⟶ D ⟶ E ⟶ F ⟶ G ⟶ cell material
   1   2   3   4   5   6

Feedback inhibitory loop

Fig. 2.4 Feedback inhibition of a hypothetical pathway.

pathway is controlled by the free concentration within the cell of the molecule which is the normal product of the pathway. High external concentrations of amino acids, for example, lead to corresponding levels being reached within the cell, and this in turn switches off the endogenous synthesis of that amino acid.

REPRESSION AND INDUCTION

Like feedback inhibition, repression is usually exerted by the product of the pathway but in this case, instead of inhibiting the *function* of a *single* enzyme, the effect is to block further *synthesis* of *all* the enzymes of the pathway. Thus whereas feedback inhibition has an immediate effect in reducing biosynthesis, repression acts more slowly (but has a more lasting effect) by reducing the total synthetic activity of the culture as the cells grow and the existing supplies of the biosynthetic enzymes are diluted out in the increasing population. A comparison of the effect of feedback inhibition and of repression on the kinetics of biosynthesis by a growing culture are shown graphically in Fig. 2.5. A fall in the endogenous concentration of the repressor molecule leads to a de-repression of synthesis of the biosynthetic enzymes—but these, of course, take a little time to reach the level in each cell that existed before repression commenced.

Another phenomenon that affects the expression of biosynthetic pathways is induction. In this case the level of expression of a pathway is increased in response to the presence of a low molecular weight molecule called an inducer. The inducer may be the precursor of a pathway. One of the most studied inducible enzymes in bacteria is the β-galactosidase synthesized by many strains of *Escherichia coli*. This enzyme is capable of hydrolysing many β-galactosides, including lactose, to yield free galactose, which is then further metabolized, in much the same manner as glucose can be by the pathways shown in Fig. 2.3, to provide energy and material for biosynthesis.

As the effect of both induction and repression is to modify the expression of genes, these types of regulation involve alterations in the rate of protein synthesis. The molecular processes underlying these types of regulation will be discussed further when the mechanism of protein synthesis has been described in detail (p. 29).

Biosynthesis of macromolecules

General points

One feature common to the synthesis of all types of macromolecule, is that one of the reactants must provide the necessary energy for the polymerization step. Thus in carbohydrate synthesis energy must be provided for glycoside bond synthesis, with pro-

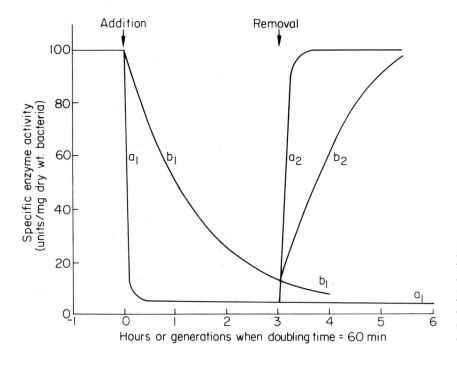

Fig. 2.5 The change of enzyme activity with time in a culture following the addition of (a_1) an inhibitor of enzyme activity, and (b_1) a repressor of enzyme formation. The second part of the graph shows the effect of removal of the compounds (a_2 and b_2).

teins the energy is needed for peptide formation, and with nucleic acids the polymerization step is the formation of a sugar–phosphate ester. In all cases the low molecular weight component of the reaction is the molecule activated to provide the energy. Thus monosaccharide nucleotides are the precursors of polysaccharides; amino acid precursors of proteins are in the form of mixed acid anhydrides with phosphoric acid; purine and pyrimidine bases for nucleic acid synthesis are in the form of nucleotide triphosphates.

The complexity of the biochemical processes underlying the synthesis of macromolecules is closely related, as may be expected, to the complexity of the macromolecules themselves. At one end of the scale, the biosynthesis of those polysaccharides that consist of a single type of monosaccharide joined together in a long chain involves, apart from the activation step, only a single enzyme acting repeatedly. Even with polysaccharides consisting of alternating monosaccharide units, biosynthesis can occur by the alternate action of two enzymes. But at the other end of the scale, the structure of proteins and nucleic acids is so complex that much more sophisticated methods of macromolecular synthesis must be used. In these cases it is not sufficient just to use the sequential action of enzymes on their own to ensure accurate synthesis of the product. Rather the enzymes have to act under the specification of a second molecule, often called a *template*, that carries information to ensure that the various enzymes operate in the correct place and in the correct order. Further details of these processes are given in the relevant sections below.

Biosynthesis of polysaccharides and related molecules

All polysaccharides are synthesized by the extension of a pre-existing polysaccharide chain by one monosaccharide unit (X) at a time, and the added monosaccharide unit enters the reaction in an activated form as a monosaccharide nucleotide, usually as the uridine-diphosphate derivative (UDP.X) but sometimes with other purine or pyrimidine nucleotides. The synthesis occurs according to the following generalized reactions:

$$\ldots X.X.X.X.X.X.X.X.X$$
$$(n \text{ residues}) \quad +$$
$$\text{uridine-diphosphate-}X$$
$$= \ldots X.X.X.X.X.X.X.X.X.X.X$$
$$(n+1 \text{ residues}) \quad +$$
$$\text{uridine-diphosphate}$$

In the case of polysaccharides involving two alternating types of monosaccharide (X and Y) the generalized reaction occurs in two steps:

Step 1
$$\ldots X.Y.X.Y.X.Y.X.Y.X.Y \quad +$$
$$\text{uridine-diphosphate-}X$$

$$= \ldots X.Y.X.Y.X.Y.X.Y.X.Y.X \quad +$$
$$\text{uridine-diphosphate}$$

Step 2
$$\ldots X.Y.X.Y.X.Y.X.Y.X.X \quad +$$
$$\text{uridine-diphosphate-}Y$$

$$= \ldots X.Y.X.Y.X.Y.X.Y.X.Y.X.Y \quad +$$
$$\text{uridine-diphosphate}$$

Not all polysaccharides consist of single unbranched chains of monosaccharide units. Some have mixtures of monosaccharides with *N*-acetyl-amino sugars and some of monosaccharides and uronic acids (Chapter 1). Such polymers are synthesized essentially as shown above, but the intermediates are the nucleotide derivatives of the relevant *N*-acetyl-amino sugars or uronic acids.

Considerable doubt exists as to the method of synthesizing branched polysaccharides. The chemical reactions involved are similar to those required for chain extension, but, as yet, there is little clear idea of how the branch points are inserted at specific points in the structure, if indeed they do occur specifically. In certain cases it has been shown that addition of a *primer molecule*, consisting of a small preformed piece of the polysaccharide to be synthesized greatly increases the rate of polysaccharide synthesis; and it may be that branch points are inserted into polysaccharides by the operation of some crude form of template, the position of the branching in the primer influencing the insertion of the branches in the new material.

Biosynthesis of lipids

Since some lipids have a more complex structure than polysaccharides the synthetic mechanisms may be correspondingly more complex. With the synthesis of phosphatidic acids and the neutral triglycerides (Chapter 1) the situation is fairly straightforward. For these compounds the precursor of the pathway is phosphoglycerol and the synthetic steps are as follows where R_1 and R_2 are long-chain fatty acids.

Step 1

$$CH_2OH$$
$$CHOH \qquad + R_1CO.SCoA \rightarrow$$
$$CH_2OPO_3H_2$$

$$CH_2O.OC.R_1$$
$$CHOH \qquad + CoA.SH$$
$$CH_2OPO_3H_2$$

Step 2

$$CH_2O.OC.R_1$$
$$CHOH \qquad + R_2CO.SCoA \rightarrow$$
$$CH_2OPO_3H_2$$

$$CH_2O.OC.R_1$$
$$CHO.OC.R_2 + CoA.SH$$
$$CH_2OPO_3H_2$$

Step 3

$$CH_2O.OC.R_1$$
$$CHO.OC.R_2 + R_1CO.SCoA \rightarrow$$
$$CH_2O.PO_3H_2$$

$$CH_2O.OC.R_1$$
$$CHO.OC.R_2 + H_3PO_4 + CoA.SH$$
$$CHO.OC.R_1$$

In each of these reactions the energy for bond formation comes from the splitting of the —CO.SCoA bond of the substituted fatty acid.

The precursor of phosphatidyl inositol, phosphatidyl serine and phosphatidyl ethanolamine (see Chapter 1 for structures) is a phosphatidic acid of the type formed in Step 2 above. This phosphatidic acid is first converted to the cytidine-diphosphate (CDP) derivative and this is then converted to the inositol derivative by the reaction:

CDP-diglyceride + inositol →
    phosphatidyl inositol + CMP

and phosphatidyl serine is formed similarly:

CDP-diglyceride + serine →
    phosphatidyl serine + CMP

Phosphatidyl ethanolamine is formed from phosphatidyl serine by the loss of CO_2.

Biosynthesis of nucleic acids

The biosynthesis of DNA and RNA is basically similar although some difficulties are introduced by the fact that DNA contains two complementary

deoxyribonucleotide polymers while RNA is usually single-stranded, at least in the period immediately following synthesis (Chapter 1). In both molecules synthesis occurs by a step-wise extension of the molecule, the overall chemical change involved in an extension by one unit being as is shown in Fig. 2.6. In DNA the low molecular weight

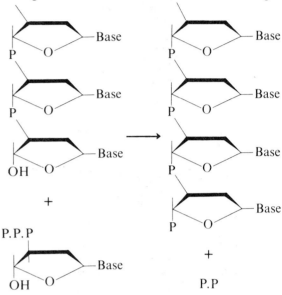

Fig. 2.6 Basic reaction leading to the increase in length of a polynucleotide chain, as found in RNA and DNA, by one unit.

precursors (which also provide the energy for the chemical bonds involved in the extension) are the *deoxyribo*nucleotide triphosphates of adenine, guanine, cytosine, and thymine, whereas *ribo*-nucleotide triphosphates of adenine, guanine, cytosine, and uracil are involved in RNA synthesis.

Although the reactions in Fig. 2.6 show how the nucleotide backbone of RNA and DNA can be extended, they do nothing to show the means whereby the correct nucleotide bases are inserted in their correct positions in the growing chain (for the sequence of purine and pyrimidine bases along the backbones of RNA and DNA is of the greatest possible significance to the organism (see Chapter 1 and p. 25). In practice DNA uses its double-stranded nature to provide its own information as to the order of insertion of bases, while RNA obtains this information by using one of the two strands of DNA. Thus DNA provides its own template, while RNA uses a DNA template.

The insertion of the correct bases into the two growing chains of a DNA molecule occurs as shown in Fig. 2.7. The process depends greatly on the 'base

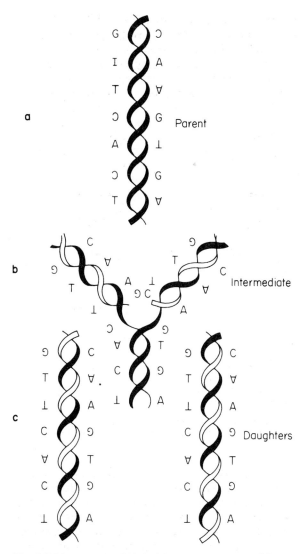

Fig. 2.7 Biosynthesis of DNA. (**a**) parental structure. (**b**) Intermediate stage with the two new strands being formed on the complementary old strands. (**c**) Final product consisting of two replicas of the original. Note that each replica consists of one 'new' (white) and one 'old' (black) strand.

strand *B*. When this step-wise process is complete, the end product is two double-stranded DNA molecules each with one 'old' and one 'new' strand (Fig. 2.7c). This type of DNA replication is called *semi-conservative* replication and is carried out by an enzyme known as the DNA-primed DNA polymerase, or often just DNA *polymerase* for short.

Synthesis of RNA bears many similarities to DNA synthesis except that the precursor molecules are different (see above) and that the base sequence of a DNA strand, rather than the RNA itself, provides the template for the insertion of the correct base in the correct position in the growing RNA chain. For this purpose, it is assumed that the base-paired double strand of DNA must open at least temporarily, in a manner analogous to that occurring in DNA synthesis.

The base pairing used for RNA synthesis on a DNA template is the same as that found in DNA, save that DNA adenine pairs with RNA uracil, rather than thymine as would be the case in DNA synthesis. This difference reflects, of course, the presence of uracil but the absence of thymine in normal RNA.

Biosynthesis of proteins

Protein synthesis is best considered in three sections corresponding to the formation of the primary, secondary, and tertiary structure of the molecule (Chapter 1). The synthetic processes involved in these three steps are very different: the first consists of the means whereby the amino acids are activated and inserted to form a single polypeptide chain in which each component amino acid occupies its correct position in the chain. By far the majority of the experimental evidence available concerns this step. The second concerns the synthesis of the —S.S— bridges that link the different polypeptide chains together, and the third is the process whereby the newly synthesized structure is folded up to make the biologically functional molecule.

Biosynthesis of the polypeptide chains

Although up to twenty different amino acids may be found in bacterial proteins, the bond linking them all is the same—the peptide bond. Consequently, two problems exist in the formation of the primary structure of proteins:

1 How is the peptide bond synthesized?
2 How is the order of the twenty different component amino acids in the polypeptide chain specified?

pairing' that exists between the complementary bases in the DNA double-strand. As the original molecule always contains an adenine residue opposite a thymine and a cytosine opposite a guanine (Fig. 2.7a), disruption of the base pairing and separation of the two strands allows the synthesis of two new DNA strands using the base sequence of the old strands as a guide (Fig. 2.7b). In this way the new strands are built up, strand *a* having complementary base pairing with the parent strand *A*, and strand *b* with complementary base pairing to

As is the case with the formation of all types of bonds between two molecules in biological systems, activation of one component of the reaction mixture is necessary for bond formation to occur. In principle, *although it must be stressed that the exact details are different in practice*, the synthesis of the peptide bond takes place according to the following basic reaction:

$$\ldots HN.CHR_1.CO.X + H_2N.CHR_2.COOH$$
$$= \ldots HN.CHR_1.CONH.CHR_2.COOH + X.OH$$

where X is a derivative that activates the carboxyl group of the amino acid, thus facilitating combination with the —NH_2 group of the second amino acid to form the peptide bond.

In practice the situation is made slightly more complex by the fact that the incoming amino acid for peptide bond synthesis is already activated for the formation of the *next* peptide bond. So the reaction sequence is as follows, where the symbol ':' indicates an activated bond.

Step 1
$$\ldots HN.CHR_1.CO:X_1 + H_2N.CHR_2.CO:X_2$$
$$= \ldots HN.CHR_1.CONH.CHR_2.CO:X_2 + X_1.OH$$

Step 2
$$\ldots HN.CHR_1.CONH.CHR_2.CO:X_2 +$$
$$H_2N.CHR_3.CO:X_3$$
$$= \ldots HN.CHR_1.CONH.CHR_2.CONH.CHR_3.$$
$$.CO:X_3 + X_2.OH$$

Step 3
$$\ldots HN.CHR_1.CONH.CHR_2.CONH.CHR_3.$$
$$.CO:X_3 + H_2N.CHR_4CO:X_4$$
$$= \ldots HN.CHR_1.CONH.CHR_2.CONH.CHR_3.$$
$$.CONH.CHR_4.CO:X_4 + X_3.OH$$

and so on.

Thus the activation energy is derived from the activated state of the last amino acid of the growing chain and not from the activated incoming amino acid, and the chain is synthesized from the amino to the carboxyl terminus.

The step-wise reactions shown above account for the synthesis of all the peptide bonds of the polypeptide chain. There are, however, two special steps that must be considered: the first and the last. The first step—namely the formation of a dipeptide from the first amino acid—cannot occur unless the amino group of the first amino acid is substituted. The first amino acid of a growing polypeptide chain is always the substituted amino acid *N*-formyl-methionine (Fig. 2.8). In this molecule the free —NH_2 group of methionine is blocked with a

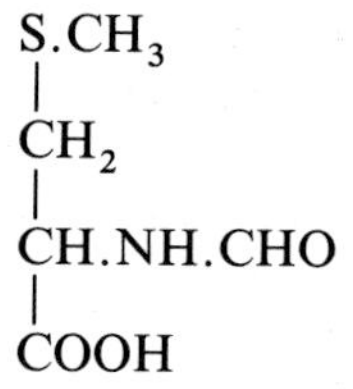

$$S.CH_3$$
$$|$$
$$CH_2$$
$$|$$
$$CH.NH.CHO$$
$$|$$
$$COOH$$

Fig. 2.8 Structure of *N*-formyl-methionine.

—CHO residue. After the polypeptide chain has been synthesized and liberated from the synthetic site, enzymes exist that split off the whole *N*-formyl-methionine residue from the chain, so the ultimate *N*-terminal amino acid of a bacterial polypeptide chain is the second residue in the order of synthesis.

Less is known about the mechanism of chain termination. Probably all that happens is that no amino acid is presented for insertion and the chain, as synthesized up to that point, is liberated from the synthetic site; but this matter will be discussed in more detail later.

THE ORDERING OF AMINO ACIDS IN POLYPEPTIDE SYNTHESIS

Although these reactions allow the synthesis of a polypeptide chain they do nothing to decide the order of insertion of the different amino acids in their correct position in the chain. This step is achieved by allowing polypeptide synthesis to occur on a molecular template. In the case of RNA synthesis, it may be recalled, the molecular template used was a DNA strand of complementary base sequence—but in protein synthesis the operation of the template is a great deal more complex. In this case the template is a single strand of RNA, and the order of insertion of amino acids is determined by the order of the purine and pyrimidine bases along the polyribophosphate backbone of this RNA. Thus an essentially linear molecule, a strand of RNA, specifies the structure of a second essentially linear molecule, the polypeptide chain.

Since there are only four different types of purine and pyrimidine bases in RNA, but 20 different types of amino acid in protein, it is clear that the base sequence of individual purines and pyrimidines along the RNA strand cannot unambiguously determine the order of insertion of amino acids into the growing polypeptide chain. For specification to be unambiguous, a sequence of at least three bases is needed for each amino acid; but on the other hand with this number there are too many possibilities and there would be more than one group of three bases available for each amino acid. In fact, if there is a 'direction of reading' of the bases (that is if the base order adenine–guanine–cytosine for example can be distinguished from the sequence cytosine–guanine–adenine) then there will be 64 distinct triplets of purines and pyrimidines available to specify the insertion of 20 amino acids.

Experimental investigation of this problem, the so-called feat of 'cracking the genetic code', shows that 61 of the 64 available triplets (or RNA *codons* as they are called) 'code' for the insertion of individual amino acids during protein synthesis, while the remaining three triplets are concerned with bringing the synthesis of a given polypeptide chain to an end. The number of different codons that specify the insertion of each type of amino acid is different. Three of the amino acids can be specified by any of six different triplets; five of the amino acids each have four specific codons; one has three; nine have two, while two are specified by a single codon only. The full list of triplet attributions is given in Fig. 2.9. The codon AUG inserts

2nd base	U	C	A	G	
1st base					3rd base
	Phe	Ser	Tyr	Cys	U
U	Phe	Ser	Tyr	Cys	C
	Leu	Ser	CT	CT	A
	Leu	Ser	CT	Trp	G
	Leu	Pro	His	Arg	U
C	Leu	Pro	His	Arg	C
	Leu	Pro	Gln	Arg	A
	Leu	Pro	Gln	Arg	G
	Ileu	Thr	Asn	Ser	U
A	Ileu	Thr	Asn	Ser	C
	Ileu	Thr	Lys	Arg	A
	Met	Thr	Lys	Arg	G
	Val	Ala	Asp	Gly	U
G	Val	Ala	Asp	Gly	C
	Val	Ala	Glu	Gly	A
	Val	Ala	Glu	Gly	G

Fig. 2.9 The key to the genetic code. The RNA bases adenine, guanine, cytosine, and uracil are referred to as A, G, C, and U. The abbreviation CT stands for 'chain termination'—see text.

methionine unless it is the first triplet in the message when *N*-formyl-methionine is inserted instead.

The overall process of protein synthesis as described so far is often summarized by saying that there is *colinearity of gene and protein*. Thus three bases on one of the DNA strands determine the nature of three bases in the RNA, and these, in turn, determine which amino acid is inserted into the growing polypeptide chain. Furthermore this arrangement accounts for the fact that a point mutation—that is a change in the nature of a single base in the DNA—may cause the change of one amino acid for another at a particular point in the polypeptide sequence of protein.

Because of its role as a transmitter of information, the RNA molecule that acts as the intermediate between the gene and the polypeptide chain is usually known as *messenger* RNA or m-RNA. Apart from describing its function graphically, this name helps to distinguish this type of RNA from others of different function in the cell.

The polypeptide product of a gene is synthesized initially as a long straight-chain molecule but after synthesis this structure folds up (p. 29). Generally the polypeptide products of a single gene appear to fold up in the same way and this is a property of the amino acid sequence of the polypeptide and is not imposed by the action of another gene distinct from the one that was responsible for ordering of the amino acids in the polypeptide chain.

THE ROLE OF TRANSFER RNA (T-RNA)

The colinearity principle shows how the order of the amino acids in a polypeptide chain is specified by the base sequence of a gene, but does little to explain how, in molecular terms, three purine or pyrimidine bases can specify the insertion of a single amino acid in a growing polypeptide chain. This task is performed by another class of RNA molecules, quite distinct from m-RNA, which are known as transfer RNAs (or t-RNAs).

Structure of t-RNAs Although a large number of different types of t-RNA are found in microbial cells (there is often more than one to every codon), all are variants of a common basic structure. Fig. 1.17 shows, as an example, the structure of one of the alanine t-RNAs from yeast. All have a single polyribophosphate chain of about 75 residues, and all contain a similar terminal sequence of C.C.A-OH.

Unlike most other types of RNA molecule, t-RNAs contain a range of atypical substituted purines and pyrimidines, such as dimethylguanine, dihydrouracil, methylcytosine and the like. Some transfer RNAs even contain thymine, a base normally found only in DNA. Another unusual feature of these molecules is that parts of the structure are double-stranded, even though the RNA concerned consists of a single polyribophosphate chain. The double-stranded regions are caused by the chain doubling back on itself and forming internal complementary base pairs.

Function of t-RNAs   As mentioned previously, the function of t-RNA molecules is to interact both with the amino acid and with the RNA codon on the messenger RNA so that the order of bases on the 'message', which is a copy of that of the gene, enables the amino acid to be inserted into its correct position in the growing polypeptide chain. The two parts of the basic t-RNA structure involved in these functions can be identified with some certainty.

The part responsible for recognizing the RNA codon is called the *anticodon*. It consists of a sequence of three bases in the t-RNA sequence that are complementary in hydrogen bonding properties to the three bases of the RNA codon. Thus in the alanine t-RNA whose structure is shown in Fig. 1.17, the anticodon is the base sequence IGC at residues 36–38 of the structure, and these three bases interact by complementary base pairing with GCU, one of the RNA codons for alanine. The part of the t-RNA molecule that interacts with the amino acid is the final adenine residue of the —$CpCpA$.OH sequence at the 3′OH terminal of the molecule. The amino acid interacts with this sequence to form a mixed acid anhydride with the phosphate residue on the terminal adenine (Fig.

$$(tRNA).CCA.OH + Amino\ acid + ATP$$

$$= (tRNA).CCA\text{-}Amino\ acid + AMP + PP$$

Fig. 2.10 Activation of an amino acid by the formation of a complex with t-RNA.

2.10) in all t-RNA molecules. This bond serves two purposes: first, it attaches the amino acid to the t-RNA molecule, but secondly its nature ensures that the amino acid is held in an activated state so that a peptide bond may be formed with another amino acid under suitable circumstances. ATP is needed for the reaction.

Synthetase enzymes   Since all t-RNAs end in the sequence $CpCpA$.OH and consequently lack specificity in this part of their structure, some means must be found to ensure that each amino acid is loaded on to its correct t-RNA. This process is carried out by a series of *synthetase enzymes*, one for

each type of t-RNA, that have specificity both for the amino acid residue and for some portion of the t-RNA molecule that does not include either the anticodon, or the —$CpCpA$.OH terminal sequence. Investigation of the nature of such enzymes has shown that as well as ensuring that the correct amino acid is loaded on to a given t-RNA molecule, it is also the catalyst responsible for the reactions summarized in Fig. 2.10.

THE OVERALL MECHANISM OF PROTEIN SYNTHESIS

Now that the multifaceted structure of the t-RNA molecule has been described in some detail, it is possible to give a complete account of the process of polypeptide bond formation. This is shown diagrammatically in Fig. 2.11. The process involves two essentially separate series of reactions that meet with the involvement of the t-RNA molecule. The first series achieves the activation of the amino acids in the form of loaded t-RNA molecules. All the 20 types of different amino acid found in protein are activated in this way and each is loaded on to one of a number of distinct t-RNAs specific for the amino acid in question.

The second series of reactions involved consists of the formation of messenger RNA on a template of DNA. This process results in the mobilization of the information inherent in the base sequence of the gene and its dispatch in the form of a series of messenger RNA molecules, to the site of protein synthesis in the cell.

These two series of reactions impinge when the activated amino acid–t-RNA complexes are provided for the formation of the polypeptide chain, the order in which they are used being determined by the order of coding triplets in the messenger RNA (Fig. 2.11). After the loaded t-RNA has been drawn to its correct position for peptide synthesis by the interaction of codon and anticodon, and after the peptide bond has been formed by the reactions shown in Fig. 2.12, the free t-RNA is released from the site of peptide synthesis until it has been recharged with another amino acid residue. In this way the amino acids are inserted into their correct locations in the growing polypeptide chain, starting with an *N*-formyl-methionine residue and continuing until one of the codons UAA, UAG, or UGA is encountered in the messenger RNA. At this point no amino acid is inserted and the polypeptide chain is released complete.

As far as is known, all polypeptide chains in the bacterial cell are synthesized by this means, all the information required for determining the amino

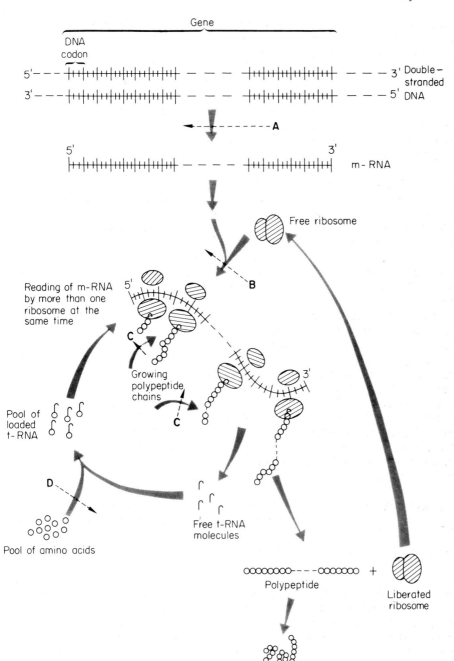

Fig. 2.11 Diagrammatic representation of the overall process of protein synthesis. The dotted arrows indicate the sites of inhibition of certain antibiotics. (**A**—actinomycin D, rifamycin; **B**—streptomycin, erythromycin, tetracycline; **C**—puromycin, chloramphenicol; **D**—some structural analogues of amino acids.)

acid sequence and length of the structure being stored ultimately in terms of the base sequence of the DNA.

RIBOSOMES

As might be expected from the complexity of the reactions involved, protein synthesis does not occur in free solution in the bacterial cells but occurs on the surface of structures—they might even be thought of as organelles—known as ribosomes (see Chapter 8). Ribosomes are composed of protein and RNA (a different type from both t-RNA and m-

RNA) in approximately equal proportions. We are still not sure of all the functions of the many proteins and enzymes present in ribosomes, nor is the role of the RNA part of the structure at all clear. What is certain is that the RNA messenger is dispatched from its site of synthesis on the DNA template to the ribosome, and it is on the surface of these structures that the interaction with the loaded t-RNA molecules occurs. The nascent polypeptide chain remains attached to the ribosome—at least at one point on its length—until a chain-terminating triplet is encountered, when the complete polypeptide is liberated free into the cytoplasm of the cell.

In practice it appears that a single messenger RNA molecule can be used a number of times before it is inactivated or destroyed. This can be shown experimentally by the kinetics of protein synthesis and also by the fact that it is possible to isolate messenger RNA molecules with more than one ribosome attached at intervals (see Fig. 2.11). Both this observation, and the size of the message in relation to the size of the ribosome, suggest that the part of the message actively involved in peptide synthesis moves in relation to the ribosome; and where there are a number of ribosomes attached to a single messenger, these structures follow one another along the RNA at regular intervals each producing a polypeptide chain. The structures in which a series of ribosomes are found attached to a thread of messenger RNA are called *polysomes*, and isolation of ribosomes from living cells often gives preparations in this form, particularly if care is

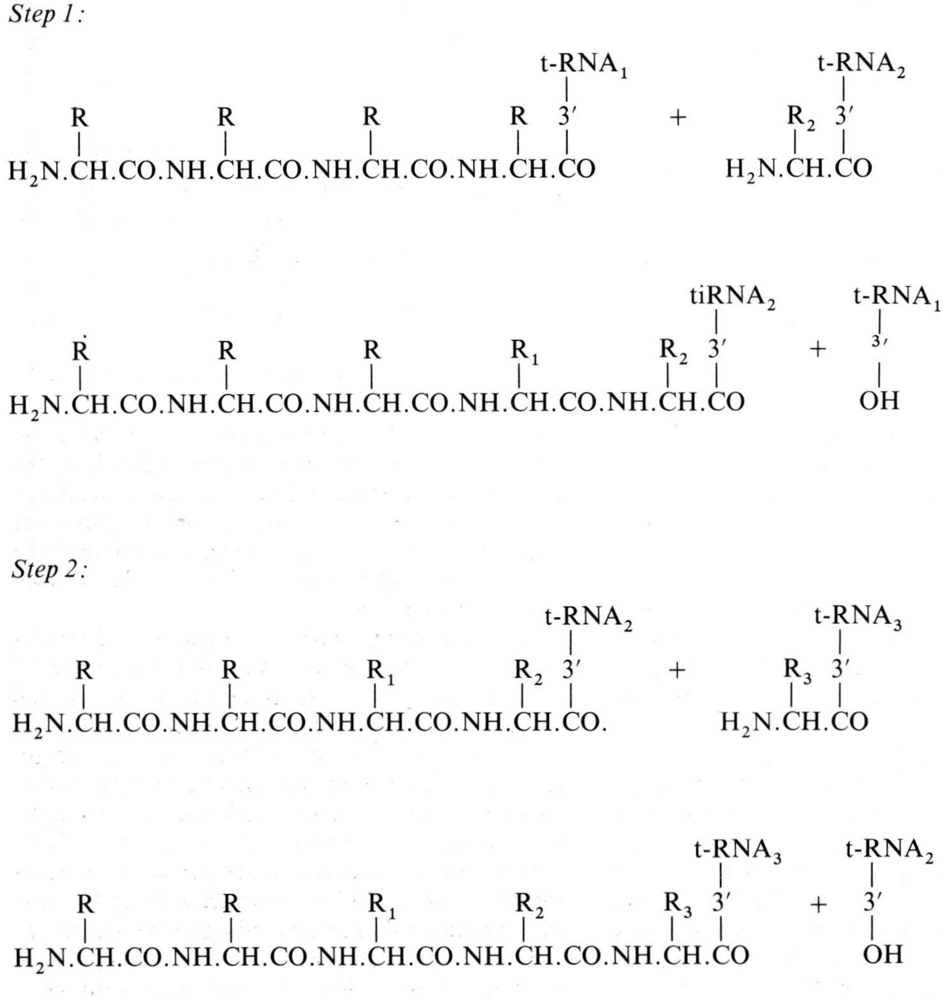

Step 1:

$$
\begin{array}{cccc}
 & & & \text{t-RNA}_1 \\
 & & & | \\
R & R & R & R \quad 3' \\
| & | & | & | \quad | \\
\text{H}_2\text{N.CH.CO.NH.CH.CO.NH.CH.CO.NH.CH.CO}
\end{array}
\quad + \quad
\begin{array}{c}
\text{t-RNA}_2 \\
| \\
R_2 \quad 3' \\
| \quad | \\
\text{H}_2\text{N.CH.CO}
\end{array}
$$

$$
\begin{array}{ccccc}
 & & & & \text{tiRNA}_2 \\
 & & & & | \\
R & R & R & R_1 & R_2 \quad 3' \\
| & | & | & | & | \quad | \\
\text{H}_2\text{N.CH.CO.NH.CH.CO.NH.CH.CO.NH.CH.CO.NH.CH.CO}
\end{array}
\quad + \quad
\begin{array}{c}
\text{t-RNA}_1 \\
| \\
3' \\
| \\
\text{OH}
\end{array}
$$

Step 2:

$$
\begin{array}{cccc}
 & & & \text{t-RNA}_2 \\
 & & & | \\
R & R & R_1 & R_2 \quad 3' \\
| & | & | & | \quad | \\
\text{H}_2\text{N.CH.CO.NH.CH.CO.NH.CH.CO.NH.CH.CO.}
\end{array}
\quad + \quad
\begin{array}{c}
\text{t-RNA}_3 \\
| \\
R_3 \quad 3' \\
| \quad | \\
\text{H}_2\text{N.CH.CO}
\end{array}
$$

$$
\begin{array}{ccccc}
 & & & & \text{t-RNA}_3 \\
 & & & & | \\
R & R & R_1 & R_2 & R_3 \quad 3' \\
| & | & | & | & | \quad | \\
\text{H}_2\text{N.CH.CO.NH.CH.CO.NH.CH.CO.NH.CH.CO.NH.CH.CO}
\end{array}
\quad + \quad
\begin{array}{c}
\text{t-RNA}_2 \\
| \\
3' \\
| \\
\text{OH}
\end{array}
$$

Fig. 2.12 Molecular interactions involved in the lengthening of a polypeptide chain by two units.

taken not to destroy RNA in the isolation procedure.

Formation of the secondary and tertiary structure of proteins

The reactions described in the foregoing sections are involved solely with the synthesis of the polypeptide chain using the information provided by the gene. The problem of how the polypeptides fold up to form active proteins and how the —S.S— cross bridges are formed still remain to be discussed.

It seems likely that folding occurs spontaneously as the growing polypeptide chain is liberated from the ribosome. Thus the process probably starts at the —NH₂ terminus of the protein as soon as the terminal N-formyl-methionine has been removed. At this point the synthesis of the carboxyl terminus may not yet have taken place.

The —S.S— cross bridges also probably form spontaneously once the relevant —SH residues in the protein have been brought into their correct juxtaposition by the folding process.

The cross bridges are formed by the reaction

$$R_1.SH + R_2.SH \rightleftharpoons R_1\text{-}S\text{-}S\text{-}R_2 + SH$$

and this occurs easily with little net energy change.

Undoubtedly a further stage is important in the biosynthesis of most proteins, at least when the process is considered in the frame-work of a growing cell. A simple example is the synthesis and arrangement of flagellin in bacterial flagella. Practically nothing is known about how the molecules are arranged to give the flagellum its characteristic properties and molecular pattern.

Regulation of protein synthesis

Basically there are two ways of influencing the level of gene expression: either the activity of existing enzymes can be modified, or their rate of synthesis can be altered. Only those methods affecting synthesis will be discussed here since effects on activity have already been described on p. 17.

In the preliminary discussion of enzyme regulation (see p. 17 *et seq.*) it was pointed out that the rate of synthesis of a gene product was under two types of control—one that 'set the pace' of gene expression and which was unaffected by the composition of the growth medium, and a second (induction or repression) in which changes in the composition of the medium could influence the rate of synthesis of the product of individual genes or groups of genes.

Messenger RNA occupies a key intermediate position between the gene and its final protein product. Consequently any alteration, either in the rate of formation of this molecule, or in its 'translation' into protein, will, in turn, influence the rate of expression of the gene, and it is in this region of the overall process of protein synthesis that the regulatory factors exert their effect rather than by limiting the availability of amino acids, energy, or any other molecule.

The 'pace-setting' of gene expression is known to be mediated by a region—the *promoter* region—on the DNA (Fig. 2.13). This region is involved in

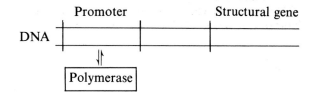

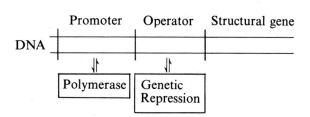

Fig. 2.13 Pace-setting and regulation of gene expression.

recognizing RNA polymerase, the enzyme responsible for synthesizing messenger RNA on the DNA template provided by the gene. A high affinity of the RNA polymerase for a particular promoter ensures efficient transcription with consequent high protein yield, while a poor affinity leads to lower levels of expression.

The biochemical basis of induction and repression has been worked out in detail for a number of microbial enzymes, notably by Jacob and Monod for β-galactosidase synthesis in *Escherichia coli*. In all cases the effect of the low molecular weight effector molecule (whether inducer or repressor) is to alter the rate of RNA synthesis, but unlike the pace-setting arrangement already described, both induction and repression are achieved by influencing the passage of RNA polymerase along the gene rather than by altering its binding to the promoter.

Inducers and repressors do not exert their effects directly on the DNA, but indirectly through a protein molecule which has the properties of recognizing both the low molecular weight effector molecules themselves, and also a specific region of the DNA known as the operator (Fig. 2.13). In both inducible and repressible systems the binding of this protein to the operator blocks transcription. In induction the effect of the low molecular weight inducer is to inhibit binding to the operator (thus allowing transcription), while in repressible systems the low molecular weight repressor must be present for this protein to block synthesis. Overall, therefore, inducers stimulate protein synthesis by removing a block to transcription, while repressors interact to block transcription.

In many bacterial cells induction and repression do not necessarily affect single genes, but may show coordinated effects on a number of genes which specify proteins of related function. A common example of such groups of genes are those specifying the individual enzymes of a biosynthetic pathway. When genes are grouped together in this way they are known as an *operon*; and commonly such operons express a single message which carries the transcripts of all the genes concerned. This mechanism has the advantage that coordinate control can easily be exercised by regulating the production of this multi-component message. On the other hand if the message is to be translated into the necessary number of separate proteins, then it must include some information as to where the individual polypeptide chains begin and end. In practice this is carried out by the existence of 'stop' and 'start' signals at appropriate points on the message so that separate polypeptide chains are formed as the ribosomes pass along the message (see Fig. 2.11).

Mucopeptide biosynthesis

Compared with the efforts that have been made by experimental workers in the field of protein synthesis, relatively little has been done to elucidate the mechanism of mucopeptide synthesis. Nevertheless, quite a lot is known about the latter process.

As might be expected from the discussion of the mechanism of polysaccharide synthesis above, the polysaccharide chain of mucopeptide is built up by the interaction of the relevant monosaccharides after their activation to form the uridine diphosphate derivatives. Therefore this part of the synthesis is, in principle, identical with the two sets of reactions shown in Steps 1 and 2 (p. 21), for the synthesis of a polysaccharide of alternating subunits. However, the process of mucopeptide synthesis is unusual in that the peptide side-chain is added to the muramic acid residue before the substituted muramic acid is used for polysaccharide synthesis; moreover, the peptide is built up stepwise and not added to the muramic acid as a preformed peptide. As a consequence the biosynthesis of even a small part of the great sack-shaped molecule of mucopeptide (see p. 14) involves many steps, as follows, where UDP is uridine diphosphate, GlcNAc is *N*-acetyl-glucosamine, and MurNAc is *N*-acetyl-muramic acid:

Step 1 Addition of alanine.

$$\text{Uridine-diphosphate-MurNAc} + \text{L-ala} = \begin{array}{c} \text{Uridine-diphosphate-MurNAc} \\ | \\ \text{L-ala} + \text{H}_2\text{O} \end{array}$$

Step 2 Addition of glutamic acid.

$$\begin{array}{c} \text{UDP-MurNAc} \\ | \\ \text{L-ala} \end{array} + \text{D-glu} = \begin{array}{c} \text{UDP-MurNAc} \\ | \\ \text{L-ala} \\ | \\ \text{D-glu} \end{array} + \text{H}_2\text{O}$$

Step 3 Addition of lysine.

$$\begin{array}{c} \text{UDP-MurNAc} \\ | \\ \text{L-ala} \\ | \\ \text{D-glu} \end{array} + \text{L-lys} = \begin{array}{c} \text{UDP-MurNAc} \\ | \\ \text{L-ala} \\ | \\ \text{D-glu} \\ | \\ \text{L-lys} \end{array} + \text{H}_2\text{O}$$

Step 4   Addition of *two* molecules of alanine together as a dipeptide.

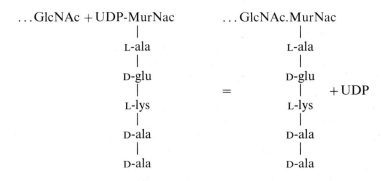

Step 5   Formation of the C1 → C4 glycoside bond.

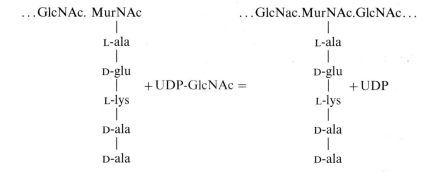

Step 6   Formation of the C4 → C1 glycoside bond.

...GlcNAc. MurNAc

    |
   L-ala
    |
   D-glu   +UDP-GlcNAc =
    | |
   L-lys
    |
   D-ala
    |
   D-ala

...GlcNac.MurNAc.GlcNAc...

    |
   L-ala
    |
   D-glu   +UDP
    | |
   L-lys
    |
   D-ala
    |
   D-ala

At this stage it is worth noticing the following points:

1 As mentioned above, the UDP-MurNAc is loaded stepwise with amino acids before the glycoside bond is formed at Step 5.

2 The last two D-alanine residues are added as a dipeptide rather than singly.

3 The final polypeptide side chain on the muramic acid has the overall composition L-ala-D-glu-L-lys-D-ala-D-ala; that is with an extra D-alanine at the end of the chain when compared with the final structure in the complete molecule (see p. 15).

Step 7   Formation of the cross bridges (see p. 32). At this stage two units as produced by Step 6 are joined together by a pentaglycine bridge. The bridge is formed between the ε-amino group of a lysine residue and the carboxyl group of the *penultimate* D-alanine of the muramic acid side chain. In the process the terminal D-alanine from one side chain is liberated.

Step 7 Formation of the cross bridges.

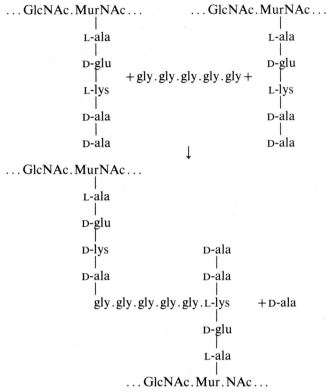

Obviously the reactions outlined in these seven steps could lead to a large molecule—particularly if the cross linking occurs between many chains to form a three-dimensional structure rather than to form a linear or sheet-like molecule. However, practically nothing is known of the detailed steps by which the prototype molecule resulting from Step 7 is interlinked to form the final macromolecule. In particular, for example, it is not known at what stage in the overall formation of the mucopeptide that the cross-linking reaction occurs, beyond the fact that it occurs after the formation of considerable lengths of polysaccharide chain. Nor is it clear whether all the side chains to the muramic acids are cross-linked by bridges to other side chains or whether some remain free. It is undoubtedly true that some of the polypeptide side chains to the muramic acid residues are free since an enzyme exists in the cell that can remove the terminal alanine residue from the chain rather than catalyzing its cross linkage to another side chain. The overall reaction catalyzed by the enzyme is:

...GlcNAc.MurNAc...
|
L-ala
|
D-glu
|
L-lys   D-ala
| |
D-ala   D-ala
| |
gly.gly.gly.gly.L-lys   →
                 |
                 D-glu
                 |
                 L-ala
                 |
             ...GlcNAc.MurNac...

...GlcNAc.MurNAc...
|
L-ala
|
D-glu
|
L-lys   D-ala
| |
D-ala   D-ala
| |
gly.gly.gly.gly.L-lys   +D-ala
                 |
                 D-glu
                 |
                 L-ala
                 |
             ...GlcNAc.MurNAc...

Attempts have been made by some workers to gain some insight into the overall process of cell-wall synthesis by using fluorescent antibodies to stain newly synthesized wall. These techniques have given conflicting evidence to date. Some species seem to form their cell walls outward as a band originating at the cell division furrow while others seem to lay down new material over the whole surface of the bacterial cell. But even if these results are taken to be a true representation of the direction of wall synthesis, the antisera are never sufficiently well fractionated to act only against a single wall component, such as mucopeptide, and consequently the fluorescent antibody approach is more likely to allow one to follow the overall synthesis of cell wall material rather than any single component within it.

Macromolecular synthesis in the context of bacterial cell duplication

In the foregoing sections we have discussed the molecular processes involved in the biosynthesis of many types of macromolecule that are important in cell duplication, but there has been little attempt to fit these reactions into their context in cell replication itself. In the following section, therefore, some attempt will be made to give an account of what is known about this aspect of macromolecular synthesis. The account will inevitably be incomplete, partly because the problem has only recently become accessible to experimental attack and partly, as implied earlier, because not enough is known of the problems involved in synthesis *in vitro* to be able to study the process in a growing cell.

Of the macromolecules that have been considered so far, most is known about DNA replication in relation to cell division. When any cell divides, the problem of providing each daughter cell with one copy of the genetic complement from the parent can be considered as having two parts:

1 How are the copies produced?
2 How are the copies distributed so that one copy goes to each daughter cell:

The present hypothesis invoked to explain this process was originally proposed by Jacob, Brenner and Cuzin. They proposed that there is a point of contact, known as the *replicator*, between the DNA and some point on the cell membrane. The replicator has the properties both of holding the DNA double thread in its correct spatial relationship with the cell as a whole but also with initiating the replication of the DNA molecule which occurs as

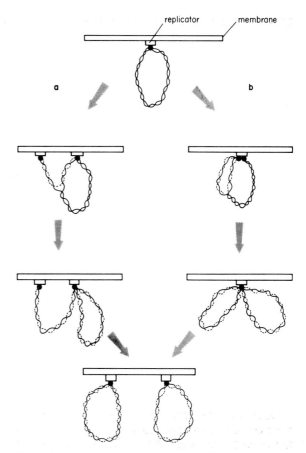

Fig. 2.14 Two methods for the replication of DNA at cell division and the distribution of the two copies to the daughter bacteria. (a) Replicator divides at the commencement of DNA replication. (b) Replicator divides at the end of DNA replication. Pre-existing DNA strands are shown as solid lines; newly synthesized DNA as dashed lines.

described on p. 32. The replicator also, it is suggested, is responsible for distributing the two copies of the DNA, one copy to each daughter cell. This last step can be achieved in one of the two ways shown diagrammatically in Fig. 2.14. Either the replicator divides when the replication of the DNA double strand is complete, thus giving the pattern shown in Fig. 2.14b, or the replicator divides as DNA replication commences as shown in Fig. 2.14a. Whichever occurs, the net effect is the same: one copy of the DNA is distributed to each daughter cell. At the present time it seems that the mechanism shown in Fig. 2.14b is correct.

The underlying mechanism of DNA replication and distribution as it occurs in growing cells may lead to some mutations of bizarre phenotype. One of the most dramatic is the formation of DNA-less bacteria. In this case the failure of DNA replication means that cell division procedes normally but that there is no duplicate copy of the DNA to pass to one of the daughter cells. Thus, in this mutant, a string of DNA-less bacteria are formed as the culture grows with a single DNA-containing cell at one end of the chain. It is interesting to note in passing that the DNA-less cells are quite capable of carrying out many of the physiological functions attributed to normal cells; what they cannot do is replicate.

The other macromolecule whose synthesis has been followed to some extent in the whole cell is mucopeptide. Here two basic patterns of synthesis seem to occur. In many Gram-negative bacteria (most of the studies have been done in *Salmonella typhimurium*) the new material is inserted more or less continuously into the old wall structure so that the new material is distributed over the whole cell surface. With streptococci, however (and this may apply to all Gram-positive bacteria), the new material is laid down on each side of the division furrow, thus producing a clear line of demarcation between 'old' and 'new' mucopeptide on the cell surface.

In summary, therefore, it is true to say that a great deal is known of the detailed biochemical steps in the synthesis of many types of bacterial macromolecule but that our knowledge of how these synthetic process are organized both in space and time to lead to the formation of a new bacterial cell is still rudimentary. Perhaps one of the most interesting speculations concerned with bacterial cell growth, and one that has a profound bearing on the various topics discussed here, is the possibility that the position of the various macromolecules in the cell itself provides information, in addition to that carried in the DNA, that assists in inserting newly synthesized macromolecules into their correct position in the developing cell. If this is so, then the possession of the genetic information carried in a cell is no use without a bacterial cell to interpret it. Which came first—the information or the cell?

Energy metabolism in micro-organisms

Whenever molecules are synthesized from more simple precursors, energy is required, and conversely the degradation of complex molecules releases energy that can then be used for other purposes. Degradative changes (or *catabolism* as it is called) may therefore provide the necessary energy for the biosynthesis (or *anabolism*) of macromolecules and their precursors.

In practice, the readily available energy stored in a chemical molecule is that which is liberated when chemical linkages in the molecule are broken. However, not all such ruptures make the same amount of energy available. For example the hydrolysis of peptide bonds

$$R.CO\text{-}NH.R + H_2O \rightarrow R.COOH + R.NH_2$$

releases relatively little energy (about 0.5 kilocalories/mole hydrolyzed), while the hydrolysis of glycosides (e.g. the linkage joining sugars in polysaccharides such as amylose—see p. 3) produces about 3.0 kcal./mole for each bond split.

A few types of molecule contain bonds that are particularly energy rich. Examples are the terminal —P.P— bonds in ATP (about 8.0 kcal./mole for each phosphate split off) and the phosphate ester link in phospho-*enol*-pyruvate (13 kcal./mole for each phosphate released).

Since bacterial growth involves the synthesis of complex molecules from more simple precursors, a positive net input of energy is needed for bacteria to grow and consequently the bacteria themselves will have to degrade a sufficient number of energy-yielding bonds to provide the necessary biosynthetic potential. Or, to put the matter in another way: the anabolic activity of heterotrophic bacteria must be balanced by an equivalent amount of catabolism.

This requirement for energy means therefore that growing heterotrophic bacteria must take up molecules from the environment which are broken down either immediately or following some limited metabolic interconversion. In many bacterial species these molecules are commonly monosaccharides—notably glucose, but with others amino acids and fatty acids are used for this purpose. In all cases, however, the result is the same: energy is made available for biosynthetic (anabolic) purposes by the catabolic destruction of the molecules.

While many of the metabolic chemical interconversions in bacterial cells are clearly anabolic, and others are catabolic, there exist some pathways whose direction may change depending on the physiological state of the bacteria and others where the metabolic interconversions yield a pool of molecules which may, depending on the circumstances, be either used for biosynthesis or broken down. In view of their dual role, such pathways are known as *amphibolic* pathways.

The alternative fate of some molecules is an important characteristic of bacterial metabolism since it allows bacteria to store up potential energy when external circumstances are favourable, but to catabolize these molecules when they are not.

Glucose provides a good example. Under conditions of excess nutritional availability, glucose tends to be polymerized in many microbial species to produce a polysaccharide, which then constitutes a store of glucose residues. The energy for this process will come from excess energy available. If external nutrients are at a premium, however, glucose is metabolized, ultimately to produce CO_2, H_2O and energy, and under these conditions the polysaccharide stores are also mobilized by reforming glucose as a result of hydrolysis.

Since the molecular changes that either require or liberate energy are so numerous in bacteria, it is clearly an advantage to have a few types of molecule which can be used to channel the energy molecules which, as it were, constitute the energy currency of the cells. In practice micro-organisms (like many living systems) use primarily ATP for this purpose.

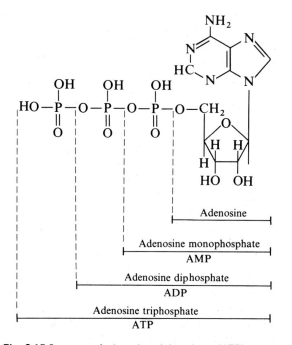

Fig. 2.15 Structure of adenosine triphosphate. (ATP)

This molecule (Fig. 2.15) consists of a triphosphate residue attached to a purine nucleotide. Hydrolysis of either the terminal or the penultimate bonds liberates about 8.0 kcal./mole split.

$$ATP \rightleftharpoons ADP + P \quad G^1_0 = -7.4 \text{ kcal./mole}$$
$$ATP \rightleftharpoons AMP + PP \quad G^1_0 = -7.6 \text{ kcal./mole}$$

In view of these reactions, the $ATP \rightleftharpoons ADP + P$ and the $ATP \rightleftharpoons AMP + PP$ interconversions can be used either as a potential source of energy or as an energy receptor. One example of each will suffice. In the formation of acetoacetyl-CoA (an important metabolic intermediate in fatty acid metabolism) from 2 acetyl-CoA molecules, one ATP is converted to $ADP + P$. Thus the energy released by this change is used to drive the formation of the bond between the two acetyl-CoA molecules. The energy of formation of this bond is 6.6 kcal./mole, while the ATP hydrolysis yields about 7.4. Thus there is energy to spare and the reaction is driven in the direction of synthesis.

In the second example glyceraldehyde-3-phosphate is oxidized to 1,3-diphosphoglyceric acid in a series of reactions. This oxidation is achieved by adding oxygen (derived ultimately from water), but in the presence of inorganic phosphate the product is not the free acid but the mixed acid anhydride involving the —COOH of the phosphoglycerate and one of the hydroxyls of phosphoric acid:

$$\begin{array}{l} CHO \\ | \\ CHOH \\ | \\ CH_2O\text{-phosphate} \end{array} \xrightarrow[-2H]{+H_2O + \text{phosphate}} \begin{array}{l} CO\text{-phosphate} \\ | \\ CHOH \\ | \\ CH_2O\text{-phosphate} \end{array}$$

In the 1,3-diphosphoglycerate formed in this way the potential energy in the CO-phosphate bond is enough to form ATP from ADP while that in the CH_2O-phosphate bond is not. Thus in the presence of ADP the following reaction occurs:

$$\begin{array}{l} CO\text{-phosphate} \\ | \\ CHOH + ADP \\ | \\ CH_2O\text{-phosphate} \end{array} \rightarrow \begin{array}{l} COOH \\ | \\ CHOH \\ | \\ CH_2O\text{-phosphate} \end{array} + ATP$$

Thus the oxidation of the aldehyde group, followed by the molecular rearrangement in the resulting molecule allows the free energy generated by this metabolic change to be passed to ATP for use elsewhere in the cell.

Although some ATP is generated by metabolic interconversions of this type in bacterial cells—the so-called *substrate-level phosphorylation* reactions—by far the majority of the ATP generated in the cell comes from the conversion of hydrogen atoms to H_2O. This is the process known as *oxidative phosphorylation*, and approximately three ATP are generated for every pair of hydrogens oxidized to form water. So wherever two hydrogens are removed from a chemical compound in a bacterial cell in the course of metabolic conversions, these two hydrogens can yield 3 ATP once oxidation to water is complete.

The conversion of these hydrogen atoms to water is achieved by the cytochrome system of the bacteria working in conjunction with other hydrogen carriers. Thus the hydrogens—which are never liberated as free hydrogen, but are immediately used to reduce an appropriate carrier—are ultimately oxidized to water in a series of steps, and the three ATP molecules are liberated one at a time as the process takes place. The earlier hydrogen carriers in this sequence undergo oxidation and reduction by the removal or uptake of hydrogen, but the cytochromes do not act in this way. These molecules—and a number of different types are involved in the process—carry a Fe atom at their active centres and with these molecules the oxidation/reduction transition occurs in terms of a Fe^{3+}/Fe^{2+} change. Be that as it may, the final stage in the transfer is the reduction of cytochrome from the Fe^{3+} to the Fe^{2+} state and the coupled reduction of oxygen to form water (Fig. 2.16).

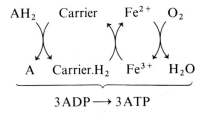

$$3\,ADP \longrightarrow 3\,ATP$$

Fig. 2.16 The oxidation of hydrogen ions to form water in micro-organisms.

Earlier in this chapter we stressed the economic advantage to the cell of having regulatory effects operating against a number of protein products at once (p. 17 *et seq.*), and also the advantage of having single enzymes serving multiple functions (p. 26). The presence of ATP and of the cytochrome system will oxidize hydrogen regardless of the chemical reactions from which it arises.

In view of these arrangements we can now see how important respiration is to living systems. Since the process ultimately oxidizes C atoms to CO_2 and H atoms to water, these reactions (particularly the latter) are the main source of ATP for biosynthesis. If oxygen is not available, the cell must rely on

substrate level phosphorylations for its energy, and for this to be equally effective quantitatively, many more molecules must be metabolized. Alternatively the bacteria must seek a source of energy other than the metabolism of small organic molecules, which is characteristic of heterotrophs. Perhaps the most efficient alternative source is light energy (*photosynthesis*) but other bacteria use the chemical interconversion of small inorganic molecules—the process known as *chemoautotrophy*.

Bacterial photosynthesis

Photosynthetic bacteria—like green plants—use light energy rather than organic molecules as a source of energy. As a consequence all the polymerized carbon in the bacteria can be built up from CO_2 by a reaction which is basically

$$CO_2 + H_2X - \xrightarrow{\text{light}} (CH_2O) + H_2O + 2X$$

where (CH_2O) denotes cell material. That is, a source of reducing power (H_2X) reduces the CO_2 to a form that can ultimately be used for biosynthesis.

In green plants the source of reducing power is water, and the general equation is in the form:

$$CO_2 + H_2O - \xrightarrow{\text{light}} (CH_2O) + H_2O + O_2$$

Oxygen gas is evolved. Most bacteria, however, use H_2S as a source of reducing power, so oxygen is *not* evolved, but sulphur is precipitated:

$$CO_2 + H_2S - \xrightarrow{\text{light}} (CH_2O) + H_2O + 2S$$

Alternatively a small organic molecule can be used as a reductant:

$$CO_2 + \text{succinate} - \xrightarrow{\text{light}} (CH_2O) + H_2O + \text{fumarate}$$

but, again, no oxygen is liberated. This then is the clear distinction between bacterial photosynthetic systems and those of green plants: *bacteria produce no oxygen while green plants do.*

As with all complex biochemical interconversions the process occurs in stages. Photosynthetic bacteria use light energy for the formation of ATP and the accumulation of reducing power by trapping the light in pigments similar to, but not identical with, those found in green plants (chlorophyll and carotenoids). The ATP and the reducing power are then re-used in further reactions to reduce CO_2 in a series of reactions which do not involve light. Probably the key step is the 'fixation' of CO_2 according to the reaction

$$
\begin{array}{c}
\text{H} \\
| \\
\text{H—C—O—phosphate} \\
| \\
\text{C} = \text{O} \\
| \\
\text{H—C—OH} + CO_2 \\
| \\
\text{H—C—OH} \\
| \\
\text{H—C—O—phosphate} \\
| \\
\text{H} \\
\end{array}
\quad
\xrightarrow[\substack{\text{ribulose}\\ \text{diphosphate}\\ \text{carboxylase}}]{}
\quad
\begin{array}{c}
\text{H} \\
| \\
\text{H—C—O—phosphate} \\
| \\
\text{HO—C—H} \\
| \\
\text{COOH} \\
\\
+ \\
\\
\text{COOH} \\
| \\
\text{HO—C—H} \\
| \\
\text{H—C—O—phosphate} \\
| \\
\text{H} \\
\end{array}
$$

ribulose 1,5-diphosphate

2 molecules of 3-phosphoglycerate

The phosphoglyceric acid generated in this way is used as the prime source of all the organic molecules needed by the cell. In fact the phosphoglycerate is also used to regenerate ribulose-5-phosphate, the substrate into which CO_2 is fixed (see Fig. 2.17). CO_2 fixation therefore occurs as part of a cyclic series of reactions, the turning of the cycle being driven by energy derived from light.

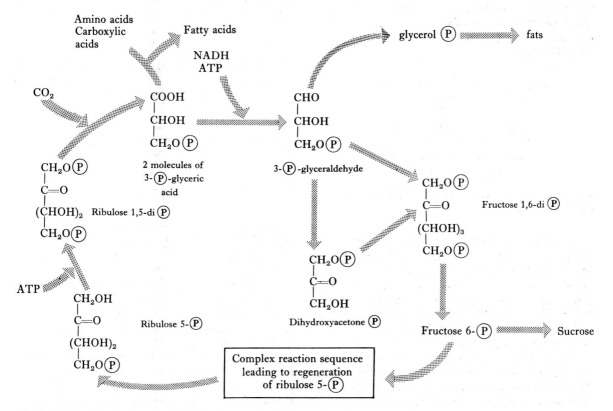

Fig. 2.17 A simplified representation of the photosynthetic carbon reduction cycle.

Chemoautotrophic bacteria

Just as photosynthetic bacteria use light energy and a hydrogen donor to reduce CO_2 to form macromolecular building blocks, so other bacteria—known as *chemoautotrophs* or *chemolithotrophs*—use energy derived from chemical interconversions of small molecules for biosynthesis.

Perhaps the simplest of these, at least in terms of the reaction used to provide energy, are the hydrogen bacteria such as *Hydrogenomonas*. These organisms use atmospheric oxygen to oxidize the hydrogen which is dissolved in small quantities in water, and the resulting energy release is used to generate ATP. Other types of bacteria can also use hydrogen as a source of reducing power, but the oxidizing agent is not oxygen. *Desulphovibrio desulphuricans* for example uses sulphate as the oxidant, and the product is water $+ H_2S +$ energy.

Apart from the bacteria that use hydrogen as a source of reducing power, there are three main types of chemolithotrophs: sulphur bacteria, iron bacteria and nitrogen bacteria. Table 2.1 shows the chemical reactions used by a number of these organisms to generate ATP. In each case the oxidation yields energy which is converted to ATP and then used for biosynthesis.

Table 2.1 Reactions used to provide energy for biosynthesis by some chemolithotrophic bacteria

Organism	Energy source	Oxidant	Products
Hydrogenomonas	Hydrogen	Oxygen	Water
Desulphovibrio sp	Hydrogen	Sulphate	$H_2S +$ Water
Nitrobacter	Nitrite	Oxygen	Nitrate + Water
Nitrosomonas	NH_4^+ ion	Oxygen	Nitrite + Water
Thiobacillus thio-oxidans	Sulphur	Oxygen	Sulphate + Water
Thiobacillus ferro-oxidans	Fe^{2+} ion	Oxygen	Fe^{3+} ion + Water

Summary

In this chapter we have outlined very briefly in chemical terms the ways in which bacteria grow and how the energy for the process is provided. Many species (the heterotrophs) take in simple organic molecules and treat them in two ways: some molecules are used as building blocks to make the necessary macromolecules for cell growth, while the others are degraded to provide the necessary energy. A few molecules have a fate which depends on the metabolic state of the organism at the time and the external environment in which it finds itself.

Other bacteria do not derive their energy from the breakdown of simple organic molecules but use either light energy—the photosynthetic autotrophs—or energy derived from the interconversion of small inorganic molecules—the chemolithotrophs. In this case there is no need to take up organic molecules from the environment since energy can be made available without them. This enables bacteria of this type to penetrate ecological niches containing inorganic salts but no carbon source save CO_2, and in fact such bacteria can grow and multiply on CO_2, NH_4 ions, water and a supply of small inorganic molecules to provide the essential elements, provided a system for producing ATP for biosynthesis is also available.

The chemical diversity of bacterial metabolism shows how extensively bacteria have evolved. In some cases the need for metabolic enzymes has been sacrificed at the expense of having to take up molecules preformed from the environment, and in some pathogenic bacteria this specialization has reached the point that they cannot grow outside the organism that they are infecting. At the other extreme a totipotent synthetic apparatus within the cell allows bacteria to grow in chemical environments that are far too spartan for organisms which do not derive their energy requirements from light or simple inorganic chemical reactions.

Books and articles for further study (Chapters 1 and 2)

General references:

LEWIN, B. (1974). *Gene Expression—1*. John Wiley, New York.

SZEKELY, M. (1980). *From DNA to Protein*. Macmillan, London.

WATSON, J. D. (1977). *The Molecular Biology of the Gene*, 3rd Edn. W. A. Benjamin, New York.

WEISSBACH, H. and PESTKA, S. (1977). *Molecular Mechanisms of Protein Biosynthesis*. Academic Press, New York.

Nucleic acids

CRICK, F. H. C. (1954). Structure of hereditary material. *Scient. Am.*, **191**, No. 4, 54.

DAVIDSON, J. N. (1976). *The Biochemistry of the Nucleic Acids*, 8th Edn. Chapman and Hall, London.

DU PRAW, J. (1971). *DNA and Chromosomes*. Holt, Reinhart and Winston, New York.

SÖLL, D., ABELSON, J. N. and SCHIMMEL, P. R. (1980). *Transfer RNA – Biological Aspects*. Cold Spring Harbor Laboratory Press, Cold Spring Harbor.

STEWART, P. R. and LATHAM, D. S. (1977). *The Ribonucleic Acids*, 2nd Edn. Springer Verlag, Berlin.

Ribosomes

BRIMACOMBE, R. (1978). Ribosomes, In *28th Symposium of the Society for General Microbiology*, Cambridge University Press, Cambridge.

RICH, A. (1966). Polyribosomes. *Scient Am.*, **209**, No. 6, 44.

STÖFFLER, G. and WITTMANN, H. G. (1977). Primary structure of proteins within the *E. coli* ribosome. In *Molecular Basis of Protein Synthesis* (Ed. by H. Weissbach and S. Pestka). Academic Press, New York.

Genetic code

CLARK, B. F. C. (1977). *The Genetic Code*, Studies in Biology no. 83, Edward Arnold, London.

CLARK, B. F. and MARCKER, K. A. (1968). How proteins start. *Scient. Am.*, **218**, No. 1, 36.

CRICK, F. H. C. (1962). The genetic code. *Scient. Am.*, **207**, No. 4, 66.

CRICK, F. H. C. (1966). The genetic code. III. *Scient. Am.*, **215**, No. 4, 55.

NIRENBERG, M. W. (1963). The genetic code. II. *Scient. Am.*, **208**, No. 3, 80.

WOESE, C. R. (1967). *The Genetic Code*. Marcel Dekker, New York.

YCAS, M. (1970) *The Biological Code*. North Holland Publishing Co., Amsterdam.

Regulation of gene expression

DU PRAW, J. (1971). *DNA and Chromosomes*. Holt, Reinhart and Winston, New York.

KENNEY, F. T., HAMKALO, P. A., and FAVELUKAS, G. (1973). *Gene Expression and Regulation*. Plenum Press, New York.

Chapter 3

Genetics of micro-organisms

Bacterial inheritance

Introduction

The structure of DNA and how this molecule can give rise to enzymes and other proteins of specific structure in biological systems was described in detail in the previous chapter. In essence, the order of purine and of pyrimidine bases along the poly-nucleotide backbone of the DNA carries the essential information for the cell to make specific proteins by the process of protein biosynthesis. Some of the proteins made in this way then give rise to the other molecules—both large and small—that go to make up the microbial cell; so the information needed to make a new bacterial cell resides in the cell's DNA, and primarily in the sequence of nucleotides along that molecule.

However, this is not the only role of the DNA in the cell. It is also responsible for passing on the necessary information to make a new cell from one generation to the next; that is, DNA both carries the information necessary to make a daughter cell from the one in which it is functioning at a particular time and also passes the information on to successive generations.

The way in which DNA replicates distribute the product molecules to the daughter cells has also been dealt with in the previous chapter. In this chapter we are concerned with the way in which the DNA is organized in the cell, and how that organization facilitates its role both as a molecule coding for proteins and also as a source of heritable information.

A further concern in this chapter is the ways in which changes in the hereditary information may occur. Broadly this may take place in two ways: on the one hand chemical alterations may take place in the DNA molecule itself—particularly in the nature of its nucleotide sequence. Such changes are called *mutations*. Their consequences generally are to alter the structure of the protein coded for by that gene but also to change the nature of the heritable information.

The second source of change is when some exogenous DNA enters a cell and interacts with the resident DNA to give rise to a hybrid molecule. Such interactions are known as *recombination events* and may produce exceptionally big changes in the nature of the heritable information passed on by a cell to its descendants. Such interactions occur both in prokaryotic and eukaryotic cells. In summary, therefore, bacterial inheritance will be discussed under the following main headings.

1. Organization of the hereditary information
2. Changes in the nature of the hereditary information—mutation
3. Changes in the nature of the hereditary information—recombination
4. Mechanisms of gene exchange.

The treatment of each of these topics here will be of the briefest. An enormous amount is now known about the organization and behaviour of DNA in microbial cells—the subject generaliy known as Microbial Genetics—and for a detailed account a more advanced text must be consulted.

The organisation of the hereditary information

Chromosomes and extrachromosomal elements

DNA is a molecular thread consisting of two strands which run in opposite directions to one another (see Chapter 2). In bacteria the majority of the DNA in the cell is organized as a single molecule known as *the chromosome*—even though the properties of such structures are different in many ways from the chromosomes of nucleated cells. The bacterial chromosome—which normally is about one millimetre long and which has a molecular weight of about 2×10^9—is covalently closed; that is, the ends of the polynucleotide threads are covalently joined to make a continuous circular molecule. DNA in such a configuration is said to be in the covalently closed, circular (or CCC) form

Linear double stranded DNA

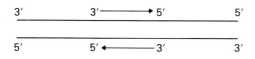

CCC DNA

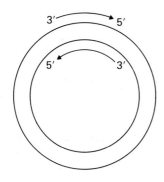

Fig. 3.1 Linear double-stranded and covalently closed, circular (CCC) DNA.

(Fig. 3.1). In bacteria and in those bacterial viruses that contain DNA the nucleic acid, the DNA is present as a single copy, and is not complexed—as is the case in nucleated cells—with basic proteins such as histones.

Although the chromosome is always the largest single piece of DNA in a bacterial cell, it is not necessarily the only one. Most bacteria carry other pieces of DNA, either as *extrachromosomal elements* (or *plasmids*, as they are often called) or in the form of bacterial viruses (bacteriophages p. 126) carried in a quiescent form by the cells.

Plasmids are closed circular molecules like the chromosome; indeed the only difference that is easy to see between these two structures is size. While the chromosome has a molecular weight of about 2×10^9, plasmids range in size between about 10^6 and 10^8. Unlike the chromosome, however, plasmids are not always present as single copies. In some bacteria, for example, multiple copies of the same plasmid may exist in a single cell, while in others single copies of a number of distinct plasmids may occur.

Bacteriophage DNA may often be incorporated as part of the bacterial chromosome, and when this is the case additional discrete DNA molecules cannot, of course, be found. Other bacteriophages are carried independently of the chromosome, and in this case their structures may be indistinguishable from plasmids. Certainly their size is similar and in practice the only way to distinguish phage DNA from other pieces of extrachromosomal CCC DNA in bacteria is to determine the nature of the information implicit in their DNA base sequence; and that can only be done by appropriate biological experiments.

Replicons

The fact that a piece of CCC DNA exists in a bacterial cell—and this applies equally to the chromosome and to plasmids—implies that the piece of DNA in question is able to replicate and segregate as the culture grows. Such independently replicating pieces of DNA in bacteria are known as *replicons*.

Apart from their relative sizes, the chromosomal replicon may be distinguished from bacteriophages and from plasmids by an important practical consideration. Extrachromosomal elements may sometimes fail to replicate or be distributed to daughter cells. This leads to loss from the bacterial cell. Normally this process does not occur frequently, although great instability of some extrachromosomal elements may be encountered on occasion.

Genotype and phenotype

Since both the chromosomes and the extrachromosomal elements are double-stranded DNA, both carry genetic information which adds to the total carried by the cell. The totality of this information is commonly referred to as the cell's *genotype*, and this term is used in contrast to *phenotype*, a word that describes the characters actually expressed by the cell at any instant in time. In bacteria, the majority of the DNA in the cell is expressed at all times, the only exceptions being information which is repressed or otherwise 'switched off'. Normally this unexpressed information does not amount to more than about 10 % of the total. This fact also serves to distinguish prokaryotes (p. 176) from eukaryotes (p. 135)—particularly higher eukaryotes such as vertebrates—where a very large amount of the total genotype may be unexpressed, particularly when the organism is in its adult form. This 'switching off' of DNA in higher organisms is often claimed to play an important role in cell differentiation in these forms.

Genes and cistrons

The biosynthetic relationship between DNA and protein in the cell means that certain regions on the DNA carry the necessary information to order the amino acids along the polypeptide chain of a single protein, and such lengths of DNA are referred to as *genes*. This relationship is summarized in Fig. 3.2.

Originally it was thought that all the DNA of bacterial cells was always expressed as protein at all times, but it is now known that this is not so. Certain regions of the DNA do not specify the synthesis of proteins but are the targets at which effector proteins bind directly to the DNA. Normally this interaction leads to some form of regulatory control on the expression of the DNA. Examples are the promoter regions at which RNA polymerase binds to the DNA (p. 29) and the operator regions, stretches of DNA where repressor molecules responsible for regulating the expression of inducible and repressible proteins bind (p. 29).

Regions that act in this way are commonly called *regulatory genes* to distinguish them from genes responsible for specifying protein products—genes normally known as *structural genes*. Unfortunately, however, this distinction between genes on the basis of their biosynthetic properties is not absolutely clear cut. Certain genes with regulatory function (see for example the *i* gene involved in regulating β-galactosidase biosynthesis in *Escherichia coli* (p. 29) also specify a protein product. Thus structural genes always specify protein products, but there are two types of regulatory gene: those that do specify a protein product and those that do not.

In practice many enzymes in bacteria contain more than one polypeptide chain and consequently the gene that specifies their synthesis has to contain the appropriate number of sections to code for the individual chains. Such sections are called *cistrons*. This aspect of gene organization is summarized in Fig. 3.3.

The term cistron arises from the nature of the genetic tests used to show their presence, a test based on the phenomenon of *genetic complementation* (Fig. 3.4). For this test to be carried out, it is necessary to establish a genetic situation in which two copies of the gene being investigated are present in the same cell. This situation is unstable and tends to break down to the state where only one gene is retained, but nevertheless the double system can be established in certain cases (see below). When such a situation is set up and the two genes are both unmutated, then the cell will normally make twice as much of a normal protein product, regardless of whether that particular protein consists of a single polypeptide chain or more than one (Fig. 3.4a). If on the other hand one of the two genes is altered so that it makes an inactive product (that is the gene is mutated) then two protein products are produced by the cell when it has two genes: one active and the other inactive (Fig. 3.4b).

It is when the two genes are both mutated so that neither individually makes an active product that complementation can occur; and then only when the proteins concerned consist of two polypeptide chains. Let us look at this phenomenon in more detail. If both genes are damaged, but the damage in both cases lies in the part of the gene coding for the same polypeptide chain, no active enzyme can be formed (Fig. 3.4c). However, if the damage lies in one part of the first gene but in the other part of the second, active protein may emerge by the cell's using the appropriate parts of the two gene products. Thus active polypeptide A, specified by gene A can complement active component B, specified by gene B, to produce the active protein AB (Fig. 3.4d). Thus the presence of cistrons in a gene can be

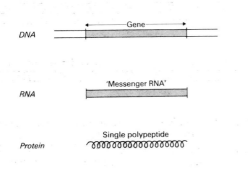

Fig. 3.2 The relationship between genes and proteins. Here one gene codes for a protein of one polypeptide chain.

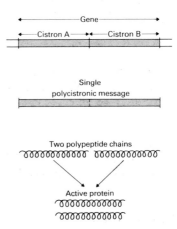

Fig. 3.3 The arrangement of cistrons in a gene coding for a protein of two polypeptide chains.

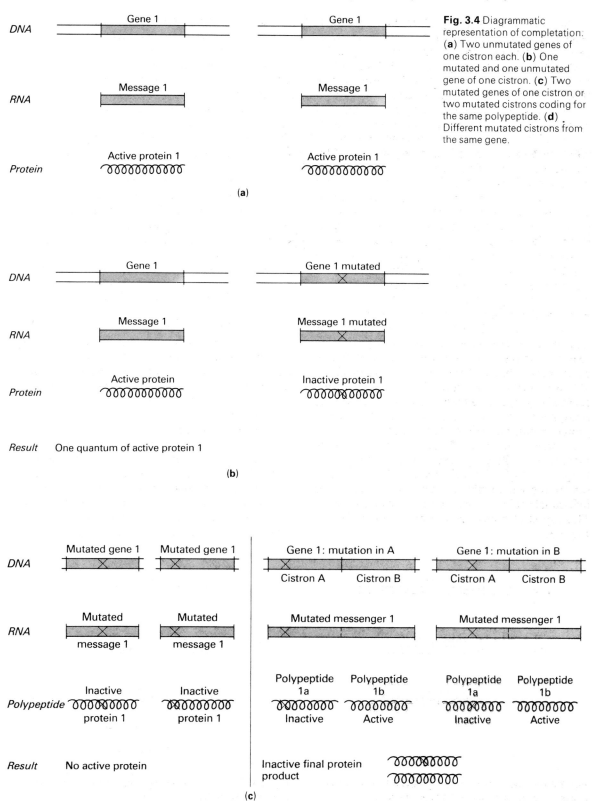

Fig. 3.4 Diagrammatic representation of completation: (**a**) Two unmutated genes of one cistron each. (**b**) One mutated and one unmutated gene of one cistron. (**c**) Two mutated genes of one cistron or two mutated cistrons coding for the same polypeptide. (**d**) Different mutated cistrons from the same gene.

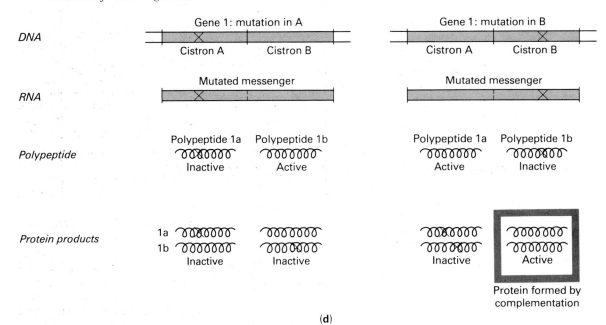

(d)

inferred when complementation occurs to produce an active product; but it is an important proviso of the test that complementation should be detected only under circumstances in which recombination (see below) between the two genes concerned can be ruled out as a source of the active protein product.

Operons

In bacterial cells, certain groups of genes, and commonly groups of genes whose protein products have related physiological function, are under co-ordinate control, and such groups are called *operons*. An example already cited in Chapter 2, was the group of three genes of lactose system of *Escherichia coli* whose expression is under the control of a single set of regulatory determinants. There are many such examples.

One of the properties of operons is that the DNA which comprises the genes concerned is commonly transcribed to form a single *polycistronic messenger RNA*; that is a single RNA molecule is made that reflects the base sequence of the whole operon (Fig. 3.5). A consequence of the formation of such a

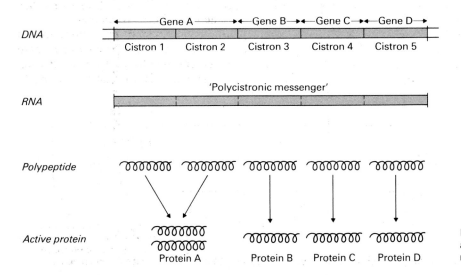

Fig. 3.5 Relationship between a gene and a polycistronic message.

single product is that a mechanism must exist whereby the single piece of information is translated into the appropriate number of separate polypeptide chains, which then group themselves to form the appropriate number of independent proteins (see Chapter 2).

In summary, therefore, the chromosome of bacteria contains a succession of genes (about 10 000 in all) arranged either singly or in groups (operons). The individual genes and operons are controlled in their expression by regulatory mechanisms of variable complexity, but the net result is that the great majority of genes in the bacterial chromosome are expressed as protein products at all times. A similar situation exists for non-chromosomal replicons. One cannot be dogmatic as to whether most genes consist of multiple cistrons or not. Certainly many bacterial proteins consist of more than one type of subunit, but such an arrangement is far from universal.

Mutation

As long as the base sequence of their DNA does not change, bacteria will 'breed true' from generation to generation. Any change in base sequence, however, will alter the informational content of the DNA and this in turn is likely to produce heritable changes in the structure of at least ohe bacterial protein.

Such *mutations* in DNA base sequence may be of a number of different kinds. Changes in the nature of a single base are called *point mutations*, removal of sections of the DNA are known as *deletions*, and the removal of a piece of DNA from one position on a replicon to another position on the same replication, or to a position on another replicon in the same cell, is known as a *transposition*. A special situation, in which the sequence of the DNA is altered either by adding or removing a single base pair is known as a '*frame-shift mutation*'.

Point mutations

Point mutations themselves are of two types: *transitions* and *transversions*. In the former a pyrimidine is replaced by a pyrimidine or a purine by a purine, while with transversions pyrimidines are replaced by purines or vice versa. The way in which a triplet of bases in the DNA codes for an equivalent triplet in the messenger RNA and how this RNA triplet—or codon—then specifies the insertion of a specific amino acid into the growing polypeptide chain has already been described in Chapter 2. Thus individual base triplets in the DNA code for the insertion of individual amino acids.

Even though there is this direct relationship between DNA sequence and amino acid insertion, this does not mean that changes in base sequence inevitably produce alterations in the amino acid sequence of the protein. The genetic code is redundant—that is certain amino acids are coded for by more than one triplet, and therefore a change of base sequence in the gene does not necessarily change the amino acid that is specified. An example of how this may occur is shown in Fig. 3.6. On the other hand, if the base change is one that does alter the nature of the amino acid inserted in the protein, then this change persists in a heritable manner in the bacterial culture as long as its effect is not lethal.

Point mutations may be caused by a variety of agents known as *mutagens*. Many of these agents are chemical compounds which produce a direct change in the chemical nature of the DNA. But ionizing radiations are also potent mutagens, and in this case the chemical changes in DNA sequence—which are inescapable if mutation is to occur—are an indirect effect of the radiation interacting either with a component of the DNA itself or with other molecules in the immediate environment.

One of the characteristics of point mutations is that they can *revert* by a further change in base sequence. Sometimes the DNA returns to its original structure, and this is an example of *back-*

DNA Codon	———— A A A ————	———— A A G ————
	———— T T T ————	———— T T C ————
RNA Codon	———— A A A ————	———— A A G ————
Amino Acid	———— Lysine ————	———— Lysine ————

A = Adenine;   T = Thymine;   G = Guanine;   C = Cytosine

Fig. 3.6 Demonstration that change in base sequence does not inevitably produce an altered protein product.

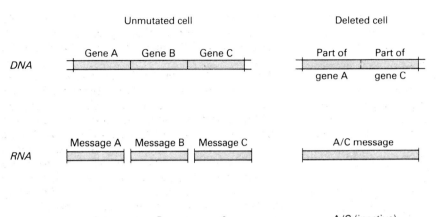

Unmutated cell   Deleted cell

Fig. 3.7 The consequence of genetic deletion on polypeptide and protein structure.

mutation to parental type. This process leads to the production of the original protein, and the back-mutated cells are indistinguishable from the original parent. A more complex step is the generation of a further 'forward' mutation that leads to the reappearance of properties similar to, but not identical with, the original parent. Such a change is called a reversion to a '*pseudo-wild-type*'. Not all changes in amino acid sequence of proteins inactivate them. In some cases, activity can be reduced or destroyed by the first mutation and then partially restored by the second.

Deletions

Deletions are mutations in which a piece of DNA is removed from a replicon and the gap closed by joining together the ends generated by the process (Fig. 3.7). The effect of deletions is to remove certain genes completely from the genome and to damage the two genes that span the gap, since the remaining half of the gene on the left of the gap will be joined with the residue of the gene on the right.

Deletions are caused by chemical agents and by irradiation. Exactly how they arise is still obscure but they probably occur as a consequence of aberrant DNA replication. They do not revert spontaneously and cannot normally be reverted by further mutagen action.

Transpositions

It has been known for many years that genes in higher organisms can change their location in a genome with consequent effects on phenotype. Only recently, however, has it become clear that an analogous process also commonly occurs in bacteria. Indeed it may be one of the most important genetic events which shapes evolution in bacteria.

Transposition in prokaryote genomes is characterized by the transfer of a defined stretch of DNA from one replicon to another, a process which does not result in the loss of the material from the donor. Thus the process duplicates genetic information. The unit of DNA which moves during transposition is now known as a *transposon*. The overall process of transposition is summarized in Fig. 3.8.

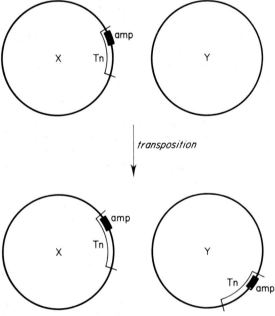

Fig. 3.8 Summary of the transposition process. In this case a transposon (Tn) specifying resistance to β-lactam antibiotics (amp) transposes from plasmid X to plasmid Y. Note the duplication of material.

Initially it seemed that transposon-specified transposition in prokaryotes was a feature peculiar to the evolution of antibiotic resistance, but this is now known not to be the case. Transposons specifying resistance to many bacterial characters unrelated to antibiotic resistance—such as toxin production, metabolic enzymes, metal ion resistance—are now known. Nevertheless it is primarily the antibiotic resistance transposons which have been the most fully investigated.

All transposons have a number of important common features, and in all cases the amount of DNA involved is more than that required to code for the particular phenotypic character which is transposed. Thus the common ampicillin resistance transposon of enteric bacteria (TnA) has enough DNA to code for about four times as much protein as is involved in the β-lactamase enzyme itself. In fact it is now clear that transposons are highly specific DNA sequences which both contain a DNA sequence coding for a product of selective value, and also for the necessary enzymes to catalyse the transposition of the transposon to other replicons. In this way, therefore, transposons are complementary in effect to self-transmissible bacterial plasmids: the transposons catalyse the transfer of genetic information between replicons (and among these will be self-transmissible plasmids) whereas the transmissibility of plasmids catalyses the exchange of information between bacterial cells (see p. 49). Under these circumstances it is obvious that the transposon/plasmid interaction must have great evolutionary impact.

Those transposons whose organization has been examined in detail all seem to share two important features: first they all seem to be limited at their extremes by inverted repeats—sequences at one end of the transposable unit which are repeated *in reverse order* at the other. Secondly they contain sequences which code for specific enzymes which catalyse the transposition process. In the case of

TnAs there are two such enzymes: one relatively small and the other relatively large. The overall organization of the ampicillin transposon TnA is shown in Fig. 3.9 as a typical example of this sort of unit. In this case the 'inverted repeats' are 38 nucleotide pairs in length but in other transposons the analogous structures are considerably longer.

One important feature of transposition, is that there seems to be little site specificity to decide whereabouts in a recipient replicon the transposon will come to rest. Certainly there seems to be no requirement for sequence homology as is the case for classical recombination—see below. Thus many transposition events would seem *a priori* to be potentially lethal, since insertion into the centre of an essential gene cannot be excluded. Laboratory studies show that lethal transpositions are, in fact, far from rare.

The fact that transposon specified transposition is potentially a lethal event argues that bacteria must have some means of minimizing its impact. Although all transposons are mutagenic, their impact often seems less in naturally-occurring strains than in specially selected laboratory bacteria. Almost certainly most prokaryotes do have means for 'immunizing' (at least partially) their replicons against the arrival of transposons, though the precise mechanisms involved are still obscure. Overall it seems clear that transposition does occur in nature at a rate sufficiently high to be an important source of evolutionary change, without commonly being so prevalent as to undermine the genetic integrity of the bacteria—probably the majority that carry transposable elements.

Recombination

The previous section has described how the structure, and therefore the informational content, of DNA can be modified by chemical means. The other method to achieve the same end is to replace

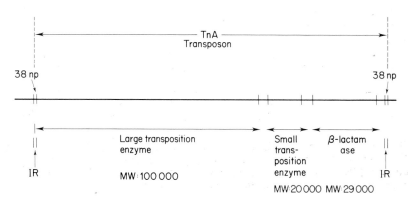

Fig. 3.9 Structural organization of the ampicillin resistance transposon TnA. IR, inverted repeat sequence; MW, molecular weight; np, nucleotide pairs. The whole TnA transposon is about 4500 np in length.

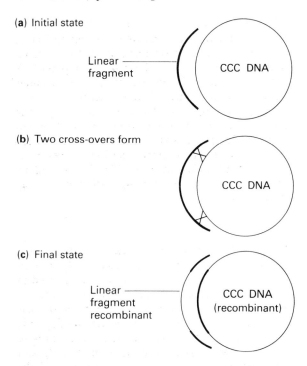

(a) Initial state

Linear ——— CCC DNA
fragment

(b) Two cross-overs form

CCC DNA

(c) Final state

Linear ——— CCC DNA
fragment (recombinant)
recombinant

Fig. 3.10 Recombination of a linear double-stranded DNA fragment with a double-stranded CCC DNA molecule. (*Note* all DNA strands are *double* strands.)

one region in the molecule by a different piece of DNA. This is the process known as *recombination*. Until recently it was a process which could only be carried out in living systems, but now *in vitro* recombination has been achieved. This has given rise to the exciting prospects of '*Genetic Engineering*'.

For natural recombination to be possible in bacterial cells, two conditions must be fulfilled: first, the cell in which recombination is occurring must contain two appropriate pieces of DNA; and secondly, there must be some base sequence homology between these pieces. The homology need not extend over the whole of the two molecules, but it must be enough for chemical interaction to occur in the required manner.

The overall effect of recombination depends on the state of the DNA molecules that are involved. For the recombinant configuration to survive, one at least of the interacting pieces must be a replicon, that is it must be in a CCC form. The other piece may, however, be either CCC or linear; but the outcome of the process is substantially influenced by which types of DNA are involved.

If linear DNA interacts with CCC DNA, two

breakages and reunions, both involving both strands of the DNA, are needed to insert a piece of DNA in to the circular replicon (Fig. 3.10). Such breakages and reunions are known as *cross-overs*. Note that a single cross-over will only give a linear product, and this will not replicate efficiently (p. 31 and Chapter 2). Even with two cross-overs residual pieces of linear DNA are generated which cannot replicate, and these cannot become part of the stable hereditary content of the bacterial population. This type of recombination is involved, for example, when a piece of linear DNA is transferred to a bacterial cell by transformation (see later).

The other type of recombination occurs when two CCC molecules interact. In this case one cross-over gives rise to a single circular molecule containing all the DNA originally present in the two interacting replicons (Fig. 3.11). A further cross-over will then generate two circles, each containing material from both parental plasmids. As a result

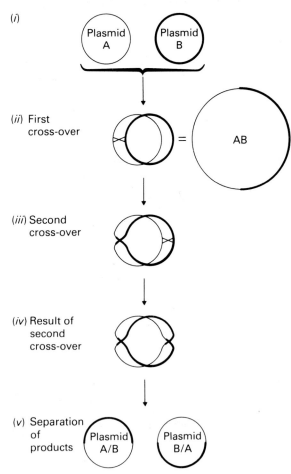

(i)

Plasmid A   Plasmid B

(ii) First cross-over   = AB

(iii) Second cross-over

(iv) Result of second cross-over

(v) Separation of products   Plasmid A/B   Plasmid B/A

Fig. 3.11 Recombination of circular replicons.

therefore, information may be exchanged between the initially interacting replicons (Fig. 3.12).

In practice cross-overs only occur (except by rare exceptions known as *illegitimate* recombinations) at points where base-sequence homology exists between the interacting molecules. But since this homology does not need to extend over the whole length of the interacting structures, recombination provides a means of introducing foreign stretches of DNA into a bacterial replicon. In the example shown in Fig. 3.12, for example, two regions of homology span a heterologous region and allow the insertion of the heterologous material into the resident replicon.

Recombination is a process of crucial importance to the evolution of bacterial cells, since it allows substantial pieces of foreign DNA to become incorporated into bacterial replicons. This allows the acquisition of additional bacterial characteristics which have proved their worth in other organisms, a process never possible by point mutation or deletion.

Recombination between pairs of DNA molecules, one a replicon and one a linear molecule, allows the order of genes along a piece of DNA to be deduced. This process is known as 'mapping'. A textbook of microbial genetics should be consulted for the methods involved in this process.

It should perhaps be mentioned that the transposition process (see p. 46) is a form of recombination. Indeed it was initially classed as an example of 'illegitimate recombination' processes. It is distinct from classical recombination in prokaryotes in that it leads to gene duplication, not simply to gene exchange.

In vitro recombination

A major development in the technology of manipulating DNA in the test tube has recently allowed recombinant DNA molecules to be constructed in the test tube. The techniques involve the isolation of intact replicons—that is CCC DNA—and their conversion to a linear form by the action of specific endonucleases. If the correct enzyme is chosen, many plasmids can be opened at one point only, and the result is a linear molecule containing all the information in the original circular structure (Fig. 3.13).

The endonucleases used for this purpose do not usually cut straight across the DNA double helix, but rupture the two strands of the molecule a few base pairs apart. Thus the linear molecule that is

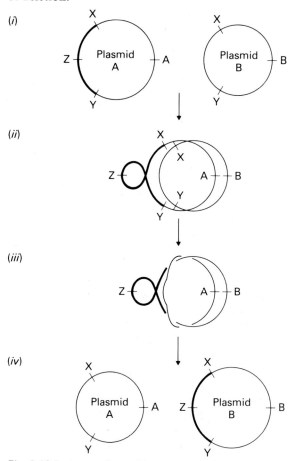

Fig. 3.12 Exchange of gene Z between two plasmids.

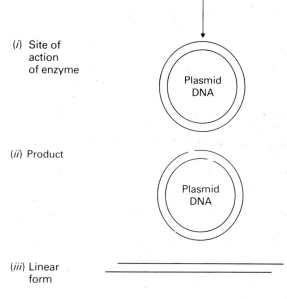

(i) Site of action of enzyme

(ii) Product

(iii) Linear form

Fig. 3.13 Action of restriction endonuclease as a plasmid containing a single susceptible site.

generated has single-strand tails attached to the ends of the linear double-stranded regions (Fig. 3.13).

If the same endonuclease preparation is now used to generate fragments of DNA from another un-related DNA source, linear fragments are produced once again, and these will each now carry tails of the same sequence as those on cut replicon because of the specificity of the enzyme involved (Fig. 3.14).

If the two preparations of DNA are now mixed in the presence of DNA ligase—an enzyme capable of joining together split fragments of DNA—some artificial recombinants will be formed in which a fragment of foreign DNA has been enclosed in the gap generated in the plasmid replicon. Note that this is not a process that leads to reciprocal exchange of DNA. Moreover, since these recom-binants are now in the CCC form they will replicate once inserted into an appropriate bacterial cell. This step is achieved by transformation (see later) and one now has the situation whereby a laboratory constructed recombinant DNA has become a herit-able part of a living system.

This approach has now been used to form a large number of 'unnatural' genetic linkages in the labor-atory, some involving only prokaryotic genes but others constructed from both pro- and eukaryotic sources. Even viral and prokaryotic genes have been recombined in this way. The technique has clearly attracted enormous interest both for the opportun-ities it gives to investigate fundamental biological problems, but also for the considerable commercial potential of some of the new constellations of genes constructed in this way.

Gene exchange between bacteria

Introduction

The way in which bacterial DNA is organized and carried in the bacterial cell has already been des-cribed in the earlier sections of this chapter. The ways in which this DNA—be it chromosomal, plasmid or bacteriophage—can be transferred be-tween bacteria will now be discussed. The only exception to this is the process of normal phage infection which is described in Chapter 7.

DNA may be transferred between bacteria by the three distinct processes known as *Transformation*, *Conjugation* and *Transduction*. In the first the DNA passes between the bacteria as a naked molecule, and the nucleic acid can be isolated in active form from the medium separating the bacteria. In con-jugation the DNA passes in a protected form during cell-to-cell contact between the organisms. This process is complex and surface appendages on the donor bacterium play an essential part in the process. During transduction, as mentioned above, the DNA is protected by the protein coat of the virus which acts as the vehicle taking the DNA from one cell to another.

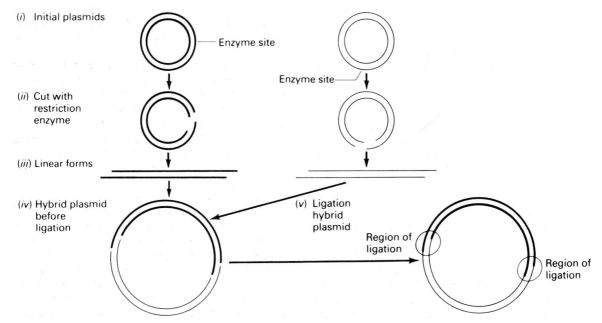

(*i*) Initial plasmids

Enzyme site

Enzyme site

(*ii*) Cut with restriction enzyme

(*iii*) Linear forms

(*iv*) Hybrid plasmid before ligation

(*v*) Ligation hybrid plasmid

Region of ligation

Region of ligation

Fig. 3.14 Artificial recombination involving two distinct DNA molecules.

Transformation

Transformation studies were the first to prove that all the information necessary to make bacterial proteins was carried as DNA since chemically pure deoxyribonucleic acid was able to transfer the ability to make a cell product from one bacterial cell to another. The DNA involved in this transfer process is liberated from the donor by lysis and then taken up by the recipient, but only when the latter is in a receptive phase of growth—usually known as the *competent state*.

The highest frequency of transformation is normally obtained when the transforming DNA is double-stranded and circular. Under these circumstances the incoming DNA can usually form a replicon immediately on reaching the interior of the recipient bacterial cell, and its inheritance is therefore assured without any involvement with the resident DNA of the recipient. This type of DNA transfer is therefore particularly relevant to plasmid transfer since the DNA concerned in this case is usually in the form of small circular structures which readily survive independently of the chromosome in the recipient bacteria. Occasionally the transferred DNA cannot replicate in the recipient, and under these circumstances transfer is said to be 'abortive'.

Although CCC DNA is transferred with relatively high efficiency, both linear double-stranded and linear single-stranded DNA can be transformed, the former more effectively than the latter. In these two cases, however, the incoming DNA has to integrate into a replicon already present in the recipient if it is to survive to become part of the hereditary information of the recipient cell line. For this purpose recombination between incoming and resident DNA is necessary (p. 48). For the step to be successful two cross-overs between the incoming DNA and the resident replicon will be necessary, and for this to occur there must be the appropriate regions of homology between the two DNA molecules. The process of transformation of double-stranded linear DNA is summarized in Fig. 3.14, the chromosome of the recipient being the source of replicative potential for the survival of the incoming DNA in this case.

If single-stranded DNA is used for transformation, an additional step is necessary before survival of the information carried by the single strand of DNA is assured. The step is the conversion of the molecule to the double-stranded form, the complementary strand being synthesized in the recipient by DNA polymerase (see Chapter 2) already present in the recipient bacteria.

The one aspect of transformation that greatly limits the efficacy of the process in transferring DNA between a wide range of living species is the need to find homology in the recipient DNA. Normally these conditions are only met in closely related bacteria and the process therefore tends to be species specific. The main exception to this is the transfer of bacterial plasmids. The self-replicating abilities of the plasmid DNA allow it to survive without recombination, and plasmid transformation can therefore occur over a much wider range of organisms than transformation of linear DNA fragments.

There has been much discussion as to how widespread transformation is in Nature. In practice the presence of nucleases in the medium surrounding many bacteria probably pose a severe limit to the survival of naked DNA in a bacterial growth medium, and the general view seems to be that transformation cannot be quantitatively an important means of gene transfer in the wild state.

Conjugation

As implied by its name, conjugational transfer of DNA requires cell-to-cell contact by the participating bacteria. Indeed the process is often referred to as bacterial mating. During the process, the DNA passes from the donor to the recipient in a protected form. As far as is known the molecule always passes as a single-stranded linear molecule, and consequently the synthesis of the complementary strand to produce double-stranded DNA is always essential for the survival of the DNA in the recipient.

In practice bacterial plasmids are by far the most common form of DNA to be passed from cell to cell by conjugation, and these pieces of DNA are normally present in bacterial cells in a CCC form. Thus the overall transfer process requires the conversion of a CCC molecule in the donor into a linear single-stranded form for transfer, and then the conversion of this structure back to the CCC form in the recipient.

The first step in the donor bacterium, after the mating bridge has formed between donor and recipient is therefore to run off a single-strand copy of one of the two circular DNA strands that constitute the CCC plasmid in the donor. This occurs by a special method of DNA replication that is distinct from that normally involved in the replication of daughter plasmids for distribution at cell division. The details of this process are complex and for a description an advanced textbook of microbial genetics should be consulted. Once the linear single-stranded DNA has been transferred to the recipient, and the complementary strand synthesized to make the double-stranded linear form, the newly formed plasmid circularizes to form a CCC form in the recipient. Once in this state the plasmid can replicate in the recipient and thereafter become part of the hereditary information of the cell.

The cell-to-cell contact necessary for DNA transfer by conjugation is catalyzed by the presence on the surface of the donor bacterium of hair-like structures known as sex-pili. Their synthesis is specified by the plasmid itself, and consequently plasmid DNA capable of undergoing transfer by conjugation normally carries the necessary information to make the donor cell produce sex-pili; that is transferable plasmids are infectious and themselves carry the necessary genes to ensure their own infectivity. The conjugational transfer of a plasmid is summarized in Fig. 3.15. One must stress that this sequence has been somewhat over-simplified. A

great deal is now known about the minutiae of conjugational transfer, but for a full description one should consult an advanced text.

Normally bacteriophage DNA is not transferred by conjugation, since this type of DNA is capable of specifying the synthesis of the specific phage vector which then achieves its transfer. Commonly therefore conjugational transfer is confined to bacterial plasmids. It must not, however, be assumed that all plasmids can specify their own transfer. These elements may be classified into two categories: *conjugative plasmids*, which can specify their own transfer, and *non-conjugative plasmids*, elements that rely on other vectors, such as bacteriophage, or helper plasmids for their transfer from cell to cell.

Usually the pieces of DNA transferred by conjugation are relatively small and can form an intact replicon in the recipient without need for recombination. As a rare event, however, larger pieces of DNA—notably fragments of the chromosome—may pass by conjugation, and recombination is then essential for the survival of the transferred material. This situation commonly occurs when a bacterial plasmid—and the notable example is the bacterial fertility factor, F—becomes itself incorporated into the bacterial chromosome. When this occurs the ability to make the cell that carries the plasmid infectious is not necessarily impaired, and as a consequence the whole chromosome becomes an infectious 'plasmid'—even though it may be as much as 100 times as large as a normal conjugative plasmid. Once this has occurred DNA transfer occurs by normal conjugation, and the whole chromosome begins to pass to the recipient cell.

Conjugational transfer of the chromosome, when it occurs in this way, has two main features that set it apart from the equivalent transfer of a bacterial plasmid. First, the length of DNA that has to pass in this way is very long compared with that of the plasmid and the process tends to break down. This leads to the transfer of fragments. Secondly, since it is very unlikely that the whole of the chromosomal replicon will be transferred intact, recombination of the fragment with the recipient chromosome is necessary for the incoming DNA to reach a replicative form. This process is summarized diagramatically in Fig. 3.16.

The fact that fragments of chromosomal DNA can be transferred by conjugation in this way has led to an important method of determining the order of genes on the bacterial chromosome. If a chromosomal mating is established, and the process interrupted at intervals, the amount of DNA that passes will reflect the duration of the mating

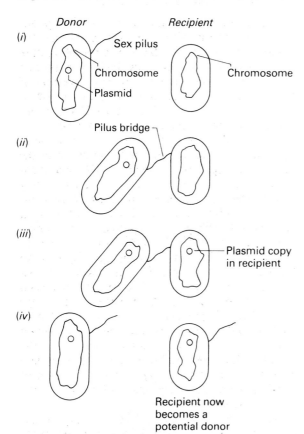

Donor   Recipient

(i) Sex pilus / Chromosome / Plasmid / Chromosome

(ii) Pilus bridge

(iii) Plasmid copy in recipient

(iv) Recipient now becomes a potential donor

Fig. 3.15 Transfer of a conjugative bacterial plasmid by conjugation (mating).

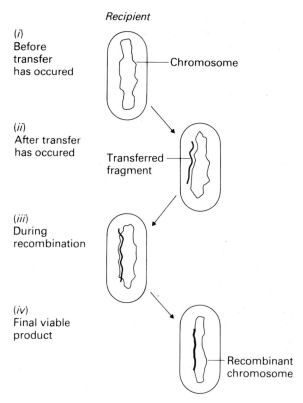

(*i*)
Before
transfer
has occured

Recipient

Chromosome

(*ii*)
After transfer
has occured

Transferred
fragment

(*iii*)
During
recombination

(*iv*)
Final viable
product

Recombinant
chromosome

Fig. 3.16 Chromosomal transfer: diagrammatic summary.

process. Since the chromosomal DNA passes to the recipient in a linear fashion, graded durations of mating, followed by analyses of the nature of the genetic information has been transferred allows one to determine the order with which genetic information occurs on the chromosome. Moreover this approach is facilitated by the observation that a given plasmid always starts the transfer of chromosomal DNA from a particular point on the chromosome. This point is known as the *initiation* site. The technique of chromosome mapping by this procedure is known as *interrupted mating*, and it has been used extensively over the last few years to establish the order of genes on the *E. coli* chromosome.

Some bacterial plasmids are capable of catalyzing transfer of chromosomal DNA which is not as comprehensive as that just described. The more limited mechanism is one where small regions of the chromosome—often only a few genes—may be transferred and become established in the recipient. In practice two versions of this mechanism are known, one where the transferred piece of DNA becomes covalently bound into the plasmid during the process, and the other where no such covalent linkage can be detected. When the chromosomal

DNA becomes part of the plasmid, two routes of survival for the chromosomal fragment are possible: either it remains in the extrachromosomal state in the recipient, or it becomes incorporated in the recipient chromosome by recombination. The situation in which a chromosomal gene becomes part of a plasmid is commonly encountered with the fertility factor F. A classic example is the incorporation of the chromosomal lactose genes (*lac*) from *E. coli* to produce the so-called F'.*lac* (pronounced F-prime *lac*) factor.

Transduction

The properties of bacterial viruses (bacteriophages) and the way in which they propagate themselves, are described in Chapter 7. Here we are concerned solely with the way in which these elements can transfer bacterial DNA between susceptible cells—the process known as transduction.

Transduction of chromosomal fragments

The normal developmental cycle of virulent phage (see Chapter 7) involves a number of phases, the outcome of which is a large number of mature phage particles each carrying their own specific DNA. The transduction process is a diversion of this normal developmental cycle to one in which bacterial DNA, rather than phage DNA, is incorporated into the maturing phage particle.

The critical stages of phage development, as far as transduction is concerned, are those in which the phage stops replication of the host DNA and turns the metabolic activity of the cell towards making phage DNA and phage-coat proteins. During this stage the DNA of the bacterial host is broken down and the cell contains many linear chromosomal fragments. When the phage comes to the stage of wrapping up DNA in the developing phage coat, to form mature phage particles, therefore, one occasionally finds errors in which the phage wraps up a piece of bacterial DNA in place of the phage DNA. For this process to occur the fragment of bacterial DNA has to be the length that would normally be incorporated (usually about 1% of the total length of the chromosome. Too large a piece cannot be accommodated, and too small a fragment does not allow the phage coat to form round the DNA satisfactorily.

Once bacterial DNA is located in a phage in this way, however, it is treated by the phage as its 'correct' DNA and consequently infection of the chromosomal fragment into a new host. The phage particle is therefore the vehicle used by the bacterial DNA in achieving transfer from one bacterial cell to another.

Once the bacterial DNA finds itself in the new host, the problems that face it are essentially those encountered by transforming DNA (p. 48). Transduced fragments will normally be in the double-stranded linear form but will not have the necessary information to circularize and form an independently replicating unit on their own account. Recombination with a resident replicon will therefore be necessary, and for this to occur at a reasonable frequency homology with the resident replicon is needed. If this is not available, the DNA will be unlikely to recombine at a reasonable frequency and will be destroyed. If, however, recombination is possible, the incoming DNA will be incorporated into the resident replicon by two cross-overs, exactly as described in Fig. 3.14.

The efficiency of transduction as a means of gene transfer is greatly affected by a number of factors. First, many bacteria are immune, or even merely insensitive, to phage infections, and consequently the exogenous DNA cannot be injected, or if injected is immediately destroyed. Thus the range of bacterial species accessible by transduction is relatively small, and is normally confined to members of the same or closely related species. Secondly, transduction is inefficient if the phage is too 'virulent' for the recipient, that is the process occurs poorly when the incoming phage kills and lyses the recipient bacteria rapidly. Usually transduction is at its most efficient when the transducing phage lysogenizes the recipient—see Chapter 7.

Plasmid transduction

As was the case with transformation, there are a number of differences between the transfer of chromosomal and plasmid material by transduction. In practice there are two of importance. First, transduction of a plasmid leads to the insertion of a piece of DNA which can itself form a replicon in the recipient. As a consequence recombination is not needed for survival, and the frequency with which the process occurs is thereby enhanced. In particular, the replicative independence of plasmids means that they can sometimes be transduced to strains and species that do not readily accept chromosomal fragments.

The other point is that many plasmids are just about the correct size to be successfully wrapped up in the developing phage particle. Whether this is a situation that has arisen because it has selective

advantage for the bacteria or the phage is unclear at present, but transduction can often be a successful method of plasmid transfer even for those plasmids that actually carry the necessary information to make them self-infectious.

Gene transfer in natural populations

Although much is now known of the ways in which genetic information can pass between bacterial cells in the laboratory, we know remarkably little about the extent to which these processes occur in nature. Perhaps the best studied examples concern the transfer of antibiotic resistance, but even in these cases our knowledge is extremely patchy. The main problem in investigating natural events is that one always has to infer what has happened after the transfer event has come to light. In the case of antibiotic resistance the thing that usually calls attention to the transfer is the emergence of a new line of resistant bacteria in a clinically significant situation.

A typical sequence of events that implies that gene transfer—probably by conjugation—plays an important role in the evolution of bacterial populations came from a study of the emergence and spread of resistance to carbenicillin—a penicillin specifically developed for the treatment of *Ps. aeruginosa* infections—in bacterial strains infecting patients in a Burns Unit in Birmingham. The sequence of events was as follows. Carbenicillin was used for two or three years in the Unit without any resistant *Ps. aeruginosa* occurring in burned patients. In 1969, however, a number of distinct *Ps. aeruginosa* strains, all resistant to carbenicillin, appeared in the Unit over a short period of time. A plasmid was shown to be responsible for the resistance, and it was easy to demonstrate that the plasmid was the same in all the strains regardless of the cultural characteristics in the strain that carried them.

As soon as the selection pressure on the resistant strains was removed by stopping the use of carbenicillin, the resistant strains disappeared, and this situation persisted as long as carbenicillin was not used in the Unit. Immediately this antibiotic was used again, however, resistant *Ps. aeruginosa* reappeared and remained prevalent until carbenicillin use was stopped once more. Through the two phases of this outbreak of resistance the plasmid concerned was the same, but the host pseudomonads differed widely in their detailed properties.

Investigation of the phenomenon showed, contrary to initial expectations, that the plasmid conferring resistance to *Ps. aeruginosa* had not disappeared completely from the Burns Unit between the outbreaks of carbenicillin resistance. It could still be found in strains of *Klebsiella aerogenes* harboured in the patients and the staff in the Unit, but the plasmid in question conferred resistance to ampicillin as well as to carbenicillin in *K. aerogenes*. Thus ampicillin use in the Unit—which was not discontinued when carbenicillin therapy was stopped—provided the necessary selection pressure to maintain the plasmid in the enteric bacteria of the gut; and transfer of the plasmid *Ps. aeruginosa* was only detected when carbenicillin was used for therapy. Ampicillin is not effective against *Ps. aeruginosa*, and therefore this penicillin does not supply the same selection pressure as carbenicillin.

This sequence of events shows that resistance plasmids may persist in reservoirs and break out in unsuspected directions as soon as novel selection pressures are applied.

This pattern of events—which is effectively microbial evolution in action—should *not* be regarded as limited solely to the transfer of antibiotic resistance. The exchange of many types of plasmid-carried genes may occur and new cell-lines will then be established if appropriate selection is applied. This then is one of the ways in which additional genetic information may pass between bacterial populations, the restrictions being, first, whether a vector capable of passing the DNA between the strains is present, and secondly, whether the new information can express itself in a self-replicating form in the recipient.

Clearly gene transfer between bacteria is easiest when the gene concerned is part of a plasmid. However, as we have seen, not all genes are present in bacteria in this form. Nevertheless, plasmids probably do act as an intermediate in the transfer of chromosomal DNA between bacterial cells. We know that as a rare event, they can integrate into the

chromosome and thus initiate the transfer of the whole chromosome to the recipient bacterium. On other occasions chromosomal genes will become part of a plasmid by illegitimate recombination (p. 48) and these plasmid recombinants can then be transmitted to a recipient during conjugation. A final stabilizing step would be the transposition of the gene from the plasmid to the chromosome in the recipient, another step which might be 'illegitimate'. Perhaps the most important point to note, in conclusion, is that evolutionary events in bacteria need only be very rare to have an impact when the pressures are there. Such are the numbers of bacteria and the time available to them for gene exchange, that the rarest events are still significant on an evolutionary time-scale.

In view of the genetic plasticity of bacteria, the question arises as to how *E. coli*, for example, remains recognizable as a discrete species, and how isolates of this species can be recognized as such with reasonable certainty when obtained from a variety of different sources. The answer probably lies in the shaping influence of selection pressure. Only those genes whose products give positive benefit to the bacterial cell and to the population of which it is part will survive, and since this is so, the fact that bacteria are speciated argues that organisms are subjected to patterns of selection pressure, patterns that reflect individual ecological niches. Thus the speciation of bacteria arises from a balance between the diversifying effects of genetic variability and the conservative influence of a relatively constant environment.

While this balance is relatively stable for some species and speciation is correspondingly clear-cut, this is not always so. The boundaries between some species are notoriously indistinct, and this may well arise from a less clear pattern of selection pressure on the one hand, and from a greater ability of the bacteria concerned to exchange genetic information on the other.

The genetics of eukaryotic micro-organisms

Introduction

Eukaryotic micro-organisms are here used to demonstrate and clarify the basic principles of genetics first worked out with higher plants and animals. We

have introduced specialized words since these are often useful in mastering complex ideas. More detailed treatment of the subject will be found in the reading list.

Reproduction of genetic material

The nucleus

In contrast to the relatively small and unprotected nucleic acid genomes of prokaryotes (p. 40), the chromosomal DNA of higher organisms is complexed with a basic protein (histone) which probably has both regulatory and protective functions. Between cell divisions the chromosomes are isolated from the cytoplasm in the *nucleus* surrounded by the *nuclear envelope*. This *interphase* nucleus also contains the enzymes necessary for chromosome replication and DNA transcription and, in addition, the transcription products: messenger, transfer and ribosomal RNAs. The ribosomal RNA accumulates in the *nucleolus* before passing with other nuclear products through pores in the nuclear envelope into the cytoplasm (Chapter 9, p. 180).

Mitosis

Genome replication in prokaryotes is a continuous process which may be out of phase with cell division. In the more complex cells of higher organisms, where the DNA is organized into more than one chromosome, there is usually a regular sequence of replication—the *mitotic cycle*. Immediately after the mitotic division there is usually a period (G1) when RNA and protein, etc. are synthesized. Then in the synthetic (S) phase the DNA and histone of the chromosomes doubles. Following S there is another growth phase, G2, and only then is the cell capable of a further mitosis.

In mitotic *prophase* the chromosomes become visible as double strands (*chromatids*) which are attached to each other only at a point of constriction, the *centromere*. At the end of prophase the nuclear membrane breaks down and a spindle is formed. The centromeres become attached to the spindle fibres and come to lie on the equator of the cell at *metaphase. Anaphase* is marked by the sudden division of the centromeres, whose daughters, with their attached chromatids, travel or are pulled towards opposite poles of the spindle. At *telophase* the daughter chromatids coalesce and nuclear envelopes reform around the two new nuclei. In this way identical genomes are passed to the two daughter cells.

Meiosis

Underlying the many different life cycles of eukaryotes there is a basic alternation of *haploid* cells containing one set of genetic information and *diploid* cells containing two sets. *Meiosis* is the process by which one diploid nucleus gives rise to

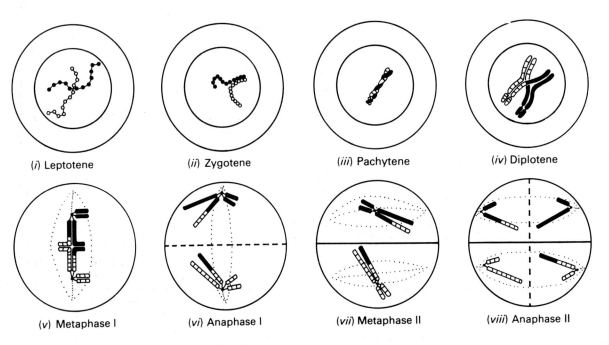

(*i*) Leptotene   (*ii*) Zygotene   (*iii*) Pachytene   (*iv*) Diplotene

(*v*) Metaphase I   (*vi*) Anaphase I   (*vii*) Metaphase II   (*viii*) Anaphase II

Fig. 3.17 Genetically important stages of meiosis. See text for details.

four haploid nuclei. Fusion of two haploid nuclei results in formation of a *zygote* and restores the diploid condition.

The meiotic process differs from mitosis in two ways—there are two successive nuclear divisions without intervening synthesis of DNA, and the nuclei produced are not identical to each other or to the parent nucleus. Because of the complexity of the first meiotic prophase it is convenient to divide it into a number of stages. When the chromosomes first become visible in *leptotene*, they appear as single beaded strands although DNA synthesis is complete. During *zygotene*, homologous chromosomes from the two parental strains (only one pair is shown in Fig. 3.17) come to lie together. This process is called *synapsis* and is facilitated by a molecular 'zip', the *synaptonemal complex* (Figs. 3.18 and 3.19), which can be seen only in the electron microscope. In *pachytene* when pairing is complete, the pairs of homologous chromosomes are so closely associated with each other that they can no longer be distinguished in the light microscope. The beaded appearance is lost and each pair is now referred to as a *bivalent*.

In *diplotene* the synaptonemal complex breaks down, the bivalents open out, and each chromosome is seen to consist of two chromatids; at various points along the bivalent pairs of chromatids are seen to have exchanged partners. These exchanges or *chiasmata* (sing. *chiasma*) are the products of breakage and reunion of chromatids during pachytene, at which time there is a small amount of DNA synthesis, and they are the cytological expression of *genetic crossing-over* (p. 61). The homologous undivided centromeres repel each other and but for the chiasmata would pull the bivalents apart.

In *diakinesis* the nuclear envelope breaks down and a spindle is organized. The centromeres become attached to spindle fibres and jockey for position until at *metaphase I* the individual bivalents are positioned with their chiasmata more or less on the

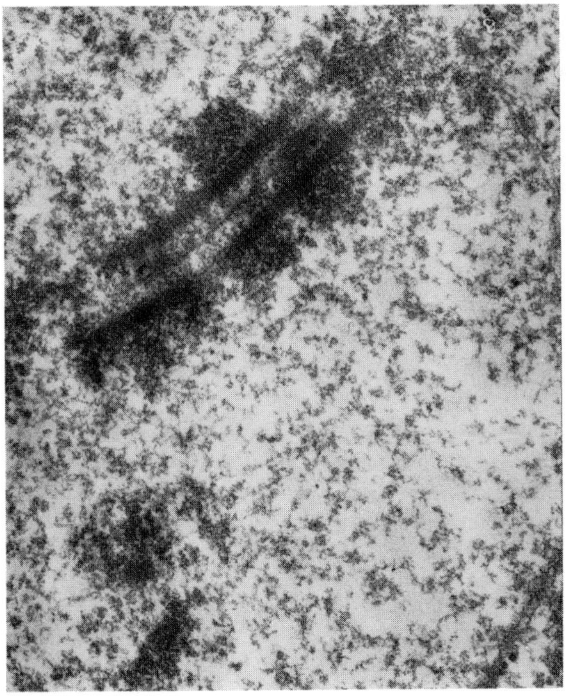

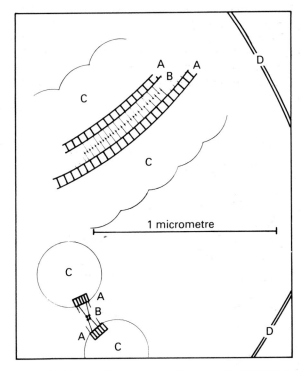

Fig. 3.18 Electron micrograph of part of a pachytene nucleus of *Xylaria polymorpha* (Ascomycotina), showing longitudinal and transverse sections through the synaptonemal complex (SC). (Photograph courtesy of Dr A. Beckett, Bristol University.) Many higher organisms have more than 100 times as much DNA per chromosome as *Xylaria*. However the dimensions of the 'zip' are the same throughout the eukaryotes although the larger chromatin masses may meet around the SC.

Fig. 3.19 Diagram of the electron micrograph Fig. 3.18. Note the paired masses of chromatin (C) held together by the synaptonemal complex composed of one central (B) and two lateral (A) elements. (D) is the nuclear envelope. The bivalent in longitudinal section probably terminates at the envelope but the end is twisted out of the plane of the section.

equator and their centromeres randomly orientated on either side. At *anaphase I* the attraction between paired chromatids lapses and the still undivided centromeres pass to opposite poles of the spindle to give two *telophase I* nuclei each with the haploid number of undivided chromosomes. Note that as a result of crossing-over the sister chromatids of these chromosomes (or half bivalents) are not genetically identical.

At this stage there may be a short *interphase*, but more usually the chromosomes pass directly to *prophase II* and two new spindles are organized. This time the centromeres come to lie on the equator of the spindle by *metaphase II* and divide normally at *anaphase II*. The daughter chromatids pass to the four poles to form four haploid nuclei.

Although superficially akin to mitosis, this second nuclear division differs from it fundamentally because of the random orientation of centromeres at both metaphases and because of the crossing-over between non-sister chromatids which took place during prophase I in the synaptonemal complex. These two factors ensure that all four products of meiosis are different. In the Euascomycetes (p. 197), which have been important in many genetic investigations, the *tetrad* is arranged linearly and each member undergoes a further true mitosis to form an ascus containing a row of eight ascospores.

Mutation

Nuclear and extranuclear mutation

The phenomenon of mutation is fundamental to genetics; it is a prerequisite of the evolution and, thence, the recognition of new genes. A mutation can be defined as any permanent change in the quality or quantity of an organism's heritable information. Not all this information is confined to the chromosomes; for example both plastids and mitochondria are known to have genetic continuity and contain their own DNA. The existence of nuclear genes controlling plastid development, however, indicates that this specific DNA is not sufficient for plastid autonomy (see p. 68).

Since the genome is replicated and divided exactly between daughter cells, chromosome mutation can be detected almost immediately in haploid organisms and will appear in specific proportions in diploids following meiosis and fertilization (p. 60). On the other hand, if one plastid mutates in a cell containing twenty, a number of cell generations must elapse before a cell is formed which by chance

contains none but mutant plastids. There is no reason to suspect that the mutational events in extranuclear particles are in any way different from those within the chromosomes, but the technical difficulties of their isolation and characterization have so far led to concentration of research on nuclear genes.

Most spontaneous mutations occur at a frequency of the order of 10^{-7}–10^{-9} per cell generation, and their behaviour can usually be interpreted as a change in the nucleotide sequence specifying the phenotype studied. Some genes, however, mutate at much higher frequencies, and it is sometimes necessary to invoke ingenious explanations for these *unstable genes*.

Chromosome mutation

The better-understood mechanisms of nucleic acid mutation are described on p. 45. Analogous changes also occur in eukaryote chromosomes, for example inversions, duplications, deletions and translocations of whole chromosome segments were described decades before DNA was identified as the genetic code. These *aberrations* can all be placed in one of two categories: those which affect the viability of the cell's immediate progeny, and those with a delayed effect on the organism's fertility. The molecular mechanisms of mutation are almost certainly similar in the chromosomes of bacteria and those of higher organisms, but the complex structure of the latter makes it difficult to recognize any alteration apart from gross structural changes.

VIABILITY

A break in a chromosome has no effect on the cell containing it so long as no division takes place. At division, however, those chromosome fragments which lack a centromere usually fail to reach the telophase nuclei, which therefore lose some genetic information. In a diploid nucleus, the other homologue may be able to supply the lost information and there may be no immediate effect, but in haploids loss of information can lead to early death.

FERTILITY

Exchange of chromosome parts within or between chromosomes usually has no immediate physiological effect on the cell and its mitotic progeny since there is no change in the total information content. At meiosis, however, crossing-over between paired aberrant chromosomes frequently leads to the formation of unbalanced gametes with consequent reduction of fertility.

SPONTANEOUS MUTATION

If large numbers of progeny from a single source are examined it is always possible to find spontaneous mutants, but one cannot be certain that these have not resulted from mutagenic effects in the background. Since it is possible to select strains with high mutation rates, and since some genes are more likely to mutate than others, this is clearly not a simple situation. Apart from the mutagenic background, there are cellular control mechanisms, such as DNA repair enzymes, which could affect mutability in different strains.

Biochemical mutants

Prototrophs and auxotrophs

Prototrophic (effectively = wild-type, symbol +) micro-organisms can often be grown on simple media, e.g. *Chlamydomonas* (p. 70) requires only inorganic salts to grow normally in liquid culture or on agar plates in the presence of light; *E. coli* will grow on a similar medium supplemented with an energy source such as glucose, while *Neurospora crassa* requires the vitamin biotin (p. 74) in addition.

Auxotrophic mutants are unable to grow on the minimal media which will support prototrophs because they cannot synthesize particular compounds. For example, *Neurospora* mutants which cannot synthesize arginine will only grow on arginine-supplemented medium. Such mutants are given the symbol *arg*; there are similar abbreviations to designate alleles conferring requirements for other amino acids and vitamins.

Auxotroph selection

It is simple to pick out a prototrophic mutant in an auxotrophic population: only the mutant will be able to produce a colony on a suitable selective medium. It is less easy, however, to select auxotrophic mutants from prototrophs because the prototrophs will grow just as well as the mutants, if not better, on supplemented medium.

One way of solving this problem depends on the fact that penicillin kills only growing bacteria. If a suspension of *E. coli* prototrophs is treated with a mutagen and then grown in minimal medium containing a lethal concentration of penicillin, all the cells which are capable of growth will be killed. After a suitable delay the remaining bacteria, including any auxotrophs present which cannot grow because of lack of various supplements, are centrifuged down, washed, resuspended in penicillin-free medium, and poured on to supplemented agar plates. The resultant colonies are tested and classified as requiring various amino acids or vitamins for normal growth.

Similarly, mutants of mycelial organisms such as *Neurospora* can be isolated by the filtration technique. Treated spores are suspended in liquid minimal medium and allowed to grow for a short time. The suspension is then filtered; the ungerminated auxotrophs pass through the filter, while the germinated prototrophs are retained and discarded.

Auxotroph classification

The original investigations of biochemical mutants, which led to the formulation of Beadle and Tatum's 'one gene—one enzyme' hypothesis were carried out by laboriously examining the growth requirements of thousands of unselected *N. crassa* spores. Lederberg, however, developed a *replica plating* technique; this was originally used to demonstrate the spontaneous nature of mutation of *E. coli* to phage resistance in the *absence* of phage, but it also provides an easy way of classifying mutants of organisms which grow as discrete colonies, e.g. Maize Smut (*Ustilago maydis*).

For replica plating a suspension of treated bacterial or fungal cells is poured on to a master plate containing medium supplemented with those growth factors which will allow the desired mutants to grow. After incubation, a piece of sterile velvet mounted on the end of a cylindrical block is pressed gently on to the plate; this 'replicator' picks up a few cells from each colony on the master plate, and it is pressed successively on to a series of plates containing minimal medium alone or with various supplements. After further incubation, colonies appear on these replica plates in the same pattern as on the master plate; absence of a particular colony from a plate is presumptive evidence that it requires the supplement missing from that plate.

Biochemical pathways

One of the first biosynthetic pathways to be worked out was that of arginine. Arginine-requiring mutants of *N. crassa* were found to fall into a number of groups including:

1 Those which grew on ornithine, citrulline, or arginine.
2 Those which grew on citrulline or arginine, but not on ornithine.
3 Those which grew only on arginine.

This suggested that arginine is the product of the sequence:

Precursor $\xrightarrow{1}$ Ornithine $\xrightarrow{2}$ Citrulline $\xrightarrow{3}$ Arginine,

and that the three groups of mutants are deficient in the enzymes mediating reactions 1, 2, and 3 respectively. The enzymes were subsequently isolated and shown to catalyze these reactions. Some mutants can be shown to accumulate excess precursors before the blocked reaction—for example, adenine-requiring mutants of yeast produce a red pigment which distinguishes them from the colourless wild-type colonies. Numerous biochemical pathways have since been elucidated in this way.

Genetic recombination and mapping

Meiosis and Mendel

THE PRINCIPLE OF SEGREGATION

When two haploid yeast cells fuse, one containing a chromosome bearing the gene A mediating, say, the citrulline → arginine reaction and the other with its homologous chromosome bearing the mutant *allele a*, the diploid (*Aa*) will be able to grow on minimal medium unsupplemented with arginine. Thus in the presence of the *dominant* gene *A* we can no longer detect the presence of the *recessive* allele *a*. After meiosis, however, two of the haploid cells produced will contain *A* and the other two *a*, so that in the haploid generation only half the cells will be able to grow on minimal medium. In other words *A* and *a*, although associated together in the diploid, segregate independently and unchanged into the haploid cells produced by meiosis.

Mendel originally deduced this principle of segregation, in 1865, by observing the products of fertilization in pea plants, which are diploid. We can observe it directly in yeast however, which has a haploid generation. Following random fusion of haploid yeast cells carrying the two genes, yielding diploids, we can distinguish between the genotypic ratio of 1 *AA*:2 *Aa*:1 *aa* and the phenotypic ratio of three prototrophs to one auxotroph.

THE PRINCIPLE OF INDEPENDENT ASSORTMENT

If we now consider a second gene pair *B* and *b*, mediating, say, biotin production versus requirement, carried on a *second* pair of homologous chromosomes, we again get the genotypic ratio of 1 *BB*:2 *Bb*:1 *bb* and the phenotypic ratio of three biotin producers to one biotin requirer, following random fusion of cells carrying either *B* or *b*. When we observe the inheritance of both sets of genes, however, we find that the haploid progeny of the *AaBb* diploids includes four types of cell in approximately equal numbers:

1 Grow on minimal medium; hence genotype *AB*.
2 Require biotin; hence genotype *Ab*.
3 Require arginine; hence genotype *aB*.
4 Require both arginine and biotin; hence genotype *ab*.

This indicates that the two factors *A* and *B* segregate at random, independently of each other.

Chromosome mapping

THE EXCEPTION: LINKAGE

We have considered the inheritance of two pairs of genes carried on different pairs of chromosomes. If we look at the haploid progeny of an *AaBb* diploid in which both pairs of genes are carried on the *same* pair of chromosomes, however, then we find that the principle of independent assortment no longer holds good. *A* and *B* still segregate from *a* and *b*, but no longer at random.

If *A* and *B* were originally on one of the parental chromosomes and *a* and *b* on the other, then *Ab* and *aB*, the *recombinants*, will only be formed if crossing-over takes place between *A* and *B*. In Fig. 3.20, where *A* and *B* have been placed at equal distances along the chromosomes, this means that if there is only a single chiasma per bivalent (p. 56) then only 1/3 of the tetrads will include the recombinant haploids *Ab* and *aB*. There will therefore be a significant excess of the parental types (*AB* and *ab* derived from the original haploids) and the two genes are said to be *linked*. Further, if a number of genes are found to be linked in this way, they are said to constitute a *linkage group*, implying that they all belong to the same chromosome. (18 linkage groups have so far been identified in yeast but this does not necessarily mean that there are 18 pairs of chromosomes, since if genes in different linkage groups can be shown to be linked the number of groups can be reduced.) Note that the two recomb-

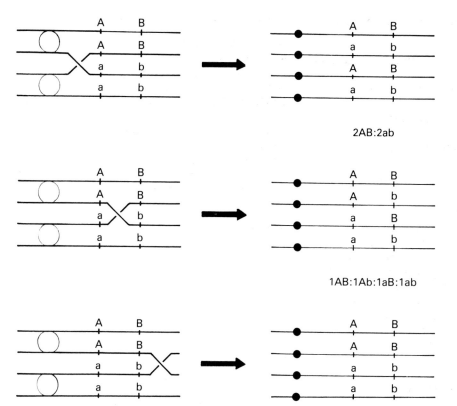

Fig. 3.20 Relationship of chiasma position to the segregation during meiosis of two linked genes, *A* and *B*.

inant classes *Ab* and *aB* will be present in approximately equal numbers, since both parents contribute similar haploid chromosome complements to the cross.

RANDOM SPORE ANALYSIS

The information gained from the study of linked genes can be used to determine the order and degree of separation of the genes along the chromosome. Thus in Fig. 3.20 one third of the tetrads are 1 *AB*:1 *Ab*:1 *aB*:1 *ab* while the other two-thirds total 4 *AB*:4 *ab*; that is, *one-sixth* of the haploids are recombinant. This 1/6 expressed as a percentage gives a map distance of 16.67 units (known as *centimorgans*) between *A* and *B*. If we then find a third gene *C* to be 12 units from *B* and 29 units from *A* (i.e., 58% of tetrads include the results of chiasmata between *A* and *C*), we can deduce that the linear order of the genes is:

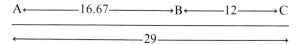

GENE ORDER

Although this method of genetic mapping is slightly more complex than we have made it appear, it is possible to build up a linkage map of all the genes in a chromosome in this way, using the so-called *three-point test cross*, on the principle that the probability of recombination between any two genes is proportional to the distance between them.

INTERFERENCE

The three-point test cross reveals occasional double cross-overs, which are the products of independent chiasmata occurring in the two regions *AB* and *BC*. Now, the chance of two occurring together in the same bivalent is the product of the chances of either event occurring singly—16.67% × 12%, i.e. 2%, in this example. However, the actual number of double cross-overs observed is often significantly lower than the expected number as calculated in this way, since a chiasma in one region is found to reduce the chance of another occurring in an adjacent region. This phenomenon, known as *interference* is explained by the mechanical re-straints imposed on crossing-over by the rigidity of the chromatids.

Interference is usually expressed as the *coincidence*, i.e., the ratio of expected to observed double cross-overs. As we should expect, the degree of interference usually increases as the distance between cross-overs decreases.

Negative interference is the opposite of interference, namely that a chiasma in one region favours a second in an adjacent region. Negative interference is detected in mapping very closely linked genes and in *intra-gene* maps.

Maps of single genes

The classic work of Pritchard with *Aspergillus nidulans* demonstrated that mutational sites within a single gene could be mapped. He used a number of independently induced mutations of an adenine synthesizing gene, *ad*. Two such mutants, *ad-8* and *ad-11* were crossed, together with outside markers, one, yellow conidia (*y*), only 0.22 centimorgans from *ad*. Just over three-quarters of a million ascospores were plated out on a medium not containing adenine and therefore selective for one class of recombinants which could arise between *ad-8* and *ad-11*, namely +, +, since the parental strains and the reciprocal recombinant class, *ad-8*, *ad-11*, all needed adenine for growth. 365 such adenine-independent progeny grew, giving a recombination percentage of only 0.04, or allowing for the unde-tected reciprocal recombinant class, a distance between *ad-8* and *ad-11* of 0.08 centimorgans.

Detailed analysis of some of these recombinants indicated that in several a second cross-over had occurred between *ad* and *y*, in fact such double cross-overs were 20 times more frequent than expected on a random basis. Hence strong negative interference was evident over these very short distances.

Using further mutants, Pritchard was able to map the relative positions and distances between 7 mutational sites in the *ad* gene. Such intra-gene maps have been constructed in many organisms and have been of value in understanding gene action and its control.

Tetrad analysis

So far, the mapping procedures described have relied on analysis of the genotypes of random spores. In ascomycetes such as *Neurospora* and *Sordaria* where linear tetrads are formed the pro-ducts of individual meioses can be isolated mechanically and analyzed. When spore-colour mutants are studied this analysis can be carried out by direct microscopic observation of the tetrads and we can use these observations to determine the map distance of the gene from the centromere. For example, the *asco* mutant of *Neurospora* has colourless instead of black spores, and the distribution of black and colourless spores in the asci of a hetero-zygote depends on whether or not a chiasma has occurred between the gene and the centromere. Thus if we find that 20 second division asci occur for every 80 first division asci, we know that a chiasma occurs between the centromere and the gene in 20 % of the asci (Fig. 3.21).

Note that in tetrad analysis we are counting crossover asci, *not* crossover spores containing crossover chromatids. This is important to re-member because in a second division ascus only *half* the spores contain chromosomes derived from the crossover chromatids. Because of this, if we wish to equate gene-centromere distances obtained in this way with the map distances obtained by random spore analysis, we must divide the percentage of second division asci by two. In this example the map distance between the gene and the centromere is:

$$\frac{\text{Second division asci}}{\text{Total asci}} \times \frac{100}{2} = \frac{20}{100} \times \frac{100}{2}$$
$$= 10 \text{ units.}$$

Similarly, if we obtain linkage data from tetrad analysis, the scores must always be divided by two to equate them with those obtained from random spore analyses.

Gene conversion

HYBRID DNA FORMATION AND THE MECHANISM OF RECOMBINATION

Studies of crosses of the spore-colour mutant grey (*g*), in *Sordaria fimicola*, with wild-type together with outside markers (*m–1* and *m–2*) one either side of *g* revealed visually exceptional asci where the +/*g* segregation was non-Mendelian with ratios of 5 + :3g, 3 + :5g, 6 + :2g or 2 + :6g or abnormal spore arrangements known as 'aberrant 4 + :4g' asci, while tests showed that the outside markers in such asci showed normal Mendelian segregation and spore arrangement.

Fig. 3.22 demonstrates that the 'aberrant 4 + :4g' asci could arise by exchange of DNA segments from one chain in a maternal DNA duplex with one chain in a paternal duplex. The two duplexes thus formed

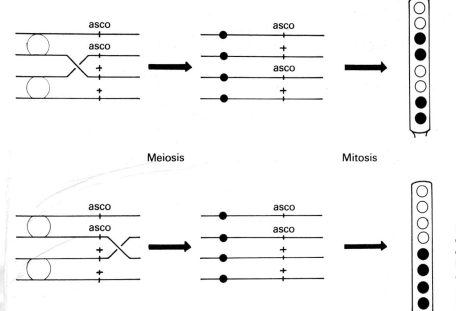

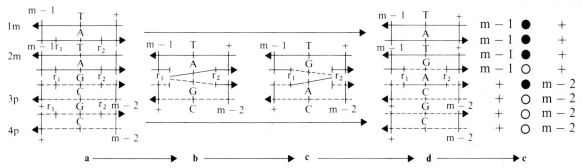

Fig. 3.21 First and second division segregation of chromosomes of the *asco*/+ heterozygote of *Neurospora crassa*. Each of the four products of meiosis divides again mitotically to give four pairs of ascospores.

would be composed of *hybrid DNA* and if the segment exchanged contained one or more mutant sites, then *heterozygous DNA* carrying mis-matched base pairs at the mutant site(s) would result.

Fig. 3.23 indicates that a *correcting enzyme system* may operate on mis-matched base pairs. If it fails to operate, aberrant 4:4 asci result but if correction occurs then non-Mendelian ratios result, the actual ratio depending on the *direction of*

correction. Hence the correcting system can convert one allele into another i.e. the occurrence of *gene conversion* which gives the appearance of non-reciprocal recombination. If the duplexes of all four chromatids are involved in hybrid DNA formation over the same segment, then 7+:1m, 1+:7m, 8+:0m or 0+:8m asci can result and these have been observed in *Ascobolus*.

Several models have been proposed to account

Fig. 3.22 Model demonstrating the mechanism of hybrid DNA formation. Note the following stages: (**a**) Early pachytene configuration. The 4 chromatids, 2 of maternal (m) origin carrying the marker *m-1* and 2 of paternal (p) origin and carrying the linked marker *m-2*, result from the cross of wild-type with black spores denoted by the base-pair A–T with the spore colour mutant, *grey*, denoted by the base-pair G–C. Two recombinators in the sense strands are denoted by r_1 and r_2. (**b**) Breakage of the sense strands at r_1 occurs in chromatids 2 and 3 followed by reciprocal exchange of single DNA chains and re-fusion at r_1. (**c**) Breakage at r_2 now occurs followed by reciprocal exchange and re-fusion. A segment of hybrid DNA is thus formed between r_1 and r_2 in both chromatids and the segments are heterozygous since the two duplexes carry mis-matched base pairs, T–G in chromatid 2 and A–C in chromatid 3. (**d**) The four products of meiosis are shown. (**e**) The four products of meiosis undergo mitosis to give an 8-spored ascus. The rules of semi-conservative replication apply and so the 8 DNA chains in (**d**) form A–T, T–A, G–C or C–G base pairs on mitotic replication. This gives rise to an aberrant 4 +:4g ascus but the outside markers show a normal Mendelian arrangement. Spores 4 and 5 (m-1, g, + and +, +, m-2 respectively) would, in a random spore analysis, be scored as recombinants each arising from a double cross-over!

Late pachytene configuration	Action of correcting enzyme system	Type of ascus
1 {T / A} 2 {T / G} 3 {A / C} 4 {G / C}	2 and 3 not corrected	Aberrant 4+:4g ● ● ● ○ ● ○ ○ ○
	only 2 corrected, G → A	5+:3g ● ● ● ● ● ○ ○ ○
	only 2 corrected, T → C	3+:5g ● ● ○ ○ ● ● ○ ○
	only 3 corrected, C → T	5+:3g ● ● ● ○ ● ● ○ ○
	only 3 corrected, A → G	3+:5g ● ● ● ○ ○ ○ ○ ○
	2 and 3 corrected, G → A / C → T	6+:2g ● ● ● ● ● ● ○ ○
	2 and 3 corrected, T → C / A → G	2+:6g ● ● ○ ○ ○ ○ ○ ○
	2 and 3 corrected, G → A / A → G / or / T → C / C → T	normal 4+:4g ● ● ● ● ○ ○ ○ ○ ● ● ○ ○ ● ● ○ ○

Fig. 3.23 Possible modes of action of the correcting enzyme system on mis-matched base-pairs in heterozygous DNA resulting from hybrid DNA formation between two of the four chromatids. Note that certain directions of correction can restore 4:4 asci with normal spore arrangements, and if the enzyme system fails to act, 4:4 asci with aberrant spore arrangements result.

for hybrid DNA formation and recombination. Fig. 3.22 shows a simple model where breaks at spaced, pre-determined, short, coded regions in the DNA sense strand, called *recombinators*, occur followed by reciprocal exchange of single DNA chains and progressing to the next recombinator giving polarity and strong negative interference between successive recombinational events, if two or more mutant sites are located in the segment.

All models suffer from the difficulty of indicating how a chiasma involves the breakage at identical molecular points of not only the sense and nonsense strands in a maternal duplex but also in a paternal duplex, followed by reciprocal exchange and re-fusion. Thus the mechanism of recombination remains partly unsolved.

Breeding systems

Mating systems

Whitehouse distinguished between *morphological heterothallism* where sexual reproduction could only occur between two thalli which differed in the morphology of their sex organs and *physiological heterothallism* where the two thalli were morphologically similar but differed genetically in their *mating-type* such that, although potentially hermaphroditic, each was incapable of self-fertilization.

The simplest sexual incompatibility system is found in Algae such as some *Chlamydomonas* spp. and in some Zygomycetes (p. 196) and Ascomycetes (p. 197), where mating-type is controlled by a single gene (mt) with only two alleles i.e. mt^+ and mt^- or in *Neurospora* designated 'A' and 'a'. Only thalli of opposite mating-type can undergo sexual reproduction.

In the Basidiomycetes (p. 198) more complex systems are found. The simpler condition is where multiple alleles at one locus exist, denoted by A_1, A_2, A_3, etc. Sexual reproduction can only occur if the two alleles in the two thalli differ, hence this condition is termed bipolar. In other of these higher fungi, multiple alleles at two, unlinked loci (denoted by '*A*' and '*B*') exist and compatible matings only occur if all four alleles are different e.g. $A_1B_2 \times A_2B_3$, i.e. tetrapolar fungi.

Further research has shown that the *A* or *B* or both loci may be *compound*. Thus in *Schizophyllum commune*, the *A* locus is composed of sub-units, designated *Aα* and *Aβ*, between which recombination can occur. A gene for *p*-amino benzoic acid synthesis has been discovered located between *Aα* and *Aβ* indicating that these are separate but closely linked genes and so three loci are concerned with mating in this species.

Heterokaryon formation and the parasexual cycle

Although meiosis has not been observed in the imperfect fungus *Aspergillus niger*, Pontecorvo has suggested that a *parasexual cycle*, which allows gene recombination, might operate in this species.

If the two strains, one carrying the biochemical marker *m–1* and the other a linked, biochemical marker *m–2*, are of the same mating-type and conidia of each are mixed and inoculated on minimal medium, neither strain can grow alone but hyphal fusions occur in the germinating conidia and 'cross-feeding' takes place between the gene products of the two, different, genetic types of nuclei in a common cytoplasm and wild-type growth of mycelia results. Such 'hybrid' mycelia are called *heterokaryons* and in this case a 'forced' heterokaryon.

Occasionally haploid nuclei fuse, assisted by *d*-camphor vapour, to form *heterozygous diploid nuclei* and diploid sectors can arise in the culture. In these, non-meiotic haploidization occurs (*ca.* 10^{-3} per mitosis) and some of the resulting haploid, uninucleate conidia contain neither marker and so grow as wild types on minimal medium although homokaryons. Further tests on other conidia indicate that they carry both markers. This indicates that *mitotic crossing-over* had taken place in the heterozygous diploid nuclei giving reciprocal recombinants.

In nature this parasexual cycle could occur between different genotypes in any of the species of the Fungi Imperfecti and so give rise to an important source of variability due to mitotic recombination. Mitotic crossing-over has also been an important tool in genetic studies of both ascomycetes and basidiomycetes and parasexual systems have recently become important in the assignment of human genes and linkage groups to particular chromosomes.

Heterokaryon incompatibility

Gene loci have been discovered which either cause lethality in fusions between the hyphae of two strains or cause a disadvantage to the heterokaryotic state which usually reverts to a homokaryon. Such strains are said to be *vegetatively incompatible*.

In *Neurospora crassa* some nine, non-allelic, heterokaryon-incompatibility loci (called *het* genes) are known. Some of these *het* genes cause plugs of precipitated protoplasm to form on fusion of incompatible hyphae and the cell contents in the affected region degenerate preventing heterokaryon formation.

The mating-type alleles act in addition as *het* alleles but an unlinked gene allele, *tol*, suppresses the incompatibility reaction of the *mt* alleles. The alleles at yet another locus, inhibit the multiplication of *i* nuclei in *I/i* heterokaryons which usually revert to a homokaryon.

In 19 wild strains of *Podospora anserina* over 92 % of possible crosses were either sexually or vegetatively incompatible, while in *Aspergillus glaucus* 35 wild isolates from a limited area were mostly vegetatively incompatible. In *A. nidulans* no less than 30 vegetatively incompatible groups have been found and these tend to be mixed together in nature and not geographically isolated. *Het* genes are known also in *Rhizoctonia solani* and in the Slime Mould, *Didymium iridis*.

These discoveries throw doubt on the 'universality' of the parasexual cycle but the mechanism of the *het* reaction may be of significance in the problem of the identification and rejection mechanism by unlike cells.

Population genetics

Much work has been done on the number, nature and geographical distribution of the mating-type alleles in bipolar and tetrapolar fungi in nature. The effect of the increasingly complex systems is to reduce the potential for inbreeding but to maintain the potential for outbreeding at a high level. Outbreeding has many advantages but it is nevertheless more advantageous for a population to inbreed than not to be able to breed at all. The compound nature of the *A* and *B* loci in tetrapolar species allows new alleles to arise by recombination, the recombinants being compatible with the parental alleles and these are important in isolated populations suffering from 'allele shortage'.

The *het* genes can be regarded as an isolation

mechanism whereby populations are split up into vegetatively (not sexually) isolated groups. Indeed, fungi offer much scope for further studies in these and other topics in the field of population genetics.

Extranuclear genetics

Besides the DNA of the nucleus, a relatively small amount of DNA is also found in mitochondria and chloroplasts, and contains genes for a limited number of organelle components. This DNA is circular in most organisms and is not associated with histones as is chromosomal DNA. Messenger RNA transcribed on the organelle DNA is translated on ribosomes present within the organelles. Despite this, the majority of the organelle proteins are coded for by nuclear DNA, and synthesized on the cytoplasmic ribosomes. Co-operation between nuclear and extranuclear genes is therefore required for the synthesis of complete organelles.

Isolation of extranuclear mutants

Although spontaneous extranuclear mutants have appeared in some organisms, for example, the *poky* mutant of *Neurospora crassa* (slow growing) and the *petite* mutant of yeast (respiratory deficient), study of extranuclear genetics has required the isolation and characterization of a large number of different types of mutant. One of the approaches used is the isolation of mutants resistant to inhibitors of mitochondrial or chloroplast function. For example, the antibiotics chloramphenicol and erythromycin, which inhibit bacterial protein synthesis, also inhibit mitochondrial and chloroplast protein synthesis without directly affecting cytoplasmic protein synthesis. This reflects structural differences between cytoplasmic ribosomes, on the one hand, and organelle and bacterial ribosomes on the other. Thus selection of mutants resistant to these drugs will yield some extranuclear mutants. However, not all the mutants isolated will be extranuclear, firstly because cell membrane permeability may be altered to prevent the drug from entering the cell and secondly because only a proportion of organelle components are encoded in the organelle DNA.

In addition to antibiotic-resistant mutants, other extranuclear mutants have been isolated in yeast which lack specific cytochromes and consequently lose the ability to grow on non-fermentable substrates.

Inheritance of extranuclear mutations

Extranuclear mitochondrial mutants have now been isolated in a number of micro-organisms, including the yeast *Saccharomyces cerevisiae*, the filamentous fungi *Aspergillus nidulans* and *Neurospora crassa*, and the ciliate protozoan *Paramecium aurelia*. Similarly, extranuclear chloroplast mutants have been isolated in the unicellular alga *Chlamydomonas reinhardi*. Since both nuclear and extranuclear mutants can affect organelle function, genetic tests have to be carried out to ascertain the location of the mutation.

This can be illustrated by reference to the *poky* (also called *mi–1*) mutation of *Neurospora crassa*. Strains containing this mutation grow more slowly than the wild-type and have an altered respiratory system. If the *poky* mutant is crossed with a wild-type strain, no segregation of *poky*/wild-type ascospores is seen in the asci. Instead, all the ascospores within one ascus are of the same type, either *poky* or wild-type. This contrasts with the 4/4 segregation seen for nuclear alleles (e.g. spore colour, Fig. 3.21) and suggests that the mutation is not located on a chromosome. The composition of the ascus depends on how the cross was carried out. In *Neurospora crassa*, the protoperithecium is fertilized by a single conidium (p. 194). If the protoperithecial parent is *poky* and the conidial parent is wild type, the asci formed will contain *poky* ascospores only. Conversely, if the protoperithecial parent is wild type and the conidial parent *poky*, all the ascospores will be wild type. These phenomena illustrate *uniparental transmission* which is a useful indication of the extranuclear location of the mutation. Since in *Neurospora* the progeny inherit the character through the female organ this is sometimes called maternal inheritance.

In *Chlamydomonas*, some chloroplast mutations have been characterized as extranuclear because they are inherited from the mt^+ parent only. It appears likely that in this case uniparental transmission results from degradation of mt^- organelle DNA in the zygote.

In some filamentous fungi, including *Aspergillus nidulans*, extranuclear mutations may be detected by the *heterokaryon test*. If a heterokaryon is formed (p. 192) between a strain containing an extranuclear mutation and a wild-type strain, both strains being suitably labelled with nuclear genetic markers, mitochondria and nuclei are able to mix in the heterokaryon, but genetic interaction between the nuclei is rare. When conidia are formed, some contain the nucleus derived from one 'parent' together with mitochondria from the other 'parent'.

This reassortment of nuclear and extranuclear genomes can be used to detect extranuclear mutations, since nuclear markers remain in the same combination as in the parent strains.

In *Saccharomyces cerevisiae*, strict uniparental inheritance is not observed. When an extranuclear mutant is crossed with a wild-type strain, both alleles enter the zygote. The zygote may then be propagated vegetatively as diploid cells, and segregation of extranuclear alleles is observed during vegetative growth. This does not occur for nuclear alleles. After a number of divisions, individual diploid cells become 'pure' or homozygous for any one extranuclear allele. If these diploid cells are sporulated, although nuclear alleles segregate 2:2, extranuclear alleles show a 4:0 or 0:4 pattern, since the parent diploid contains only one mitochondrial type.

Extranuclear recombination

In some organisms, recombination has been observed between extranuclear mutants. At present, no system is known for regulation of recombination which could be compared to meiosis and nuclear recombination, and once mitochondria or chloroplasts come into contact in the cytoplasm, it is possible that many rounds of recombination take place following fusion of organelles. Because of this behaviour, it has not been easy to map extranuclear genomes by the conventional genetic technique of recombination frequency. In yeast and *Aspergillus*, for instance, recombination seems to occur only at two levels, high and low. It is possible that low recombination figures indicate that two mutations are in the same gene, but it has not been possible to carry out complementation tests with extranuclear mutants to verify this. In nature, extranuclear recombination between different organisms may be very limited. In the first place, uniparental inheritance in the sexual stage results in elimination of one of the parental genomes before recombination can occur. Secondly, although extranuclear recombination could occur in heterokaryons formed between two strains of some fungi, the existence of nuclear heterokaryon incompatibility genes (p. 65) will make heterokaryon formation rare. In *Chlamydomonas reinhardi*, study of chloroplast recombination has been made possible by the use of *biparental zygotes*. These zygotes, which occur naturally at a low frequency only, permit the transmission of chloroplast genomes from *both* parents instead of just from the mt^+ parent, giving an opportunity for recombination to occur. Sager found that ultra-violet irradiation of the mt^+ gamete before mating increased the frequency of biparental zygotes, and was able to use this system to construct a circular map of chloroplast DNA showing the linkage relations of several chloroplast genes.

Cytoplasmic petite mutants of yeast and their use in mapping

One of the earliest known extranuclear mutants of yeast is the *cytoplasmic petite* (designated ρ^-). This mutant, which has lost all respiratory ability and is unable to grow unless supplied with a fermentable substrate such as glucose, appears spontaneously at high frequency (up to 10^{-2} in some strains). The mutation is not a simple one, but results from the random deletion of regions of mitochondrial DNA. Some petites contain as much mitochondrial DNA as the wild-type, though it may be repetitive, resulting from the reiteration of that DNA which remains. Other petites (designated ρ^-o) lose all detectable mitochondrial DNA. The petites which retain some DNA are useful for mapping studies, since a type of deletion mapping may be carried out.

Firstly, a physical map of the mitochondrial DNA may be made by the use of *restriction endonucleases* (p. 49). These enzymes, produced by some bacteria, cleave DNA at specific sites, and by analysis of the products of digestion, a map can be constructed showing the location of the various cleavage sites. Once a map has been constructed for the wild-type mitochondrial DNA, a large number of petites can be mapped in a similar way, and it can be seen which areas of the map have been lost as a result of the petite mutation. Mutations such as chloramphenicol resistance can then be located relative to the wild-type physical map. Petite mutants may be crossed with wild-type strains to find out which genetic loci have been lost as a result of the DNA loss. By correlation of marker loss or retention with the physical map, a conventional map may be obtained for the mutant loci. Deletion of large sectors of mitochondrial DNA is only possible because *Saccharomyces* is able to grow by fermentation alone. Most eukaryotic micro-organisms are obligate aerobes, and could not survive such a loss.

The role of organelle DNA

A combination of genetical and biochemical studies carried out on extranuclear mutants has added considerably to our knowledge of the role of organelle DNA. It now appears that the mitochondrial DNA of micro-organisms is responsible for coding mitochondrial ribosomal RNA and transfer RNA, together with some of the polypeptide sub-units of cytochromes aa_3 and b, and of the ATP-ase. The latter enzyme is the one responsible for ATP synthesis during oxidative phosphorylation.

Chloroplast DNA also codes for organelle RNA, and for one of the two sub-units of ribulose diphosphate carboxylase, the enzyme responsible for the CO_2 fixation step in chloroplasts. In addition, a number of unidentified chloroplast-membrane proteins seem to be of chloroplast origin.

How organelle DNA and special protein synthesizing systems evolved in the eukaryotic cell is not clear, but some of the proteins which they make are very hydrophobic, and it may be essential for them to be synthesized *in situ* rather than being transported into the organelles after synthesis on cytoplasmic ribosomes. The organellar systems are under fairly strict control from the nucleus, and organelle DNA replication and transcription seem to be carried out by enzymes of nuclear origin. Nevertheless, some authorities have suggested that chloroplasts and other cell organelles may have evolved from 'prokaryotic symbionts'.

Books and articles for reference and further study

Bacterial inheritance

FINCHAM, J. R. S. (1976). Microbial and Molecular Genetics, 2nd Edn. Hodder and Stoughton, London.

GOODENOUGH, U. (1968). *Genetics*, 2nd Edn. Holt, Reinhart and Winston, New York.

HARTMAN, P. E. and SUSKIND, S. R. (1969). *Gene Action*, 2nd Edn. Prentice Hall, Englewood Cliffs, New York.

HAYES, W. (1968). *The Genetics of Bacteria and their Viruses*, 2nd Edn. Blackwell Scientific Publications, Oxford.

HERSKOWITZ, I. H. (1977). *Basic Principles of Molecular Genetics*, 2nd Edn. Nelson, London.

LEWIN, B. (1976). *Gene Expression—1*. John Wiley, New York.

SRB, A. M., OWEN, R. D. and EDGAR, R. (1970). *Facets of Genetics*. Readings from the Scientific American. W. H. Freeman, New York.

STENT, G. (1971). *Molecular Genetics, an Introductory Narrative*. W. H. Freeman, San Fransisco.

WINKLER, U., RÜGER, W. and WACKERNAGEL, W. (1978). *Bacterial, Phage and Molecular Genetics*, 2nd Edn. Springer Verlag, Berlin.

WOESE, C. R. (1967). *The Genetic Code*. Harper Row, New York.

Special books on plasmids

BRODA, P. (1979). *Plasmids*. W. H. Freeman, London.

FALKOW, S. (1974). *Infectious Multiple Drug Resistance*. Pion, London.

MEYNELL, G. G. (1973). *Bacterial Plasmids*. Macmillan, London.

Genetics of eukaryotic micro-organisms

BAINBRIDGE, B. W. (1980). *The Genetics of Microbes*. Blackie, London, 193 pp. Standard Text.

BURNETT, J. H. (1976). *Mycogenetics*. Wiley, London, 375 pp. Advanced text.

CATCHESIDE, D. G. (1977). *The Genetics of Recombination*. Arnold, London, 250 pp. Specialist text.

ESSER, K. and KUENEN, R. (1967). *Genetics of Fungi*. Springer-Verlag, New York, 500 pp. Advanced text.

FINCHAM, J. R. S. (1976). *Microbial and Molecular Genetics*. 2nd Edn. Hodder and Stoughton, London, 150 pp. Simply written introductory text.

FINCHAM, J. R. S., DAY, P. R. and RADFORD, A. (1979). *Fungal Genetics*. 4th Edn. Blackwell Scientific Publications, Oxford, 448 pp. Standard text at intermediate level.

LEWIN, R. A. (ed.) (1976). *The Genetics of Algae*. Blackwell Scientific Publications, Oxford, 360 pp. Specialist monograph.

SAGER, R. (1972). *Cytoplasmic Genes and Organelles*. Academic Press, New York, 405 pp. Advanced text.

SMITH-KEARY, P. F. (1974). *Genetic Structure and Function*. Macmillan, London, 368 pp. Standard text at intermediate level.

STAHL, F. W. (1979). *Genetic Recombination—Thinking About It in Phage and Fungi*. W. H. Freeman, London, 333 pp. Specialist Text.

WHITEHOUSE, H. L. K. (1973). *Towards an understanding of the mechanism of heredity*. 3rd Edn. Arnold, London, 528 pp. Standard text at intermediate and advanced level.

WOODS, R. A. (1980). *Biochemical Genetics*. 2nd Edn. Chapman and Hall, London, 80 pp. Simply written introductory text.

Chapter 4

Nutrition and the influence of environmental factors on microbial activities

Introduction

In this chapter the main types of nutrition found amongst micro-organisms are outlined and the effects of environmental factors on microbial growth and reproduction are discussed in general terms. Exceptions to the generalizations will inevitably occur.

Though the various environmental factors are discussed individually it is important to realize that they are frequently interrelated. Thus the nutrients required may depend amongst other factors on the pH and temperature of the environment, and for photosynthetic organisms on whether light is available. Most of the investigations of the effects of environmental factors relate to one or only a few factors, and their interrelationship is often not sufficiently taken into account. Similarly there are interactions between nutritional factors, the requirement for a particular substance depending on other constituents of the medium (for example, some vitamins, p. 72).

Organisms grown *axenically* (i.e., cultures of a single species of micro-organisms) on a simple sterilized medium are unlikely to behave in the same way as they would in a complex environment such as the soil, where other organisms are present which compete for available nutrients and liberate compounds such as toxins and antibiotics. Sterilization of artificial media may further complicate interpretation by chemically altering the medium.

In addition to these problems of assessing the effect of environment and nutrition on micro-organisms there is a basic difficulty of measurement of growth and/or reproduction. The absolute measure of growth is dry-weight increase, but it may be difficult or, where autolysis occurs rapidly, impossible to separate organisms from the medium in which they are growing. Efforts to avoid these problems by measuring some other parameter, e.g. volume growth, linear growth, or increase in optical density of suspensions, have the disadvantage that the value measured may be water uptake, vacu-

olation, or pigment production rather than true growth (Chapter 5).

For the isolation of micro-organisms and the study of their gross physiology, relatively crude culture media are often used. These media are based on natural products and represent attempts to reproduce the natural environment of the organism to be studied. Autotrophs, which do not require organic nutrients, can best be grown on simple solutions of inorganic salts supplemented with soil extracts or similar materials when vitamins and minor elements are required. Among media widely used for the culture of heterotrophs (see below) are those containing meat extract, blood, milk, peptones, hydrolyzed casein, yeast extract, malt extract, or potato extract; all are, to varying degrees, crude sources of amino acids, vitamins and bases and are sometimes supplemented with sugars and mineral salts. These may be used as fluid culture media or as 'solid' media when gelled, usually with agar. Solid media are widely used for the isolation and maintenance of cultures but are less useful in work aimed at establishing the precise nutritional requirements of organisms. For this, completely defined culture media made from pure chemicals must be used. With more fastidious organisms development of such a medium may be extremely laborious, especially as other factors, such as pH, gaseous atmosphere, etc., may need to be precisely controlled to permit growth. Despite these difficulties, the investigation of nutrition has led to many fundamental discoveries in metabolism and enzymology.

Nutritional types

All organisms require for growth the elements necessary for synthesis of their cellular constituents, and an energy-generating system, which consists essentially of an electron donor and an electron acceptor. If an organism is growing under aerobic conditions oxygen will be the terminal electron

Table 4.1 Nomenclature of nutritional types

Nutritional type	Energy source	Carbon source	Examples	Equivalence
Chemotrophs				
Chemoorganotrophs	Oxidation of organic compounds	Organic compounds	Medically important bacteria, many soil bacteria, fungi, protozoa	Heterotrophs
Chemolithotrophs	Oxidation of inorganic compounds	CO_2	*Nitrosomonas, Nitrobacter, Thiobacillus*	Autotrophs
Phototrophs				
Photoorganotrophs	Light	Organic compounds	Purple non-sulphur bacteria, some algae	Autotrophs
Photolithotrophs	Light	CO_2	Purple sulphur bacteria, green bacteria, many algae	Autotrophs

acceptor, but with anaerobic organisms a variety of compounds is capable of replacing oxygen. Differences in the source of the energy, that is the electron source, are used to distinguish between various groups of micro-organisms (Table 4.1). Lithotrophs are able to grow using either the energy liberated by the oxidation of certain inorganic compounds or the energy from light (Chapter 2). The former are chemolithotrophs and the latter photolithotrophs; both can utilize inorganic sources of carbon and nitrogen. Chemoorganotrophs use organic compounds as energy and carbon sources; in many instances the same compound serves both purposes.

Fungi, most protozoa, slime moulds and many bacteria are chemoorganotrophs. The complexity of their nutritional requirements varies enormously according to the synthetic ability of the species or even the particular strain. Many fungi can grow and sporulate satisfactorily with a simple sugar, an inorganic or simple organic nitrogen source, and mineral elements. Some fungi have been cultured only on chemically undefined plant or animal extracts which contain a wide range of organic compounds. The obligate parasites of plants, such as many of the rusts and smuts (see Chapter 13), have unknown, presumably very complex, requirements and have not been cultured axenically.

Protozoa and slime moulds generally have complex nutritional requirements and many have not been grown in axenic culture or in chemically defined media. Those that have been investigated commonly require a range of amino acids and vitamins in addition to carbohydrates and in this they resemble the metazoa. A complication arises in the close relationship between protozoa and some of the green algal flagellates. In dark conditions and/or in the presence of certain chemicals in a suitable nutrient medium some of these algae may lose their chlorophyll and change from photosynthetic autotrophs to heterotrophs. This change may be the result of a mutation, or, alternatively, a phenotypic response. Other algae are also capable of heterotrophic growth in dark conditions but the range of substrates used is more restricted than that metabolized by true heterotrophs. The majority of algae are photolithotrophs

Bacteria are nutritionally the most versatile micro-organisms, ranging from photosynthetic and chemosynthetic lithotrophs to chemoorganotrophs with requirements of varying complexity. Some of the heterotrophs can grow on a single simple sugar and an ammonium salt; others, such as *Lactobacillus* spp., require a wide range of amino acids and vitamins. Many parasitic bacteria have been grown only in complex undefined media whilst a few, e.g. *Treponema pallidum*, have never been grown in artificial culture. Considerable use is made of the nutritional requirements of bacteria and yeasts for taxonomic purposes (see Chapter 8). There is, however, less need to use such criteria in dealing with filamentous fungi, algae and protozoa because of the availability of morphological characters for classification.

Nutritional requirements

Quantitatively the most important elements required by living cells are carbon, hydrogen, oxygen, nitrogen, sulphur and phosphorus. A wide range of other elements and compounds is also required in much smaller amounts.

The ability of a compound to supply one or more of the elements depends on the organism taking it in (p. 81) and then metabolizing it. Many large polymers, such as some oligosaccharides and some proteins, are not utilized by certain micro-organisms because, due to their size, they are not taken up, though within the cells enzymes are present which can degrade the polymers. However, many fungi and bacteria release extracellular enzymes which break down the polymers into transportable units. These extracellular enzymes, and even whole series of enzymes involved in particular metabolic pathways in the organism, may be inducible, that is, they are produced only in the presence of the substrate or closely related compounds (Chapters 2 and 3). The ability to produce such enzymes is one manifestation of the latent adaptability of micro-organisms to changing environments.

Even if a particular compound is normally taken up it may still not be utilized, although the micro-organism has the intracellular enzyme systems to do so, because of a deficiency in membrane transport systems. Uptake of substances (p. 81) frequently involves carriage across cell membranes by transport proteins. These 'permeases', like enzymes, are inactivated by, for example, extreme pH values and this may prevent utilization of various nutrients. Furthermore, the permease systems are specific for particular compounds or group of compounds and may be differentially affected by such factors as pH. If a variety of compounds is available only one may be taken up though all are utilizable; an order of preference for various nutrients is thus established.

Carbon sources

Carbon is the basic structural unit of all organic compounds and is therefore needed in comparatively large quantities. Lithotrophs can synthesize organic compounds from carbon dioxide or the bicarbonate ion, and no other source of carbon is required. Chemoorganotrophs need an organic carbon compound as an energy source, which also serves to provide most of the carbon required by the cell. Studies with radioactive carbon have shown that many chemoorganotrophs require, in addition, small quantities of carbon dioxide, although they cannot use it as their sole source of carbon since it cannot be further oxidized to produce energy. The range of carbon compounds used is enormous, varying from gaseous carbon dioxide through simple sugars and amino acids to lipids, polysaccharides and proteins.

The chemoorganotrophs—fungi, most protozoa and many bacteria—commonly, and often preferentially, use the simple carbohydrates, particularly D-glucose. Disaccharides, monocarboxylic acids, amino acids, lipids, alcohols and polymers such as starch and cellulose are less readily utilized, approximately in the order given. There is a great variation between organisms in the range of compounds used. Many fungi and bacteria have a large range of catabolic and anabolic enzymes and can synthesize all the substances they require from a single carbon source or, conversely, can use a large selection of substances; mixtures may be stimulatory but are not essential. Other fungi and bacteria, and protozoa generally, are more exacting, and require several carbon-containing compounds because, being metabolically deficient, they lack the enzymes to convert a single source to all the compounds that they require. These complex requirements may include several carbohydrates, a selection of organic acids and, in the case of protozoa, various steroids, phospholipids, purines and pyrimidines.

Nitrogen sources

Nitrogen occurs in cells as a constituent of proteins, which occur as enzymes as well as structural polymers, and of the purines and pyrimidines which are constituents of nucleic acids and some growth factors. Within the cell the basic units of nitrogen metabolism are amino acids, which may be obtained from the environment as such or formed from inorganic nitrogen. Some bacteria and many blue-green bacteria can 'fix' atmospheric nitrogen; that is, they can convert nitrogen gas into organic nitrogen (p. 251). Many fungi, algae and bacteria and a few protozoa can utilize ammonium salts but nitrates are less readily used and nitrite is used by only a few organisms since it is frequently toxic. Protozoa and some bacteria require more complex nitrogen sources, e.g. a selection of 10 to 15 different amino acids. Apart from this absolute requirement, many micro-organisms may grow better when supplied with a mixture of amino acids, proteins or protein hydrolysate, than with a single amino acid or an inorganic source. The mixture of amino acids may supply the organism with one which it can synthesize only slowly and which, in the absence of an exogenous supply, limits growth. In addition some protozoa may possibly require peptones or natural proteins.

Other elements

Elements other than carbon and nitrogen required by micro-organisms are supplied usually as inorganic compounds, but some organisms require certain sulphur- and phosphorus-containing organic compounds. The elements needed can be divided into two classes according to the concentration at which they are required. The macronutrients such as phosphorus, potassium, sulphur, magnesium, calcium, iron, and sodium are usually required at relatively high levels (10^{-3} to 10^{-4}M) and hence it is easy to demonstrate that growth does not occur in their absence. Micronutrients or trace elements such as manganese, copper, cobalt, zinc, and molybdenum are required only at very low concentrations (10^{-6} to 10^{-8}M) and it is an exacting task to demonstrate a requirement for them. The usual methods for removing macronutrients from a medium leave a sufficient level of trace elements as contamination to satisfy the need for them and in many instances a requirement for one micronutrient can be satisfied by another one. The distinction between macro- and micronutrients is not clear cut; certain elements such as iron lie between the two classes and may be considered to belong to either. Generally speaking, it is true to say that the macronutrient elements are incorporated into compounds of structural importance or have a physiological role such as osmoregulation and are thus required in large quantities and that the micronutrients are co-factors for enzymes.

Since all these elements are intimately involved in basic metabolism their requirement is fairly general amongst all groups of micro-organisms, though the amounts required vary. There are organisms in whose metabolism inorganic ions play a special part; for example, the iron and sulphur bacteria use large quantities of some compounds containing these elements.

The form in which elements other than carbon and nitrogen are supplied in culture and some of their uses within the cell are summarized in Table 4.2.

The function of some micronutrients is unknown, even though a definite requirement for them has been demonstrated in some micro-organisms. Thus boron is needed by algae and some bacteria, and vanadium, tungsten, chromium and gallium are sometimes stimulatory to fungi.

Vitamins

There is often confusion between the terms 'vitamin' and 'growth factor'. In this book 'vitamin' is restricted to meaning a complex organic compound required in small quantities by living organisms for the normal functioning of their physiological processes; vitamins are constituents or precursors of co-enzymes and are hence required at very low

Table 4.2 Summary of nutrient supply and use (excluding carbon and nitrogen)

Element	Usually supplied in synthetic media as	Use
Phosphorus	KH_2PO_4, K_2HPO_4	Component of nucleic acids, phospholipids and co-enzymes. (high levels for medium buffering)
Sulphur	$MgSO_4$	Component of the amino acids cystein and methionine and the vitamins biotin and thiamin
Potassium	KH_2PO_4, K_2HPO_4	Co-factor and maintenance of electrical and osmotic potentials
Sodium	NaCl	Required by some bacteria, blue-green bacteria and fungi
Calcium	$Ca(NO_3)_2$	Stability of some extracellular enzymes, bacterial and some fungal sporulation
Magnesium	$MgSO_4$	Co-factor, e.g. for nitrogenase, chlorophyll and bacterial cell wall components
Iron	$FeSO_4$	Component of cytochromes and required for the synthesis of chlorophyll (including that of purple sulphur bacteria), diphtheria toxin and haemins
Manganese	$MnSO_4$	Co-factor, e.g. for superoxide dismutase, aminopeptidase L-arabinose isomerase
Copper	$CuSO_4$	Co-factor for oxidases, tyrosinase (especially in melanin synthesis)
Cobalt	$CoSO_4$	Constituent of vitamin B_{12} complex
Zinc	$ZnSO_4$	Co-factor for alkaline phosphatase, alcohol dehydrogenase
Molybdenum	$(NH_4)_6Mo_7O_{24}$	Co-factor for nitrogenase and nitrate reductase

levels, usually only a few micrograms per litre of culture medium. Organic compounds required for growth and reproduction but not occurring in co-enzymes are regarded as growth factors. Some are required in relatively high concentrations for structural purposes and are not distinct from other nutrients (e.g. amino acids) which are specifically required by an organism unable to synthesize them.

The basic metabolic pathways and therefore the co-enzyme requirements of most organisms are very similar. There is, however, a very wide variation in the ability of micro-organisms to synthesize vitamins and consequently a variation in those that have to be supplied. Some fungi, bacteria, algae and protozoa are able to synthesize all the compounds of this type that they require; others do not have the synthetic machinery and one or more vitamins have to be supplied. Usually protozoa have a limited synthetic power and often require an external supply of several vitamins. In contrast bacteria and fungi generally require an exogenous source of only one or two vitamins. The inability to synthesize vitamins may not reflect a complete absence of the appropriate pathway and in some cases suitable intermediates satisfy the requirement as readily as the vitamin itself.

The chemical structure of the commonest vitamins required by micro-organisms and some of their uses is given in Table 4.3.

Growth factors

Many micro-organisms, particularly protozoa, require growth factors which are involved in membrane formation. Inositol is needed by many fungi, particularly yeasts, but *Actinomyces israelii* is the only bacterium known to have this requirement. Absence of inositol from media of exacting fungi often causes the formation of morphologically peculiar cells. The high level at which inositol is required is also in keeping with a structural role. Choline is required at levels similar to inositol by some organisms (e.g. pneumococci) and is also probably involved in membranes. Inositol and choline are both required by many *Mycoplasma* species which, unlike bacteria but like some protozoa, also require sterols for their membranes. Some fungi require sterols for sexual reproduction.

Many instances are known of bacteria, mycoplasmas and protozoa, and a few among fungi, requiring long-chain fatty acids. For example, oleic acid is essential to several *C. diphtheriae* strains, but this need can be satisfied by other unsaturated fatty acids. These factors are also required in concentrations greater than that of vitamins.

Several amines are known to be necessary growth factors for some organisms although their role is uncertain. Putrescine, spermidine and spermine are needed by *Haemophilus parainfluenzae* and a mutant strain of *Aspergillus nidulans*, for example, and are required at levels of 0.2–10 μg/ml. They are thought to be involved in membrane stability, but polyamines such as spermine and spermidine occur also in association with ribosomes. Glutamine and asparagine are respectively required by some *Neisseria gonorrhoeae* and *Pediococcus* strains at levels which suggest that they do not have a structural role.

Purines and pyrimidines are found to be only stimulatory to the growth of many bacteria but are essential for others. The stimulatory effect is usually not highly specific since compounds causing the stimulation can frequently be replaced by another substance of the same type. *Lactobacillus arabinosus* is stimulated by the purine guanine but other purine bases, adenine, xanthine, and hypoxanthine, may be effectively substituted for it. However, *Shigella boydii* has a specific requirement for adenine, as has *L. bifidus* for adenine, guanine, xanthine and the pyrimidine uracil. The latter compound also stimulates the growth of some *Lactobacillus* species and can be replaced, in most instances, by the related compounds uridine, uridylic acid, cytidine and cytidylic acid, and sometimes by cytosine. But again, instances are known of organisms having specific requirements such as that of *Staphylococcus aureus*, when growing anaerobically, for uracil and *L. bulgaricus* for orotic acid. Purines and pyrimidines are used to synthesize nucleic acids (p. 10), which explains why some organisms such as *Lactobacillus gayonii* are stimulated not by purines, nor purine nucleosides, but by purine nucleotides.

There are several other compounds known to be required as growth factors including amino acids, peptides and proteins and such compounds as hematin and biopterin but, in addition, there may be further unidentified substances which are required by organisms such as those bacteria, protozoa and fungi which can be cultivated only on complex undefined media.

Microbiological assay of nutrients

The fact that certain nutrients are necessary for the growth of some micro-organisms and determine the amount of growth that can take place, makes it possible to estimate quantitatively the required nutrients. Among the substances which have been estimated by this method are amino acids, vitamins, and minor elements.

Table 4.3 The chemical structure and uses of some common vitamins

Vitamin	Structure	Used As	In
Thiamin (vitamin B$_1$)	Pyrimidine / Thiazol — Pyrophosphate	Thiamin pyrophosphate	Decarboxylases, transaldolases, transketolases
Biotin		Prosthetic group of biotin enzymes	Carboxylases, deaminases
Riboflavin		Flavin nucleotides (FAD and FMN)	Oxidation/reduction processes, electron transport
Pyridoxin (vitamin B$_6$)		Pyridoxal phosphate	Amino acid transaminases, deaminase and decarboxylases

Pantothenic acid	Pantoic acid β-alanine	Co-enzyme A	Fatty acid and keto acid metabolism
Nicotinamide		Pyridine nucleotides (NAD and NADP)	Oxidation / reduction processes
Cyanocobalamin (vitamin B$_{12}$)	See Mandelstam and McQuillen (1973)	Adenosine derivative	Transmethylation and isomerisation of some organic acids
Tetrahydropteroyl-glutamic acid	Pteridine (a) *p*-aminobenzoic acid (b) Glutamic acid (c) (a) + (b) = Pteroic acid (a) + (b) + (c) = Pteroyl glutamic acid		Transfer of 1-C units
Lipoic acid			Hydrogen and acyl-group transfer

Water

Most micro-organisms contain about 90% water by weight. Spores and other resistant structures contain considerably less. Water is the solvent in which all metabolic reactions occur and through which, in most instances, all exchange of substances with the environment takes place.

The amount of water available to micro-organisms cannot be measured simply as the total amount in the system since some will be associated with solutes and other molecules in such a way as to be unavailable. The amount of water bound in this way affects the vapour pressure (P) of a solution and the ratio of this to that of pure water (P_0) gives a measure of the amount of available water (a_w) in the solution

$$a_w = \frac{P}{P_0} = \frac{\text{Relative humidity}}{100}$$

Micro-organisms grow in the range $a_w = 0.63$–0.99. Bacteria, algae and protozoa are active only at the top end of this range, and only a few xerophilic fungi (i.e., those able to grow at very low available water levels) are able to grow at a_w values of down to 0.63. Although no micro-organism can grow without a relatively high level of available water, many have stages in their life cycle, when spores or cysts are formed, which enable them to survive long periods of desiccation.

The response of organisms to water and aqueous solutions is of great practical importance. Desiccation provides an efficient means of preventing biodeterioration and of keeping organisms in culture collections (p. 94).

Osmotic pressure

Micro-organisms have a higher internal concentration of solutes than the surrounding aqueous medium and, therefore, have a higher osmotic pressure; for example, Gram-positive bacteria have an internal osmotic pressure of about 20 atmospheres and Gram-negative ones of 5–10 atmospheres. The internal osmotic pressure may be reduced to a minimum by converting all low molecular weight substances not immediately involved in metabolism into insoluble storage products such as lipids and polymeric carbohydrates (p. 156). However, the excess osmotic pressure over that of the environment causes water to enter the cell which would swell and burst if no mechanism of osmo-regulation existed. In algae, fungi and most bacteria a more or less rigid cell wall exists surrounding the protoplast which prevents its bursting. In filamentous fungi, however, the hyphal tip has only a very thin wall where it is being laid down and a decrease in the osmotic pressure of the environment can cause bursting of this tip. There is evidence that in fungi and bacteria the osmotic pressure of the cytoplasm is adjusted according to the environmental osmotic pressure in such a manner that a very large pressure difference between the two is prevented. In protozoa, because there is no rigid cell wall to contain the protoplast, water is continuously withdrawn from the cytoplasm and is periodically expelled into the environment by the contractile vacuole or similar mechanism. This removal of water against the osmotic gradient requires metabolic energy. Solute uptake by some protozoa may be limited so that excessive differences in osmotic pressure between the cell and the environment do not occur, and the solute concentration is often less in protozoa than in bacteria and fungi.

The effects of changes in the external osmotic pressure of micro-organisms have not been extensively investigated but a distinction must be made between immediate effects and the long-term effects on growth at a high osmotic pressure. Bacteria (Fig. 8.21) and fungi are plasmolyzed if transferred to a medium of high concentration from one of a lower osmotic pressure, but the protoplast does not always completely separate from the cell wall. In all these instances, unless the change is drastic, normal growth follows a return to solutions of normal osmotic pressure. Fungi may show temporary or permanent cessation of apical growth and production of lateral branches below the apex when subjected to osmotic shock. Similarly bacteria may undergo morphological variation; changes have been observed in both directions between bacillary and coccal forms when grown under conditions of high osmotic pressure.

Solutions of sugar or salt at high osmotic pressure are used in food preservation, although some organisms such as the osmophilic fungi, particularly some yeasts, and halophilic bacteria and fungi which grow in solutions with high solute concentration may still cause spoilage (Chapter 15).

Hydrostatic pressure

Some protozoa, fungi, and bacteria can live under conditions of considerable hydrostatic pressure, for example, in the depths of the oceans, either in suspension or in the sediments. Viable bacteria have been recovered from environments where the depth of water exceeds 10000 metres and the pressure is up to 1140 atmospheres. It is not certain whether these organisms are active at these depths but it is known that not all of these isolates are barophilic (i.e., grow better at or require high pressures). It is suggested that some isolates grow at 1 atmosphere but not at the high pressures of the environment from which they were isolated. Various bacteria, yeasts, and viruses can withstand pressures up to 1200 atmospheres for at least a few minutes but not even deep-sea bacteria can reproduce at 1500 atmospheres and prolonged exposure even to pressure of 100–600 atmospheres inhibits the normal growth of most bacteria. The effect of growth at just tolerable pressures is generally to increase cell size and especially to cause the formation of long, bizarre filamentous cells. The influence of pressure has been only slightly studied at a metabolic level but it is known to affect bacterial luminescence, several enzyme reactions, sulphate reduction, and pseudo-podium formation in amoebae. The reasons for the effects of pressure are far from clear and no coherent pattern of response is apparent. In many instances insufficient consideration has been given to related effects, particularly of temperature and other environmental and nutritional factors.

Oxygen and carbon dioxide

Obligate aerobes are micro-organisms incapable of growth without molecular oxygen which is used as their terminal electron acceptor. Obligate anaerobes use some alternative substance and may actually be killed by exposure to oxygen. There are several explanations of oxygen toxicity but it has recently been found that all aerobic bacteria examined contain the enzyme superoxide dismutase (SOD) which catalyzes the reaction:

$$O_2^- + O_2^- + 2H^+ \rightleftharpoons H_2O_2 + O_2.$$

It has therefore been suggested that aerobes are able to survive because this enzyme destroys the super-oxide free radicle which is generated by many biochemical reactions involving molecular oxygen and which is relatively long lived and has great potential for causing cellular damage, by oxidation of —SH groups and lipid peroxidation for example. Conversely it is thought that the absence of this enzyme from most anaerobes may be an important contributory factor in their oxygen sensitivity. Low levels of SOD have been found in a few anaerobes but this does not invalidate the suggestion, for insufficient data are available to conclude whether there is any correlation between levels of SOD with the spectrum of aerotolerance found among the anaerobes. Facultative anaerobes are capable of growing in the presence or absence of air.

The microaerophiles grow at lower concentrations of oxygen than required by most aerobes but some organisms often included in this group are really misnamed and, in fact, need a high carbon dioxide concentration (i.e. they are capneic, e.g. *Brucella abortus*) rather than a low oxygen tension. Carbon dioxide requirement is probably a universal feature of micro-organisms and is not confined to lithotrophs and a few others as was once thought. This requirement is often difficult to demonstrate as the quantity needed may be so low that it can be supplied by metabolically derived carbon dioxide, but it has been shown that the removal of this gas from bacterial cultures inhibits growth and that at least some bacteria and protozoa die when deprived of it for too long. In contrast, too high a concentration of carbon dioxide adversely affects growth. In the fungi, for example, some normally filamentous forms become yeast-like (an aspect of dimorphism) and some Mastigomycotina form resistant sporangia.

Algae are normally aerobic organisms but a few species, such as *Scenedesmus* sp. are known to be capable of anaerobic autotrophic, but apparently not heterotrophic growth. Thus under both aerobic and anaerobic conditions, these algae require carbon dioxide. Fungi too are mainly aerobes though some such as the yeasts, are facultative anaerobes. Vegetative growth and asexual reproduction can take place under anaerobic conditions but sexual reproduction almost invariably requires at least some oxygen. Bacteria and protozoa show much the same range of diversity with respect to gas requirement. They range from obligate aerobes such as *Pseudomonas* and the protozoa which inhabit the better aerated bodies of water through the facultative protozoa and bacteria such as *Escherichia coli*, the microaerophilic bacteria such as *Lactobacillus casei*, the chemoorganotrophic capneic bacteria, to the obligate anaerobes such as *Clostridium* and the protozoa which occur in the gut of metazoa In an environment unfavourable with respect to gases, growth of these organisms slows and ceases and motile bacteria become non-motile, a point of practical importance when examining for motility.

Movements in response to chemical substances

Movement of the whole organism in response to chemical stimulus is called chemotaxis. A change in direction of growth of filamentous fungi and algae

...ot motile is called chemotropism. Motile ...anisms may move towards or away from ...;: positive chemotaxis and negative chemo...xis respectively. Generally compounds with inhibitory or deleterious properties cause a negative response, but if a substance inhibits motility it may cause an accumulation of cells within its effective area without actually attracting the organisms. Compounds which cause a positive response are sometimes nutrients but chemicals not required for nutrition are involved in the initiation of aggregation of myxobacteria (p. 164) and dictyostelids (p. 210) and in the attraction of the motile gametes of some micro-organisms. Similar chemotropic responses may occur along chemical gradients in hyphal anastomoses or prior to the fusion of progametangia in Zygomycotina.

Recently considerable progress has been made in understanding chemotaxis of *E. coli*. This depends upon the possession of a chemosensor which is specific for a compound or group of structurally related compounds. These are invariably components of transport systems, for example the periplasmic galactose-binding protein and one component of the bound phosphotransferase system located in the cytoplasmic membrane are known to be involved. There are at least twelve chemosensors for attractive compounds and about eight for repellents.

In the absence of a gradient of a substance eliciting a chemotactic response *E. coli* alternately swims in more or less straight lines for about one second and tumbles for about a tenth of a second. Where a gradient of an attractant exists and the bacterium is moving towards the higher concentration the intervals between tumbles increase and *vice versa* when moving in the opposite direction. Swimming occurs when all the flagella operate together in a bundle as a single organelle; tumbling occurs when the direction of rotation of one or more of the individual flagella is reversed and the bundle ceases to exist, the individual flagella acting independently. How the organism senses concentration gradients and how this information determines the direction of rotation of the flagellum are but two of the many outstanding questions.

An account of chemotaxis in slime moulds is given on p. 211.

Physical factors

Temperature

Temperature affects the rate of all processes occur-ring in micro-organisms and it may determine the type of reproduction, the morphology of the organism and the nutrients required. For any particular organism the minimum and maximum temperatures are the lowest and highest at which growth occurs, and the optimum temperature is that at which the growth rate (or any other process) is the greatest. Their values vary between different strains of a single species and with the stage in the life cycle, the age of cultures and the nutritional status of the medium. Metabolic processes obey normal chemical laws, the rate of reaction increasing with increasing temperature, until a point where denaturation of the thermolabile enzymes involved in catalysis of the reactions begins to reduce the rate. The determination of optimum temperatures may be complicated by a time factor; initially the growth rate may be rapid but subsequently it may decrease as proteins are slowly denatured. Thus, the optimum temperature should be considered as the temperature which allows the best sustained growth. Enzymes vary in the temperature at which they are denatured and as the temperature is raised, enzyme systems may be inactivated one by one. Those systems necessary for reproduction may be the most sensitive and therefore may be destroyed before those required for growth, thus explaining the wider temperature range for vegetative growth than for sporulation.

It is common practice to assign organisms to one of three categories according to their temperature relationships though there is considerable overlap between the classes. The majority of micro-organisms are mesophiles and grow at moderate temperatures (Table 4.4). Among all the main groups of micro-organisms there are however thermophils, which are able to grow at, or have a requirement for, elevated temperatures, though the range over which they grow is not wider. Conversely, there are psychrophilic organisms with low optima, maxima and minima (Table 4.4). How thermophils deal with the problem of denaturation of enzymes and how psychrophiles maintain a sufficiently rapid metabolic rate to grow is uncertain. It has been suggested that thermophiles may produce enzymes more rapidly than mesophils, thus countering thermal denaturation, and it is known that many of their enzyme proteins are particularly thermostable. Isolated enzymes from psychrophiles do not show differences from comparable enzymes from mesophiles and it is probable that the ability of psychrophiles to transport solutes across the cell membrane efficiently at low temperatures due to the peculiarities of its

Table 4.4 Temperatures (°C) for vegetative growth of representative organisms

Organism	Minimum	Optimum	Maximum
Thermophiles			
Bacillus stearothermophilus	30	55	75
Synechococcus eximius	70	79	84
Mesophiles			
Escherichia coli	10	37	45
Streptococcus faecalis	0	37	44
Neisseria gonorrhoeae	30	36	38
Saccharomyces cerevisiae	1	29	40
Fusarium oxysporum	4	27	40
Chilomonas paramecium	10	28	35
Phychrophiles			
Pseudomonas fluorescens	−8	20	37
Candida scottii	0	4–15	15
Chlamydomonas nivalis	−36	0	4

membrane fatty acid composition accounts for their ability to grow under these conditions.

Hydrogen ion concentration

There are many ways in which the hydrogen ion concentration (pH = $-\log_{10}[H^+]$) of the environment may affect micro-organisms and some of these have been mentioned earlier in the chapter. The net effect of pH acting on these various factors is expressed by the resulting growth and reproduction of the micro-organisms. The nutritional status of the environment is most markedly affected by pH both by altering the adsorption and solubility of ions and the dissociation of molecules, thus determining their availability to the micro-organisms and suitability for transport across the cell membrane (p. 81). In artificial culture growth-limiting pH conditions frequently arise as a result of the micro-organism's own metabolic activities (p. 85).

Micro-organisms grow over a pH range usually with a fairly well defined optimum which, in contrast to the temperature optimum, lies approximately at the middle of the range permitting growth. The minimum pH for growth is generally about pH 2.5 and the maximum about pH 8–9; the optimum varies widely but is frequently between pH 5 and pH 7.5. Bacteria such as *Thiobacillus thiooxidans* and *Acetobacter* spp. and some imperfect fungi (*Penicillium* spp. and *Aspergillus* spp.) are capable of growth at the very low pH values of between pH 0 and pH 2 and some *Bacillus* spp. and Cyanobacteria can grow at pH 11.

Living cells are very well buffered internally against pH changes, and environmental values have to be extreme before the intracellular pH is much affected. Extracellular enzymes are of course directly influenced by the environmental pH.

Light

Light of particular wavelengths from the range between the near ultra-violet (250 nm) to the near infra-red (1100 nm) affects micro-organisms in four main ways. It provides the energy for photosynthesis, it causes oriented response (tropic and tactic movements), it may be required for or may stimulate sporulation in fungi, and it can have deleterious or lethal effects. Responses are mediated by photosensitive compounds which are transformed by absorption of light energy; this effect is translated into the observed response by light-independent reactions ('dark' reactions). Of the four responses mentioned, photosynthesis, the tropic and tactic movements, and fungal sporulation are dealt with here. The deleterious or lethal effects are discussed in Chapter 6.

The pigment primarily involved in photosynthesis is chlorophyll but others such as carotenoids and phycobilins also can absorb light and transfer this energy to chlorophyll. The structures of the pigments are not the same in all micro-organisms and the different types have different absorption spectra. Thus the wavelength of light supplied will determine which of a mixture of organisms can photosynthesize. In some photosynthetic organisms, light may induce effects not directly attributable to photosynthesis, for example sterol content of *Scenedesmus* is raised as the light intensity is increased.

For most non-photosynthetic organisms light is unnecessary or deleterious but sporulation of some fungi, both sexual and asexual, is often initiated or increased by radiation in the ultra-violet or blue spectral regions, and zonation of a colony may occur when it is grown in alternating periods of light and dark; bands of dense sporulation develop as a result of the periods of illumination. Light may also influence pigment production itself; some normally pigmented fungi and mycobacteria fail to produce any pigment or produce reduced amounts if grown in the dark.

The phototactic behaviour may result either from a response to light or to its absence: for example the purple bacterium *Rhodospirillum rubrum* accumulates in the light zones because organisms which have crossed from the light to the dark zone stop and reverse their direction, whereas those crossing from a dark to a light zone carry on. The reason for the reversal of direction of movement is not certain but it is known that the action spectra for phototaxis and photosynthesis are the same. Algae often respond tactically to light, accumulating in regions of light provided the intensity is not too great. In *Euglena* the photoreceptor is not the photosynthetic pigment since the action spectra for photosynthesis and phototaxis do not correspond; only the light from the blue region of the spectrum is effective in inducing movement. *Euglena* responds to light of lower intensity than do phototactic bacteria and responds both positively and negatively to light. Both these types of response occur also in the blue-green algae, dinoflagellates and the diatoms, depending on the wavelength and intensity of the light, and this is responsible for the diurnal movements of some phytoplankton and benthic-forms (p. 265).

Photosynthetic rate, and hence growth rate in autotrophs, is roughly proportional to light intensity, until saturation point is reached. The light intensity which is saturating depends on other environmental factors such as temperature and mineral nutrition and previous exposure levels.

Spore-bearing structures in fungi often respond to unilateral illumination by directional growth or increase in growth rate; they are phototropic. They usually grow towards light (positively phototropic) which in the natural environment will increase the likelihood of spores being released away from the substrate. Thus stipes of Agaricaceae, asci and perithecial necks of some Ascomycotina, sporangia of many Zygomycotina and the conidiophores of a few Deuteromycotina are all positively phototropic.

Gravity

In fungi geotropism is demonstrated mainly by spore-bearing structures, though it may be modified by phototropic effects (see above), which are usually dominant. Thus sporangia of some Zygomycotina are negatively geotropic (i.e., grow away from the earth). The higher Basidiomycotina orientate in response to gravity very precisely so that the gills or pores (p. 193) are vertical, allowing the spores borne on them to fall freely and be efficiently dispersed.

The reactions of protozoa and algae to gravity have not been fully investigated. It has been suggested that some free-living amoebae may respond positively, maintaining their position in the bottom sediments of natural water, whilst free-swimming forms may react negatively remaining in the surface layers of water. The position within the water is, however, controlled by many other factors such as light and chemical gradients (see above) and many show no reaction to gravity.

Mechanical stimuli

Motile bacteria with flagella arranged at one end of the cell and some ciliate protozoa exhibit a complex avoidance behaviour on contacting an object in their path; a series of movements involving backing off, turning at an angle and proceeding again is repeated until the organism has moved around the obstruction. In contrast bacteria which have peritrichous or amphitrichous flagella (p. 141) merely reverse their direction of motion on coming into contact with an obstruction: algae, too, usually have negative thigmotactic responses. Some protozoa which live on surfaces of stones etc. in water have specialized cilia to hold on with and show positive thigmotaxis.

In fungi few tactile responses are fully documented. In natural conditions one example of the tactile response is fruit body formation induced in some hypogeous (underground) fungi by contact with plant roots and tactile responses are also thought to play a part in the penetration of some antheridia into the oogonium during fertilization (e.g. *Achyla* sp., p. 196). The formation of appressoria in some plant pathogens (e.g. *Botrytis*) may be stimulated by contact with the host plant surface. Some Deuteromycotina may be stimulated to sporulate by very drastic tactile stimuli such as cutting or scraping the mycelium, though whether this is a response to touch or to metabolites released into the medium from killed cells is in doubt.

Some bacteria align themselves according to the orientation of the molecules in the substrate. Probably the best known examples of this phenomenon, *elasticotaxis*, is the alignment of *Cytophaga* cells with the cellulose micelles in cellulose fibres. Other myxobacteria are known to arrange themselves along lines of substrate orientation induced by stress in agar.

Uptake and translocation of nutrients

The rate and method of nutrient uptake depends on the nature of the nutrient, the environment of the cell and the structure and properties of the cell envelope. The cell wall of actively growing organisms is generally regarded as freely permeable to many but not all small molecules and the cytoplasmic membrane as the principal permeability barrier of vegetative cells.

Movement of molecules across membranes may occur as a result of diffusion gradients, electrical potential gradients, or most commonly carrier-mediated transport mechanisms which move substances against such gradients and bring about their accumulation in the cytoplasm.

The diffusion of substances from a high to a low concentration takes place in both directions across the membrane. The most common substances moving on concentration gradients are gases such as carbon dioxide and oxygen. The movement of gases is not generally controlled by the cell, except as a result of the establishment of gradients by metabolic processes. Water also moves across the membrane by *passive diffusion* from more dilute conditions to more concentrated ones and is under cellular control as a result of adjustment of the amounts of solute.

In addition to simple diffusion, uptake may be by *facilitated diffusion* which is mediated by a carrier. For example yeasts take up sugar by this means which does not depend on metabolically derived energy. The characteristics of such uptake are that the carrier may be saturated, is solute specific and that it takes place at a rate in excess of that expected for the diffusion of hydrophilic molecules like sugars and some amino acids across the hydrophobic cytoplasmic membrane. The energy necessary is derived only from the existence of a gradient of the solute and therefore such systems mediate export or import according to the prevailing solute concentration on either side of the membrane.

To concentrate a solute within a cell, against a solute gradient, energy must be supplied to a carrier system with the properties described above; the process is then one of *active transport* (although this term is sometimes reserved for uptake processes by which the solute is accumulated unchanged within the cell). Energy may be coupled in one of three ways. Firstly, the accumulation of some sugars, some amino acids and some ions in a number of bacteria is achieved by coupling with part of the electron transport system. Secondly, energy may also be coupled indirectly through ion *symporters*. These are usually H^+, Na^+, or other monovalent cations and it is their movement under the influence of the energy-generated electrochemical gradient which achieves the uptake of non-electrolytes as in the Na^+-melibiose co-transport system of *Salmonella typhimurium* for example. Thirdly, in group translocation systems, such as the phosphoenolpyruvate (PEP) sugar phosphotransferase system by which uptake of many sugars and related compounds is achieved in a wide range of bacteria, the energy is derived from PEP and the sugar is found as the phosphorylated derivative within the cell.

In no instance of carrier-mediated uptake is there a certain explanation in molecular terms of the way in which the compound taken up is actually moved across the membrane, but it is presumed that some conformational or rotational change in the carrier is involved.

In many Gram-negative bacteria periplasmic binding proteins exist which are involved in transport of some but not all amino acids and sugars, but their relationship with carriers and their function is uncertain. These binding proteins are also chemotaxis receptors (p. 78).

Further methods by which nutrient substances cross the cell membrane are pinocytosis and phagocytosis. These are particularly important in some heterotrophic protozoa, myxomycetes and dictyostelids (p. 210). When a food particle comes into contact with the cell membrane the latter invaginates, surrounding the particle; this process is called *phagocytosis*. Eventually the invagination is occluded and the food vacuole so formed moves into the cytoplasm. *Pinocytosis* is a similar phenomena involving the enclosing of liquids rather than particles. Enzymes are secreted into the food vacuole and the products of digestion presumably cross the vacuole membrane by one of the systems indicated above.

Fungi may translocate nutrients through the mycelium. Functionally, the hyphae consist of open tubes: if septa are present they normally have at least one pore in them (p. 187) and thus the cross-walls are not thought to limit translocation under normal conditions. Movement occurs towards growing and sporulating regions from the older parts of the colony. The rate of movement is too great for diffusion to be solely responsible and the main translocation method is protoplasmic streaming.

Books and articles for reference and further study

AINSWORTH, G. C. and SUSSMAN, A. S. (eds.) (1965). *The Fungi*, an advanced treatise, Vol. 1. The Fungal Cell. Academic Press, New York, 748 pp.

AINSWORTH, G. C. and SUSSMAN, A. S. (eds.) (1966). *The Fungi*, an advanced treatise, Vol. II. The Fungal Organism, Academic Press, New York, 805 pp.

AINSWORTH, G. C. and SUSSMAN, A. S. (eds.) (1968). *The Fungi*, an advanced treatise. Vol. III. The Fungal Population. Academic Press, New York, 738 pp.

DAWES, I. W. and SUTHERLAND, I. W. (1976). *Microbial Physiology*. Basic Microbiology Vol. 4. Blackwell, Oxford, 192 pp.

HALL, R. P. (1967). Nutrition and Growth of Protozoa. In *Research in Protozoology*, Vol. I, ed. Chen, T-T, Pergammon Press, Oxford, pp. 388–404.

HAMILTON, W. A. (1975). Energy coupling in Microbial Transport. In *Advances in Microbial Physiology* 12, (ed.) Rose, A. H. and Tempest, D. W., Academic Press, New York and San Francisco, 1–53.

HOLZ, G. C. (1964). Nutrition and Metabolism of Ciliates. In *Biochemistry and Physiology of Protozoa*, Vol. III (ed.) Hutner, S. H., Academic Press, New York, pp. 199–242.

KITCHING, J. A. (1957). Some factors in the life of free living Protozoa. In *Microbial Ecology*; 7th Symp. Soc. Gen. Microbiol., Cambridge University Press, pp. 259–286.

KOTYK, A. (1975). Modes of coupling of energy with transport, in *Biomembranes: Structure and Function* (ed.) Gardos, G. and Szasz, I., Vol. 35 Proceedings of the Ninth FEBC Meeting North-Holland/American Elsevier, pp. 65–97.

LEWIN, R. A. (ed.) (1962). *Physiology and Biochemistry of Algae*, Academic Press, New York and London, 929 pp.

MANDELSTAM, J. and MCQUILLEN, K. (1973). *Biochemistry of Bacterial Growth*. Blackwell Scientific Publications. Oxford, London, Edinburgh and Melbourne. 582 pp.

MORRIS, J. G. (1975). The Physiology of Obligate Anaerobiosis. In *Advances in Microbial Physiology* 12. (ed) Rose, A. H. and Tempest, D. W., Academic Press, London, New York and San Francisco, pp. 169–246.

PARKINSON, J. S. (1981). Genetics of bacterial chemotaxis. In *Genetics as a Tool in Microbiology*; 31st Symp. Sec. Gen. Microbiol. Cambridge University Press, pp. 265–290.

SIMONI, R. D. and POSTMA, P. W. (1976). The Energetics of Bacterial Active Transport. *Ann. Rev. Biochem.* **44**, 524–554.

SLEIGH, M. A. (1973). *The Biology of Protozoa*. E. Arnold, London, 315 pp.

SMITH, J. E. and BERRY, D. R. (1976). *The Filamentous Fungi*. Vol. 2. Biosynthesis and Metabolism, Edward Arnold, London, 520 pp.

STEWART, W. D. P. (ed.) (1974). *Algal physiology and Biochemistry*. Botanical monographs 10. Blackwell Scientific Publications, Oxford, 989 pp.

Chapter 5

Growth of micro-organisms in artificial culture

The Oxford English Dictionary lists several meanings for the word grow and it is commonly used to describe increases in amounts of various objects, clouds, crystals and even of abstractions such as love, as well as living material. For our purposes growth may be defined as an increase in mass of the whole or part of a living organism by the synthesis of macromolecules. Organisms can grow by increase in mass of their constituent cells followed sometimes by cell division and an increase in cell numbers. More complex organisms can grow also by inclusion of organized interstitial cellular products. In higher plants and animals growth may be accompanied by differentiation of cells to form particular structures or organs. Thus growth of living material represents the production of order out of chaos, materials being incorporated in an orderly sequence under the control of the cells' genetic material.

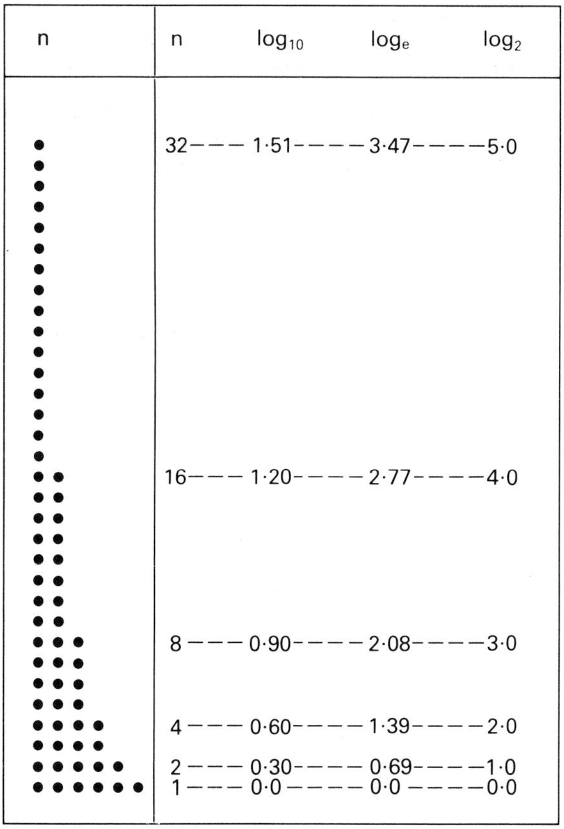

Growth of unicellular organisms

Many unicellular organisms, including most bacteria, the fission yeasts and some algae, increase in number by growing until almost exactly twice their original size and then dividing into two, usually precisely equal, halves. Provided the environment is suitable the process is repeated so that all the organisms continue to grow and divide. Clearly, this form of growth and division, known as *binary fission*, enables one cell to become successively 2, 4, 8, 16, 32, 64, etc. with no theoretical limit. If cell divisions occur at equal time intervals and all cells divide, which happens in practice, then the logarithm of cell numbers plotted against time will yield a straight line, giving rise to the expression *logarithmic growth*. Bacteria have long been known to increase in numbers in this way, but that they do so is merely a fact of arithmetic and does not give any information concerning the growth process (Fig. 5.1).

For a living cell to be growing exponentially (logarithmically) the increase in the mass of the cell during any small interval of time must be exactly proportional to the size of the cell at that time. This implies that all the freshly created material must have contributed to the growing process from the moment of its formation.

Fig. 5.2 compares the size achieved by a cell growing at a constant rate (arithmetically) and one growing fully exponentially. It should be noted that in both the increase in cell numbers will be exponential, but where cell enlargement is itself exponential the increase in numbers will be greater.

These arguments hold equally well if mass increases rather than number increases are studied,

n	n	$\log_{10}$	$\log_e$	$\log_2$
	32 — — 1·51 — — — 3·47 — — — — 5·0			
	16 — — 1·20 — — — 2·77 — — — — 4·0			
	8 — — 0·90 — — — 2·08 — — — — 3·0			
	4 — — 0·60 — — — 1·39 — — — — 2·0			
	2 — — 0·30 — — 0·69 — — — —1·0			
	1 — — 0·0 — — — 0·0 — — — —0·0			

Fig. 5.1 A diagrammatic representation of the doubling of bacterial numbers by binary fision from a single cell to 32 cells represented in various logarithmic notations. Logs to base 2 give a direct indication of the number of cell division cycles.

Fig. 5.2 A representation of the size of a cell originally of one unit length after growth for a standard time and at a standard growth rate if the cell was growing (**a**) arithmetically (at a rate which remained constant irrespective of cell size); (**b**) at a rate recalculated once halfway through the growth period to be proportional to the cell size at that time; (**c**), (**d**), (**e**) as (**b**) but with the rate recalculated 9, 19, and 99 times during the growth period; and (**f**) with the growth rate continually recalculated to always be proportional to the cell size.

and it will never be immediately obvious from inspection of a growth curve whether an apparent exponential increase in mass of a population of cells results from each cell having grown exponentially, or was from a regular periodic doubling in numbers of cells which individually grew at a constant linear rate.

Because of the small size of bacteria the exact nature of their growth has not been fully elucidated. Precise measurements of a cell growing under the microscope should determine whether its growth is linear or exponential but accuracy is difficult when approaching the limit of resolution of the light microscope and better methods, based on the size distribution of cells in a homogeneous growing population have been used. Obviously in a population of cells growing linearly there will be equal numbers of cells of all sizes whereas in a similar random sample of cells growing exponentially there will be fewer large ones.*

Studies like this have revealed that different types of cells grow in different ways. Small rod-shaped bacteria do grow exponentially, whereas long filamentous bacteria, fission yeasts and fungi tend to grow at a constant linear rate, and spherical bacteria probably sigmoidally.

Numerical representation of growth of unicellular organisms

If an experiment is performed in which the numbers of cells present in a growing culture is determined at

intervals and the results recorded in graphical form, a straight line is obtained if the numbers are expressed as logarithms. To display the growth of cells as logarithms is not only a convenience but often a necessity, since otherwise the paper size would become unmanageable or the scale unsuitably small. Moreover, since we have seen that bacterial numbers increase according to the scale of 2, a logical system is to use logarithms to the base 2 to express the numbers ($\log_2 2 = 1$, $\log_2 4 = 2$, $\log_2 8 = 3$, $\log_2 16 = 4$, etc.). In this way each increment of 1 in the ordinate scale indicates a cell division and the doubling of the population (Fig. 5.1). However, if total mass measurements are made on a culture of cells which are growing truly exponentially (compound interest), the logical base for the logarithms is the exponential (e) (e = 2.718282 approx., that is the mass to which one cell would grow by true exponential growth in the same time as the same cell would take to double its size by linear growth). For this reason it is usual to plot curves of bacterial mass against time by means of Naperian logarithms ($\log_e$) to express the mass. Inevitably, this leads to some confusion in the beginner's mind since he finds that a cell's specific growth rate representing the slope of the mass growth curve in (Fig. 5.3b) is not the simple reciprocal of the doubling time (t_d) for the numbers of cells in a culture (Fig. 5.3a) but rather:

$$t_d = 1/\mu \times C$$

where C is a constant for converting $\log_2$ to $\log_e$ (i.e. $\log_2 e$).

From the slope of the curves of mass ($\log_e$) plotted against time, and of numbers ($\log_2$) against time, four terms can be derived, only two of which are commonly used:

Mass ($\log_e$) vv time	Numbers ($\log_2$) vv time
M/t = specific growth rate	N/t = specific doubling rate (never used)
t/M = 'ething time' (never used)	t/M = doubling time

* This, of course, is because cells growing exponentially spend more time growing from size 1 to size $1\frac{1}{2}$ than in growing from size $1\frac{1}{2}$ to size 2 prior to dividing.

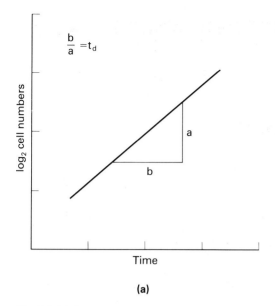

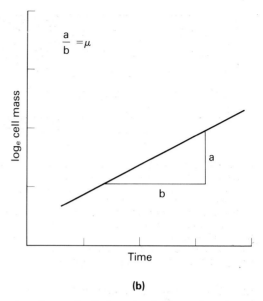

(a)   (b)

Fig. 5.3 (**a**) A graphical representation of doubling time, calculated during the exponential phase of growth and (**b**) a similar graphical representation of the specific growth rate.

It is now possible to define a number of terms used in microbial growth studies—the following are precise definitions and it must be remembered that some of the terms are often regularly used loosely by many bacteriologists who should know better! *The specific growth rate* often abbreviated to (μ)* is the rate of growth per unit organism.

$$\mu = \frac{dm}{dt} \cdot \frac{1}{m}$$

The growth rate (written μ_g) is the rate of total mass increase (mass growth rate) or total number increase (number growth rate) of a whole culture. It is not constant but increases as the numbers of dividing cells in the culture increases.

$$\mu_g = \frac{dm}{dt}$$

Note: The growth rate = specific growth rate × cell mass

$$\mu_g = \mu_s \times m$$

The generation time (t_g) is the time taken by a particular cell, from its formation by cell division to its own fission. Clearly this must be measured for a particular cell by observation under the microscope. *The mean generation time* (tgm) is the arithmetic mean of a number of observations of individual

generation times of cells growing in the same culture during the same growth period.
The doubling time is the time taken for the number of cells in a population to double. It need not be identical with the mean generation time since this would presuppose, *inter alia*, that all the cells in the culture were actively dividing, and yielding viable offspring.

All the above definitions referring to numbers usually include only *living* cells, whereas those defining quantities refer to *total cells mass*, living and dead, as these are the parameters most usually measured. In practice, a growing culture has a high percentage viability so the differences are negligible.

Factors affecting growth rate

The age of the culture, its genetic constitution and a number of external factors, such as temperature, pressure, light and pH, influence growth rate and the type of growth. These are considered in Chapter 4. The nature and concentration of the food supply also have a profound effect.

Clearly, if either carbon or nitrogen is unavailable in the environmental medium the cell cannot synthesize fresh cell material and no growth can occur, but it is now generally accepted that there is no threshold concentration for either carbon or nitrogen below which growth cannot occur. It has been argued that in any nutrient environment only one substance can be limiting for growth at any one time, and such an assumption greatly simplifies the consideration of the relationship between specific growth rate and substrate concentration. The

---

* The conventions used here (μ_g for the overall growth rate and μ for the specific growth rate) are to clarify existing confusion where μ has been used indiscriminately for both.

curves in (Fig. 5.4 and 5.5) show the relationship between specific growth rate and substrate concentration, the former increasing to an asymptote representing the maximum specific growth rate for this system. Further increases in the substrate concentration do not substantially increase the growth rate; indeed at very high concentrations many substrates become inhibitory to growth and the specific growth rate may then fall to zero. The curves shown in (Fig. 5.4 and 5.5) are representative of the equation

$$\mu = \mu_{max} \left(\frac{(S)}{K_m - (S)} \right)$$

where μ and μ_{max} are the actual and maximum possible specific growth rates, (S) the concentration of limiting substrate and K_m the Monod constant (defined as the concentration of substrate at which the growth rate is exactly half maximum), named after Jaques Monod (1910–1976) who first recognized this application of the equation. The equation defines an organism's specific growth rate

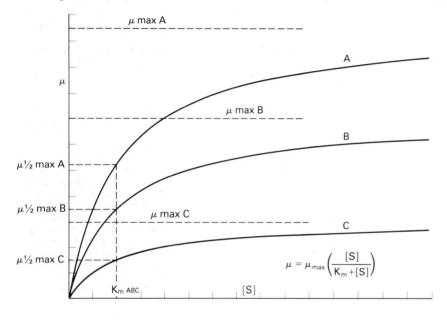

$$\mu = \mu_{max} \left(\frac{[S]}{K_m + [S]} \right)$$

Fig. 5.4 The Monod growth equation relating specific growth rate to substrate concentration, maximum growth rate (μ_{max}) and the Monod constant (K_m) for 3 bacterial species **A**, **B** and **C** having equal K_ms but different maximum growth rates.

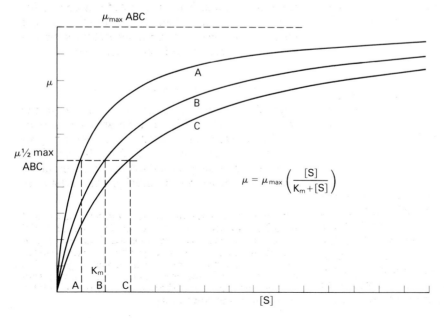

$$\mu = \mu_{max} \left(\frac{[S]}{K_m + [S]} \right)$$

Fig. 5.5 The Monod growth equation relating three bacterial species **A**, **B**, and **C** having the same maximum growth rate but different Monod constants (K_ms).

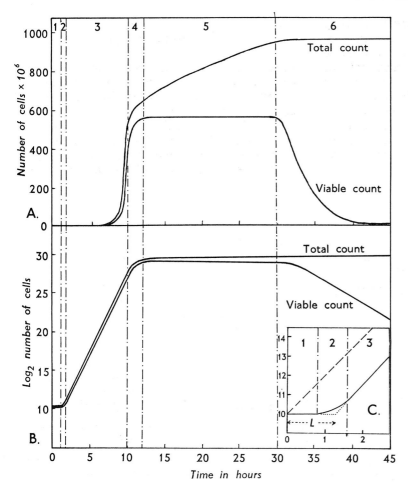

Fig. 5.6 Phases of growth. (**A**) A typical growth curve of *Escherichia coli* in a static 5 ml volume of nutrient broth at 30°C. Both total and viable counts are plotted on an arithmetic scale against time and the phases of growth are indicated by numbers, viz.:
1. Lag phase; 2. Acceleration phase; 3 Exponential phase; 4. Retardation phase; 5. Stationary phase; 6. Phase of decline. (**B**) The same data as in **A** are redrawn by plotting numbers of cells as logarithms to the base 2, against time. This graph clearly demonstrates the exponential nature of phase 3 of the growth curve. A unit increase on the logarithm to the base 2 axis indicates a doubling of the population. The doubling time is the time required for this unit increase. (**C**) Graphical derivation of lag time (data as in **B**). The solid line represents the observed viable growth curve; the broken line the 'ideal' curve if no lag in growth occurred. Lag time, L, can be obtained by extrapolation of the exponential portion of the growth curve. It may be seen that the lag time, derived in this way, is of longer duration than the lag phase (Phase 1).

under a given set of circumstances and may enable prediction of the outcome in cases of competitive interaction between species.

Phases of growth

In batch culture, growth proceeds through a series of arbitrarily defined phases. The shape of the resulting growth curve and the usually accepted limits of these phases is given in Fig. 5.6.

LAG PHASE

When medium is inoculated with organisms the prevailing conditions are usually different from the cells' previous environment. As a result, there usually follows a period of 'adjustment' during which only limited increase in numbers occurs. Clearly, during this time the theoretical curves for growth and number increase would be expected to differ since there must be a period of cell growth

prior to division. This undoubtedly accounts for some of the observed lag, but the situation is more complicated since, if mass is measured, a similar period of reduced growth rate is observed.*

THE EXPONENTIAL OR LOGARITHMIC PHASE

After four or five divisions in the lag phase the majority of cells will be growing with very similar specific growth rates resulting in uniform generation times and a constant doubling time. So long as these conditions persist the culture remains in the exponential phase, but, eventually, after perhaps six

---

* It is not clear exactly what causes these delays in onset of maximum growth rate, but over the first few microbial divisions 3 changes can be observed. Namely, a progressive increase in the specific growth rate, a reduction in the mean generation time of each succeeding generation and a reduction in the proportion of cells showing very long generation times. These three effects presumably contribute to the observed lag.

to eight further divisions the specific growth rate becomes reduced.

THE PHASE OF RETARDATION

Early this century a Czechoslovakian microbiologist, Oscar Bail, recognized that the exponential phase ended and growth of a culture ceased completely long before the nutrients in the medium were exhausted. His experiments showed that if a fully grown culture was filtered to remove the living cells, or the suspended cells were killed by heating *in situ*, the same tube of medium would, if reinoculated, grown an equal crop of the same organisms. Attempts by Bail to increase the maximum population by artificial means failed and high density suspensions died rapidly. Bail termed this population the M-concentration, the maximum concentration of cells achieved in a culture where adequate nutrients still existed. Bail's observations have often been confirmed, but no really satisfactory explanation has been presented and it seems that the cause of cessation of growth may be the result of limited transport of material to the cell—a hypothesis not far removed from Bail's suggestion of a minimum 'living space' for each organism.

The decline in the specific growth rate, from whatever cause, initiates the phase of retardation of growth and eventually leads to the stationary phase where no further growth or cell division occurs unless accompanied by the death of an equal number of cells.

STATIONARY AND DECLINE PHASES

Eventually, and again from largely unknown causes, the numbers of live cells start to decline and the culture may eventually become sterile. There seems, however, little doubt that the accumulation of toxic metabolic products and the ageing of individual cells contributes to death.

Cell age

Some micro-organisms that divide by fission do so by budding-off newly formed sections, and, in these cases, it would not be difficult to regard the mother cell as composed of older cell material than the bud. In the case of bacterial binary fission, however, where each half of the dividing cell is indistinguishable, it is now generally believed that the cell contents are completely mixed prior to cell division and each new cell is physiologically identical. An opposing view, that the original cell of the inoculum preserved its identity and had the greatest age of all cells in the culture was for many years held by the Russian Bacteriological School. It supposed a hierarchy of cells of differing ages within a population which would have suggested that older cells would die first, producing a gradually increasing rate of death in a culture. In fact, the reverse is observed; cells die at a decreasing rate, proportional to the numbers present which suggests a mandatory random incidence of death, equally probable in any cell. To speak of a 'young' cell in a growing culture infers a cell which has just divided—likewise an old cell would be one that is about to divide. In stationary phase cultures, however, cells can, of course, attain substantial ages. Referring to cultures, however, the term 'old culture' would infer one in or approaching the stationary or decline phase, whilst a 'young culture' is understood to have been more recently inoculated. These terms, though used loosely, should not be confused.

Growth of bacteria on solid media

The growth of bacterial colonies on solid media is often characteristic of the species or even the strain of organism (Fig. 5.7). The conditions affecting the shape of the colony and its rate of growth are obviously complex. For instance, the limitation of available nutrients diffusing from the medium to and into the colony are complicated by conditions such as the solid media, aqueous surface film and atmosphere, as well as the thickness and diameter of the colony. With prolonged growth the proximity of neighbouring colonies, the thickness of the medium and the availability of oxygen and carbon dioxide are important factors. Death and autolysis of cells in the centre of an older colony or dispersed throughout it, temperature, pH, and osmotic conditions will all influence colonial growth. Mathematical expressions for viable counts of colonies, or their size (diameter or thickness) are of limited validity.

Multiplication of bacteria on surfaces like talc, powdered glass, chalk or precipitates can take place when they are suspended in water containing very low concentrations of nutrients, (e.g., dissolved ammonia and carbon dioxide), owing to the absorption of solutes on to these surfaces producing an increased local concentration.

Growth of filamentous organisms

Growth in length

Some filamentous organisms, such as the filamen-

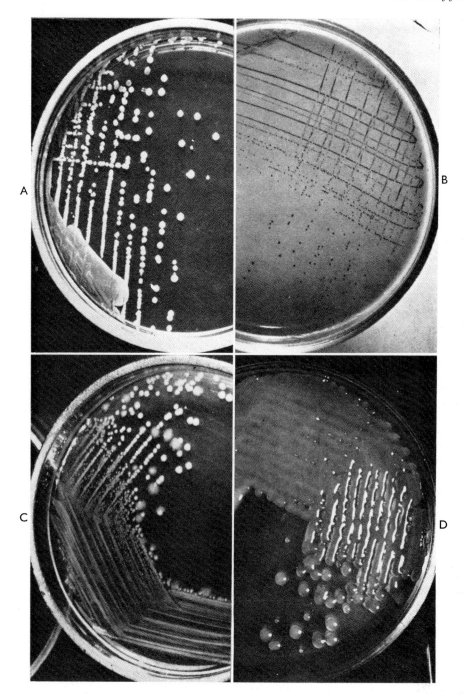

Fig. 5.7 Colonies of different bacteria grown for twenty-four hours at 37 °C (normal size). (**A**) *Staphylococcus albus* on nutrient agar. (**B**) *Streptococcus faecalis* on bile lactose agar with neutral red added as indicator. The acid produced by fermentation of the lactose results in the indicator changing to a strong pink colour which is taken up by the growing colonies. Most streptococci produce small colonies irrespective of the media on which they are grown. (**C**) *Shigella sonnei* on nutrient agar showing smooth (S) and rough (R) colonies. A rapid rate of mutation from S→R is typical of this organism. (**D**) *Klebsiella pneumoniae* on lactose nutrient agar. Vigorous production of capsular polysaccharides results in mucoid colonies. (A. H. Linton)

tous or thread-forming iron and sulphur bacteria, some green algae (e.g., *Spirogyra* and other members of the Zygnemaceae) and some blue-green bacteria (e.g., *Nostoc* and *Anabaena*) are little more than chains of individual organisms. Any cell of such a filament is potentially capable of binary fission and hence the growth in length of the chain is merely the sum of the growth of the individual component cells.

Fungal hyphae, the filaments of *Streptomyces* and related genera, and those of many algae, increase in length only by the extension of a zone just behind the tip of the hypha. This has been most studied in the fungi. In such a growth pattern under constant conditions new primary wall material is constantly being produced at the tip of the hypha and soon after formation this primary wall loses its extensibility. Although growth in length of the hypha takes place only over a narrow subapical zone it is clear that the cytoplasm in a much longer part of the hypha must increase (grow) in volume in order to supply the advancing tip. For *Neurospora* (p. 198) it has been calculated that at least a 12 mm portion of the hypha must be involved in such duplication of the cytoplasm. Translocation of nutrients and organelles is thus essential to apical growth of a hypha.

It has been generally accepted that under constant conditions, established fungal hyphae grow at a uniform rate, although a few apparent exceptions have been claimed. Branches occur by extension of areas of wall which have either retained or regained their plasticity. The pattern and frequency of branches is controlled by the genetic constitution of the species, by environmental factors, and by the phenomenon of apical dominance. The mechanism of the latter is not fully understood, but the result is to favour growth of the parent hypha and to limit that of the branches.

Secondary growth of the individual cells of complex structures

Young cells are usually hyaline, thin-walled and non-vacuolated. Vacuoles develop and granules of reserve foods and drops of oil commonly become more conspicuous as the cell ages. The structure of the cell wall is also subject to change. The wall of a fungal hypha is thin and elastic at the growing tip but becomes inextensible behind the zone of elongation, and may thicken through the deposition of secondary wall material. In spores and in the individual cells of the complex fruit-bodies of the higher fungi, rigidification of the wall may not take place uniformly all over the cell; this may result in extension of the cell in some directions and not in others, often giving complex shapes. A study of the stages in development of the basidia and basidiospores of certain Basidiomycotina (p. 198) showed that the final shapes could be explained as the result of the combined effects of the pressure of the surrounding cells during development, and of the rate and site of hardening of the cell wall. Cells, such as the asci of most hymenial Ascomycotina (p. 197), which develop in a closely packed single layer or hymenium and are thus subjected to equal lateral pressure, tend to be cylindrical or clavate; those developing under conditions of equal pressure from

all directions, such as the irregularly distributed asci of the Plectascales, are usually globose.

Secondary thickening of fungal cell walls is most striking in many spores, and in the peripheral cells of many sclerotia (p. 190), rhizomorphs (p. 190) and fruit-bodies. This is often associated with deposition of a black pigment, usually assumed to be melanin. Some spores remain thin-walled and are usually viable for only a limited time. Others develop thick walls and may become dormant. An entirely new layer or layers may be laid down inside the original spore wall from which it may differ in structure and presumably also in chemical composition. In some spores, e.g. the zygospores of *Rhizopus* (p. 189) and allied genera, the original wall may be sloughed off and replaced by several distinct layers differing from it and from each other. The cell walls (or particular wall layers) of fungus spores are usually inextensible, largely impermeable to water and dissolved salts and often differ markedly in fine structure from the walls of vegetative hyphae. On germination these inextensible walls break to allow emergence of the germ tube, the wall of which is continuous either with the innermost layer of the spore wall or with a new inner layer formed just prior to germ-tube emergence. The ornamentation which is such a conspicuous feature of the spores of many fungi may arise in a number of ways according to the species, but always involves deposition of secondary material.

Pigments develop in the cytoplasm or wall of many micro-organisms, notably in spores and reproductive structures. All fungal spores are colourless when first formed but many become pigmented later. The basidiospores of some species of agarics do not become fully pigmented until some hours after attaining full size and are not shed until pigmentation is completed. Maturation thus continues after full size is attained.

Growth of colonies of filamentous organisms

GROWTH OF FUNGAL COLONIES ON SOLID MEDIA

If a plate of a suitable agar medium is inoculated at the centre with fungal spores or mycelium, a colony of circular outline is produced within a few days and will continue to grow at a rate dependent on the isolate used, the nature of the medium, and environmental factors. Initially there is a lag period while spores in the inoculum germinate, or broken or damaged hyphae in a mycelial inoculum recover. Radial growth then commences. Very young mycelia whose *total* hyphal length is only a few millimetres are *undifferentiated* in the sense that all their hyphae are physiologically and morphologically equal. Growth of such mycelia is exponential (see p. 83), graphs of the logarithm of *total length* of hypha plotted against time being linear. However in a matter of only hours *differentiation* sets in as some hyphae assume the status of leading hyphae, spreading more or less radially, while others constitute lateral branches of first, second or higher orders which occupy the spaces between. Once differentiation commences growth of the colony ceases to increase exponentially, its acceleration diminishing. The rate of extension growth of the leading hyphae, which equals the rate of radial growth of the fungus colony, becomes constant and generally remains so unless *staling* occurs. Staling is usually due to the harmful effects of the accumulation of the colony's own metabolites and is manifested by the radial growth rate falling, even to zero. Staling is commoner on rich than on poor media, and at supraoptimal than at suboptimal incubation temperatures.

The radial growth rate of the differentiated fungal colony is generally independent of the depth of the agar substrate. That a constant radial growth rate should be maintained by a differentiated colony requires explanation since one might expect the growth rate to go on rising as the biomass in the colony increases. It can be shown, by severing marginal hyphae at different distances back from their tips, that the extension rates of the apices are unaffected by the presence of mycelium lying more than a few millimetres back from the margin; actual distances range from 0.4 to about 10 millimetres in different species. Severing the hyphae at shorter distances than these lessens their extension growth rate. There is evidently a peripheral growth zone in which the hyphae contribute materials to the apical extension of the colony's leading hyphae. A relationship between the width of this peripheral growth zone (W), the specific growth rate (μ, see p. 85) of the hyphae, and the radial growth rate of the colony (R) is supported by experimental data, namely;

$$R = \mu W$$

μ was estimated from the kinetics of growth of the organisms in submerged culture and W by severing colony margins till the least width which did not lessen the marginal advance was found. Calculated and observed values of R agreed well. Since radius generally increases at a uniform rate it is likely that the width of the peripheral growth zone remains uniform, but the factors which govern and determine its width are still obscure. The width is

characteristic of the strain of fungus, and while it is influenced by some environmental factors (e.g. nutrient concentration) it is unaffected by others (e.g. temperature). Consequently the use of radial growth rate to measure the response of growth of a mould to an environmental influence is most informative if the width of the peripheral growth zone is unaffected. When this condition is satisfied, this way of studying growth is convenient because it is non-destructive; successive measurements of the same colony are possible. If the size of a Petri dish limits the range of measurements, long horizontal tubes half-filled with solid culture medium and inoculated at one end can be used instead.

GROWTH OF FUNGAL COLONIES IN LIQUID MEDIA

Filamentous fungi grow in liquid media with one or other of three principal forms. If the culture consists of an unshaken body of nutrient solution the mould usually forms a thick buoyant mat with submerged and aerial portions, and commonly sporulates on the latter. If the culture is shaken, stirred, or agitated by forced aeration, sporulation is generally prevented, the vegetative fungus growing either in a *filamentous* form, the hyphae fragmenting as they reach a certain order of size, or a *stromatic* or *pellet* form in which the fungus grows as numerous separate ball-shaped colonies each consisting of radially extending mycelium and reaching diameters of several millimetres. The growth form depends on the species, the size of the inoculum, and the conditions. Since the hyphal fragments in the filamentous form generally remain in an undifferentiated state, growth in the presence of excess nutrients is exponential (p. 83). However, because of the differentiation that occurs in the pellets in the stromatic growth form, growth of the individual pellets, and of the culture as a whole, quickly ceases to increase exponentially. As the pellets grow they individually assume constant radial growth rates, and, provided that their number remains constant, growth of the biomass in the culture follows the cube root law:

$$M^{\frac{1}{3}} = Kt + M_0^{\frac{1}{3}}$$

where M = the biomass after time t
   M_0 = the biomass at time zero
and K = a constant (*not* the specific growth rate)

Diffusion of oxygen into pellets limits their growth rate at a size depending on their texture. Because of this impediment to oxygen access, pellets become very heterogeneous physiologically.

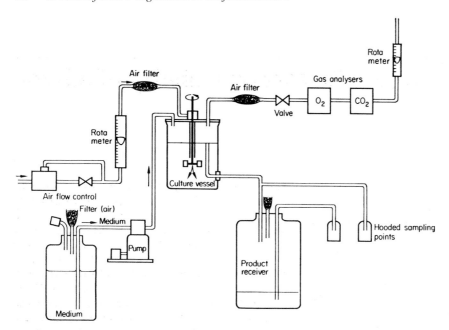

Fig. 5.8 Continuous culture apparatus. (After Herbert, D. (1958). *Recent Progress in Microbiology. VIIth International Congress for Microbiology.* Almquist and Wiksell, Stockholm, 382pp)

The interpretation of dry-weight data on the growth of filamentous fungi is sometimes complicated because not only does the mycelium grow by autocatalytic processes whereby cell components such as proteins and nucleic acids increase exponentially, but also sometimes by assimilatory processes which lead to the accumulation of, for example, polysaccharides. The *replicatory* and *assimilatory* processes may vary in their importance at different stages in the growth of a culture.

CONTINUOUS CULTURE

The chemostat is a system of cell culture in which nutrient medium is continuously added and spent culture continuously removed from the culture vessel (Fig. 5.8). In the steady state which can be thus achieved the viable population of cells is constant, the removal of cells being balanced by new cell growth. Continuous cultures of this kind are found in many natural circumstances such as the alimentary canal. The technique has been adapted by industry and in the laboratory for the continuous production of micro-organisms (e.g. yeasts) or for products (e.g. beer, p. 351).

Books and articles for reference and further study

BRODY, S. (1945). *Bioenergetics and Growth*, Reinhold Publishing Co., New York, 1023 pp.

BURNETT, J. H. (1976). *Fundamentals of Mycology*, 2nd edn.; Edward Arnold, London, Chapters 3 and 4.

DEAN, A. R. C. and HINSHELWOOD, C. (1966). *Growth, Function and Regulation in Bacterial Cells*, Clarendon Press, Oxford, 439 pp.

ERRINGTON, F. P., POWELL, E. O. and THOMPSON, N. (1965). Growth characteristics of some Gram-negative bacteria. *J. gen. Microbiol.*, **39**, 109–123.

GUNSALUS, I. C. and STANIER, R. Y. (1962). *The Bacteria*, Vol. IV. The Physiology of Growth, Academic Press, New York, 459 pp.

HERBERT, D. (1958). Some Principles of Continuous Culture, in *Recent Progress in Microbiology*, VIIth International Congress for Microbiology, Almqvist and Wiksell, Stockholm, pp. 389–396.

MANDELSTAM, J. and MCQUILLEN, K. (1968). *Biochemistry of Bacterial Growth*, Blackwell Scientific Publications, Oxford and Edinburgh, 540 pp.

MEADOW, P. and PIRT, S. J. (eds.) (1969). *Microbial Growth*, 19th Symp. Soc. gen. Microbiol., Cambridge University Press, 450 pp.

MONOD, J. (1942). *Recherches sur la Croissance des Cultures Bacteriennes*, Hermann et Cie, Paris, 210 pp.

MONOD, J. (1949). The growth of bacterial cultures. *Ann. Rev. Microbiol.*, **3**, 371.

NOVICK, A. (1955). Growth of Bacteria. *Ann. Rev. Microbiol.*, **9**, 97–110.

PAINTER, P. R. and MARR, A. G. (1967). Inequality of mean interdivision time and doubling time. *J. gen. Microbiol.*, **48**, 155–159.

PIRT, S. J. (1975). *Principles of microbe and cell cultivation.* Blackwell Sci. Pub., Oxford, London, Edinburgh and Melbourne, 274 pp.

POWELL, E. O. (1956). Growth rate and generation time of bacteria with special reference to continuous culture. *J. gen. Microbiol.*, **15**, 492–511.

TEMPEST, D. W. and HERBERT, D. (1965). Effect of dilution rate and growth-limiting substrate on the metabolic activity of *Torula utilis* cultures. *J. gen. Microbiol.*, **41**, 143–150.

Chapter 6

Influence of external factors on viability of micro-organisms

Survival in nature

In nature many agents may cause the death of micro-organisms, and many are employed by man for the destruction of harmful species.

Organisms grow only within certain limits of pH, temperature, osmotic pressure, and other environmental conditions (p. 85). Outside these limits they may survive but be unable to grow and reproduce; further departure from optimal conditions may lead to death. Death may result also from desiccation or starvation. Predators, parasites or competitors which produce antibiotics or other toxic products may also cause death of micro-organisms. An organism may cause its own destruction by producing toxic metabolites, a process known in mycology as 'staling', or by inducing other unfavourable changes in the environment (p. 88).

Micro-organisms differ widely in the ability to survive unfavourable conditions, and different phases of the same species vary greatly in their sensitivity. A state of reduced metabolic activity (hypobiosis), whether brought about artificially or occurring naturally, greatly increases an organism's ability to survive unfavourable conditions.

Hypobiosis

Hypobiosis can be brought about artificially by low temperatures, lack of oxygen, loss of water, and exposure to high salt concentrations, or by a combination of these. With the exception of low temperature, these factors are probably also involved in natural hypobiosis in resting bodies such as the endospores of bacteria, the sclerotia and chlamydospores of fungi, and the cysts of protozoa. In most instances thick cell walls help to maintain appropriate concentrations of oxygen and water.

In a state of hypobiosis, organisms are able to survive for long periods in the absence of nutrients, and have a greatly increased resistance to lethal agents. For example, the amoeboid phase of *Naeg-leria gruberi* is destroyed by drying for 15 seconds, freezing to $-30°C$ for 45 minutes or heating to $50°C$ for 2 minutes. Cysts of this protozoan, however, will survive drying for 23 months, freezing for $3\frac{1}{2}$ months, or heating to $50°C$ for 30 minutes. Spores of *Bacillus anthracis* have been shown to be viable after 31 years in the dry state and those of *Clostridium sporogenes* after 46 years in alcohol. Dried fruit-bodies of the fungus *Schizophyllum commune* have survived 34 years in a vacuum and retained the ability to resume spore production when moistened.

The circumstances that determine the production of resting bodies are not fully understood. Starvation is a frequent cause, as with many protozoa which produce cysts when the supply of bacteria on which they feed is exhausted. In contrast, bacteria may produce endospores on rich media under certain circumstances. The germination of protozoan cysts may be induced by a wide variety of agents e.g. cysts of *Schizopyrenus russelli*, a soil amoeba, by the presence of various amino acids; that of *Colpoda duodenaria*, a ciliate, by a variety of factors which are thought to stimulate activity of the tricarboxylic acid cycle.

Hypobiosis has been subdivided into *dormancy*, in which metabolic activity although greatly reduced is detectable, and *cryptobiosis* (sometimes *anabiosis* or *latent life*) in which metabolic activity is absent or undetectable. There is no doubt that a wide variety of organisms can survive a complete cessation of metabolism. For example, many algae, spores of bacteria and of some fungi and fragments of lichens have survived exposure for 2 hours to a temperature a fraction of a degree above absolute zero. It has been calculated that at $-200°C$ simple chemical reactions would be eight million times slower than at $20°C$ so at even lower temperatures the complex reactions involved in metabolism cease completely. The absence of chemical activity at these temperatures, however, ensures the maintenance of structural integrity and hence the resumption of metabolism on return to normal conditions. One method of inducing hypobiosis, namely freeze-drying, is now widely employed for preserving cultures of bacteria for long periods. Cell suspensions are rapidly frozen, by means of 'dry ice' (solid carbon dioxide) in alcohol and are then dried under high vacuum. It is usually necessary to include in the suspension protective agents such as glucose, ascorbic acid, gelatin, serum, or glutamate. The way in which these substances act is still not clear. The dried suspensions are maintained under vacuum or in an inert gas such as nitrogen to avoid oxidation to which they are very susceptible in the dried state.

Senescence and natural death

The life of most higher organisms, if other mishaps are avoided, is terminated by a gradual decline in metabolic activity (*senescence*) ending in death. This, however, is unusual in micro-organisms, in which the individual commonly divides into two daughter cells. In this, micro-organisms resemble the unspecialized cells of higher organisms under favourable conditions (such as in tissue culture or in an actively growing tissue) rather than individual multicellular organisms.

In fungi a common example of natural death is provided by basidiomycete fruit-bodies which die on completion of spore discharge, i.e., when their function is fulfilled. In *Neurospora crassa* a 'natural death' mutant is known in which the vegetative growth of a colony ultimately ceases; rejuvenation can be obtained by crossing or by heterocaryosis with a normal strain. In the suctorian protozoan *Tokophyra infusionum* which reproduces by budding, a definite limited life-span occurs, the length of which may be modified by nutritional factors, excess food in general shortening life.

Lethal agents

A wide variety of physical and chemical agents bring about the death of micro-organisms. These agents vary in efficiency and each is influenced by (1) many environmental factors and (2) the nature of the micro-organisms exposed to them.

Factors influencing the activity of lethal agents

Certain environmental factors influence the activity of both physical and chemical agents. Since the rate of killing is approximately logarithmic (p. 84), the time taken to kill all organisms is dependent upon the initial load of organisms present. The pH of the environment also influences the efficiency of many lethal agents. Heat, for instance, is more effective at acid pH, a property which makes it possible to sterilize acid fruit juices at lower temperatures than neutral solutions (p. 344). The activity of many acidic and basic agents is very dependent upon pH since, in general, unionized molecules enter cells more readily than do ions; consequently the concentration attained within the cell is greatest when the pH outside the cell is one at which the agent is in an undissociated state. Some chemical agents, such as mercuric chloride, whose action depends on the ionized mercuric ion, are more effective in aqueous than in alcoholic solution and the presence of other salts may suppress ionization of the molecule.

The rate of killing by chemical agents is dependent on the concentration of the agent and the temperature, the activity usually being greater as the temperature is raised (Fig. 6.1).

Many chemicals are most effective in a pure inorganic aqueous environment. Chlorine, for instance, is effective as a disinfectant for the purification of drinking water when present in a few parts per million (p. 334). The activity of such chemicals, however, may be greatly reduced in the presence of organic matter such as serum, pus, milk, faeces, etc. Usually the concentration of the disinfectant is lowered by chemical interaction with organic matter; sometimes the latter protects micro-organisms against the lethal agent. Lethal chemicals which are not readily inactivated by organic matter and have the additional property of low toxicity for mammalian tissue, are used in the treatment of infectious diseases (p. 103).

Under normal conditions, most non-sporing micro-organisms are susceptible to physical and chemical agents. A few notable exceptions occur; for instance, while mycobacteria are as sensitive to heat as other non-sporing micro-organisms, they are much more resistant to chemical agents by virtue of their protective waxy envelope. This selective resistance to chemicals facilitates their isolation from mixed floras in pathological materials. Fungal spores are usually more resistant than non-sporing micro-organisms but bacterial endospores are the most resistant forms of life under natural conditions. High temperatures are required to kill endospores and these structures, together with mycobacteria, are susceptible to only a few chemical agents, including formaldehyde, glutaraldehyde, ethylene oxide and iodine, and then only under certain conditions. Few disinfectants are active against viruses; halogens and phenol are generally active and some viruses are inactivated by

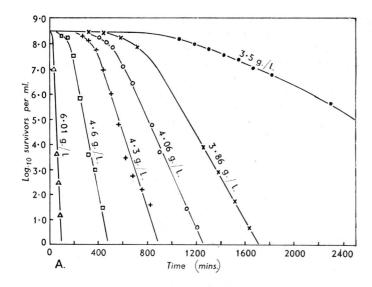

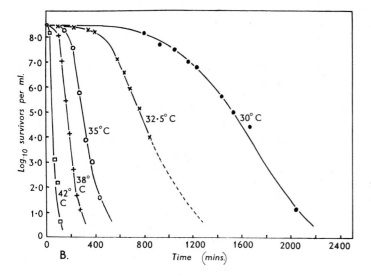

Fig. 6.1 The effect of phenol on *Escherichia coli*. (**A**) The relationship between the logarithms of survivors and time, for *E. coli* when exposed to various concentrations of phenol at 35 °C. (After Jordon, R. C. and Jacobs, S. E. (1944). *J. Hyg., Camb.*, **43**, 275.) (**B**) The relationship between the logarithms of survivors and time, for *E. coli* when exposed to 4.6 g phenol per litre at various temperatures. (After Jordon, R. C. and Jacobs, S. E. (1945). *J. Hyg., Camb.*, **44**, 210.)

quaternary ammonium salts and xylenols (p. 102). A number of compounds able to inactivate viruses *in vitro* have little action on them in tissue cells because of the close association of viruses with host cells. In outbreaks of foot and mouth disease more reliance is placed on cleansing with sodium carbonate solutions than on viricidal action. Fungi are not necessarily killed by substances lethal to bacteria but many chemicals are fungicidal and some may be specifically active against particular groups of fungi (p. 334). Griseofulvin, for example, is selectively active against dermatophyte infections of animals and is used in the treatment of these.

The choice of lethal agent for use in a particular situation therefore depends on the materials to be treated, the environmental factors and whether or not sterility is required. The use of heat sterilization, whilst being the most readily controlled, is limited to heat stable materials; chemicals, on the other hand, cannot be used for the sterilization of food or culture media because of the residual chemical's adverse effects.

Definition of terms

The antimicrobial agents discussed in this chapter

are used in a number of ways and the terminology to describe these processes requires careful definition. *Sterilization* is an absolute term indicating the complete destruction or removal of all micro-organisms, including the most resistant bacterial spores. *Disinfection*, usually used of chemical agents, is the killing of many harmful micro-organisms without necessarily achieving sterility. In practice, disinfectants by killing important pathogenic organisms often render certain objects safe for use, e.g. clinical thermometers. Disinfectants which have a sufficiently low toxicity to be used safely on skin and mucous membranes are termed antiseptics. Antimicrobial chemicals which possess a very low toxicity are used as *chemotherapeutic* agents in the treatment of infectious disease in animals and plants and include the important group of naturally produced antibiotics.

Some antimicrobial agents are mainly employed as killing agents (*bactericidal* or *germicidal*), whereas others inhibit growth and multiplication (*bacteriostatic*). Killing agents are essential for sterilization, but many of the useful chemotherapeutic agents are bacteriostatic. These inhibit multiplication of the infective agent while the body's defences destroy the micro-organisms already present.

Dynamics of disinfection

Exposure of a population of micro-organisms to a lethal agent does not normally result in instantaneous death but an exponential decrease in the number of viable cells in the population, that is, in each successive unit of time a constant fraction of the micro-organisms survives. Plotting the logarithm of the number of surviving organisms against time often results in a straight line graph (Fig. 6.1A), the curve being similar in form to that expressing the course of a unimolecular reaction. This, however, is true only under certain limited conditions such as when a disinfectant is present in excess and the rate of disinfection is sufficiently rapid. Departure from the exponential relationship can be caused by a variety of factors; for instance, the death rate may be initially slow but soon accelerates, reaching a constant value which is maintained until most organisms are killed, with the exception of a few organisms which may persist and die much more slowly than the majority. Plotting values under these conditions of survival against time result in sigmoid curves (Fig. 6.1B) rather than straight lines.

Two measures of the effectiveness of a lethal agent are commonly used. These are the LD_{50} (i.e., the dose needed to kill 50% of the individuals) or, alternatively, the dose which kills the entire population. The LD_{50} is particularly useful for comparing the potency of different lethal agents since it can be determined with ease and accuracy and allows for differences in susceptibility of individual cells. However, in practice it is often necessary to ascertain the dosage needed to kill all the organisms, a measurement that is of less scientific value, since it is dependent on the incidence of abnormally resistant individuals, but of great medical importance.

Physical agents

Many physical agents have an adverse effect upon the viability of micro-organisms. Some lethal agents, e.g. heat and certain radiations, are widely used in sterilization procedures. Others, such as cold, desiccation and those producing cell damage, kill a proportion of a population of cells but cannot be relied upon to produce sterility. The mechanical process of filtration is also used in sterilizing fluids.

Heat

High temperature is one of the most reliable and widely used techniques available for killing micro-organisms. Unlike disinfection by chemicals no toxic residue remains when sterilization is completed and its use is only limited where materials to be sterilized may be damaged by the temperatures used or adversely affected by moisture.

The temperature that is lethal for a particular

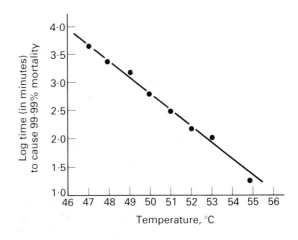

Fig. 6.2 The lethal effect of different temperatures on *Escherichia coli* at ph 7. (After Jordon, R. C., Jacobs, S. E. and Davies, H. E. F. (1947). *J. Hyg., Camb.*, **45**, 333.)

micro-organism depends on the time of exposure (Fig. 6.2). One practical term used is the 'thermal death time' which is the time required to kill a culture at a specified temperature. In a moist environment the thermal death time for non-sporing bacteria ranges from several hours at 47°C, one hour at 60°C, to 5 minutes at 70°C, none being able to survive more than a few minutes at 80°C. In practice heating a culture at 80°C for 10 minutes in the presence of water kills all non-sporing bacteria and the majority of fungal spores. More resistant spores, such as the ascospores, found in certain strains of *Byssochlamys fulva*, resist 86–88°C for 30 minutes. Some bacterial endospores are killed by this temperature but the majority are killed in the same time only by temperatures above 100°C. The resistance of bacterial spores constitutes the major problem in heat sterilization.

The majority of viruses are inactivated by exposure to temperatures between 50 and 60°C for 20 minutes but a few, e.g. the serum hepatitis virus, are inactivated only by higher temperatures such as 100°C for 20 minutes.

MOIST HEAT

It is important to distinguish between moist and dry heat processes of sterilization. Moist heat is the more efficient lethal agent, sterilizing at a lower temperature and in a shorter time. Death of micro-organisms is the outcome of coagulation and denaturation of the structural proteins and cellular enzymes and this occurs more readily in the hydrated than in the dry state. Irrespective of the material being processed moist heat sterilization takes place in an aqueous environment.

Exposure to temperatures of the order of 63°C for 30 minutes or 72°C for 20 seconds are used in processes of pasteurization, as for milk (p. 348) to kill non-sporing pathogens and beverages (p. 351) to terminate fermentations. Raising the temperature to boiling point also kills many bacterial spores but several hours exposure are required to kill the more heat-resistant spores and sterility cannot be ensured. However, boiling is widely used to render instruments safe for use in some medical and dental work, but since sterility is not guaranteed, this treatment is inadequate for surgical instruments. The addition of 2% sodium carbonate to the boiling water gives as effective killing in 10 minutes as boiling in pure water for several hours. The corrosive action of the ·alkali on plated metal (e.g. surgical instruments), however, limits the use of this technique.

Steam is a particularly effective lethal agent since it provides a very rapid process of heat transfer. As steam condenses on reaching a cold object, it not only gives up its latent heat of vapourisation but also the partial vacuum so created sucks more steam to the same object. It is essential that pure moist steam is used, i.e. all air must be discharged from the equipment (Fig. 6.3) and the steam must not be superheated else the process of heating becomes virtually a less efficient dry heat process.

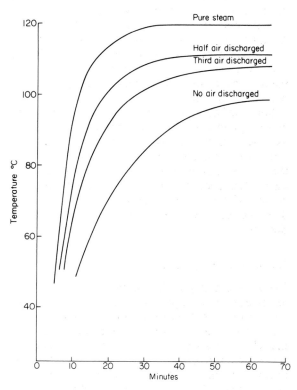

Fig. 6.3 Temperatures of steam–air mixtures in an autoclave. Steam at 15 lb per square inch was introduced into a chamber completely or partially evacuated and the temperatures within the chamber were determined at intervals. (Data, from Underwood, W. B. (1941). *A textbook of sterilization*, American Sterilizer Co., Erie, Pa.)

Sterilization by steam is carried out in an autoclave. A range of pressures can be employed; the sterilizing times in Table 6.1 are recommended for high vacuum autoclaves and include the minimum time required for sterilization plus a safety margin (see M.R.C. Report, 1959). In autoclaves where air is removed by downward displacement, longer times are required especially when permeable materials are being sterilized. Under these conditions exposure for 25 minutes at 121°C (i.e. at 15 lb per

Table 6.1 Equivalent minimum sterilizing time/temperatures with moist and dry heat

Moist heat			Dry heat	
lb./sq. in. (pure steam)	Temperature °C	Sterilizing time	Temperature °C	Sterilizing time
0	100	20.0 hours		
5	109	2.5 ,,		
10	115	50.0 minutes		
15	121	15.0 ,,	120	8.0 hours
20	126	10.0 ,,		
30	134	3.0 ,,	140	2.5 ,,
			160	1.0 ,,
			170	25.0 minutes
			180	10.0 ,,

in^2) is adequate; with bacteriologically clean materials, shorter times (e.g. 20 minutes) or, lower temperatures (e.g. 109°C) are used.

The sterilization of contaminated heat labile materials, such as woollen blankets and medical equipment containing plastic components, etc., has proved a problem but the use of steam at sub-atmospheric pressures (e.g. 80°C approx. −8 p.s.i.) has proved effective. The penetration and efficiency of heat transfer by steam makes this technique superior to others at the same temperature and provides a disinfection (pasteurization) process from which dry, wrapped articles can be obtained. To achieve a sterilizing process, formalin can be injected into the apparatus during the heating process and this is carried by the steam into all exposed parts of the equipment and all spores are destroyed after 2 hours. The formalin is finally removed under vacuum and residual traces neutralized with ammonia.

DRY HEAT

Sterilization by dry heat has limited application. The higher temperatures and longer exposure times (Table 6.1) are often damaging to many materials. It is used mainly for the sterilization of materials which must be kept dry, e.g. anhydrous fats, oils, certain surgical instruments, powders, certain glassware, etc. Usually a hot air oven is used, the heating of which takes place by heating elements in the base or sides, the air being circulated by a fan. The process of heat transfer to the objects is by convection, radiation and conduction. Time must be allowed (1) for heat penetration to bring all objects to the required temperature, (2) for the object to be held at a standard temperature to kill all spores, and (3) for the oven to cool before opening. The entire process takes several hours and is therefore considerably longer than the moist heat processes.

Non-sporing organisms are killed by exposure to 100°C dry heat for 1 hour, fungal spores at 115°C for 1 hour and the majority of bacterial spores (excluding a few highly resistant spores) at 160°C for one hour.

INFRA-RED RADIATION

Infra-red radiation is used for the sterilization of particular materials such as all-glass syringes. This is a form of dry-heat sterilization. The materials to be sterilized are exposed to infra-red radiation as they pass on a belt through the equipment. Temperatures of 180°C are rapidly attained with greater certainty than in a hot air oven and exposure for not less than 10 minutes is adequate to sterilize. Under vacuum, infra-red radiation can attain temperatures of 280°C and subsequent introduction of nitrogen avoids oxidation of the materials on cooling.

Cold

Chilling and freezing bacterial cultures brings about the death of a proportion of the cells, this proportion varying with the cooling procedures used. Rapid freezing to low temperatures, however, often results in a state of hypobiosis (p. 93) leading to prolonged survival. The death rate during *cooling* or *chilling* depends on several factors including the rate of cooling, the nature of the suspending medium and the growth phase of the organism. For instance, cold shock caused by the rapid cooling of young cultures of *Escherichia coli* from 37°C to 4°C produces a 95% reduction in viable cells whereas similar cultures slowly cooled to 4°C over a period of 30 minutes show no loss in viability. The greatest loss in viability occurs with actively metabolizing cells.

Freezing causes further physical and metabolic damage to the cells. The death rate is found to be greatest at temperatures just below freezing point compared with still lower temperatures. In addition to mechanical damage from ice crystals, forming in the cell, the major effect is the precipitation of coagulable cellular proteins which is much greater at $-2°C$ than at $-20°C$ or lower. More organisms survive sub-freezing temperatures if they are protected by organic substances such as peptone, glucose, sucrose, glycerol, sodium glutamate, etc. Dependent upon the presence or absence of these substances the number of survivors under frozen conditions decreases with storage time.

Some moulds and yeasts are more resistant to freezing than are most bacteria but bacterial spores are virtually unaffected. Some viruses are inactivated at low temperature; but others including bacteriophage, are preserved by holding at $-60°C$. A few protozoa can be preserved by freezing.

Desiccation

When dried some loss in viability occurs with all micro-organisms. Bacterial spores are least affected but vegetative cells demonstrate rapid and considerable loss in viability during the drying process followed by reduced loss under storage. Drying is less destructive under vacuum than in air, as is also storage in an inert atmosphere which guards against the lethal effects of oxidation. A residual water content of 30–40% is considered to be the most harmful.

The period of survival after natural drying varies between organisms. Some delicate pathogens survive for a few hours only; others remain viable for weeks or months especially if they are protected by body fluids, such as serum, and dried on fabrics, as blankets and clothing.

Spray-drying used for the preparation of dried milk and egg powders is not a means of sterilization and both pathogenic and non-pathogenic organisms survive. Care must be taken therefore to safeguard against bacterial multiplication in these products after reconstitution prior to cooking.

Visible and ultra-violet radiation

Light can bring about chemical change, and hence biological damage, only if it is absorbed; light that passes through the cell without being absorbed has no effect. Visible light, that is, electromagnetic radiation of wavelength 400–750 nm is absorbed by relatively few of the compounds present in non-photosynthetic organisms, and therefore has little harmful effect, and the same is true of ultra-violet radiation of wavelength (300–400 nm). Ultra-violet radiation of wavelength less than 300 nm, on the other hand, is strongly absorbed by proteins and nucleic acids, in which it brings about chemical change, with the result that relatively small doses of such radiation will cause chromosome damage, genetic mutation or death. Higher doses are required to cause inactivation of enzymes.

If micro-organisms are treated with various dyes (e.g. erythrosin) they become sensitive to damage by visible light. Such dyes are said to possess photodynamic action. The germicidal effect of sunlight is due largely to the ultra-violet component, which is only a few per cent of the total. The lower limit of wavelength of the ultra-violet received at the earth's surface is about 290–300 nm, but is dependent on the clarity of the atmosphere and on the altitude and latitude. It will be appreciated, therefore, that the germicidal power of sunlight varies greatly.

The main ultra-violet-induced lethal photoproduct formed in DNA is the pyrimidine dimer in which adjacent pyrimidine bases (thymine or cytosine) in the same DNA strand dimerize. The hydrogen bonds to the purine bases on the opposite strand are thus broken and the DNA becomes distorted in that region. As a result DNA replication past the dimer is slowed down to the point where the presence of a few dimers is lethal to the cell.

In some micro-organisms the harmful consequences of exposure to short wavelength ultra-violet radiation can be partly averted by prompt exposure of treated organisms to visible light, an effect known as photoreactivation. Such micro-organisms, e.g. *E. coli* and baker's yeast, possess photoreactivating enzyme. The photoreactivating enzyme combines mechanically with the irradiated DNA (but not with unirradiated DNA), the UV lesion acting as the substrate. The complex absorbs light, producing repaired DNA with subsequent liberation of the enzyme. The enzyme–substrate complex does not dissociate in the dark. It has recently been calculated that *E. coli* possesses about 25 photoreactivating enzyme molecules per cell.

Penetration by UV is poor but it has uses in the laboratory where cabinets used for carrying out inoculations can be conveniently sterilized by UV from a quartz mercury vapour lamp left on continuously. Similar sources may be used for the disinfection of indoor atmospheres (p. 337) and for the treatment of water supplies for special purposes in some hospitals.

Atomic radiation

There are two distinct types of atomic radiation: electromagnetic waves of the same class as light waves, but produced by high voltage apparatus and having much shorter wavelengths and far higher energies (X-rays); and subatomic particles travelling at very high velocities, produced by the breakdown of radioactive elements such as cobalt 60 (γ-rays). The high speed subatomic particles, released by radioactive decay or accelerated by means of such instruments as the cyclotron, include charged particles such as β-rays (high speed electrons), protons, α-particles (helium nuclei) and uncharged particles such as neutrons. The effect of all these atomic radiations on matter, living or otherwise, is to cause ionization. An electron is knocked out of one molecule and is gained by another. As a result, both molecules become ions of high chemical reactivity capable of causing biological damage. The different types of atomic radiation differ in the number and distribution of the ionization they cause, and in their penetrating power. A major cause of radiation damage in living tissue is probably the production of free hydroxyl and hydrogen radicles, an indirect result of the ionization of water.

There are vast differences in sensitivity of different organisms to atomic radiation, micro-organisms being much more resistant than most higher organisms. Thus the LD_{50} for most mammals, including man, is less than 1000 roentgens (r), whereas for *Escherichia coli* it is 10000 r, yeast 30000 r, *Amoeba* 100000 r, Paramecium 300000 r, and for *Micrococcus radiodurans*, the most resistant vegetative bacterium known, 750000 r. The severity of radiation damage depends to a considerable extent on conditions at the time of irradiation. Thus the sensitivity of cells is considerably reduced by anaerobic conditions at the time of irradiation. Recently it has been found that pre-treatment with certain chemicals such as β-mercapto-ethanolamine, gives some protection against subsequent irradiation. The effects of irradiation damage may be modified by post-irradiation treatment. For example, if irradiated *E. coli* cultures are subsequently kept at 18°C far more bacteria survive than at higher or lower temperatures. It is thought that at this temperature the processes which repair radiation damage are more effective than those extending the damage.

Since vegetative bacteria present a more or less identical target to radiation, the amount of radiation damage caused does not vary much from species to species. Thus differences in radiation sensitivity, which vary enormously from species to species, must be due to differences in the ability of each bacterial species to modify such damage, and in fact, can be explained largely by the absence or presence of a DNA dark-repair mechanism, and if present, of its efficiency. The dark-repair mechanism operates by excizing damaged single-stranded regions of DNA and repolymerizing the missing bases. Unlike photoreactivation which operates only for UV damage, dark-repair mechanism will repair DNA damaged by UV and ionizing radiation, nitrogen mustards, and alkylating agents, etc.

In dealing with the effects of both short-wavelength ultra-violet and atomic radiation it is important to distinguish between genetic and physiological effects. If severe, both may cause death, but whereas a threshold value exists for physiological damage, there is apparently no minimum dose required to cause mutation.

Radiation, as a means of sterilization, is finding an increasing number of applications. Package

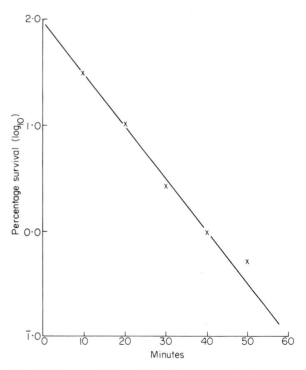

Fig. 6.4 The rate of killing of *Klebsiella pneumoniae* by ultrasonic waves, 700000 cycles per second. (Data from Hamre, D. (1949). *J. Bact.*, **57**, 279.)

irradiation of syringes, catheters, petri dishes, surgical gloves, and sutures is widely used, but certain plastics (e.g. cellophane) are damaged by radiation. Careful precautions are essential to ensure that no materials receive an overdose of radiation since this causes some plastics to become discoloured and brittle. The use of radiation in the sterilization of foods is limited by its effect on their appearance and flavour.

Cellular disintegration

Micro-organisms can be broken up by ultrasonic vibrations or mechanical agitation. Ultrasound, which includes inaudible sound waves with vibrations of the order of 700 000 cycles per second, is only effective in liquid suspension and, apart from heating effects, the lethal action is attributed almost entirely to the physical destruction of cells. Efficiency is related more to the intensity of the wave emission than to its frequency. Damage to cells is attributed mainly to 'gaseous cavitation'; it is presumed that the cell walls are weakened by the

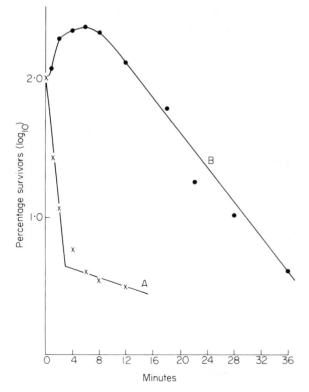

Fig. 6.5 The rates of inactivation of vaccinia virus by ultrasonic waves, 20 000 cycles per second. (**A**) the virus is suspended in buffered saline; (**B**) the virus is suspended in 20 per cent inactivated rabbit serum. (T. J. Hill)

creation of minute cavities resulting from the formation of small gas bubbles in the fluid under the alternating pressures produced by the sound field. As with other lethal agents the order of destruction in general follows an exponential rate (Fig. 6.4).

All actively living cells including micro-organisms are susceptible to ultrasound. Gram-negative bacteria are generally more susceptible than Gram-positive bacteria and rod-shaped organisms more than cocci. Bacterial spores are highly resistant. Some workers report viruses to be resistant to ultrasound whilst others consider them to be susceptible (Fig. 6.5). The different finding may in part be related to the size and shape of the virus particles. Ultrasound is used in virology to disrupt tissue cells to release intracellular viruses and to break up aggregates of virus; this may explain the initial increase in pox-forming units in the serum suspension in Fig. 6.5B.

Cellular disintegration can also be brought about by mechanical agitation in the presence of abrasives (e.g. carborundum), sand or glass beads. Alternatively shearing forces developed by forcing a fluid or frozen suspension through a narrow opening bring about cell disintegration. These techniques are often used to produce cell-free enzymes for biochemical studies since the preparations are not subject to heating as with ultrasonics.

Filtration

Filtration, not itself a lethal agent, is used widely in industry and laboratory for the sterilization of heat-labile fluids. By this process organisms are removed from fluids and may be separated from their soluble products of metabolism, e.g. exotoxins. Filters are made from many materials, including unglazed porcelain, diatomaceous earth (kieselguhr), asbestos fibres, sintered glass, and synthetic membranes of cellulose nitrate (collodion) and cellulose acetate. All types are available in a range of grades of different pore size and are selected according to the nature of the fluid to be filtered and the size of particle which is to be removed. In addition to the mechanical action of filters other factors must be considered such as the effect of pressure exerted to force the fluid through and the electric charges on the filter and on the suspended particles, which may result in their electrical adsorption to the filter.

In the past, filters were used to remove micro-organisms of bacterial size or larger, and 'sterilization by filtration' was regarded as rendering a fluid bacteria-free and not virus-free. Membrane filters are now made in a wide range of pore size including many much smaller than the smallest bacterium. The finer ones are used for ultra-filtration and measure from 5 nm in size. These can be used for filtering fluids free of viruses and by a choice of different membranes of known pore size it is possible to estimate with reasonable accuracy the virus particle size according to which membrane just fails to retain the virus.

Chemical agents

Disinfection by chemicals is a much more complex subject than sterilization by physical agents. Many chemicals can cause the death of micro-organisms but their modes and rates of action are extremely diverse. Some are general protoplasmic poisons damaging all living cells; others, in particular the antibiotics (p. 105), are highly specific, being relatively non-toxic to mammalian tissues and some micro-organisms but active in high dilution against specific ones, a feature termed selective toxicity. Chemical disinfectants may be bactericidal at certain concentrations but bacteriostatic at high dilution; these differing activities are considered to be the result of quite separate and unrelated modes of action and not simply of dilution. Thus mercuric ions in low concentrations interfere with enzyme action by combining with —SH groups, a reaction which can be reversed or prevented by adding other —SH containing compounds such as glutathione; in high concentration mercuric ions irreversibly denature cell proteins.

The majority of disinfectants are general poisons, their lethal action being the result of their capacity to coagulate, precipitate or otherwise denature both structural and essential enzymes of tissue and micro-organismal cells. Groups of chemicals which do this are considered below. Because of their general toxicity some are useful only as general disinfectants; others, with milder toxicity, can be applied as antiseptics to living tissue.

HALOGENS

Halogens or compounds which release halogens, e.g. hypochlorites, chloramines, hypobromites, are strongly bactericidal by the oxidation of proteins and similar substances. In consequence of their vigorous chemical activity, solutions are readily inactivated by combination with organic matter, and this limits their usefulness (p. 103). An alcoholic solution of iodine (one per cent) is a useful skin disinfectant but may cause reactions in sensitized individuals.

ALKYLATING REAGENTS

Heterocyclic compounds, e.g. ethylene oxide, as well as methyl bromide and lactones, e.g. β-propiolactone, are effective bactericides through their ability to alkylate bacterial structures. The gas ethylene oxide effectively kills bacterial spores and can be used for the sterilization of heat-labile materials and equipment. To avoid explosion, the gas is mixed with an inert gas under pressure at a standard humidity and temperature. This method is widely used in industry for sterilizing plastic goods and, because ethylene oxide is extremely penetrating, it is the method of choice for long plastic coils (as in dialysers) and for equipment with finely threaded bolts.

PHENOLIC COMPOUNDS

This group of disinfectants, which includes phenol, lysol (a solution of cresols in soap), and xylenols, are intensely toxic by virtue of protein denaturation. Proteins are precipitated by one to two per cent phenol. Their activity is not greatly reduced by organic matter but none are capable of killing bacterial spores. Thus most vegetative cells of bacteria are killed by 1% phenol in 5–10 minutes at 20°C, but anthrax spores survive 24 hours in 5% phenol. Newer phenolic preparations such as 'clear soluble phenolics' (e.g. hycolin, stericol) are as efficient as lysol but much safer.

Halogenated phenols are much less toxic than phenols but their activity is severely depressed by organic matter since the halogen contributes to the activity. This group includes 'Dettol', a chloroxylenol, and 'Hexachlorophane'. Hexachlorophane is effective against staphylococci but has recently been restricted in use because of reports of brain damage in animals.

Chlorhexidine ('Hibitane') is a very useful disinfectant with a wide spectrum of activity and very low tissue toxicity but it is relatively easily inactivated. It is used as an alcoholic solution for skin disinfection and, in aqueous solution, as a general antiseptic, either alone or mixed with cetrimide (a quaternary ammonium compound). This mixture ('Savlon') is less easily inactivated than chlorhexidine alone.

ALDEHYDES

Formaldehyde, which is usually marketed as a 40% aqueous solution (formalin), even when diluted to 1%, is able to kill bacterial spores and tubercle bacilli. In contrast to most disinfectants it can therefore produce complete sterility but is highly toxic. As a moist gas or solution its penetration into organic matter and crevices is slow but it is highly efficient for sterilizing smooth surfaces. Suitable dilutions have been used to treat footwear of persons suffering with fungal infections, such as athlete's foot, and to treat wool prior to sorting in order to kill any spores of anthrax bacilli which may be present.

Other useful aldehydes include a 2% aqueous solution of glutaraldehyde; it is less irritant than formalin but more rapidly bactericidal.

ALCOHOLS

Alcohols kill vegetative bacteria rapidly but have no action on spores. Their activity requires the presence of water and 70% gives optimal activity, absolute alcohol being relatively inactive. Isopropyl alcohol is slightly more effective than ethyl alcohol and is preferred as a skin disinfectant.

ACIDS AND ALKALIS

Mineral acids and alkalis produce their main lethal effects through their hydrogen and hydroxyl ions respectively, although their characteristic ions may also contribute. The more highly dissociated molecules therefore produce the greatest effect. Hydrogen ions are more effective than hydroxyl ions. Organic acids often owe their activity to the undissociated molecules.

HEAVY METALS AND THEIR SALTS

All the heavy metals are antibacterial and antifungal to some degree. Silver and copper exhibit these properties at minute concentrations, a property referred to as oligodynamic activity. The salts and organic complexes of mercury, tin, silver and, to a lesser degree, copper are all actively lethal agents. When ionized in aqueous solution the metal ions combine with and precipitate cell proteins.

SURFACE ACTIVE COMPOUNDS

These include cationic, non-ionic, and anionic compounds, of which the most important are the cationic *quaternary ammonium compounds* (QACs). Their activity has been variously attributed to denaturation of cell protein, inactivation of enzymes and disruption of the cell membrane as a result of their surface-active properties. It is likely that inactivation of enzymes concerned with energy-producing metabolism is the most important mechanism of inhibition. They are not highly active but by virtue of their 'surface wetting' properties they are very useful cleaning agents and are sometimes used in combination with antiseptics, e.g. chlorhexidine ('Hibitane'), to make the latter more effective. Gram-positive organisms are generally more sensitive than Gram-negative ones but quaternary ammonium salts are not active against *Mycobacterium tuberculosis*, bacterial spores, viruses or fungi.

DYES

Prior to the introduction of sulphonamides and antibiotics, the triphenylmethane dyes (e.g. methyl violet, brilliant green, etc.) and the acridines (e.g. acriflavine and proflavine) were widely used in the treatment of wounds and burns because of their high antiseptic action and low toxicity. Dyes are more active against the Gram-positive than Gram-negative bacteria and crystal violet, which is fungistatic, has been used against mycotic skin infections. An important feature of the acridines is their sustained activity in the presence of serum, but the discoloration of the skin by these compounds has resulted in their being superseded by more acceptable antiseptics.

Selective toxicity

Some antiseptics are selectively active against Gram-positive organisms and most are able to kill Gram-positive organisms at a lower concentration than Gram-negative organisms. The wider the spectrum of activity the greater the usefulness of an antiseptic.

The property of selective toxicity is most marked among the antibiotics; each has its own spectrum of activity (Table 14.1). Consequently the range of microbial diseases which can be treated by each is essentially restricted.

In addition to the importance of selective toxicity between micro-organisms the relative toxicity of an antibacterial agent for the host compared with its toxicity for the micro-organism is of great practical importance. This is considered below.

Chemotherapy

This term was coined by Paul Ehrlich who defined it as 'the use of drugs to injure invading organisms without injury to the host'. Toxicity to the host is usually the limiting factor which determines whether or not a chemical can be used as a therapeutic agent. Frequently a substance which is harmful to micro-organisms is harmful to the cells of the host and therefore the choice of a chemotherapeutic agent depends on the difference in degree between these two actions. Ehrlich devised the following index of the usefulness of a chemotherapeutic agent in which the higher the index, the more useful the agent will be.

Chemotherapeutic Index
$$= \frac{\text{Minimum Curative Dose}}{\text{Maximum Tolerated Dose}}$$

The anti-malarial drug quinine and a few other effective chemotherapeutic agents were known for centuries, but a rational approach to the discovery of new agents was not possible until it was appreciated that infectious disease was caused by micro-organisms. In the late nineteenth century the differential action of some dyes in staining bacteria without staining the surrounding tissues suggested to Ehrlich the value of searching for toxic compounds having a greater affinity for micro-organisms than for mammalian tissues. Ehrlich undertook a systematic study of organic arsenical compounds in the hope of finding one which retained the effectiveness of inorganic arsenic in killing the causal organism of syphilis, *Treponema pallidum*, while being less toxic to man. He finally succeeded, discovering a compound arsphenamine, named Salvarsan or 606 (since it was the 606th compound he tested) which was used for many years in the treatment of syphilis.

The two principal approaches now employed in the search for new chemotherapeutic agents are the synthesis of chemical analogues of known metabolites, and the screening of the many naturally produced antibiotics of soil organisms for their suitability as chemotherapeutic agents.

METABOLIC ANALOGUES

Essential metabolites, whether substrates or coenzymes, if present in small amounts in a cell or tissue, can be antagonized by substances known as metabolic analogues. Each analogue exerts its antagonism by occupying and blocking the enzyme sites used by the metabolite. The similarity must be in the chemical configuration, dimensions and electron distribution since the active sites on the enzymes are known to be highly polarized. Many examples of metabolic analogues are known but one example only will be considered here.

The sulphonamides and folic acid antagonists The discovery by Domagk (1935) that certain azo dyes containing an aromatic sulphonamide group had a better chemotherapeutic index against streptococcal septicaemia in mice than any previously known, was destined to alter the whole outlook in chemotherapy. Tréouel, Nitti, and Bovet (1935) showed that the reduction product of the dye, *p*-aminobenzene sulphonamide, was active both *in vitro* and *in vivo*. Since then a large number of sulphonamide compounds have proved to be effective against virulent streptococci, pneumococci, neisseria and, to a lesser extent, salmonellae and shigellae.

Woods (1940) observed the specific antagonism of *p*-aminobenzoic acid (PABA) to sulphonamide action and Fildes (1940) proposed the theory that sulphonamides inhibit bacterial growth by blocking an enzyme system associated with the use of PABA as an essential metabolite. Many other antagonists to sulphonamides exist and not all can be explained by this theory. Nevertheless, the primary and most important effect on most susceptible micro-organisms appears to be that of interference with the incorporation of PABA into folic acid during its enzymic synthesis. The similarity in structure and absolute size of the sulphonamide molecule to that of PABA is one of the best examples of a structural analogue interfering with an enzyme controlled synthesis (Fig. 6.6).

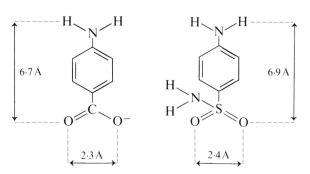

p-aminobenzoic acid (anion)   *p*-aminobenzene sulphonamide

Fig. 6.6 The structural similarity between *p*-aminobenzoic acid and *p*-aminobenzene sulphonamide.

Some microbial parasites cannot use external sources of folic or folinic acid since these compounds cannot penetrate the cell wall; they must be therefore synthesized within the cell from PABA, pteridine, and glutamic acid (Table 4.3). Sulphonamides, by virtue of their structural similarity to

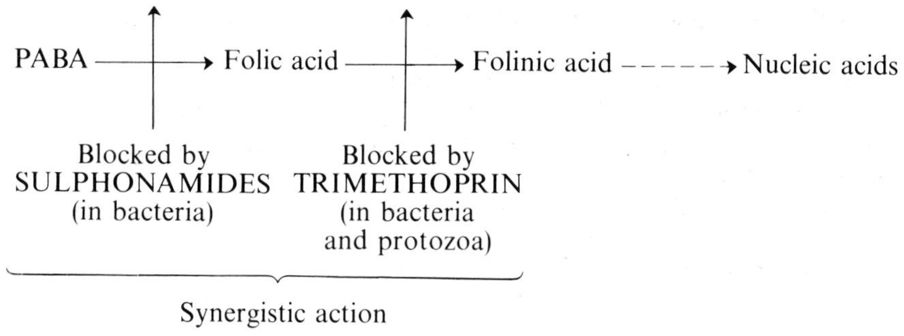

PABA ———→ Folic acid ———→ Folinic acid ----→ Nucleic acids

Blocked by
SULPHONAMIDES
(in bacteria)

Blocked by
TRIMETHOPRIN
(in bacteria
and protozoa)

Synergistic action

Fig. 6.7 Sites of inhibition by analogues in the metabolic pathway leading to nucleic acid synthesis.

PABA act as competitive inhibitors of this synthesis and therefore have powerful bacteriostatic effects on organisms dependent on PABA. These include many pathogenic bacteria, toxoplasma, and plasmodia. In the cell folic acid is converted to folinic acid by the bacterial enzyme dihydrofolate reductase. This conversion can be blocked by a number of diaminopyrimidines (e.g. trimethoprim and pyrimethamine) which have an affinity for this enzyme. Consequently these compounds are effective chemotherapeutic agents against certain bacteria and protozoa, and a combination of sulphonamide and trimethoprim, co-trimoxazole, is becoming increasingly important in modern chemotherapy. The function of folinic acid is to act as a co-factor for enzymes concerned in the synthesis of purines and pyrimidines used in nucleic acid synthesis but none of these later stages are sensitive to sulphonamides (Fig. 6.7). Organisms which can absorb folic acid from an external source are insensitive to sulphonamide.

Most vertebrates can satisfy their requirements for folic acid either from their diet or from organisms in their intestinal tract which are able to synthesize it; alternatively they may be supplied with folinic acid in their diet. Any slight effect of sulphonamides on mammalian enzymes is counteracted by excess folic and folinic acids supplied to the cell.

ANTIBIOTICS

Antibiotics are chemical agents produced by many micro-organisms which are harmful to other microbial species. The clinically useful antibiotics are produced by certain actinomycetes and by some species of *Penicillium* and *Bacillus*. Many hundreds of antibiotics have been isolated but relatively few are clinically useful because of an unsatisfactory antimicrobial spectrum of activity, or excessive toxicity *in vivo*. The mechanisms of some antibiotics are considered below, the spectra of activity of

clinically useful ones in Chapter 14 (p. 334) and their commercial production in Chapter 16 (p. 362).

The action of antimicrobial agents including antibiotics

To be effective, antimicrobial agents have ultimately to inhibit an essential chemical reaction in the cell and the biosynthetic interconversions, described in Chapter 2 provide many excellent targets for such compounds. In practice inhibitors exist for practically every interconversion that occurs in the cell although the potency of the inhibition varies widely from compound to compound and reaction to reaction.

Broadly speaking all inhibitors can be classified under two main headings although there are a few compounds, notably those that interfere with ribosome function, that may not fall into this general scheme. The two main groups are:

1 Inhibitors that interfere with the function of enzymes by interaction at the enzyme active centre (Type A).
2 Inhibitors that interfere with nucleic acid metabolism by becoming incorporated as part of a nucleic acid molecule and thus disrupting its normal replicative activity (Type B).

In bacterial cells the majority of reactions that can be blocked effectively by antibiotics form part of either the protein/nucleic acid or cell wall biosynthetic systems. At our present stage of knowledge, the steps in the biosynthesis of lipids do not seem to be so susceptible to inhibition, but this may merely reflect the fact that we know a good deal less about the details of this process than about protein and cell wall synthesis. A variety of steps in energy metabolism of the cell are also susceptible.

The steps in protein and nucleic acid biosynthesis that may be blocked by various inhibitors are indicated in Fig. 2.11. Thus RNA polymerase is inhibited both by actinomycin D and by rifamycin, while the ability of the ribosome to catalyze the translation of m-RNA is impaired by streptomycin, tetracycline, erythromycin, and the other macrolide antibiotics. Streptomycin, tetracycline, and the macrolides all inhibit ribosome function in different ways but the detailed mechanism of their action is, as yet, unclear. Whereas tetracycline and erythromycin seem to block polypeptide synthesis completely, streptomycin (under some conditions at any rate) can allow translation of the message but many errors appear in the resulting polypeptide chain due to the insertion of incorrect amino acids as the chain grows.

Puromycin and chloramphenicol are also potent inhibitors of protein synthesis but in these it is the chemical reactions involved in polypeptide chain extension that are inhibited. Of the two antibiotics, the action of puromycin has been completely elucidated while that of chloramphenicol is not yet completely understood. Puromycin is incorporated in place of an amino acid at the growing point of the polypeptide chain and this process blocks further chain extension. It also causes the detachment of the incomplete chain from the ribosome; thus in practice the antibiotic leads to the synthesis of many incomplete, and therefore inactive, protein molecules in the inhibited cell.

Apart from the reactions shown in Fig. 2.11, a number of metabolic interconversions indirectly related to, but essential for, polypeptide synthesis can be readily inhibited. The biosynthesis of the nucleotide precursors of both RNA and DNA can be blocked very effectively by some antibiotics (e.g. azaserine and diazo-norleucine) and the loading of amino acids on to the relevant t-RNAs can sometimes be impaired severely by structural analogues of the amino acid concerned. For example, L-5-methyl tryptophan blocks the loading of L-tryptophan on to tryptophan-t-RNA in *Escherichia coli*.

Another biosynthetic process indirectly related to protein synthesis is DNA replication. This process can be effectively inhibited by a number of molecules of Type B (see above). Effective examples of such compounds are ethidium bromide, proflavine, mitomycin, and many members of the acridine series of dyes.

The other major series of biochemical reactions in bacteria inhibited by therapeutically effective antibiotics is cell wall synthesis. The overall process consists of at least seven steps (see Steps 1–7, p. 30) and a number of clinically important antibiotics act by blocking one of these reactions. The penicillins and cephalosporins inhibit Step 7 (see p. 32), that

is, they block the formation of the cross bridges between the two adjacent side chains on the *N*-acetyl-muramic acid residues. It is thought that the four membered β-lactam ring of these antibiotics plays a crucial part in their inhibitory action because the —CO–N— bond of the lactam ring opens in the course of their action to block irreversibly a key chemical group at the active centre of the enzyme catalyzing Step 7.

The overall reaction is probably:

$$\text{ENZYME—X—R} + \begin{array}{c} | \;\; | \\ -\text{C—C—} \\ | \quad\;\; | \\ \text{C—N—} \\ \!\!\!\!\!/\!\!/ \\ \text{O} \end{array} + H_2O \rightarrow$$

$$\begin{array}{c} | \;\; | \\ -\text{C—C} \\ \!\!\!\!\!\!/ \qquad | \\ \text{ENZYME—X—OC} \quad \text{N—} + \text{R—OH} \\ \qquad\qquad\qquad | \\ \qquad\qquad\qquad \text{H} \end{array}$$

where —X–R is the key group of the enzyme blocked by the antibiotic.

In addition to penicillins and cephalosporins, the antibiotics vancomycin and bacitracin inhibit cell wall biosynthesis. In neither case has the precise mode of action of the inhibitor been worked out, but it seems likely that some reaction between steps 3 and 6 are blocked in both cases.

Just as inhibitors of the biosynthesis of the precursors of RNA and DNA may exert a lethal effect on cell growth, so inhibitors of the biosynthetic precursors for cell wall synthesis can be effective antibiotics. Step 4 in the cell wall biosynthetic pathway involves the addition of D-alanyl–D-alanine, to the growing side chain of the *N*-acetyl-muramic acid residues (see p. 31). The D-alanyl–D-alanine dipeptide required for this step is formed in the cell as a separate and preliminary step according to the following reaction:

D-alanine + D-alanine = D-alanyl–D-alanine + H_2O,

and this reaction is sensitive to the antibiotic D-cycloserine. The antibiotic is thought to exert its action because it is a close structural analogue of D-alanine. This ensures that the inhibitor is drawn to the active centre of the enzyme that normally handles D-alanine as a substrate but, unlike D-alanine, the cycloserine interacts effectively with the enzyme and thus blocks its ability to form D-alanyl–D-alanine from its metabolic precursors

Cell wall synthesis in many moulds may also be effectively inhibited by antibiotics. As would be expected, because of the different chemical constitution of bacterial and fungal cell wall structures, antibiotics active against bacteria do not inhibit cell wall synthesis in fungi.

Recently it has become apparent that there are some antibiotics that can affect enzyme action indirectly by influencing the chemical environment in which they operate. Most common among this group of antibiotics are those that destroy the permeability properties of bacterial membranes so that certain ions leak out of the cell. Examples of such molecules are the actins and alamethicin, antibiotics which cause a specific leakage of potassium ions from susceptible bacteria. Appropriate concentrations of potassium ions within the cell are essential for the effective functioning of many essential systems in bacteria and hence for survival.

Methods of testing disinfectants and antibiotics

The testing of disinfectants

No completely satisfactory test for disinfectants has been devized which takes into account the effects of organic matter, time, temperature and toxicity for host tissues. The Rideal–Walker test compares the activity of a disinfectant with that of phenol against a culture of *Salm. typhi* under standard conditions; these conditions may be modified, e.g. by the addition of organic material. More recently, 'in-use' and 'capacity use–dilution' tests have been introduced in which the conditions of testing resemble more closely the situation in which the disinfectant is being used.

The assay of antibiotics

The potency of samples of antibiotics is assessed by bio-assay. As with the bio-assay of nutrients (p. 73), the activity of the sample of unknown potency is compared with that of a standard preparation against a test organism known to be sensitive to the antibiotic under test. In contrast to bio-assays of nutrients, which measure the amount of growth, those of antibiotics measure inhibition of growth. The bio-assay of antibiotics is usually carried out by serial dilution or agar diffusion techniques.

SERIAL DILUTION METHODS

A series of dilutions of both the standard antibiotic preparation and the test sample are prepared in a fluid nutrient medium. The tubes are inoculated with a standard suspension of the test organism and incubated. The greatest dilutions of the standard and of the sample which completely inhibit growth is noted and from these the potency of the test sample is estimated. This measures the bacteriostatic concentration of the drug; the bactericidal level may be determined by sub-culturing from each tube and determining which concentration of the drug kills the test organism. Defects of the test are that it is laborious, requires strictly aseptic conditions (a chance contaminant may completely invalidate results) and is liable to be influenced by the growth of resistant mutants which have a selective advantage in the presence of the antibiotic. Serial dilutions are now mainly used for the assay of antibiotics which diffuse too slowly for agar diffusion methods to be applicable.

AGAR DIFFUSION METHODS

In this bio-assay the antibiotic diffuses from a confined source through nutrient agar seeded with the test organism whilst the plate is incubated. A concentration gradient of the antibiotic is thus set up and this results in a zone of inhibition being formed around the antibiotic reservoir. Under constant conditions of temperature, medium, and inoculum size, the size of the zone of inhibition is proportional to the concentration of the antibiotic at the source (Fig. 6.8). In this bio-assay the size of zones produced by a range of dilutions of a standard preparation are plotted and the concentrations of antibiotic in various dilutions of the sample which give zones of comparable size are read from the graph. With antibiotics of small molecular size the relationship between the square of the zone diameter and the logarithm of the antibiotic concentration is strictly linear (Fig. 6.9).

Bacteriocins

The bacteriocins are a class of bacterial products which exercise a very specific bactericidal activity; they have some relationship with bacteriophages and possibly also with sex factors (p. 51). They were discovered and first studied in the intestinal Gram-negative organisms but many other bacteria, both Gram-negative and Gram-positive, produce them. They take their specific names from the organisms producing them, e.g. colicins from *Escherichia coli*, pyocins from *Pseudomonas aeruginosa* (*pyocyanea*), megacins from *Bacillus megaterium*.

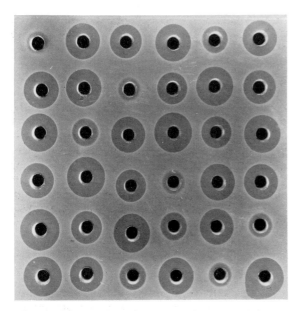

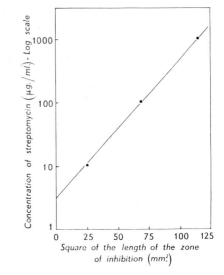

Fig. 6.9 A standard assay curve obtained with *Klebsiella pneumoniae* and streptomycin by the agar diffusion technique. A straight line relationship results by plotting logarithms of the antibiotic concentrations against the corresponding values of the square of the inhibition-zone diameters. The assay of antibiotic solutions of unknown potency can be determined by reading from the standard curve the concentrations corresponding to the square of the inhibition-zone diameters produced by these solutions. (After Linton, A. H. (1958). *J. Bact.* **76**, 94.)

Fig. 6.8 A streptomycin assay plate using the agar diffusion technique. Nutrient agar is seeded with the test organism (*Klebsiella pneumoniae*) and the antibiotic is placed in the cup, cut in the agar with a sterile cork borer, from which it diffuses. The size of each zone of inhibition is proportional to the concentration of the antibiotic in the cup. In the assay plate demonstrated, six different dilutions are used, three of which are standard solutions of known concentration, the other three are dilutions of the unknown solution being assayed. These are randomly distributed throughout the plate, each row and each column including all six dilutions. The average of six zone measurements for each dilution is determined and this cancels variations which may occur in different parts of the plate. (A. H. Linton)

All true bacteriocins are proteins but some may become associated with the lipopolysaccharide of the cell wall of the producer and form a complex which can be split only by drastic methods. Some bacteriocins, particularly pyocins, in electron micrographs resemble bacteriophage tail structures (Chapter 7). They may be considered as products of defective phage genomes, able to code for only part of the phage structure. The initiation of bacteriocin synthesis kills the producer cell.

Like bacteriophages they are adsorbed to specific receptors on, or possibly under, the bacterial surface from which they exercise a lethal action by a mechanism which has not, so far, been elucidated. The target of the lethal effect varies between bacteriocins. They may degrade the cell's DNA, or inhibit DNA and RNA synthesis, or interfere with protein synthesis. Unlike bacteriophages they are not regenerated in the cells to which they have been adsorbed and, as a rule, they do not lyse them in the process of killing.

While adsorption of the bacteriocin 'particles' takes place very rapidly, there is an interval between adsorption and the irreversible bactericidal action during which the cell may be rescued by inactivation of the bacteriocin already adsorbed to the cell with trypsin.

Most, though not all, bacteriocins are good antigens and evoke in rabbits the formation of neutralizing antibodies (p. 329). Precipitating antibodies are produced also, but it is not entirely clear whether they react with the bacteriocin itself or with an associated component.

The potential or actual synthesis of bacteriocin is due to genetic determinants, the bacteriocinogenic factors, which resemble prophages (p. 131) but are probably not integrated into the bacterial chromosome. Cells carrying such factors are immune to the effect of the corresponding bacteriocin.

As with a prophage, the synthesis of the lethal product specified by a bacteriocinogenic factor is repressed in most cells harbouring the factor and is active in only a few of them; but the synthesis of bacteriocin can be induced, like the synthesis of phage components, by UV irradiation, mitomycin C and other agents. The massive amount of bacteriocin produced after induction is due mainly to a great increase in the proportion of organisms producing it rather than to an increase in the amount produced by individual cells.

Some colicinogenic factors (genetic determinants for colicins) can be transferred from producer to non-producer strains. The factor for colicin 1b (col 1b), acts as a sex factor directing the formation of pili (p. 51) which allow cells to conjugate and thus to pass on the col factor. Other col factors are not able to cause formation of their own pili, but rely for transmission on col 1b, the F factor, or on transduction by bacteriophages (p. 53). Whether there is any mechanism for the transfer of bacteriocinogenic factors other than col factors is not known.

Bacterial typing by inhibition patterns

The identification of bacterial species of medical importance is often aided by tests which depend on the inhibitory effect of certain chemical compounds on that species selectively e.g. the inhibition of *Strept. pneumoniae* by ethyl hydrocuprein. In epidemiological and research studies, the differentiation of strains or types within a species can sometimes be achieved by a characteristic pattern of sensitivity to a standard set of antibacterial substances or agents.

Resistogram typing

Methods have been reported and others are being studied in which drops of solutions of a range of chemical substances (e.g. malachite green, boric acid) are placed in a grid pattern on a culture plate seeded with the test organism. After incubation, the pattern of inhibition can be observed. Such methods have been used for typing *Pseudomonas aeruginosa*, *Proteus* spp. and *Candida* spp., and are useful research tools where many variations can be introduced to suit a particular requirement.

Colicin typing

The specificity of action of bacteriocins has been utilized for bacterial typing, and encouraging results have been obtained with *E. coli*, *Shigella sonnei*, *Proteus* spp. and *Pseudomonas aeruginosa*, although, at present, the methods are not standardized between different laboratories as with phage typing (see below). Fig. 6.10A illustrates a typical typing plate in which 12 test strains of *E. coli* are exposed to two of a standard set of colicin-producing cultures.

Phage typing

The success of phage typing depends on the very

Fig. 6.10

(**A**) *Colicin-typing of E. coli.* Colicin-producing strains are first grown for 48 hours (2 per plate) and then the growth removed (shown by dark streaks). Test strains are then inoculated across the plate and, after incubation, the pattern and extent of inhibition noted.

(**B**) *Phage-typing of a strain of Staph. aureus.* The strain to be typed is carpeted on to the surface of a nutrient agar medium and specific phages are spotted in the centre of the appropriate squares of the grid. The phage-type is determined by the pattern of sensitivity obtained.

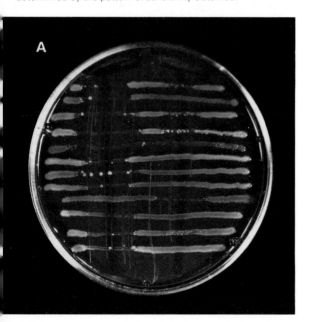

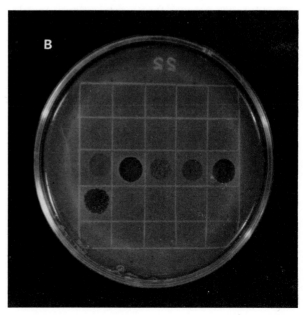

high specificity of phages for their host bacteria. Phage-typing methods were first introduced in 1925 to distinguish strains of salmonellae. As a result of subsequent development of the methods, the type distribution of *Salmonellae typhi* is now known for most of the civilized world and routine typing contributes greatly to our knowledge of the epidemiology of typhoid fever. The phages employed for *S. typhi* are active only against freshly-isolated strains possessing the Vi-antigens and the method is termed 'Vi phage typing'.

Phage typing is widely used for distinguishing strains of *Staph. aureus*. This enables the sources of a particular phage type of staphylococcus responsible for an outbreak of sepsis, to be traced and eliminated. Staphylococcal phage typing presents quite different problems from the typing of salmonellae. Whereas there are only a few hundred typhoid carriers in England, there are many millions of staphylococcal carriers and thus a much greater variety of types are found among staphylococci than typhoid bacilli. Despite the fact that staphylococcal phages are not as stable as those of *S. typhi*, the method does permit a fairly reliable identification of individual strains which is very valuable in short-term investigations.

Phage typing of *Staph. aureus* is technically more difficult than the typing of salmonellae; the staphylococcal phages are derived from lysogenic strains of *Staph. aureus* and selected on the basis of their range of host specificities. Twenty-two phages are included in the basic set of typing phages; the phages are numbered by international agreement and are divisible into four groups of antigenically related phages. Almost all strains derived from human sources belong to groups I, II or III. Any one strain of *Staph. aureus* is likely to be susceptible to several phages, usually of the same antigenic group. Some strains of *Staph. aureus* are resistant to all the phages used and cannot be typed but new phages are added to the basic set from year to year, thus increasing the range and value of the typing method.

The phages are prepared for use from filtrates of infected susceptible host cultures known as propagating strains. The filtrates are purified and then titrated to determine the dilution required to just produce confluent lysis when spotted on to a culture plate seeded with the host bacterium; this dilution is referred to as *Routine Test Dilution* (or RTD) and is the dilution normally employed in the typing method.

A culture plate is seeded with the staphylococcus to be typed, then each of the phages of the standard set is dropped on to the agar surface in a standard pattern, often with a grid marked on the culture plate (Fig. 6.10B). After drying, the plates are incubated, usually at 30°C. The phage type of the test strain is then designated by the numbers of the phages which have produced lysis, under the standardized conditions of the test, e.g. phage type 84/85 is a strain of *Staph. aureus* which has recently caused several serious outbreaks of hospital infection, and which is lysed only by phages 84 and 85. Some strains which are not lysed by phages at their routine test dilution, may be typable when more concentrated phage suspensions are employed (usually $1000 \times$ RTD).

Phage typing is a very valuable method but since it is an empirical procedure, it is open to many sources of error. For example, if phage lysates are too concentrated, non-specific lysis may occur. Moreover, since most staphylococcal propagating strains are lysogenic, some harbouring as many as five different temperate phages at a time, concentrated lysates are likely to contain both the propagated typing phage and some of the temperate phages. These temperate phages may themselves become lytic or they may alter the susceptibility of the test organism to the typing phage.

Books and articles for reference and further study

General references

ALBERT, A. (1968). *Selective toxicity*, Methuen and Co. Ltd., London, 531 pp.

COWAN, S. T. and ROWATT, E. (eds.) (1958). *The Strategy of Chemotherapy*, 8th Symposium of the Society for General Microbiology, Cambridge University Press, Cambridge, 360 pp.

CRUICKSHANK, R., DUGUID, J. P., MARMION, B. P. and SWAIN, R. H. A. (1973). *Medical Microbiology*, 12 Edn., Vol. 1. Churchill Livingstone, Edinburgh and London, 667 pp.

KAVANAGH, F. (1963). *Analytical Microbiology*, Academic Press, New York, 707 pp.

KELSEY, J. C. and MAURER, I. M. (1966). An in-use test for hospital disinfectants. *Mon. Bull. Min. Hlth.*, **25**, 180–184.

KELSEY, J. C. and SYKES, G. (1969). A new test for the assessment of disinfectants with particular reference to their use in hospitals. *The Pharmaceutical Journal*, **202**, 607–609.

NEWTON, B. A. and REYNOLDS, P. E. (eds.) (1966). *Biochemical Studies of Antimicrobial Drugs*. 16th Symposium of the Society for General Microbiology, Cambridge University Press, Cambridge, 349 pp.

REPORT, M. R. C. (1959). Sterilization by steam under increased pressure, *Lancet*, **i**, 425.

SYKES, G. (1964). *Disinfection and Sterilization*, E. and F. N. Spon, London, 486 pp.

WILLIAMS, R. E. O., BLOWER, R., GARROD, L. P., and SHOOTER, R. A. (1966). *Hospital infection—causes and prevention*, Lloyd-Luke Ltd., London, 386 pp.

Mechanisms of antibiotic action

FRANKLIN, T. J. and SNOW, G. A. (1974). *Biochemistry of Antimicrobial Action*, 2nd Edn., Chapman and Hall, London.

GALE, E. F., CUNDLIFFE, E., REYNOLDS, P. E., RICHMOND, M. H. and WARING, M. (1981). *The Molecular Basis of Antibiotic Action*, 2nd Edn., John Wiley, London and New York.

NEWTON, B. A. and REYNOLDS, P. E. (eds.) (1966). *Biochemical Studies of Antimicrobial Drugs*. 16th Symposium of the Society for General Microbiology, Cambridge University Press, Cambridge.

Part II

Form, size and life cycles of micro-organisms

Introduction

Biologists recognize that the traditional division of living things into the two 'kingdoms', plant and animal, is no longer tenable. Micro-organisms include some groups which are plant-like (e.g. the green algae), some which are animal-like (e.g. protozoa) and others which have some characteristics of both kingdoms (e.g. fungi). A third kingdom is now recognized, i.e. the *Protista*, which includes those organisms of small size and relatively lowly differentiation which have hitherto been loosely termed 'micro-organisms'.

With the exceptions of the viruses (Chapter 7), some of the true slime moulds (p. 209) and a few other examples, micro-organisms, like plants and animals, normally consist of cells. In many of them the thallus comprises only one cell, but some are multicellular, the constituent cells being similar to one another in many species, but differentiated into distinct types of cell of varied form and function in the higher algae and higher fungi. Even the most highly differentiated forms, however, lack the true tissues, formed by cell division in more than one plane, characteristic of higher plants and animals.

The study of cell organization, made possible by the electron microscope and the ultra-microtome, and by advances in biochemical techniques, has revealed fundamental resemblances and differences between various micro-organisms and offers a sound basis for their division into major groups (i.e. viruses, prokaryotes and eukaryotes) and for the further subdivision of these.

Micro-organisms of Protista may be subdivided as in the scheme below:

Classification of micro-organisms

A   Organization subcellular   VIRUSES

AA Thallus unicellular, multicellular, or plasmodial
     B   Nucleoplasm not bounded by a membrane   PROKARYOTA
         C   Chlorophyll absent, or if present of type   Bacteria (including
              different from that of plants   Actinomycetes)

         CC Chlorophyll present, together with   Cyanobacteria (Blue-
              characteristic blue-green pigment, pigments   green bacteria)
              not located in discrete plastids.

    BB Cells or plasmodia containing one or more discrete   EUKARYOTA
        membrane-bounded nuclei
     D   Cell(s) of vegetative thallus (with a few
             exceptions) possessing a cell wall(s).
         E   Chlorophyll present and located in   Algae (excluding the
              discrete chloroplasts.   'blue-green' bacteria)

         EE Chlorophyll absent.   Fungi
     DD Cell(s) of vegetative thallus lacking true
          cell walls.
         F   Thallus unicellular, remaining so.   Protozoa
         FF Thallus unicellular at first,   Slime moulds
              becoming a plasmodium or   (Myxomycetes or
              pseudoplasmodium and   Mycetozoa)
              eventually forming a
              fructification

Chapter 7

Structure and classification of viruses

General introduction

The discovery of viruses

In 1892 Iwanowski showed that the symptoms of tobacco mosaic (p. 290) could be reproduced in healthy plants by rubbing their leaves with juice from infected ones, even after the juice had been passed through a filter capable of holding back the smallest bacteria. A few years later Loeffler and Frosch showed that foot and mouth disease of cattle could also be induced by bacteria-free filtrates of fluid from the vesicles characteristic of the disease. Other plant and animal diseases were soon added to the list of those caused by these filter passing agents,

first known as 'filterable viruses' and now just 'viruses'. It is now known that many other organisms (bacteria, blue-green bacteria, fungi, mycoplasma, protozoa and insects) can be infected with viruses.

The nature of viruses

Rather than attempt a simple definition of a virus it may be more instructive to describe their major characteristics.

Viruses are distinguished by their *small size*. Only the largest e.g. smallpox virus, reach 300 nm in diameter. Many are 100 nm or less and the smallest are 20 nm in diameter.

Table 7.1 A comparison of the properties of viruses with some other micro-organisms

| | Bacteria | | | | Mycoplasmas | Viruses |
	Gram-positive	Gram-negative	Rickettsiae	Chlamydiae		
Size	ca. 1 μm diam.	ca. 1 μm diam.	ca. 300 nm	ca. 300 nm	Some have reproductive units < 150 nm	Largest are 300 nm, majority < 150 nm
Nucleic acids	DNA and RNA	DNA and RNA	DNA and RNA	DNA and RNA	DNA and RNA	Either DNA or RNA
Growth on artificial media	+	+	−	−	+	−
Intracellular replication	−[1]	−[2]	+	+	−	+
Mode of replication	Binary fission	Binary fission	Binary fission	Binary fission and ? budding	Binary fission and budding	Intracellular assembly of constituents parts (nucleic acid and protein subunits, etc)
Muramic acid	+	+ + +	+	+	−	−
Sensitivity to antibiotics used in clinical treatment[3]	+	+	+	+	+	−

[1] Some will replicate intracellularly, e.g. *Mycobacterium tuberculosis*.
[2] Some will replicate intracellularly, e.g. *Brucella abortus*.
[3] See also Table 14.1.

Most viruses have a relatively *simple structure*. All have only one of the two types of nucleic acid and the simplest, e.g. poliovirus, contain only nucleic acid (in the case of poliovirus this is RNA) surrounded by a protein coat (known as the capsid). Some infectious agents, the viroids which cause a variety of plant diseases (e.g. potato spindle tuber), have no protein capsid and consist only of low molecular weight RNA. The simplicity and regularity of virus structure is emphasized by the fact that some simple viruses can be prepared as crystals. The larger and more complex viruses often have an additional lipoprotein covering (the envelope) surrounding the capsid. Viruses are therefore acellular micro-organisms, having no nucleus, energy generating systems or organelles for motility.

Viruses are *obligate intracellular parasites*, a fact which explains why early microbiologists failed to grow them on ordinary bacteriological media. For replication viruses rely entirely on the metabolic processes of the host cell, particularly those associated with nucleic acid replication and the translation of m-RNA. Moreover, unlike other micro-organisms, viruses do not replicate by binary fission or budding but by synthesis of their component parts inside the host cell, followed by assembly into virus particles.

Some of the smallest bacteria, the Chlamydiae (e.g. the causal agents of trachoma and psittacosis) and the Rickettsiae (e.g. the causal agent of typhus) were at one time confused with the true viruses because they too are obligate intracellular parasites and are only about 300 nm in diameter. A comparison between viruses and other micro-organisms is shown in Table 7.1.

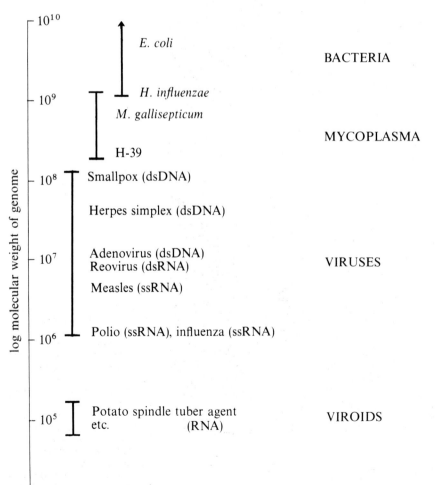

Fig. 7.1 Genome sizes of various groups of micro-organisms. ds = double strand, ss = single strand, DNA = deoxyribose nucleic acid and RNA = ribose nucleic acid.

Structure and chemical composition of viruses

The amount of nucleic acid which viruses contain varies enormously, e.g. the virion (i.e. virus particle) of influenza virus contains 1% RNA, tobacco mosaic virus 5% RNA, smallpox virus 5% DNA and some of the bacterial viruses (known as bacteriophages or phages) as much as 50% DNA. In

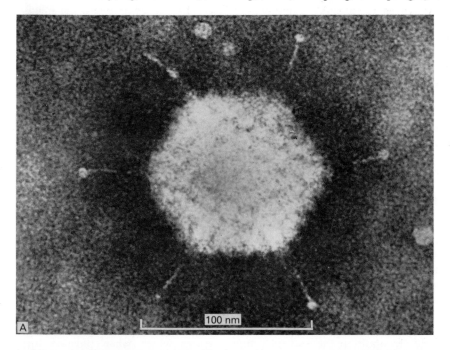

100 nm

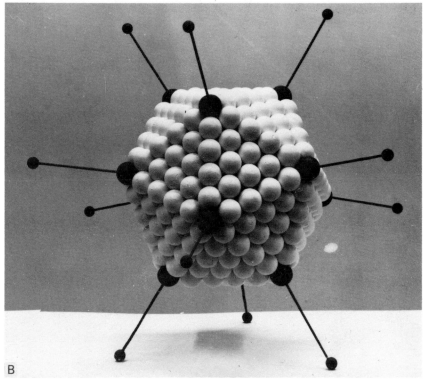

Fig. 7.2
(**a**) A virion of adenovirus negatively stained with phosphotungstate. (Valentine, R. C. and Pereira, H. G. (1965). *J. Mol. Biol.,* **13**, 13–20.)
(**b**) A model of the capsid of adenovirus. Note the distinctive structure of the capsomeres (coloured black) at the vertices of the icosahedral structure. Such capsomeres have five neighbours and are called 'pentons'; the remaining capsomeres with six neighbours are called 'hexons'. (Valentine and Pereira.)

functional terms, i.e. the molecular weight of the genome, there is also great diversity among viruses; this is shown in Fig. 7.1. Some viruses such as smallpox, therefore may have 300–400 genes whilst others like MS-2 (a small RNA phage) may only have three genes. Some, such as the viruses of polio, influenza, measles and tobacco mosaic, have the additional unique feature of possessing their sole genetic information as RNA.

The physical configuration of the genome can also vary with different viruses. Hence with DNA and RNA viruses the nucleic acid can be single- or double-stranded and in the case of DNA viruses the molecule may be linear or a covalently closed circle. Moreover, in some viruses e.g. poliovirus, the genome exists as a single molecule in the virion whilst in others, e.g. influenza virus the genome is segmented, i.e. the virion contains different segments of RNA (in the case of influenza there are 8 or 9 segments).

A further feature, particularly of some plant viruses, is that different segments of the genome are contained in different particles. Such viruses, e.g. alfalfa mosaic are known as component viruses.

The viral protein usually forms the largest part, from 50 to 90 % of the virion. One of the main functions of the protein coat or capsid, is to protect the delicate core of nucleic acid. In addition the capsid proteins may also provide specific receptors for attaching the virus to the host cell, e.g. the fibres on the capsid of adenovirus (Fig. 7.2) and the tail fibres of some bacteriophages (Fig. 7.3).

The inner nucleic acid core of some animal viruses contains very basic proteins. These probably serve to neutralize the acidic groups of the nucleic acid thereby facilitating the packaging of the genome in the small volume of the virion. A similar function is ascribed to the polyamines found in the core of some phages and plant viruses.

Besides structural proteins some viruses also contain enzymes, e.g. some phages contain the enzyme lysozyme which facilitates entry of the phage nucleic acid through the cell wall. In other viruses the enzymes are usually nucleic acid polymerases, e.g. influenza virus has an RNA-dependent RNA polymerase, smallpox virus a DNA-dependent DNA polymerase and the RNA tumour viruses have an RNA-dependent DNA polymerase (a reverse transcriptase).

Besides protein and nucleic acid, the larger and more complex viruses often contain lipid in the form of an outer envelope of lipoprotein, e.g. influenza virus contains 25 % lipid.

How are the molecular components of viruses arranged to form virions? Electron microscopy of viruses, particularly after the introduction of metal shadowing, revealed their overall morphology (Fig. 7.4), but it was only after the application of X-ray diffraction that this question began to be answered. This technique has been applied successfully to the very simple viruses which when purified form crystals or paracrystals, e.g. poliovirus or tobacco mosaic virus (TMV) respectively.

The first virus to be studied by X-ray diffraction was TMV. From such studies the capsid of TMV is known to be a hollow rod consisting of identical

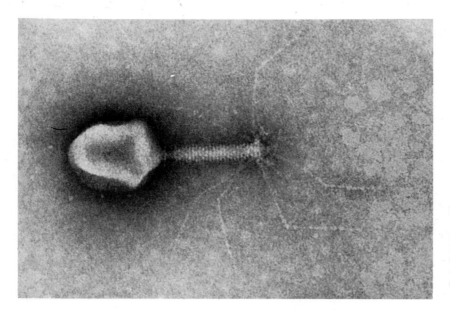

Fig. 7.3 T.4 phage negatively stained with phosphotungstate (× 280 000). Note the tail fibres radiating from the tail base plate, the striations corresponding to the contractile outer sheath of the tail and the elongated head. (E. Kellenberger.)

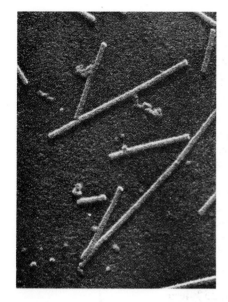

Fig. 7.4 Electron micrographs of two plant viruses illustrating the metal shadowing technique. *Left* the rod-shaped virus TMV (×33 000). *Right* the isometric virus of tomato bushy stunt. (×72 000. H. L. Nixon).

protein molecules (structure units) arranged in the form of a helix around the long axis of the virion (Fig. 7.5). The RNA of the virion also exists as a helix within that of the protein. Other rod-shaped plant viruses, including those less rigid than TMV, e.g. white clover mosaic virus (Fig. 7.6), and the rod-shaped phages, have a similar helical organization.

Although more or less spherical, some animal viruses (e.g. influenza and measles), have capsids with helical symmetry. In such viruses the helical capsid is wound up like a ball of wool inside an envelope of lipoprotein. The capsid (analogous to the tubular TMV virion) is occasionally visible in disrupted particles (Fig. 7.7).

Viruses with isometric (nearly spherical) capsids have a different structure. In these viruses the protein sub-units of the capsids are arranged as if located on the surface of a special class of delta-hedron viz. an icosadeltahedron. This is a polyhedron whose faces are all equilateral triangles and having the symmetry of an icosahedron (a twenty sided solid). Such solids are said to have cubic or icosahedral symmetry.

Negative staining electron microscopy allowed direct visual demonstration of helical and icosahedral symmetry in virus structure (see Figs. 7.2, 7.6, 7.7, 7.8, 7.9). The morphological sub-units revealed on the surface of viruses by negative staining are termed capsomeres. The number of capsomeres present varies greatly between different viruses, e.g. the virion of the phage $\phi \times 174$ has 12 capsomeres whereas that of herpes simplex virus (the virus of cold sores) has 162. In most cases a single cap-

0   10nm

Fig. 7.5 Diagram showing approximately 1/20th the length of the rod-shaped TMV virion. Note the helical arrangement of both the protein structure units (large shoe-shaped structures) and the inner RNA (beaded structure). The last turns of the protein helix have been omitted to show the position of the RNA more clearly. (After Klug and Caspar *et al.* (1962) *Cold Spring Harbor Symp. quan. Biol.*, **27**, 49.)

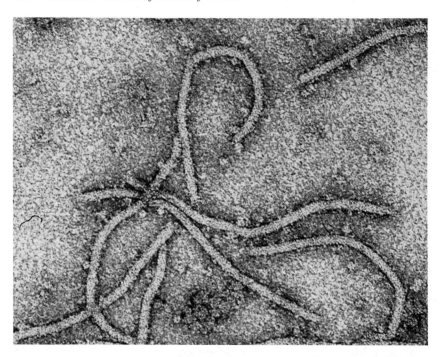

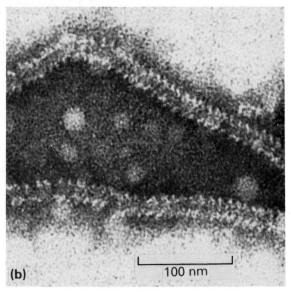

Fig. 7.6 Virus of white clover mosaic (helical symmetry), negatively stained with uranyl acetate. Note the flexuous nature of the particles. ($\times 174\,000$. R. G. Milne, Rothamstead Experimental Station.)

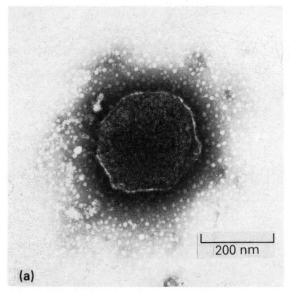

Fig. 7.7
(**a**) Virion of measles virus negatively stained with phosphotungstate (T. J. Hill).
(**b**) Purified preparation of measles virus ribonucleo-protein negatively stained with phosphotungstate. The structure of the helical nucleocapsid is clearly visible (T. J. Hill).

somere consists of more than one protein molecule.

Negative staining electron microscopy has also clearly shown the lipoprotein envelope of some viruses surrounding the nucleocapsid (nucleic acid core plus protein capsid) (Fig. 7.7, 7.10). From thin sections of virus-infected cells it can been seen that viral envelopes are derived morphologically from host cell membranes. For example in the case of herpes simplex virus, the virus acquires an envelope of host inner nuclear membrane as it passes from the nucleus into the cytoplasm (Fig. 7.11). In other viruses e.g. influenza, measles, the envelope is acquired from the cytoplasmic membrane as the virions 'bud off' from the cell surface. Host cell

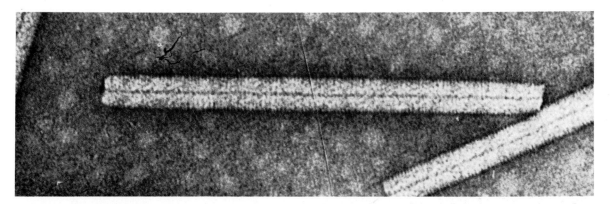

Fig. 7.8 Virions of TMV (helical symmetry) negatively stained with phosphotungstate. Note (1) the stain has filled the central hole and (2) the groove between the helix of structure units is visible as cross striations.

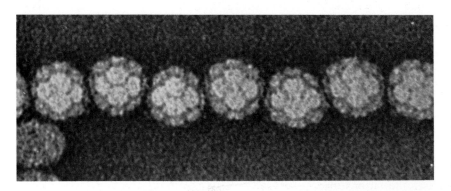

Fig. 7.9 A row of turnip yellow mosaic virions (icosahedral), negatively stained with uranyl acetate ($\times 500\,000$). Note the clearly visible capsomeres on the surface of the virions. (J. T. Finch)

membranes which give rise to virus envelopes are considerably modified by the inclusion of viral antigens. In animal viruses the envelope probably facilitates the entry of the virus into the host cell; the envelope fusing with the cell membrane thereby allowing the nucleocapsid into the cell interior.

The observation that viral capsids consist of repeated identical protein molecules or limited numbers of different protein molecules, can be explained on theoretical grounds as the most economical means of using a genome as small as those found in viruses. Also on theoretical grounds it can be shown that the most stable protein capsids i.e. in which the sub-units make identical bonds with their neighbours, can be made either as a helix or a sphere (or close approximation to a sphere, an icosadeltahedron).

In summary most viruses can be classified on a structural basis as either having icosahedral or

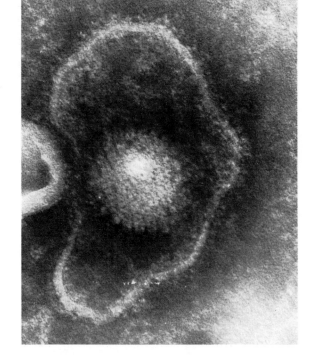

Fig. 7.10 Herpes simplex virion negatively stained with phosphotungstate ($\times 300\,000$). Note the loose outer envelope and the inner capsid with the regularly arranged capsomeres. (Mrs J. D. Almeida)

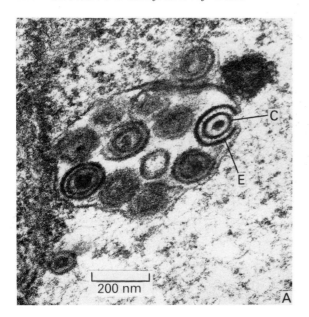

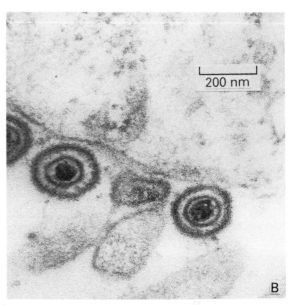

Fig. 7.11
(**A**) A capsid (C) of herpes simplex virus undergoing envelopment by budding through the inner nuclear membrane. The portion of membrane which will form the envelope (E) is thickened and darkly staining. (T. J. Hill)
(**B**) Mature, enveloped particles of herpes simplex virus. (T. J. Hill)

helical symmetry; in some cases the virions also possess a lipoprotein envelope. The two types of virion architecture are illustrated diagrammatically in Fig. 7.12.

Although most viruses fit into the helical or icosahedral categories, there are at least two groups which have a more complex structure. The poxviruses (a group including smallpox virus) are large brick-shaped or ovoid viruses. The virion has a central DNA-containing body (the nucleoid) surrounded by a series of membranes (Fig. 7.13). While the simple isometric phages have icosahedral symmetry and the rod-shaped or filamentous phages have helical symmetry, the 'tailed' phages have a more complex structure (Fig. 7.3). The head of such phages contains the nucleic acid and can be spherical, ellipsoidal or hexagonal in shape; sizes range from 25 to 100 nm. The phage tail, which may be up to 100 nm in length, is narrower than the head but varies in complexity with different phages. In the T-even phages (T2, T4, T6 all attack *Escherichia coli*) the tail has an inner tube (possibly having helical symmetry) and an outer contractile sheath.

At the head end the sheath is attached to a collar and at the opposite end to a hexagonal plate (Fig. 7.3). Fibres radiating from this plate serve to attach the phage to the host bacterium prior to infection (p. 127). The tailed phages probably represent a combination of icosahedral and helical symmetry in the structure of the head and tail respectively.

A summary of the structural feature of some animal, plant and bacterial viruses is shown in Table 7.2.

Virus replication

Viruses replicate only in living cells (p. 124). The

Fig. 7.13 Section of a vaccinia virus virion (× 200 000). Note the complex structure of membranes surrounding the inner DNA-containing nucleoid. (M. A. Epstein)

Fig. 7.12 Diagrammatic representation of sections through (A) an enveloped virus with icosahedral symmetry, (B) an enveloped virus with helical symmetry. (After Caspar, 1962)

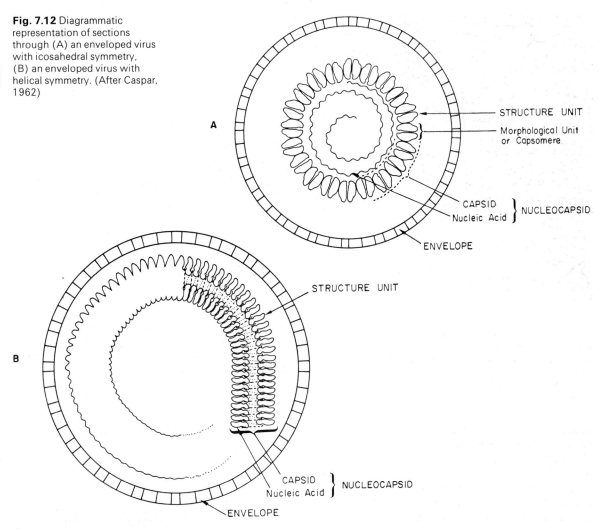

Table 7.2 The main features of the size and structure of some animal, plant, and bacterial viruses

Virus	Nucleic acid	Symmetry of capsid	Presence of envelope	Size of virion
Vaccinia	DNA	?	+	230 × 300 nm
Herpes simplex	DNA	Icosahedral	+	Non-enveloped 110 nm Enveloped 180–250 nm
Tipula iridescent virus (infects larvae of the crane fly, *Tipula paludosa*)	DNA	Icosahedral	–	130 nm
Adeno	DNA	Icosahedral	–	75 nm
T$_4$ phage	DNA	Icosahedral and helical	–	Head: 95 × 65 nm Tail: 110 × 25 nm
ϕX 174 (phage)	DNA	Icosahedral	–	30 nm
Tobacco mosaic	RNA	Helical	–	Rigid rod 300 nm long
Potato X	RNA	Helical	–	Flexible rod 500 nm long
Influenza	RNA	Helical	+	80–200 nm
Polio	RNA	Icosahedral	–	28 nm
Turnip yellow mosaic	RNA	Icosahedral	–	28 nm
MS2 (phage)	RNA	Icosahedral	–	24 nm

term 'replicate' infers that the process of virus multiplication is different from that of micro-organisms and tissue cells which divide by binary fission with or without mitotic division of their genetic components. The mode of entering the host cell varies from virus to virus. For animal viruses this is described below. The plant viruses and the bacteriophages have to overcome the additional obstacle of the cell wall in order to penetrate their host cells. In the case of phages it seems to be the general rule that only the virus nucleic acid penetrates the bacterial cell wall and the virus capsid remains outside. This is achieved either by an elegant injection device associated with the phage tail and aided by a lysozyme-like enzyme or by enzyme action alone (p. 129). In the case of plant viruses the cell wall is breached either by trauma e.g. associated with grafting or very often as a result of direct 'injection' by virus-carrying insect vectors, e.g. aphids (p. 294).

The mode of replication of viruses is considered to be similar for all and has been most completely worked out for bacteriophage (p. 127). The viral nucleic acid upon entering the cell takes over control of the cellular metabolic processes and codes for the separate synthesis of viral nucleic acid and protein which later combine to form the mature virus particle. The virus yield from a cell infected with a single virus particle varies widely but often ranges from 10 to 100 particles.

Viruses may be propagated in tissue cultures, made from animal or plant tissues, in susceptible animals, plants or micro-organisms.

REPLICATION SEQUENCE

Tissue culture studies have revealed the nature of the replication sequence which occurs in the infected host cell. This sequence includes: (1) adsorption of the virion to the cell plasma membrane; (2) penetration of the virion into the cell which, in animal cells, is by active micropinocytosis (i.e. engulfment in a manner similar to phagocytosis but on a smaller scale) by the host cell itself or by fusion of the virus envelope with the cell membrane; (3) eclipse of the virion, due to enzymically controlled decapsidation processes which release the nucleic acid 'core', intact virus being no longer found in the cell; (4) the latent period, when biosynthesis of new virus components takes place; (5) the assembly period, when immature virus forms first appear in the cell; (6) maturation of progeny virions from the infected cells by either budding-off processes (e.g. influenza virus) or cell lysis (e.g. poliovirus). With

some animal viruses the complete sequence from (1) to (7) may take as little as 6 hours.

The host cell must be capable of supporting this sequence of steps in viral replication. Many viruses have a single or limited host cell range; others may replicate in a number of different host cells but the quantity of virus produced in each cell type may differ widely.

To propagate a virus in tissue culture it is necessary first to choose the appropriate host tissue cell. Cell susceptibility is related to such factors as the ability of the virus to adsorb to the cell plasma membrane, which, in turn, depends on the presence of virus 'receptors' on the membrane, as well as the provision by the cell of the appropriate biosynthetic processes for the support and replication of the invading virus.

Laboratory cultures of animal tissue cells provide the most useful means of propagating animal viruses both for production of virus antigens and for diagnosis of virus diseases. A variety of tissue cultures are available but those including actively multiplying tissue cells are the most useful for the propagation of viruses. Monolayer cultures are widely used for the study of particular virus-cell interactions and cytopathic effects (p. 312). Tissues freshly taken from a living or recently dead animal are treated with mild proteolytic enzymes (e.g. trypsin), a process which, whilst maintaining cell viability, causes disaggregation of the cells. These cells are seeded into culture tubes or bottles containing physiologically and nutritionally balanced fluids plus appropriate antibiotics to inhibit bacterial and fungal contaminants. The cells adhere to the lower surface of the vessel when incubated at 37°C, multiply, and form a confluent tightly-packed sheet of cells within 2 to 4 days; the term monolayer is used to describe the cell sheet which is rarely more than one cell thick (Fig. 7.14a). Latent viruses, when present in the animal tissue from which the cultures were prepared, may be stimulated into active replication by either the enzymic disaggregation process or the subsequent cultivation of the cells *in vitro*, or the infection of the cultured cells with another virus. Despite this possibility, tissue culture constitutes a highly efficient *in vitro* system for the propagation of viruses, with the outstanding advantage that cultures can be prepared from most cells or tissues known to be susceptible to particular viruses. In addition, animal cells cultured *in vitro* are no longer subjected to immunity surveillance mechanisms, or to the regulatory actions of antibodies, hormones or various kinins, which operate within the animal body, though they are invariably capable of expressing cellular antiviral activity by means of interferon production (p. 331). They may also be cloned to provide homogeneous cell cultures with selected characteristics.

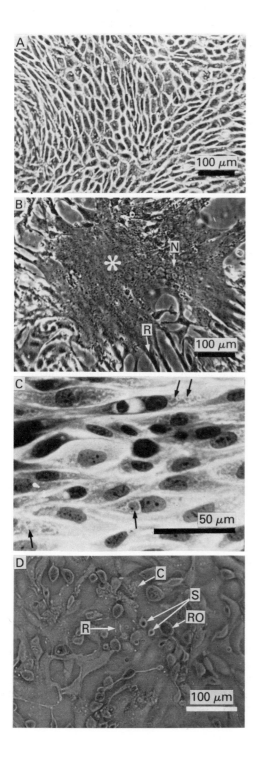

PROPAGATION OF ANIMAL VIRUSES IN EXPERIMENTAL ANIMALS

When using animals it is necessary to consider: (1) their natural susceptibility to infection (Table 7.3) or immune status to the virus; (2) the possibility of latent infection with the same or other virus (often challenge of another virus stimulates a latent virus to become active, as occurs with herpes simplex in man, the cause of the common cold sore on the lips which often erupts when the patient is challenged by a common cold virus); (3) the most suitable route of inoculation, which is usually related to the affinity of the virus for particular tissues (Table 7.3). Infection is recognized by characteristic signs and symptoms of disease.

A widely used host for animal virus propagation is the developing chick embryo. The tissues and associated membranes support the replication not only of fowl viruses but also of a small number of mammalian viruses to which the hatched chick and adult bird are not susceptible. Mammalian viruses such as vaccinia, smallpox, ectromelia and herpes simplex may be propagated on the chorioallantoic membrane, and influenza virus in the primitive respiratory tract as well as in the cells bounding the fluid-filled amniotic and allantoic cavities. The chick embryo is not susceptible to most other mammalian viruses, e.g. poliovirus, herpes zoster and foot-and-mouth, and its usefulness for the propagation of mammalian viruses is limited in this respect. Whilst maternal antibodies against fowl viruses may be present in small amounts in the yolk of the fertile egg the chick embryo does not possess an antibody-forming mechanism. Evidence of virus replication may be indicated by pock formation in the chorioallantoic membrane (Fig. 7.15), by lesions and inclusion bodies (p. 313) in particular tissues, death of the embryo, or the demonstration of the presence of viral antigen by specific serological tests

Fig. 7.14 Viral cytopathic effects in tissue culture. (A) Normal monolayer of dog kidney cells (Phase contrast). (B) Large, central syncytium (polykaryocyte, asterisked) formed by fusion of many dog kidney cells when a monolayer is infected with canine distemper virus, and showing clusters of constituent nuclei (N, arrowed) and retraction (R, arrowed) of peripheral cytoplasm (Phase contrast). (C) Intracytoplasmic inclusions (some arrowed) in dog kidney cells infected with canine distemper virus and stained by May–Grünwald–Giemsa (Bright field). (D) Extensive rounding-off (RO, arrowed), tailing-off or retraction (R, arrowed), shrinkage (S, arrowed), cytolysis (C, arrowed) and cell detachment at a focus of poliovirus infection in a monkey kidney cell monolayer (Phase contrast). (A, B, and C: G. H. Poste; D: L. W. Greenham.)

Table 7.3 The natural and experimental hosts of some common animal viruses

Virus	Natural host	Experimental host	
		Animal	Route of inoculation
Arthropod-borne Encephalitides (Arboviruses)	Man and other mammals	Mouse	Cerebral
Canine Distemper	Dog	Ferret	Nasal
Foot-and-Mouth	Cloven-hoofed animals	Guinea pigs	Hind footpads
		Suckling mice	Cerebral
Influenza	Man	Chimpanzee, ferret, mouse	Nasal
		Chick embryo	Amniotic cavity (at isolation)
			Allantoic cavity (adapted)
Smallpox (variola)	Man	Chick embryo	Chorioallantoic membrane
Vaccinia	Man	Chick embryo	Chorioallantoic membrane
		Rabbit	Skin
Mumps virus	Man	Monkey	Parotid gland
Poliovirus	Man	Monkey	Cerebral, occasionally peritoneal
Rabies	Dog, man, etc.	Mouse	Cerebral

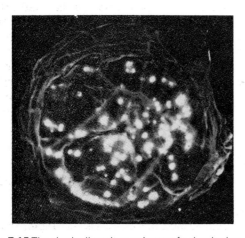

Fig. 7.15 The chorioallantoic membrane of a developing chick embryo infected with vaccinia (nat. size). After incubation for two days the membrane was removed to formal saline and examined for lesions. Plaques of necrosis resulting from virus activity are clearly seen.

(p. 326). Since one pock on the chorioallantoic membrane can arise from infection with one viral particle it is possible to assay for the number of pock-forming units in a virus suspension by this technique.

Bacteriophage

Introduction

Numerous different bacteriophages (phages) are known since almost all species of bacteria, including actinomycetes, are susceptible to infection by specific ones. Bacteriophages exhibit high specificity of action, each being able to infect only one group of closely related organisms, usually a single species (although the T phages of *Escherichia coli* will also attack *Shigella sonnei*), or occasionally only a few strains of a single species. This highly selective activity provides a very useful method for typing bacterial isolates, and phage-typing methods are widely used in modern epidemiological studies.

Phages together with their host bacteria provide a model which is convenient for the intensive study of host–parasite relationships and virus multiplication. In particular extensive studies of the T-even coli phages (T2, T4 and T6) have contributed greatly to present-day knowledge of the genetics of micro-organisms (p. 40 *et seq.*).

Demonstration of bacteriophages

Bacteriophages can be isolated from many natural sources such as faeces, sewage, and polluted water. Like all viruses, phages multiply only within living cells and since they can be seen directly only by electron microscopy, phage activity must usually be demonstrated by indirect means. When a drop of a bacteria-free filtrate of a fluid containing phage particles is added to a young broth culture of susceptible bacteria, the culture will clear visibly within a few hours due to lysis of the bacterial cells. The bacteria-free filtrate can be used to infect a second broth culture and this process may be continued indefinitely.

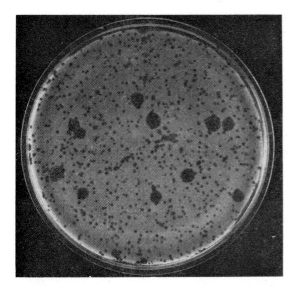

Fig. 7.16 Plaques of different size caused by the growth of at least two bacteriophages on a carpet of actively multiplying bacteria ($\times\frac{1}{2}$). Each plaque arises from the growth of a single phage particle initially inoculated on to the culture. (A. H. Linton).

The lytic activity of phage was first demonstrated on solid media where, on a culture plate seeded or carpeted with susceptible bacteria, phage growth and activity is indicated by cleared areas known as plaques (Fig. 7.16). These plaques behave like colonies; they may show differences in size and general morphology and can be 'picked off' by touching with an inoculating needle and transferred to a fresh culture of susceptible bacteria. Moreover, a single phage particle produces one plaque by multiplication and liberation of many phages which attack and lyse neighbouring bacteria. Hence, fluids containing phage particles can be titrated by plating measured volumes of suitable dilutions of the fluid with sensitive bacteria. The plaque count is analogous to plaque counting of animal viruses in tissue cultures.

The presence of phage is not always evident since latent infection may occur without bacteriolysis. Accordingly, phages are subdivided into two main groups by the relationship they establish with their bacterial hosts. Virulent or lytic phages always produce lysis of the host cells. Temperate phages may lyse the host, but sometimes their genetic material becomes associated with that of the host cell as prophage and is transferred to daughter cells on division. Thereafter a small proportion (10^{-2} to 10^{-7} per cell generation) of the progeny cells lyse to release free phage. Hence bacterial cultures which carry temperate phages are termed lysogenic and the process by which the phage enters the prophage state is known as lysogenization.

Replication of virulent bacteriophages

The stages in the infective process are similar to the multiplication cycles of other viruses, and may be considered as follows: (1) adsorption of free phage; (2) penetration of phage nucleic acid; (3) intracellular development; (4) maturation and release of new phage (Fig. 7.17).

Study of the characteristic sequence of events during a viral multiplication cycle entails making various determinations at precise times after initial infection. To do this, the infection must be synchronized; otherwise infection of different cells is spread over a long period of time during which new virus particles, released when some infected cells lyse, will infect other cells. Synchronization can be achieved by allowing adsorption of virus to bacteria for only a few minutes and then removing the unadsorbed virus. When it is desired to study the multiplication cycle in cells infected by single virus particles, 'one-step conditions' are employed; once adsorption is complete, the virus cell mixture is diluted into a large volume of nutrient broth so that progeny viruses released later are very unlikely to become adsorbed to the uninfected cells because they are so sparse.

ADSORPTION OF FREE PHAGE

For adsorption to take place, the phage particles must first collide with the cell, then bonds must be established between the two surfaces. Under optimum conditions, this process is very rapid, almost all phages being adsorbed within a few minutes. Adsorption is usually irreversible but can sometimes be reversed by varying conditions such as pH and temperature. Frequency of collision between phages and bacteria depends mainly on the concentration of virus particles and cells. When there is an excess of phage particles, as many as 300 may adsorb to one bacterial cell. Frequency of collision is also affected by the size and mobility of the virus and the mutual attraction or repulsion brought about by electrostatic forces. The formation of bonds after collision is dependent on the chemical structure of the two surfaces. As with the union between an antigen and antibody, the presence of complementary chemical groups on the two surfaces is essential, e.g. with coli T_1, amino groups on the viral surface bond with receptors containing acidic groups on the cell surface. The receptor sites of the bacterial cell render it highly specific for a small range of phages and cannot be found in the walls of bacteria resistant to those phages. The receptors may be situated in different layers of the wall, e.g. receptors for phages T_2 and T_6 exist in the lipoprotein layer of the wall whereas phages T_3 and T_4 adsorb to receptors in the lipopolysaccharide layer. Receptor substances have been separated from cell wall preparations by gentle chemical extraction and have been shown to combine specifically with phage active against the whole cell. The phage surface must also be modified before adsorption can occur; this may be by attachment of positively charged cations (e.g. Mg^{++} or in some cases, L-tryptophane. The extent of modification varies from one phage to another but each phage is specific with regard to the co-factor required for adsorption.

Some phages (T-even phages) are equipped with fibrils which extend from the base plate at the tip of the tail. These phages adsorb to receptors on the bacterial cell by attachment of their fibrils to the cell surface, leaving the virus particle in a characteristic position at right angles to the cell wall and with the head sticking out (Fig. 7.18a). The receptors for the very small phages f_1 and f_2 are only found on the surface of F^- cells. Some other phages attach specifically to flagella.

Bacterial mutation can change the nature of the chemical groups on the receptors and by so doing prevent phage adsorption, rendering the cells phage-resistant.

PENETRATION OF PHAGE NUCLEIC ACID

In the case of phages with more complex tails (e.g. T_2) the phage nucleic acid is injected into the bacterial cell usually by insertion of the tail core in the cell wall and contraction of the sheath, leaving the empty protein coat outside. The T-even phages are known to hydrolyze the mucopeptide of the bacterial cell wall by means of a phage lysozyme

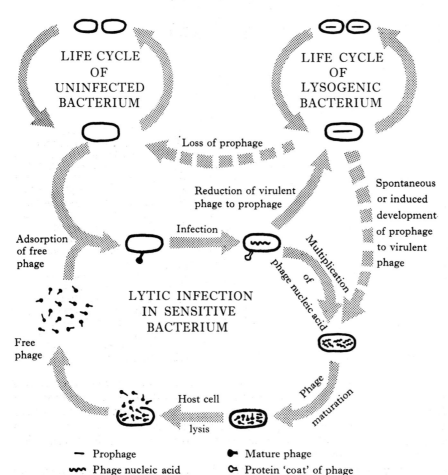

Fig. 7.17 Phage–host life cycles. (Modified from Jawetz, Melnick, and Adelberg, 1966).

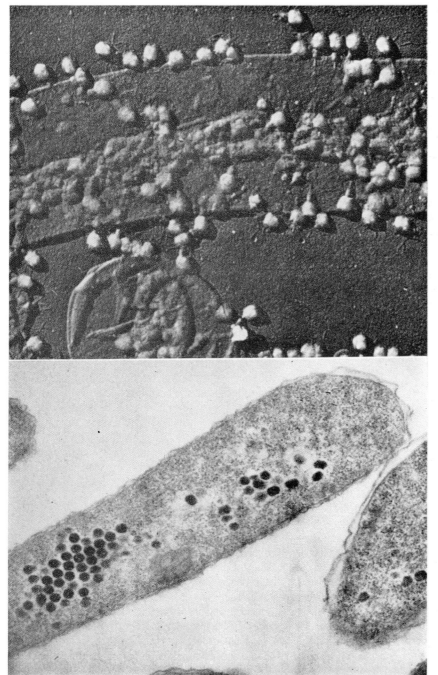

Fig. 7.18 (A) Electron micrograph of T2 bacteriophage adsorbed on to empty cell walls of *Escherichia coli* (×40 000). Each phage particle is adsorbed by its tail, the end of which has been introduced through the cell wall (E. Kellenberger). **(B)** Electron micrograph of a thin section of *Escherichia coli* infected with T2 bacteriophage thirty minutes before fixing (×50 000). Mature phage particles are visible, their polyhedral heads being well preserved. The bacterial cell wall, detached from the cytoplasm, is clearly visible. (Kellenberger, E. *et al.* (1958) *J. Biophys. Biochem. Cyt.*, **4**, 671)

which is attached to the tip of their tails. The dissolved hole may help in the insertion of the tail core in readiness for injection of nucleic acid, but such a hole is apparently not essential since some other phages which possess no lysozyme are still able to insert their tail core. Holes produced by phage lysozyme may cause a leakage of cell con-

tents; normally the holes are repaired fairly quickly but if a very large number of phages produce holes in the same cell within a short time, there may be so much leakage that the cell lyses without virus multiplication. This is known as 'lysis from without'.

Electron microscopy of T-even phages has shown

that irreversible adsorption of a phage always results in contraction of the tail sheath. The base plate and fibrils are held firmly against the cell surface and contraction of the sheath pulls the head towards the cell, causing the hollow tube or core to penetrate the cell wall reaching through to the cell membrane but probably not piercing this. Rather like primitive muscle, the sheath contains the source of energy for inserting the tube through the wall and does not depend on an energy supply from the bacterium. Other phages which are not equipped with contractile sheaths are still able to effect penetration with their tube but at a much slower rate than the T-even phages and are dependent on the host cell for the energy required to achieve entry of their tube or, in the case of the tiny spherical phages, of their capsid.

Once the cell membrane is reached, the release of phage DNA is triggered off and the contents of the head are expressed on the membrane through which the nucleic acid molecules are able to pass into the cell. The empty protein head and tail structures remain attached to the bacterial cell wall but they take no further part in the infective process.

INTRACELLULAR DEVELOPMENT

If the bacterial cells are disrupted soon after the injection of phage nucleic acid, the presence of phage cannot be demonstrated by plaque formation on a susceptible plate culture. This short period before phage can once more be demonstrated is called the eclipse phase.

Immediately after injection of phage DNA into the host cell, synthesis of a new protein required by the phage for multiplication commences. The first products, often referred to as 'early proteins', include certain enzymes which are required for later construction of the complicated molecules peculiar to the phage, e.g. kinases for the formation of nucleotide triphosphates, and a phage DNA polymerase.

Subsequently during the eclipse phase, 'late proteins' appear which include the protein sub-units for the phage head and tail structures. Construction of new phage DNA is accomplished with some components from the medium and some from degraded host DNA. Genetic recombination, in which phage particles undergo random exchange of genetic material can occur at this stage (p. 53).

MATURATION AND RELEASE OF NEW PHAGES

The assembly of the newly-synthesized components

to form a complete or mature, infectious virus particle is known as maturation. Maturation of phages involves assembly of many components and therefore occurs in several steps. Viral DNA molecules are first condensed into large particles surrounded by a membrane and resembling phage heads. These are recognizable before maturation is complete in thin sections of infected bacteria (Fig. 7.18b). Phage heads are completed by further condensation of capsid monomers around the DNA condensate, but the heads are not yet stable since premature lysis of the cells at this stage by artificial means results in the DNA escaping from the protein coat of the new head. The hollow tube and base plate appear to be assembled next, followed by the contractile sheath and finally the fibrils are added. Each step of the phage assembly is controlled by genes in the phage nucleic acid, maturation being a sequential process in which each step recognizes or requires the previous step. The assembly can be achieved in vitro by adding the components in the correct combination.

For almost all phages, synthesis of phage components and assembly of mature phages continues until the bacterial cell bursts. The cell wall is weakened by the action of phage lysozyme and the cell finally bursts releasing the newly formed infective phages into the surrounding medium. One exception to this method of release is the tiny phage M13 which appears to leak out of the cells by an unknown mechanism, without killing the host cells.

The replication cycle from adsorption to release of new phages takes from 15 to 30 minutes.

Replication of temperate bacteriophages

Unlike virulent phages, temperate phages enter into a symbiotic relationship with their hosts and do not usually produce lysis (Fig. 7.17). They gain entry to bacterial cells in the same way as virulent phages but thereafter do not interfere with the synthesis of bacterial components nor do they usually produce any visible changes in the culture.

Although temperate phages do not lyse their hosts, their presence can often be demonstrated because the occasional phage particle takes on a virulent role and as a result a very small number of cells in a culture may lyse. By plating an excess of sensitive indicator strain for the phage with a small inoculum of the lysogenic culture, any release of phage in virulent form will be clearly shown by plaques in the carpet of sensitive cells around the affected colonies of the lysogenic culture.

The prophage (p. 128) exists as one or a few particles per cell, it behaves like a host gene and is reproduced synchronously with the host nucleus, to which it is bound. The mechanism of attachment has been studied in the case of phage, where it has been found that the small circular phage chromosome is integrated into the much larger circular chromosome of the bacterium at a point where a short sequence of homologous nucleotide pairs occurs in the two chromosomes. One prophage may be substituted for another in a lysogenic culture following infection with a second temperate phage.

Infection by temperate phages confers resistance on the host bacterium to superinfection with the same or related phages. This immunity is clearly different from the resistance of certain bacteria to virulent phages through failure to adsorb these. The occasional spontaneous change in prophage which allows the production of virulent phage particles can be greatly accelerated by the use of various agents, e.g., ultra-violet light.

The phenomenon of transduction depends on the activities of temperate phages. When a temperate phage from one lysogenic culture is allowed to infect a second bacterial culture, it is possible for a small number of closely linked genes to be transferred from one bacterium to another. The transferred material remains as a stable feature in the recipient bacterium; properties such as resistance to antibiotics can be transferred in this way.

Some temperate phages can confer new properties on the host cell and this is known as phage conversion. The prophage behaves as a host gene, each cell receiving the prophage acquiring the new property. For example, virulent toxin-producing strains of *Corynebacterium diphtheriae* have been isolated from cultures of avirulent non-toxin-producing strains as a result of infection with specific phages.

Actinophages

Particular viruses attack actinomycetes, and have been found in several streptomycetes, *Actinomyces bovis* and *Nocardia farcinica*. Such viruses—'actinophages'—are abundant in the soil. Some possess a tail which, in several at least, is unlike that of the T-even coli phages in that it lacks a visible contractile sheath, tail fibres and a base plate. A terminal knob or group of prongs possibly assumes the function of the base plate, though the mechanism of attachment is obscure. In other actinophages there is no tail at all.

Virus diseases of fungi

Until the early 1960s no disease of a fungus had been proved to be due to a virus, although viruses had been suggested as the cause of certain abnormalities in fungus fruit-bodies and mycelium. Indeed several authors had cited the supposed lack of virus diseases as a character distinguishing fungi from actinomycetes.

Mushroom growers, however, had long been concerned about poor cropping and the production of abnormal fruit-bodies under conditions of intensive cultivation. It has now been shown conclusively that a number of well-known disorders of cultivated mushrooms, such as 'watery stipe' and 'die-back', are due to a virus or a complex of viruses. Electron microscopy shows that virus particles of at least two and probably three distinct types (viz. small and larger spherical particles and elongated ones) may be present in watery stipe fruit-bodies. These particles are morphologically comparable with those of certain plant viruses and distinct from those of bacteriophages. Serological tests and density-gradient centrifugation gave similar results for all the types of particle observed, but improved techniques of virus isolation are desirable. Spread of the virus along infected hyphae and through anastomoses between diseased and healthy ones has been demonstrated.

Polyhedral virus particles, 25–30 nm in diameter and containing RNA, have been isolated from a slow-growing strain of *Penicillium stoloniferum*. The viral RNA from this fungus stimulates the production of interferon (p. 331) in animals. Cultures of *P. stoloniferum* free of virus were obtained from heat-treated conidia of the infected strain. There is strong evidence that some other examples of slow growth and abnormal form of moulds are due to virus infection.

Algal viruses

Virus infections of algal cells are now recognized and are best known in the Cyanobacteria (p. 171).

The first virus of blue-green bacteria was isolated in 1963 from infected cells of *Lyngbya* cultured from a waste stabilization pond. It also infects species of *Plectonema* and *Phormidium* and hence has been coded as LPP-1. This virus is stable over a pH ranged from 5 to 11, which is similar to that within which the host *Plectonema boryanum* grows. In culture, plaques vary in size between 0.1 and 8 mm in diameter and this is considered to be due to strain variation. The virus contains double-stranded

DNA with a molecular base composition of 55% guanine plus cytosine.

The classification of viruses

Introduction

Early attempts at virus classification were based largely on the more easily observed features of virus diseases, e.g., tissue tropisms, host range, symptomatology and pathology, rather than the characteristics of virus particles, such as chemical composition and structure, which are more difficult to determine.

Increased knowledge has led to systems of virus classification based on the more basic immutable characters of the virion e.g. type of nucleic acid and fine structure. An internationally agreed system of virus classification is now being constructed on the basis of such characters (Wildy, 1971). The groups of viruses are defined by a non-phylogenetic hierarchy of characters and the system will include all types of viruses (plant, bacterial, vertebrate and invertebrate). Although the virus families and genera have been given latinized names, virologists have so far resisted the use of latinized binomial generic and specific names for individual viruses. It seems that vernacular names e.g. tobacco mosaic virus, will be used for some time to come. Some workers favour the use of a binomial nomenclature based on the present vernacular names plus an informative cryptogram as the second part. Amendments can be made to the cryptogram as new facts are obtained.

Mainly because of their medical importance and the availability of good culture systems, the viruses causing disease in vertebrates have been more extensively studied than those of other hosts. Therefore the taxonomy of these viruses is rather more advanced; a summary of the major vertebrate virus families is given below. The main characters used to define these groups are the type of nucleic acid, capsid symmetry, presence of envelope, site of replication in the cell, number of capsomeres (for virions with icosahedral symmetry) diameter of helix (for viruses with helical symmetry) and particle size. For diagnostic purposes viruses are more often identified by serological means (p. 326).

DNA-containing vertebrate viruses

PARVOVIRIDAE

Small (Latin, parvus = small), non-enveloped, icosahedral viruses, particles 22 nm in diameter with a single-stranded genome. The group includes Kilham's rat virus and the virus of feline panleucopenia. These viruses have a predilection for rapidly dividing cells.

PAPOVAVIRIDAE

Named after three of the group members: the *pa*pilloma (wart) viruses, *po*lyoma virus (causing many different tumours in mice) and *va*cuolating agent (also known as SV40, found in 'normal' monkey kidneys). The virions are non-enveloped, icosahedral and 40–55 nm in diameter. The genome is cyclic and double-stranded. Nearly all members are oncogenic (tumour producing).

ADENOVIRIDAE

First isolated from the *aden*oidal glands of man. The particles have icosahedral symmetry, are 80 nm in diameter and non-enveloped. The genome is linear and double-stranded. Adenoviruses are generally associated with respiratory diseases and often produce latent infections. Some serotypes produce tumours when inoculated into newborn hamsters.

HERPESVIRIDAE

The icosahedral 100 nm capsids are surrounded by a lipoprotein envelope (Fig. 7.12a). The genome is linear and double-stranded. The group takes its name from the virus causing herpes simplex (cold sores) in man. Other herpes viruses include varicella–zoster (chicken pox–shingles), pseudorabies and infectious laryngotracheitis. Many cause latent and recurrent infections and some, e.g. Marek's disease virus of the chicken, cause tumours.

IRIDOVIRIDAE

A group originally formed from several insect viruses causing iridescent crystals in the infected larvae. The icosahedral capsid is 190 nm in diameter and the vertebrate viruses have an outer envelope. Unlike all other DNA-containing vertebrate viruses except the poxviruses, they replicate in the cytoplasm. Members found in vertebrates include African swine fever virus and lymphocystis virus of fish.

POXVIRIDAE

The largest (230×300 nm) and most complex (Fig.

7.13) of the vertebrate viruses. Like the iridoviruses, these viruses replicate in the cytoplasm. The various genera depend largely on serological relationships e.g. viruses of the *Orthopoxvirus* genus (vaccinia, cowpox, mousepox, smallpox, etc.) are sufficiently closely related to give common neutralizing antibodies. Some poxviruses produce tumours, e.g. rabbit myxoma and fibroma, others produce more generalized infections with a rash, e.g. smallpox.

RNA-containing animal viruses

PICORNAVIRIDAE

A group of small (Gk. pico = small), 22 nm, non-enveloped, icosahedral viruses which are sub-grouped according to their acid sensitivity. The genus *Enterovirus* contains acid stable viruses such as poliovirus, the coxsackieviruses and the echo-viruses (*E*nteric *C*ytopathic *H*uman *O*rphan viruses). The genus *Rhinovirus* contains acid-labile members such as many of the common cold viruses and the virus of foot and mouth disease. Infections with picornaviruses are often inapparent—hence the term 'orphan' viruses, i.e. viruses without a disease.

Like all other RNA viruses they replicate in the cytoplasm of the host cell.

TOGAVIRIDAE

Particles are 20–40 nm with icosahedral symmetry and an outer envelope. Many members of the group are transmitted by arthropod vectors; the viruses replicate in the arthropod tissues. The proposed genera are *Alphavirus* (group A arboviruses), e.g. Eastern and Western encephalitis viruses, *Flavivirus* (group B arboviruses), e.g. yellow fever virus, *Pestivirus*, e.g. swine fever virus, *Rubivirus*, viz. german measles virus.

ARENAVIRIDAE

Pleomorphic enveloped viruses, 35–120 nm in diameter. The name (Latin, arena = sand) arises from the fact that thin sections reveal 20–30 nm granules within the virion; some of these granules are ribosomes derived from the host cell. Some members of the group cause inapparent infections in rodents, e.g. lymphocytic choriomeningitis, whilst others cause severe generalized infections in man, e.g. Lassa fever.

REOVIRIDAE

Non-enveloped, icosahedral viruses, 70–75 nm. The genome has two distinctive features; it is double-stranded RNA and fragmented into ten different segments. The genus *Reovirus* consists of viruses causing inapparent infections in the respiratory and enteric tract (*R*espiratory *E*nteric *O*rphan viruses). However, the *Orbiviruses*, e.g. African Horse Sickness virus, cause disease in man and other animals and are spread by arthropod vectors. The *Rotaviruses* (so called because of the wheel-like appearance of the capsid) form a separate genus. They cause infantile gastroenteritis in humans, mice and calves.

ORTHOMYXOVIRIDAE

The name is derived from the fact that members have receptors for certain mucins. Hence they are able to agglutinate red blood cells. The particles are roughly spherical, 80–200 nm in diameter, and have a coiled-up nucleocapsid, probably with helical symmetry, 8–9 nm in diameter. Surrounding the latter is an envelope containing the haemagglutinin and the enzyme neuraminidase. The genome is single-stranded but like the Reoviridae it consists of 8 or 9 segments.

There are three main serological subgroups; type A cause disease in man, birds, horses and pigs, while B and C have been recovered only from man.

PARAMYXOVIRIDAE

These viruses are similar in some respects to the orthomyxoviruses e.g. some produce haemag-glutination and contain neuraminidase. However, the particles and the helical nucleocapsid are larger, 100–300 nm and 18 nm in diameter respectively (Fig. 7.7b); moreover, the genome is unsegmented. The family includes the viruses of mumps, measles, Newcastle disease and canine distemper and the parainfluenza viruses.

RHABDOVIRIDAE

A group of bullet-shaped viruses (Gk., rhabdos = a rod) 70 × 175 nm in size. The virions are enveloped and have helical symmetry. Members include the viruses of rabies and vesicular stomatitis (a bovine disease). The latter, and other group members, are borne by arthropod vectors.

BUNYAVIRIDAE

Enveloped virions 90–100 nm diameter. The nucleocapsid probably has helical symmetry and the genome is divided into several segments. All these viruses multiply in and are transmitted by arthropods. Members include Bunyamwera and Rift Valley fever viruses.

CORONAVIRIDAE

Enveloped virions 80–120 nm in diameter. The envelope has petal-shaped projections giving the particle a sun-like (Latin, corona = crown) appearance. Human isolates cause common colds and gastroenteritis; similar disease patterns occur in other animals.

RETROVIRIDAE

All members carry the unique enzyme *re*verse *tr*anscriptase (hence the name, 'retro') which transcribes a DNA copy from the RNA of the virus genome. The DNA 'provirus' is inserted into the genome of the host cell. The virion structure is complex and may represent a combination of helical and icosahedral symmetry. Members include several tumour viruses of birds and mammals.

Other viruses

Although the classification of other viruses (plant, bacterial and invertebrate) is less advanced, the application of basic principles similar to those used for vertebrate viruses has enabled the formation of several taxonomic groups (see Wildy, 1971 and Fenner, 1976). At present some 16 'genera' of plant viruses have been proposed, e.g. the *Tobamovirus* group (tobacco mosaic and related viruses) and *Tobravirus* group (tobacco rattle and related viruses) and 8 families of bacteriophages.

A number of important viruses remain as yet unclassified e.g. the virus of human hepatitis B and the so called 'slow viruses' (p. 314).

Books and articles for reference and further study

General

BURKE, D. C. and RUSSELL, W. C. (eds.) (1975). Control Process in Virus Multiplication, *25th Symp. Soc. gen. Microbiol.*, Cambridge University Press, Cambridge.

CASPAR, D. L. D. (1965). Design principles in virus particle construction, in *Viral and Rickettsial Infections of Man*, edited by Horsefall, F. L. and Tamm, I., 4th Edn., Pitman Medical Publishing Co. Ltd., London and J. B. Lippincott Co., Philadelphia, pp. 51–93.

CRAWFORD, L. V. and STOKER, M. G. P. (eds.) (1968). The Molecular Biology of Viruses, *18th Symp. Soc. gen. Microbiol.* Cambridge University Press, Cambridge, 372 pp.

DALTON, A. J. and HAGENAU, F. (1973). *Ultrastructure of Animal Viruses and Bacteriophages: An Atlas.* Academic Press, London, New York and San Fransisco.

FENNER, F., MCAUSLAM, B. R., MIMS, C. A., SAMBROOK, J. and WHITE, D. O. (1974). *The Biology of Animal Viruses* 2nd Edn., Academic Press, London, New York and San Francisco.

HAHON, N. (ed.) (1964). *Selected Papers on Virology*, Prentice Hall, Inc., Englewood Cliffs, New Jersey (a collection of papers from the 18th century work of Jenner onwards, illustrating the development of virology), 313 pp.

PENNINGTON, T. H. and RITCHIE, D. A. (1975). *Molecular Virology* (in series, Outline Studies in Biology) Chapman and Hall, London.

SMITH, K. M. (1974). *Plant Viruses*. Chapman and Hall, London.

TOOZE, J. (ed.) (1973). *The Molecular Biology of Tumour Viruses*. Cold Spring Harbour Laboratory.

Reviews

Advances in Virus Research. Published annually by Academic Press.

Progress in Medical Virology. Published annually by S. Karger.

Classification

FENNER, F. (1976). The classification and nomenclature of viruses. *J. gen. Virol.*, **31**, 463.

WILDY, P. (1971). Classification and nomenclature of viruses. *Monographs in Virology*, No. 5, Karger, Basel.

Chapter 8

Structure and Classification of Prokaryotic Micro-organisms

Distinctive features of Prokaryotic cells

The cells of prokaryotes are structurally more complex than viruses but less so than those of the eukaryotes (Chapter 9). They differ from the latter principally by lacking a perforated nuclear membrane surrounding the nucleoplasm and hence a *well* defined nucleus (p. 179). They also lack organelles, such as mitochondria (p. 181), do not exhibit cytoplasmic streaming and the structure and composition of cell walls, flagella and chromatophores, where present, are different. The existence of a definite reproductive cycle in most, but not all, eukaryotes and the general absence of a precise cycle among prokaryotes is additional evidence for the concept of two fundamentally different groups of organism.

The distinctive aspects of morphology and fine structure of the various groups of prokaryotes and classification are considered together with other important characters not covered elsewhere in this book. The *Prokaryotes* include the large and varied group *Bacteria* together with the *Blue-green bacteria*. The latter and the actinomycetes historically have been studied with algae and fungi respectively. For this reason and despite their inclusion within the prokaryotes, they are considered separately after the general account of the prokaryotes.

Bacteria

Introduction

Bacteria include a great variety of micro-organisms of different shape and size. A few are as large as $5\,\mu m$ in diameter; but the majority fall into the range 0.2–$1.5\,\mu m$ in diameter; most measure one or only a few micrometres in length but some measure several hundred micrometres. These dimensions lie between those of the fungi and the viruses.

Bacteria were first described and drawn as early as 1689 by the Dutch microbiologist, Antonie van Leeuwenhoek, using simple lenses which he ground himself, but a systematic study of these small organisms did not begin until early in the nineteenth century when attempts were made to classify those now recognized as bacteria. Knowledge of the detailed structure of bacterial cells was not, however, acquired until the last two decades when the extensive use of the electron microscope coupled, often, with techniques of chemical and biochemical dissection has proved fruitful.

Morphology

Bacteria, owing to their minute size, can be examined under the light microscope only with objectives of high resolving power. With visible light no purpose is served by magnification beyond 1500 diameters since particles smaller than half the wavelength of the light used cannot be resolved. In practice only lines $0.25\,\mu m$ or more apart can be distinguished. Further useful magnification can be achieved using shorter wavelength ultra-violet light.

Most bacteria are transparent and have a refractive index similar to that of the aqueous fluids in which they are suspended. When they are examined in transmitted light under the ordinary optical microscope little detail can be seen. In order to render them more easily visible, staining with aniline dyes after killing and fixing, was introduced during the latter half of the nineteenth century. Stained preparations still provide a useful method of examining bacterial morphology. More recently phase contrast microscopy has been used for observation of living cells. This makes possible the study of structures previously seen only in fixed and specially stained preparations (p. 139).

Using these techniques bacteria are seen to fall into three morphological types—spheres, straight rods and curved rods; the curved rods being regarded as rudimentary helices. The morphology of different cells in a population arising from a single individual are not identical; variation in size and form of the individual cells occurs with age and growth conditions. This variation or pleomorphism is particularly characteristic of some organisms, e.g.

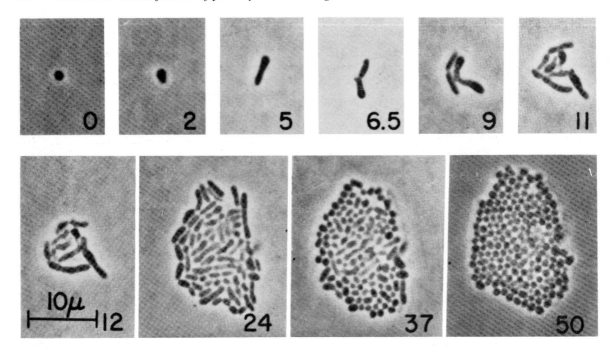

Fig. 8.1 *Arthrobacter globiformis* growing on agar plus 0.5% peptone. Numerals indicate time in hours. (Ensign, J. C. (1974) *J. Bact.*, **87**, 924)

Arthrobacter globiformis (Fig. 8.1) and *Mycoplasma* spp. (p. 168). In addition under certain, usually adverse, conditions many bacteria may change their form and structure to produce involution and L-forms (p. 138).

SPHERES

Most cocci (Greek and Latin for 'berry') approximate to true spheres but variations occur (Fig. 8.2, i–vi). *Staphylococcus aureus* and *Streptococcus pyogenes* are spherical, *Streptococcus faecalis* is ellipsoidal and *Neisseria gonorrhoeae* is kidney-shaped. Cocci may divide irregularly in one, two or three planes and the daughter cells may separate or remain attached, the resulting arrangements being useful for identification. Division in one plane gives rise to pairs of organisms (diplococci), e.g. *Streptococcus pneumoniae*, or to chains, e.g. *Streptococcus pyogenes*. Regular division in two planes at right angles to each other results in tetrads (e.g. *Micrococcus roseus*) and division in three planes to cubical packets of eight or more cocci as in the genus *Sarcina*. Irregular division in two or three planes gives the grape-like clusters characteristic of *Staphylococcus aureus*. Diameters of cocci range from 0.5 to 1.25 μm.

STRAIGHT RODS

An organism of this morphological type was formerly referred to as a 'bacillus' (Latin for 'stick'), but as it is impossible to distinguish in speech

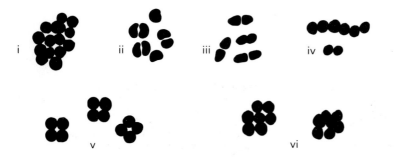

Fig. 8.2 Various morphological types of cocci drawn to approximately the same scale (i) Staphylococcus ; (ii) diplococci (e.g., *Neisseria* spp.) ; (iii) diplococci (e.g., *Streptococcus pneumoniae*) ; (iv) *Streptococcus pyogenes* ; (v) micrococci (e.g., *Micrococcus roseus*) ; (vi) *Sarcina lutea*. (A. H. Linton)

---

between the words bacillus and *Bacillus* (the generic epithet of one of the aerobic spore-forming organisms), its use as a description of morphology is now avoided and the word rod is used instead. Rods are straight or *very* slightly curved cylinders. The ratio of length to diameter varies greatly and in some instances is not much greater than unity. For example *Serratia marcescens* is 0.5 µm in diameter and 0.5–1.0 µm long, *Nitrosomonas europaea*, 0.8–0.9 × 1.0–2.0 µm and *Haemophilus influenzae* 0.3–0.4 × 1.0–1.5 µm. Others are longer, for example, *Bacillus anthracis*, 1.0–1.2 × 3.0–8.0 µm, and *Sphaerotilus natans*, 0.7–2.4 × 3.0–10.0 µm. End walls of rods are usually convex (e.g. *Serratia marcesens*, Fig. 8.3, i) but may be flat (truncate) or even slightly convex *Bacillus anthracis*, Fig. 8.3, iii), club-shaped at one end as in *Corynebacterium* (the name means clubbed rod) or tapering (e.g. *Fusobacterium fusiforme*, Fig. 8.3, v). Certain species of rod-shaped bacteria invariably occur as single cells; others, especially those which produce rough colonies (p. 89), remain attached in chains. In the infected animal *B. anthracis* occurs as single cells or short chains of two or three rods while in culture it forms invariably long filamentous chains. Chains of *Spaerotilus*, contained within a sheath, are frequently extremely long (Fig. 8.20). In *Corynebacterium* spp. (Fig. 8.3, iv) cells remain in association with each other but post-fission movements following 'snapping division' may give rise to characteristic arrangements such as the palisade or angular arrangements sometimes described as having the appearance of chinese letters. In other related bacteria the rods show true branching, either with development into filaments (*Nocardia* Fig. 8.46b.) or without (*Bifidobacterium*, Fig. 8.3 vi); some produce an extensive mycelium (*Streptomyces* Fig. 8.46g) which fragments to form spores. Occasionally rods, reproduce, not by binary or transverse fission or by fragmentation, but by producing buds on appendages (*Hyphomicrobium*, Fig. 8.3, vii). Other budding bacteria have complex shapes and cut off buds from the main body of the cell (*Ancalomicrobium* Fig. 8.3, viii).

Among the Myxobacteria some organisms exhibit simple differentiation, the vegetative rods (0.3–0.7 × 5.0–8.0 µm) rounding off to form myxospores (p. 164); others form large structures called fruiting bodies in which the cells are differentiated into stalk cells, sporangial wall cells and myxospores.

CURVED RODS

Probably the best known example of a curved rod is *Vibrio cholerae* (Fig. 8.4 i) but it is often seen as a straight rod. The relatively recently discovered *Bdellovibrio* (bdella is Greek for a leech) also exists as a small, 0.25–0.4 × 0.8–1.2 µm, curved rod in its parasitic phase but as a helix, it is 20 µm long, in its host independent stage (Fig. 8.4, ii and iii). Other rigid-walled spiral bacteria such as *Spirillum* may

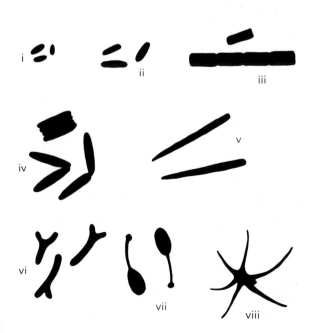

Fig. 8.3 Various morphological types of rod drawn to approximately the same scale (i) *Serratia marcescens*; (ii) *Escherichia coli*; (iii) *Bacilus anthracis*; (iv) *Corynebacterium*; (v) *Fusobacterium fusiformis*; (vi) *Bifidobacterium*; (vii) *Hyphomocrobium*; (viii) *Ancalomicrobium*. (A. H. Linton)

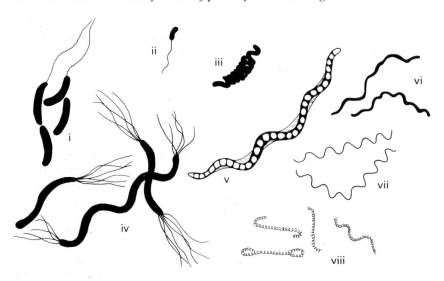

Fig. 8.4 Various morphological types of curved rods drawn to approximately the same scale. (i) *Vibrio cholerae*; (ii and iii) *Bdellvibrio*; (iv) *Spirillum*; (v) *Cristispira*; (vi) *Borrelia*; (vii) *Treponema*; (viii) *Leptospira*. (A. H. Linton)

achieve even greater lengths, up to 60 μm (Fig. 8.4, iv) and the flexible-walled spiral forms, such as *Spirochaeta*, may be even longer, reaching lengths of 500 μm. *Cristispira* is intermediate in size being about 150 μm in length (Fig. 8.4, v); *Borrelia*, *Treponema* and *Leptospira* are relatively short, from a few to 20 μm long and are all very slender, ranging in width from 0.2 to 0.5 μm, 0.9 to 0.5 μm and 0.1 μm respectively (Fig. 8.4, vi, vii and viii).

PLEOMORPHIC FORMS

Many bacteria, under certain conditions such as cold shock, the presence of specific antiserum, bacteriophage attack or the presence of antibiotics such as the penicillins, which interfere with cell wall synthesis, tend to develop an unusual form of pleomorphism involving the formation of large spherical or distorted cells, the so-called 'large-bodies'. The cells revert to their normal form if the inducing stimulus is removed after only a brief exposure. If the stimulus persists for a long time, the cells either perish or may become stabilized in the so-called L-form. L-forms consist of minute filterable bodies and non-rigid, fragile, large globules of cytoplasm and resemble *Mycoplasma* (p. 168). These may be spherical (0.125–0.250 μm in diameter) in young cultures becoming filamentous (up to 150 μm long) with age, finally forming chains of spherical bodies. They are bounded not by a typical cell wall but by a membrane (p. 156) and break up readily when touched but may be examined *in situ* by dark ground microscopy from fluid cultures or in impression smears stained by Giemsa. Rickettsias

which are also pleomorphic are in contrast bounded by a typical bacterial cell wall. They are very small spheres, rods or filaments, the smallest dimension of which is about 0.2 μm.

Another type of pleomorphism occurring among organisms not normally showing large variations in shape may be due to the action of autolytic enzymes. These are normally controlled in actively metabolizing cells but may be responsible for the frequent occurrence in old cultures of organisms of abnormal morphology. Such forms are characteristic of certain bacterial species, e.g. *Pasteurella pestis* and *Neisseria* species and are known as involution forms. These are mostly dead, but those still viable give rise to normal cells again when introduced into a favourable medium.

GRAM'S STAINING METHOD

The majority of bacteria stain readily with aniline dyes. Their differential ability to retain the stain during subsequent washing is of great diagnostic value. Gram's method is but one important example of a procedure based on this ability and because of its central importance in bacterial classification is considered here in detail; other important staining methods are described in appropriate sections of the text.

In 1884 Christian Gram developed his method for differentiating bacteria in tissue sections. Bacteria are subdivided by their reaction to this stain into those which retain it, termed Gram-positive, and those which are decolourized termed Gram-negative.

The organisms are stained with a basic dye of the triphenylmethane group (e.g. crystal violet) at slightly alkaline pH, followed by mordanting, usually with iodine in a solution of potassium iodide, but sometimes with other agents such as picric acid. The crystal violet combines with the iodine to form a complex. At this stage of the process there is no difference in appearance between Gram-positive and Gram-negative organisms, the amount of crystal violet and iodine taken up being similar. Subsequent washing of the stained preparation with a neutral (organic) solvent (usually ethanol or acetone) causes the crystal violet–iodine complex to be eluted from the Gram-negative species whilst the Gram-positive organisms remain fully stained. The difference is rendered more striking by counter-staining with another dye of contrasting colour (e.g. dilute fuchsin).

The main difference in behaviour between Gram-positive and Gram-negative organisms is thus seen to reside in the ease with which the Gram complex can be eluted. This may be due either to its being bound to wall components present in Gram-positive organisms but absent from the Gram-negative ones or to differences in the permeability of the cell wall of the two types.

Fine structure

Many of the staining techniques used in bacteriology reveal little of the internal structure of the organisms. Indeed some almost completely obscure anything but gross morphology. For this reason many special staining techniques, such as the Feulgen method for nuclear bodies, the malachite green method for endospores and the Fontana method for flagella, have been devised to reveal detail of particular structures. The use of phase contrast and dark ground microscopy has led to the acquisition of further information but by far the most powerful

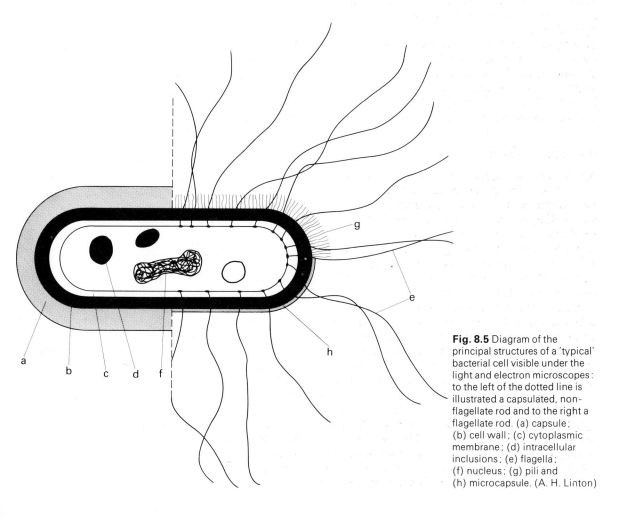

Fig. 8.5 Diagram of the principal structures of a 'typical' bacterial cell visible under the light and electron microscopes: to the left of the dotted line is illustrated a capsulated, non-flagellate rod and to the right a flagellate rod. (a) capsule; (b) cell wall; (c) cytoplasmic membrane; (d) intracellular inclusions; (e) flagella; (f) nucleus; (g) pili and (h) microcapsule. (A. H. Linton)

tool for studying bacterial anatomy is the electron microscope in which magnification of 250 000 diameters can easily be achieved.

In Fig. 8.5 the principal structures of a 'typical' bacterial cell are shown. Not all of these will occur in all organisms and some—flagella, pili and capsules, for example—will be present only under certain conditions. Some bacteria will additionally sometimes or always have other structures.

FLAGELLA

Not all bacteria are motile but motility, when it does occur, is effected in one of three principal ways. The

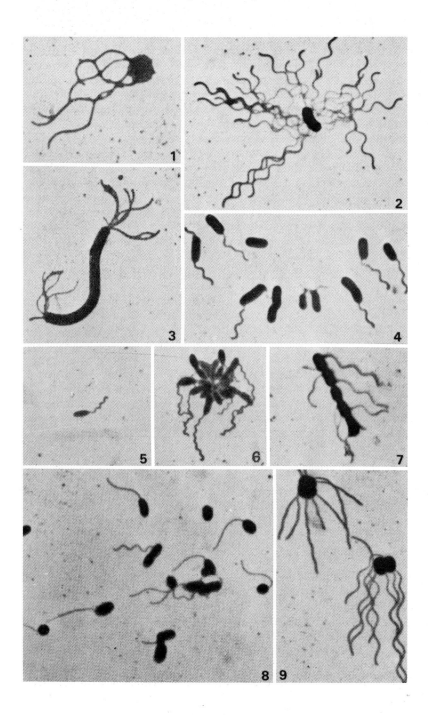

Fig. 8.6 The shape and arrangement of bacterial flagella in stained preparations.
(1) Protoplast of *Bacillus megaterium* (×3000) showing attached flagella after the cell wall has been dissolved away. (Weibull, C. (1953) *J. Bact.*, **66**, 688) (2) *Clostridium parabotulinum* (×2000) with peritrichous flagella.
(3) *Spirillum* sp. (×2000) with lophotrichous flagella.
(4) *Pseudomonas aeruginosa* (×2000) with single polar flagellum. (Leifson, E. (1951) *J. Bact.*, **62**, 377) (5) *Cellvibrio* sp. (×2000) with single polar flagellum. (6) *Caulobacter vibrioides* (×2000) showing a typical rosette of organisms each with a single flagellum.
(7) *Streptococcus* sp. (×2000) showing a flagellated chain of cocci; flagella are rare in this genus. (8) Marine organism of unknown identity (×2000) with single polar flagellum.
(9) *Sarcina* sp. (×2000). One cell, shows waves of large amplitude, the other waves of smaller amplitude. (2–9, E. Leifson)

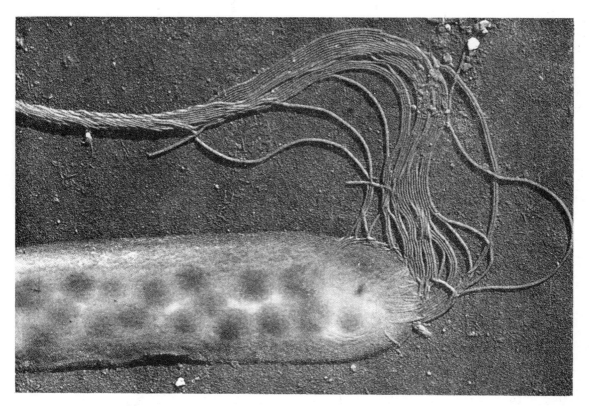

Fig. 8.7 *Spirillum* sp. (×0000). Electron micrograph showing polar flagella *apparently* originating in the cytoplasm, intertwined to form a thick bundle. The dark areas may be vacuoles. (Houwink, A. L. (1953) *Biochem. Biophys. Acta.*, **10**, 360)

most common is by means of flagella (singular flagellum; from Latin meaning 'whip'). Flagella are unbranched, helical filaments of uniform thickness, about 20 nm, throughout their length which may be many times greater than that of the bacterium (Fig. 8.6). Their size is too small for them to be seen in their natural state under the light microscope but considerable thickening by means of metal salt deposition or special staining techniques, such as those using fluorescent dyes, renders them visible (Fig. 8.6). Their motion can be observed by dark field microscopy or indirectly by watching the agitation of indian ink particles in the vicinity of flagellate bacteria mounted in soft agar to slow down their rate of movement. Dark field examinations can be greatly improved by suspending organisms in 0.5 % methyl cellulose which not only reduces the rate of movement but also induces the aggregation of the flagella in bundles sometimes sufficiently large to be seen by phase contrast microscopy. In stained preparations the wavelength and amplitude of each wave of an individual

flagellum can be determined (Fig. 8.6). Flagella of different wavelength may, unusually, be present on the same organism.

The presence or absence of flagella and their arrangement and number is a species characteristic. A single bacterium may possess from one to a hundred flagella. For example, vibrios have a single flagellum (monotrichate—from Greek trichos meaning 'hair') (Fig. 8.4, i), *Spirillum* a tuft of flagella at one or both ends (lophotrichate) (Fig. 8.4) and *Clostridium parabotulinum*, many flagella distributed around the cell (peritrichate) (Fig. 8.6, 2). Other rods may have a single flagellum or a tuft at or near one or both poles. Only a few spherical bacteria are flagellate (Fig. 8.6, 9).

It has been obvious for some time that flagella probably originate inside the cell wall, because organisms from which the cell wall has been removed (i.e. protoplasts, p. 155) retain their flagella (Fig. 8.6, 1) and electron micrographs show the proximal ends apparently well within the cytoplasm (Fig. 8.7). Until recently it has been thought that flagella originate in a vesicle in the cytoplasm. Now these are regarded as artifacts and the current view is that the basal structure is partly in the cytoplasmic membrane and partly in the wall. The basal structure is complex and different in Gram-positive

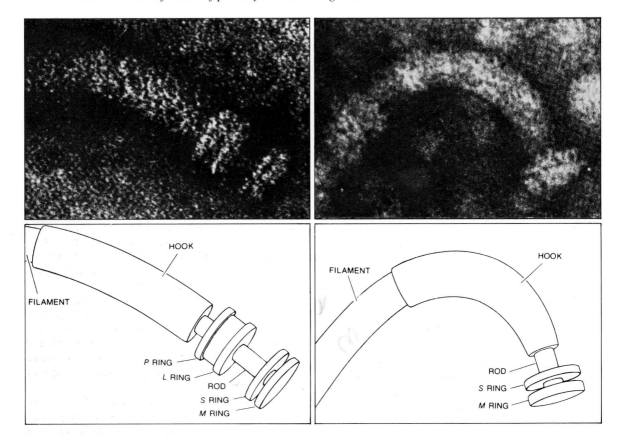

Fig. 8.8 Hooks and basal structures of bacterial flagella. The elements seen in the micrographs are identified in the drawings. In *E. coli*, a Gram-negative bacterium (*left*), four rings are mounted on a rod attached to the proximal end of the hook. The *M* ring binds to a preparation of the inner (cytoplasmic) membrane of *E. coli* and the *L* ring binds to a preparation of the outer (lipopolysaccharide) membrane of the cell wall. In gram-positive bacteria such as *Bacillus subtilis* (*right*) the *L* and *P* rings are not present; the *M* ring binds to the cytoplasmic membrane. (× 9000000). Micrographs: M. L. DePamphilis. Drawings: H. C. Berg. From *How Bacteria Swim*, H. C. Berg. Copyright © 1975 by Scientific American Inc. All rights reserved.

and Gram-negative cells (Fig. 8.8) and interpretable in terms of the differences in the structure of the cell walls of these two categories of organism (p. 149). Outside the wall is a short length of filament termed the hook (about 0.05 μm long), somewhat thicker than the major part (Fig. 8.8) and different in its chemistry as evidenced by its greater resistance to chemical digestion. The remainder, and greater part, of the filament can be seen in electron micrographs to be constructed of bead-like subunits which are usually helically arranged but in some regions may be arrayed longitudinally (Fig. 8.9). In some instances a sheath may surround the flagel-lum. This is evidently not membranous, although in some Gram-negative bacteria it appears to be a continuation of the LPS layer (p. 150). In other instances it is highly structured. Its function is unknown.

The bulk of flagellar filaments can be separated from cells by vigorous mechanical treatment and pure preparations obtained by differential centrifugation. These retain their helical configuration and when disaggregated into their component subunits by acid treatment will, on neutralization and seeding, reaggregate to reform identical structures. Analysis of the components show that more than 98 % of the total weight is protein which characteristically has a high acidic amino acid content, few aromatic amino acid residues and no cysteine. Its molecular weight is consistent with each bead visible in electron-micrographs corresponding to a single molecule of this protein, called flagellin. The flagellar proteins constitute the diagnostically important H-antigens.

Flagellin is synthesized within the cell and moves out through the hollow central core of the organelle to its tip, where self-assembly occurs. Growth of flagella takes place very rapidly, at a rate of about

0.5 μm per minute, full length being attained in 10–20 minutes.

There is now no doubt that in flagellate organisms flagella are responsible for movement. The speeds attained vary considerably, for example, 2 μm per second has been recorded for peritrichous flagellate cells and 200 μm per second for *Vibrio cholera* which is monotrichous. Directly flagella cease to beat, swimming bacteria come immediately to a stop because of the viscous forces acting upon them.

The means by which flagella cause bacterial movement was until recently uncertain. It was possible that by sequential alteration of the configuration of the component flagellin molecules a wave was propagated in the flagellum. Alternatively, and seemingly more unlikely because of the mechanical problems implicit in it, is the possibility that the whole helical structure rotated. It is not possible to distinguish between these two possibilities by direct observation of unaltered flagella but there is now good evidence that flagella rotate. It seems that the basal structure is analogous to an electric motor with the S and M rings common to both Gram-positive and -negative organisms (Fig. 8.8) acting as the stator and rotor respectively.

A less common means of locomotion is that exhibited by the spirochaetes. These helically shaped organisms (Fig. 8.3, v) move due to flexions of the cell caused by axial fibres running along the long axis of the cell but within the flexible outer envelope. In cross section it can be seen that there are two rows of fibres (Fig. 8.10), one of which is anchored near each end of the cell and which overlap and interdigitate in the mid region of the cell (Fig. 8.11). These axial filaments when isolated can be seen to have a structure identical with flagella of Gram-positive cells (Fig. 8.8) but it is not clear whether changes in the configuration of the cell surface, sometimes to the extent of forming coils, causing the cell to move, is caused by flagellar rotation.

The gliding bacteria also have a flexible cell wall and move as a result of its bending (Fig. 8.12). In this case though, a solid/liquid or a liquid/gas interface is essential for movement to occur. These organisms move relatively slowly compared to flagellate bacteria. This bending may be caused by contraction of the fibrils to be seen in close association with the cell envelope of some but not all gliding organisms. Some form of attachment to the surface over which the bacterium is moving is necessary for gliding and in *Flexibacter polymorphus* this is achieved by fibres, probably polysaccharide in nature, passing from the inner layers of the cell envelope through the wall and emerging through pores in cup-shaped elements arranged on the outer surface of the outer membrane (Fig. 8.12).

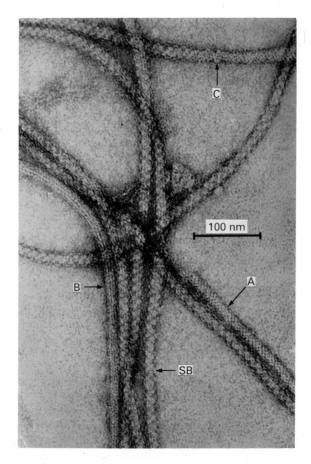

Fig. 8.9 Molecular architecture of flagella. Bacterial flagella are built of globular flagellin molecules. These are usually helically arranged (**A**), but longitudinal arrangements (**B**), and intermediate stages (**C**) may occur. **SB** is a sheathed region. (*Pseudomonas rhodos*. Lowy, J. and Hanson, J. (1965) *J. Mol. Biol.*, **11**, 293)

PILI

Pili (from Latin for hair) or fimbriae (from Latin for fringe) are very fine, straight filamentous appendages much smaller than flagella (Fig. 8.13), and are found only on certain freshly isolated Gram-negative bacteria. They measure less than 10 nm in diameter and usually less than one μm long. Pili are best demonstrated by electron microscopy in metal-shadowed specimens. The number per cell varies between one and 400 and that there may be one or

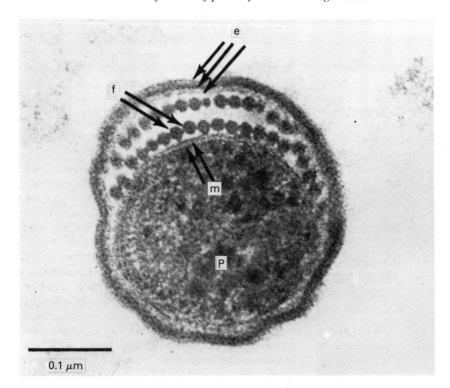

0.1 μm

Fig. 8.10 Cross section of the mid region of a spirochaete showing the two series of fibres (f) between the cell wall (e) and the cytoplasmic membrane (m). p=protoplast (Listgarten, M. A. (1964) *J. Bact.*, **88**, 1093)

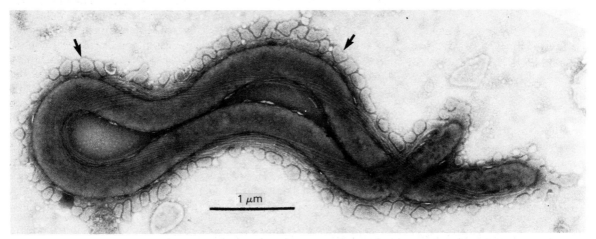

1 μm

Fig. 8.11 Interdigitation of fibres in the mid-region of a spirochaete. The outer envelope is disrupted and appears as globules (arrowed). (Listgarten, M. A. (1964) *J. Bact.*, **88**, 1091)

more morphological types present, at least ten have been recognized. These are designated by number in order of discovery or named according to their function. Pili consist of protein subunits, of molecular weight about 17000 and, like flagella, may be disaggregated into these components which, under the right conditions, reassemble to form structures indistinguishable from the original pili.

Sex pili have been studied most fully and are required for transfer of DNA between donor and recipient cells in conjugation. In some strains conjugating pairs are contiguous and, since it has been shown that phage adsorbed onto the tip of pili may later be located on the cell surface, it has been suggested that pili may, in these instances, act as retractile organelles bringing the two cells into contact. In other strains the mating pairs, though linked by pili, remain distant to one another and

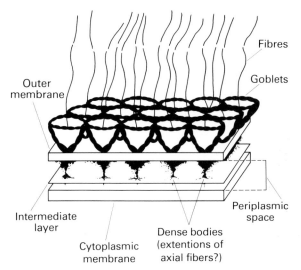

Fig. 8.12 Fine structure of cell envelope layers of *Flexibacter polymorphus*. (Ridgway, H. F. (1975) *Canad. J. Microbiol.*, **21**, 1748)

Labels: Outer membrane; Fibres; Goblets; Intermediate layer; Cytoplasmic membrane; Dense bodies (extentions of axial fibers?); Periplasmic space

recognized by their ability to haemagglutinate red blood cells from suspensions poured over them. It has also been shown that isolated pili adhere to epithelial cells and it is thought that the possession of pili is one factor contributing to the adhesion of *Neisseria gonorrhoeae* to the urethra wall, a property of importance in its pathogenicity.

CAPSULES, SLIME LAYERS AND SHEATHS

Many bacteria produce a non-living secretion of viscid material around the external surface of their cell wall. According to the amount of material produced and the degree of its association with the cell one of three terms may be applied, but it is not possible to define exactly the border line between them. If the material is closely adherent to the cell and detectable only with difficulty it is known as a *microcapsule*; if equally sharply defined but extensive and readily visible using appropriate techniques (see below) it is called a *capsule*; if copious in quantity and only relatively loosely associated with the cell it is referred to as a *slime layer*.

Few methods of capsule staining are of general application and probably the best way of visualizing them is to mount the organisms in indian ink—so-called negative staining (Fig. 8.15a). Capsules may also be rendered visible by treatment with specific antiserum which increases their refractivity (e.g. the 'Quellung' reaction, Fig. 8.15b). Examination by electron microscopy usually reveals no detail but osmium tetroxide fixed, ruthenium red stained, thin sections show that *Streptococcus pneumoniae* and *Klebsiella pneumoniae* capsules are composed of very fine fibrils which in the former are woven as a mat (Fig. 8.15c) and in the latter are arranged parallel to each other and normal to the bacterial surface (Fig. 8.15d).

here it is thought that DNA is transferred either down a hollow core, which electron micrographs suggests may be present, or by association with surface pilin. Usually one to four sex pili occur on each cell corresponding to the number of sex-factors in the cell. Sex pili act as receptor sites for male specific phages which enables them to be recognized easily in electron micrographs and to be distinguished from other pilus types (Fig. 8.14). The vesicle seen in this figure once thought to be distal, is now established to be at the proximal end of the pilus and probably to represent a membrane fragment.

Apart from the function of sex pili in DNA transfer, other types of pili cause bacteria to adhere both to one another and to foreign cells, such as red blood cells. Piliate colonies on agar plates may be

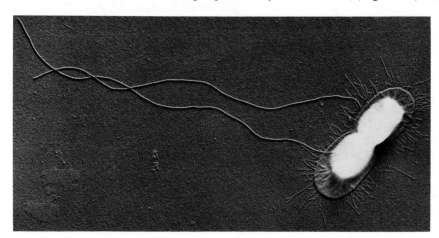

Fig. 8.13 Two flagella and about 200 pili on a dividing cell of *Salmonella anatum*. (Duguid, J. P. (1966) *J. Path. Bact.*, **92**, 107)

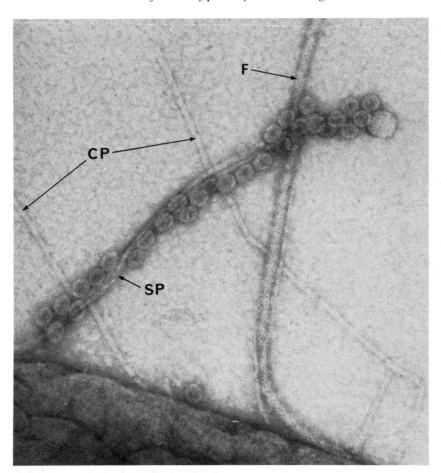

Fig. 8.14 Electron micrograph of *Escherichia coli* appendages. SP=sex pilus to which RNA phage MS2 particles are attached. CP=common pilus. F=flagellum (×000 000).

Capsules are commonly carbohydrate, and the simplest of this type, homopolysaccharides, contain only one monomer. *Acetobacter xylinum*, for example, has a capsule of cellulose (Fig. 8.16) which has only one bond type (β 1–4 linked glucose). *Leuconostoc mesenteroides* capsule also contains only one sugar but although the principal bond type is α 1–6, α 1–4 and α 1–3 linkages also occur and the structure becomes quite complex chemically (Fig. 1.3). Many bacteria produce heteropolysaccharide capsules, some of which may be chemically fairly simple, containing two sugar types and two types of glycosidic linkage as in *Steptococcus pneumoniae* type III (β 1–3 linked cellobiuronic acid residues—Fig. 1.2) and *Streptococcus* species (hyaluronic acid—Fig. 8.17). Others, such as *Streptococcus pneumoniae* type IV, which has three sugars and two bond types, form relatively complex branched carbohydrates (Fig. 1.5).

A few organisms produce polypeptide and protein capsules. *Bacillus anthracis* produces a polymer of the unusual amino acid, D-glutamic acid, linked in the 3 position (Fig. 8.18), and *Pasteurella pestis* synthesises a capsular protein. Certain halophilic organisms, when grown in suboptimal salt concentrations, even produce a DNA slime. In some instances capsules may be composed of more than one type of polymer. Thus in *Streptococcus salivarius* the exopolysaccharide consists of a levan (poly β 2–6 fructose) and another antigenically distinct carbohydrate. *Bacillus megaterium*, another example, when grown in medium rich in carbohydrate and nitrogen, produces a highly ordered capsule with polysaccharide at the poles of the cell and polypeptide in the remaining zones. When carbohydrate is absent from the medium the entire exopolymer is polypeptide (Fig. 8.19a–c). This also illustrates the effect of certain environmental conditions on capsular synthesis; broadly speaking carbohydrate capsules are produced when the medium C/N ratio is high and within certain limits in some organisms (e.g. *Streptococcus salivarius*) there is proportionality between the amount of polymer (levan) produced and the concentration of

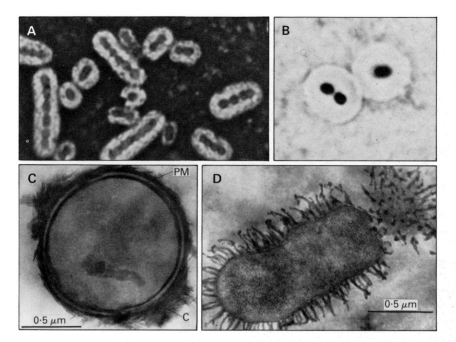

$$\left[\begin{array}{c} \text{CH}_2\text{OH} \\ \text{OH} \\ \text{OH} \end{array} \quad \begin{array}{c} \text{CH}_2\text{OH} \\ \text{OH} \\ \text{OH} \end{array} \right]_n$$

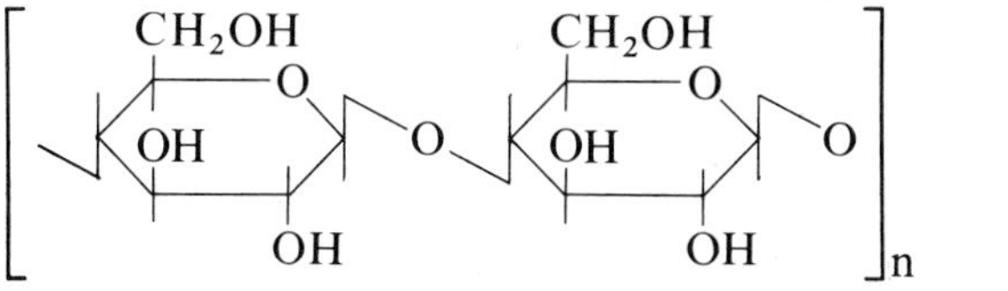

Fig. 8.16 Repeating unit of cellulose as found in *Acetobacter xylinum* capsules.

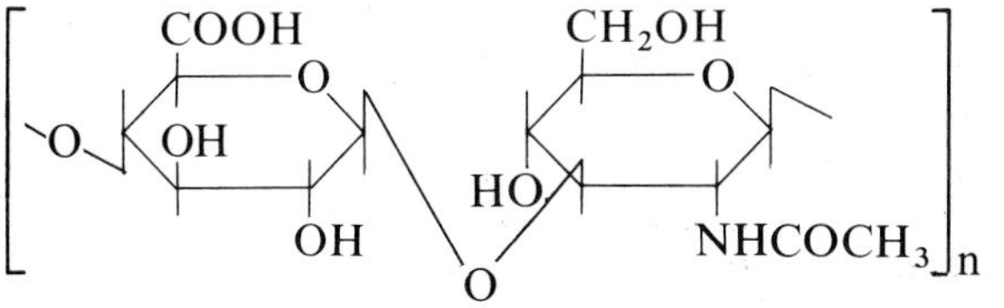

$$\left[\begin{array}{c} \text{COOH} \\ \text{OH} \\ \text{OH} \end{array} \quad \begin{array}{c} \text{CH}_2\text{OH} \\ \text{HO} \\ \text{NHCOCH}_3 \end{array} \right]_n$$

Fig. 8.17 Repeating unit found in hyaluronic acid capsule of *Streptococcus*.

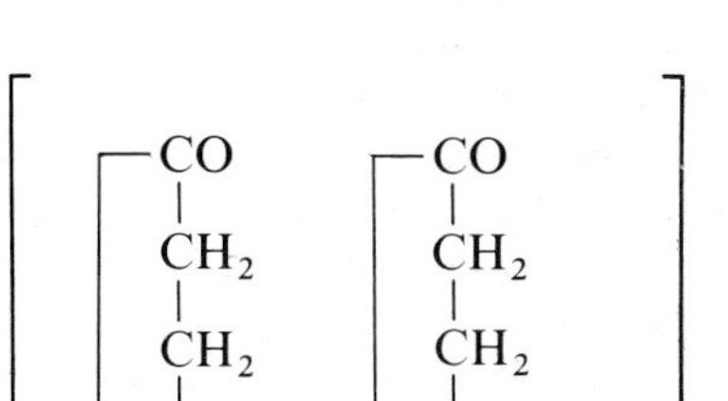

$$\left[\begin{array}{c} \text{CO} \\ | \\ \text{CH}_2 \\ | \\ \text{CH}_2 \\ | \\ \text{CHNH} \\ | \\ \text{COOH} \end{array} \quad \begin{array}{c} \text{CO} \\ | \\ \text{CH}_2 \\ | \\ \text{CH}_2 \\ | \\ \text{CHNH} - \\ | \\ \text{COOH} \end{array} \right]_n$$

Fig. 8.18 Dipeptide fragment of *Bacillus anthracis* poly-D-glutamic acid capsule.

Fig. 8.15 Bacterial capsules. (**A**) Smear of capsulated pneumococci of an uncentrifuged spinal fluid from a fatal case of meningitis. The capsules are demonstrated by negative staining using indian ink. (× 3600). C. F. Robinow).
(**B**) Capsule swelling test performed on sputum from a pneumonic patient (× 3600). (C. F. Robinow).
(**C**) *Diplococcus pneumoniae* stained with ruthenium red, dehydrated with alcohol and embedded in Epon 821. The mat-like capsule (C) surrounds each bacterium. Ruthenium red penetration is evident at the plasma membrane (PM) and into the cytoplasm.
(**D**) *Klebsiella pneumoniae* stained with ruthenium red, dehydrated with Darcupan and embedded in Araldite 502. Capsular fibrils (C) are seen at regular intervals along the wall. (Springer, E. L. (1973) *J. gen. Microbiol.*, **74**, 21)

sugar (sucrose) present. Conversely low C/N ratios favour synthesis of polypeptide capsules.

At the chemical level, exopolysacharide synthesis is usually *via* UDP-sugar derivatives but levan and dextran apparently are exceptional in being formed by transglycosylation from sucrose or similar oligosaccharides.

The role of capsules is uncertain. They are not essential structures because (1) they are not synthesized under all environmental conditions, (2) mutants exist which have lost the ability to produce them and (3) cells from which they have been removed by enzymic digestion remain viable. The presence of a capsule, however, is often associated with the virulence of pathogenic organisms (p. 310). When capsulate organisms enter the animal body they are

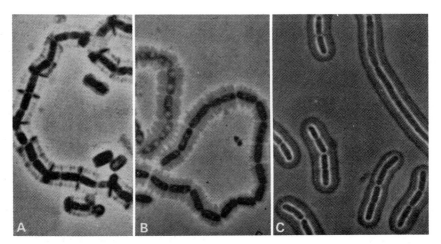

Fig. 8.19 Phase-contrast of *Bacillus megaterium* (× 2500) showing the influence of nutrition on the nature of the capsule. (**A**) Culture grown in carbohydrate-rich broth and treated with polysaccharide antibody. The polysaccharide of the capsule with which the antibody reacts, is shown to be concentrated at the ends of the organisms and at transverse septa across the capsular region corresponding to the points of cell division. (**B**) Culture grown in carbohydrate-rich broth and treated with polypeptide antibody. This reacts with polypeptide capsular material and shows that this is distributed uniformly throughout the capsule except for the limited spaces occupied by polysaccharide. (**C**) Culture grown in a carbohydrate-free medium and treated with polypeptide antibody. The uniform distribution of polypeptide materials throughout the capsule indicates the absence of polysaccharide under these conditions. (Tomcsik, J. (1956) *Ergebn. med. Grundlag. forsch.*, **1**)

able to resist phagocytosis by the white cells of the blood, i.e. they are not readily engulfed by these scavenger cells or, if they do become engulfed, they resist digestion by the intracellular enzymes and continue to multiply. Loss of capsule, not only leads to loss of virulence but is often accompanied by other changes, such as alteration of the colony structure on the surface of solid media. Capsulation is only one of many factors involved in the pathogenicity of the organism.

Other hypothetical functions involving protection are the prevention of bacteriophage attack and of dessication. It is also suggested that capsules may act as ion exchangers, adsorbing nutrients and subsequently releasing them for use by the cell.

Capsule formation may be responsible for con-

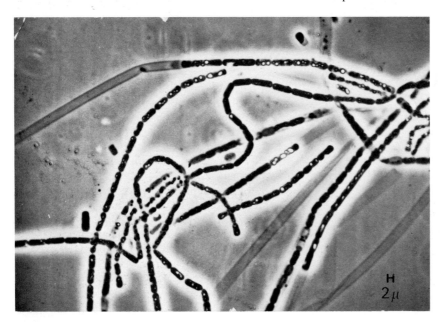

Fig. 8.20 *Sphaerotilus natans*, grown on 0.1 per cent glucose and 0.1 per cent peptone. Empty sheaths and inclusion globules of poly-β-hydroxybutyric acid can be seen (×1620). (Mulder, E. G. (1964) *J. appl. Bact.*, **27**, 151)

siderable economic loss in dairy and other food industries (p. 347). Carbohydrate-containing materials become 'ropy' when encapsulated organisms grow in them. However, some organisms, e.g. *Leuconostoc* species, are employed commercially for the production of dextran as a plasma 'extender' used in the treatment of shock resulting from blood loss, and in the manufacture of chemically cross-linked polymers used in molecular exclusion chromatography.

The chemistry of the sheaths of organisms such as *Sphaerotilus natans* (Fig. 8.20) has not been established but proteins, carbohydrates and probably lipids are present.

CELL WALLS AND APPENDAGES

The cell wall is typically a rigid structure which gives shape to the bacterium (Fig. 8.21a). This can be demonstrated by plasmolysis (Fig. 8.21b), by electron microscopy of cell wall fragments (which retain their shape when freed of their contents) and by the fact that the cellular contents assume a spherical shape when the cell wall is enzymically dissolved (Fig. 8.21c).

Pure preparations of bacterial cell walls can be obtained by mechanical disintegration of the organism, followed by differential centrifugation. Autolysis, osmotic lysis and heat shock have also been used for some cell wall preparations. Chemical and enzymic methods are not used now because

these alter the chemical composition of the wall. Traces of contaminating protein and nucleic acids are removed by washing or by treatment with appropriate enzymes such as trypsin or ribonuclease. The preparations may be assessed for purity by electron microscopy or by chemical analysis. Typical compositions are shown in Table 8.1.

The principal structural polymer of both Gram-positive and Gram-negative bacteria is peptido-glycan (mucopeptide; murein).

The broad similarity in structure of mucopeptides from Gram-positive and Gram-negative cells contrasts with the marked difference in quantity of this compound in cell walls of these two groups. In the former up to 95% of the wall may be mucopeptide and it is arranged in several layers. In the latter as little as 5% is found and, from thickness measurements of the mucopeptide region in elec-

Fig. 8.21 Bacterial cell walls and associated structures. (**A**) *Bacillus megaterium* stained with Victoria blue to reveal the cell walls and cross-wall septa. The cytoplasm is everywhere in close contact with the cell wall and the latter cannot be distinguished from the cytoplasmic membrane. (C. F. Robinow). (**B**) *Bacillus megaterium* plasmolyzed in 5 per cent KNO₃ and stained as in (**A**). The cytoplasm has separated from the cell walls thus revealing their rigid nature. (Robinow, C. F. (1953) *Exp. Cell Res.*, **4**, 392). (**C**) Bacterial protoplasts. Phase contrast photomicrographs of protoplasts of *Bacillus megaterium*. The cell walls have been dissolved away by lysozyme in the presence of an osmotic stabilizer (sucrose) leaving the living protoplasts which assume a spherical shape. (K. McQuillen)

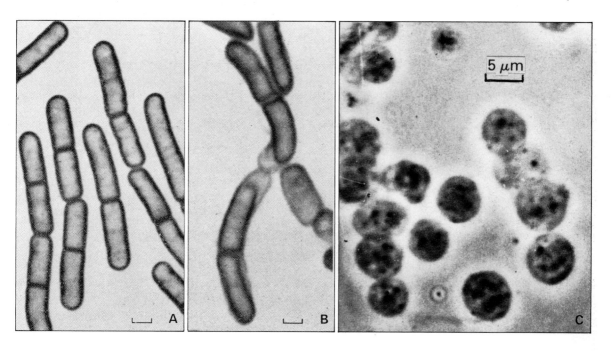

Table 8.1 Typical chemical compositions of cell walls of Gram-positive and Gram-negative bacteria

	Gram-positive	Gram-negative
Monomers		
Amino acids	~4	~18
Amino sugars	N- acetylglucoamine and N-acetylmuramic acid	
Other sugars	Few common types	Few 'exotic' types
Lipids	<2%	<20%
Polymers		
Mucopeptide	+	+
Teichoic acid or teichuronic acid	+	–
Protein	In some instances	+
Lipopolysaccharide	–	+
Lipoprotein	–	+

tron micrographs of thin sections, it is seen to be a single layer. Substantial differences also occur in the accessory polymer found in the two groups (see below).

Electron micrographs of thin sections of Gram-negative bacteria may be interpreted as illustrated in Fig. 8.22. Surrounding the cytoplasmic membrane and in intimate contact with it, is the thin mucopeptide layer. The spaces within its meshwork and immediately outside it, constitute the periplasmic region which contains for example, enzymes such as alkaline phosphatase and macromolecules in transit from their site of synthesis to the outer layers of the cell wall which is membranous in character and largely hydrophobic in nature. Typically it consists of 25% phospholipid, 25–30% lipopolysaccharide (LPS) and 45–50% protein. The lipid is mainly phosphatidylethanolamine with smaller amounts of phosphatidylglycerol and a little diphosphatidylglycerol (Fig. 1.6.). The protein can

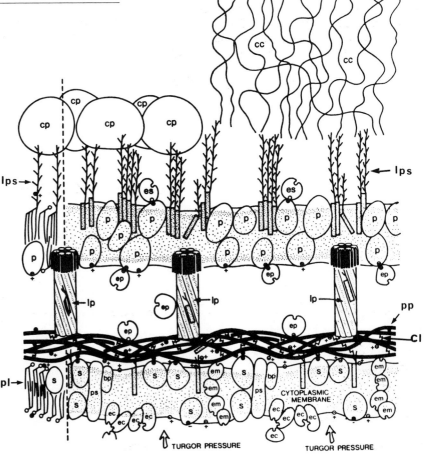

Fig. 8.22 Diagrammatic representation of the Gram-negative cell wall. +=free cation, −=free anion, ⊕=bound cation, ⊖=bound anion, cl=peptidoglycan peptide cross link, pp=peptidoglycan polysaccharide, pl=phospholipid, lps=lipopolysaccharide, cc=capsular carbohydrate, cp=capsular protein, es=enzyme localized at the cell surface, p=outer membrane protein, ep=periplasmic enzyme, lp=lipoprotein, s=cytoplasmic membrane protein, em=cytoplasmic membrane enzymes involved in synthesis of components of the surface structure, ps=permeases, ec=cytoplasmic membrane enzymes whose activity is directed towards the cytoplasm. (Costerton, J. W. (1975) *J. Antibiot. Chemother.*, **1**, 363)

be separated electrophoretically into only four or five bands, two of these at least are glycoproteins and one lipoprotein. The components of this layer are arranged mainly as a unit membrane (p. 155) with the lipoprotein molecules covalently linked to mucopeptide through their protein moieties, bridging the periplasmic space. In some instances LPS (Fig. 8.23) which is located on the outer surface with its hydrophobic lipid A (Fig. 8.24) and forming an integral part of the membrane and its polar polysaccharide projecting into the environment surrounding the cell, may also be linked covalently to protein. The LPS polytetrasaccharide terminus (Fig. 8.25) is identical with the O-antigen, important in the serological identification of certain Gram-negative bacteria. Lipid A is the component responsible for endotoxic properties (p. 311).

The walls of Gram-positive bacteria are often amorphous (Fig. 8.32), although all those so far examined contain, in addition to mucopeptide, teichoic acids. These in many instances represent a major component of the wall accounting for up to 10% of dry weight of the cell. Teichoic acids are anionic polymers which always contain either glycerol (Fig. 8.26) or ribitol (Fig. 8.27) usually combined with sugar and D-alanine and in which sugar residues may form an integral part of the polymer chain. It is thought that one molecule of teichoic acid is covalently linked, through a phosphodiester

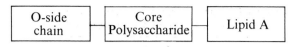

Fig. 8.23 Components of lipopolysaccharide.

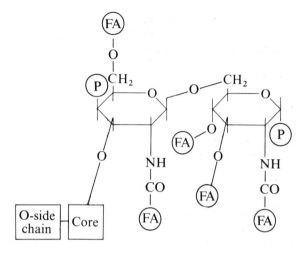

Fig. 8.24 Structure of lipid A. (FA)=fatty acid residues, e.g., lauric, myristic, hydroxymyristic and palmitic acid; (P)=phosphate.

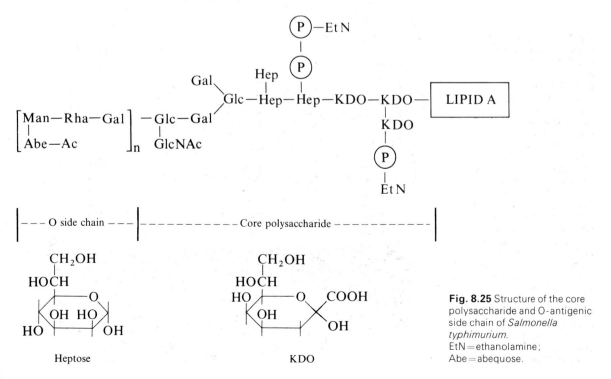

Fig. 8.25 Structure of the core polysaccharide and O-antigenic side chain of *Salmonella typhimurium*. EtN=ethanolamine; Abe=abequose.

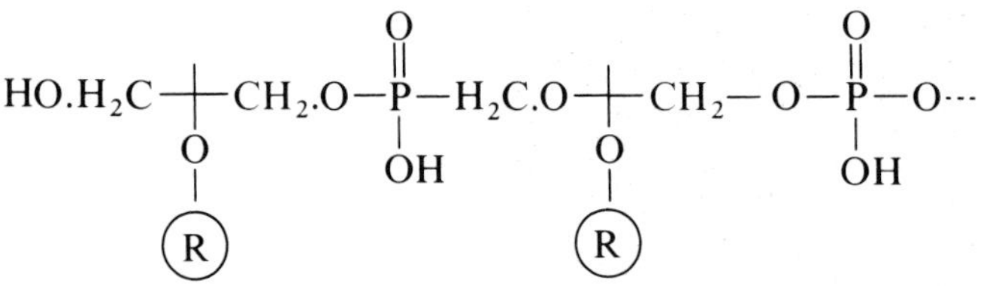

$$\text{HO.H}_2\text{C} - \overset{\underset{\underset{\text{(R)}}{|}}{\text{O}}}{\underset{|}{\text{CH}_2}} \text{O} - \overset{\overset{\text{O}}{\|}}{\underset{\text{OH}}{\text{P}}} - \text{H}_2\text{C.O} - \overset{\underset{\underset{\text{(R)}}{|}}{\text{O}}}{\underset{|}{\text{CH}_2}} - \text{O} - \overset{\overset{\text{O}}{\|}}{\underset{\text{OH}}{\text{P}}} - \text{O} \cdots$$

Fig. 8.26 Structure of 1,3-polyglycerol phosphate teichoic acid. (R) = H or D-alanine or sugar residues.

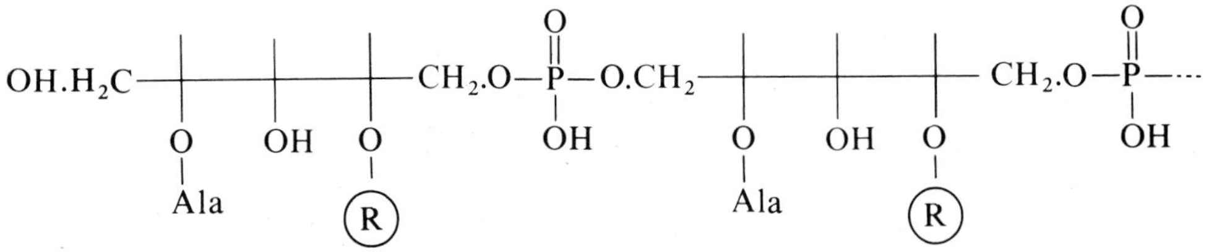

$$\text{OH.H}_2\text{C} - \overset{\underset{\underset{\text{Ala}}{|}}{\text{O}}}{|} - \overset{\underset{\text{OH}}{|}}{|} - \overset{\underset{\underset{\text{(R)}}{|}}{\text{O}}}{|} - \text{CH}_2\text{.O} - \overset{\overset{\text{O}}{\|}}{\underset{\text{OH}}{\text{P}}} - \text{O.CH}_2 - \overset{\underset{\underset{\text{Ala}}{|}}{\text{O}}}{|} - \overset{\underset{\text{OH}}{|}}{|} - \overset{\underset{\underset{\text{(R)}}{|}}{\text{O}}}{|} - \text{CH}_2\text{.O} - \overset{\overset{\text{O}}{\|}}{\underset{\text{OH}}{\text{P}}} - \cdots$$

Fig. 8.27 Structure of 1,5-polyribitol phosphate teichoic acid. (R) = sugar residues.

bridge to one muramic acid residue in each mucu-peptide glycan chain. Some evidence indicates that teichoic acid is surface located but since a radial arrangement of glycan chains can be excluded (p. 14) and since a substantial proportion of glycan chains are substituted with teichoic acid, it is probable that only some of these molecules are entirely surface located. The remainder will, at least partly, be interwoven with mucopeptide in the inner wall region (Fig. 8.28). In some organisms, according to environmental conditions, teichuronic acid is synthesized alternatively to teichoic acid. Teichoic acids may be involved in the regulation of autolytic enzymes, in the binding of cations and, especially in the case of the membrane types (p. 152), in the maintenance of the correct levels of magnesium ions for membrane stability and the activity of certain membrane-associated enzymes.

Other examples of matrix materials found in Gram-positive bacteria are the immunologically active C polysaccharide and M protein of *Streptococcus* and the proteins of *Staphylococcus*.

The outermost layer of the walls of both Gram-negative and Gram-positive bacteria is often a regular array of subunits of greater (e.g. *Flexibacter*, Fig. 8.12) or lesser (e.g. *Spirillum*, Fig. 8.29); *Bacillus polymyxa*, Fig. 8.30) complexity.

The cell wall and prostheca (cell wall bound appendage) of the budding bacteria are apparently similar to those of typical Gram-negative bacteria. Other bacteria, such as *Gallionella*, excrete an

acellular appendage, distinct from the wall, consisting of microfibrils of organic material aggregated to form a flat twisted ribbon which is often impregnated with iron hydroxide.

Rickettsiae, although obligate intracellular parasites, also have a rigid cell wall in which mucopeptide has been identified.

CYTOPLASMIC MEMBRANES

The cytoplasmic membrane lies within the cell wall and is related to it in Gram-positive bacteria by membrane- or lipo-teichoic acid, of the GTA type (p. 152), whose fatty acid residues (Fig. 8.31) are envisaged as being intercalated with the cytoplasmic membrane lipids and whose hydrophilic portion is thought to be associated with the inner region of the wall. Under normal osmotic conditions the wall and membrane are in close contact but the protoplast can be induced to shrink away from the wall under conditions of very high osmotic pressure (plasmolysis—Fig. 8.21b).

The membrane is about 7.5 nm thick, accounts for about 10% of the cell's dry weight and consists mostly of lipid (16–29%) and protein (40–75%)

Fig. 8.28 Location of wall teichoic acid in relation to mucopeptide glycan chains.

Fig. 8.29 *Spirillum* sp. (×70000). Electron micrograph of the wall of a crushed cell. The cell wall is shown curling over at the bottom of the photograph thus revealing the outer face of the wall. The top of the photograph shows the inner face of the cell wall from which part of the inner membrane has come off exposing the inner face of the outer membrane. This consists of spherical particles arranged hexagonally in a single layer. (Houwink, A. L. (1953) *Biochim. biophys. Acta.*, **10**, 360)

with very much smaller amounts of nucleic acids and carbohydrate, which may be due partly to contamination during preparation of the membrane for analysis. Although two broad categories of protein, based on the ease with which they may be extracted from membrane fractions by solvent treatment, may be recognized, in contrast to the outer membrane of Gram-negative cell wall (p. 150), there is no evidence of occurrence of structural proteins; it is probable that all cytoplasmic membrane proteins have either catalytic or transport functions. Typically about 100 different protein species, all present in about the same quantities, may be detected in cytoplasmic membrane preparations.

The predominant lipids in cytoplasmic membranes, are phospholipids, in particular, phosphatidylglycerol and/or phosphatidylethanolamine (Fig.

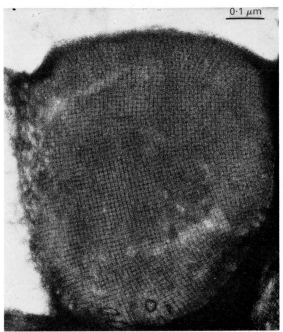

0·1 μm

Fig. 8.30 *Bacillus polymyxa* cell wall negatively stained with uranyl acetate showing rectangularly arranged globular protein units. (Murray, R. G. E. (1967) *J. Bact.*, **93**, 1509)

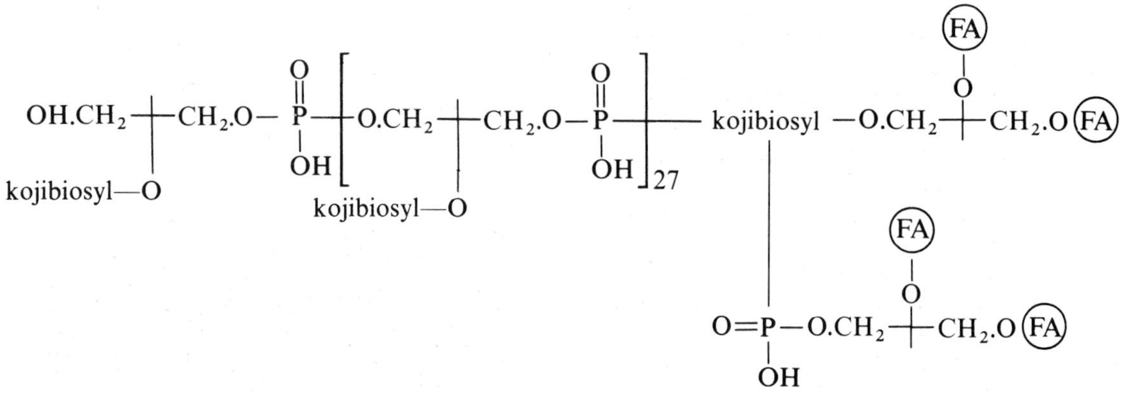

$$\text{OH.CH}_2 \overset{|}{\underset{\text{kojibiosyl—O}}{\rule{0pt}{1em}}} \text{CH}_2.\text{O} - \overset{\overset{\text{O}}{\|}}{\underset{\underset{}{\text{OH}}}{\text{P}}} \left[\text{O.CH}_2 \overset{|}{\underset{\text{kojibiosyl—O}}{\rule{0pt}{1em}}} \text{CH}_2.\text{O} - \overset{\overset{\text{O}}{\|}}{\underset{\underset{}{\text{OH}}}{\text{P}}} \right]_{27} \text{kojibiosyl}$$

Fig. 8.31 Structure of the membrane (lipoteichoic acid) extracted from *Streptococcus faecalis.*

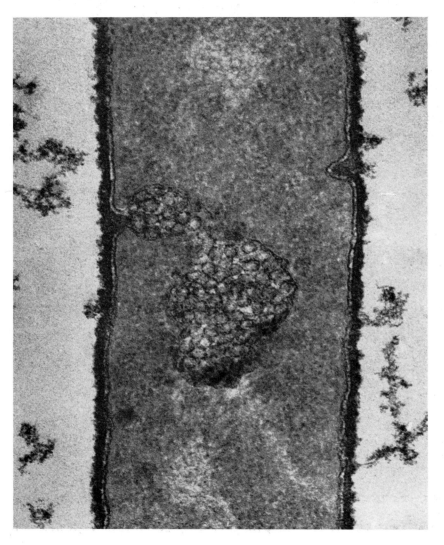

Fig. 8.32 Electron micrograph of a thin section of *Bacillus subtilis* (×170 000). The triple-layered cytoplasmic membrane is seen beneath the thicker densely stained amorphus cell wall; within the cytoplasm the pale areas are the nuclear bodies, one of which is associated with the mesosome which is continuous with the plasma membrane at the point of septum growth. (W. van Iterson (1965))

1.6). Membranes of Gram-positive bacteria characteristically contain one or both of these together with two or three more closely related compounds such as phosphatidic acid and diphosphatidylglycerol, whereas cytoplasmic membranes of Gram-negative bacteria typically contain only one or two lipids. The lipid composition is dependent on growth conditions and the fatty acid moieties are especially subject to variation although generally branched-chain, saturated types are found in Gram-positive bacteria and a mixture of saturated, unsaturated and straight chain varieties in Gram-negative bacteria. In addition, fatty acids containing a cyclopropane ring occur and lipids containing this structure, together with those having unsaturated regions in the chain and branched chains, which contribute to the fluidity of the lipid phase. In general, psychrophilic (p. 78) organisms have higher proportions of these types of lipid in their membrane than mesophiles.

In electron micrographs of fixed stained sections, the cytoplasmic membrane in most bacteria appears as two dense lines between the cytoplasm and the cell wall. This suggests that the membrane consists of three layers, outer and inner electron-dense layers each about 2 to 4 nm thick and an inner electron-transparent space about 3 to 5 nm thick (Fig. 8.32). This composite structure is called a 'unit membrane' and resembles the appearance of plasma membrane of eukaryotes (p. 176) under similar conditions. It consists of a bilayer of phospholipids in which the hydrophobic fatty acid residues are arranged perpendicularly to the plane of the membrane and form the electron transparent layer. The polar groups of the phospholipid make up the outer and inner layer and maintain upon themselves layers of protein, together forming the electron dense layers (Fig. 8.33a). Such a structure is entirely satisfactory as a basis for explanations of the permeability barrier functions of the cytoplasmic membrane but is incompatible with some of its other properties and it is now envisaged that the unit arrangement, and one in which membrane proteins exist in a lipid matrix (fluid mosaic model—Fig. 8.33b), coexist with intermediates between them. Such a dynamic structure is compatible with other functions of the membrane, such as active transport of certain solutes, synthesis and carriage of components of polymers like mucopeptide and the assembly of these outside the membrane, and electron transport.

A distinguishing feature of the prokaryotes is their lack of intracellular membrane structures. In many Gram-positive bacteria, however, invaginations of the cytoplasmic membrane occur

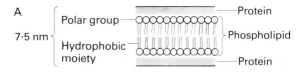

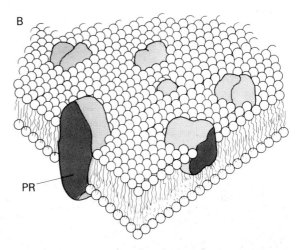

Fig. 8.33 **A**. Trilaminar 'unit' membrane model. **B**. The fluid mosaic model. PR = globular proteins randomly distributed in the membrane forming specific aggregates.

(Fig. 8.32) often of complex structure, frequently in association with a developing cross-septum. Many physiological functions have been described to these organelles called mesosomes but none have been identified certainly. There is also doubt about the different structures seen in fixed thin sections since their appearance has been shown to depend on the conditions of fixation. When protoplasts are formed from cells containing mesosome a string of bead-like vesicles may be evaginated.

The chromatophores of *Rhodospirilum rubrum*, which are vesicles about 60 nm in diameter and which contain the photosynthetic pigments of this organism, are also continuous with the cytoplasmic membrane.

Amongst the prokaryotes are found a range of types, lacking a rigid cell wall, whose outer layer is membranous in character. When the cell wall of Gram-positive bacteria is removed by treatment with lysozyme in the presence of an osmotically stabilizing agent, such as 0.2 M sucrose, the spherical product which completely lacks cell wall material is referred to as a protoplast (Fig. 8.21c). Protoplasts increase in mass in a suitable nutrient liquid medium but will not divide. When plated onto a suitable nutrient agar surface at least a proportion of the population will regenerate cell walls and divide. It is more difficult to prepare protoplasts of Gram-negative bacteria; treatment with a detergent or chelating agent, which disrupts the outer membrane, is required to allow lysozyme access to the mucopeptide. When a spherical body is produced in this way it often has residual cell wall material on its outer surface and is termed a sphaeroplast.

Some bacteria, such as *Streptobacillus moniliformis* and *Bacteroides* spontaneously form pleomorphic wall-less forms, called L-forms. Many other Gram-negative bacteria when subjected to cold shock or the action of wall-active antibiotics, for example, also produce L-forms. The first stage in their formation is the appearance of enlarged, distorted cells of the parent bacterium—the large bodies. These revert to normal vegetative rods if the inducing stimulus is removed. Continued exposure to adverse conditions leads to fragmentation to small, wall-less forms resembling *Mycoplasma* cells (see below). If these are stable they are referred to as L-forms. Most lack mucopeptide components in their outer surface structures.

Mycoplasma cells are bounded by a trilaminar membrane which, uniquely among the prokaryotes, contains sterols. In spite of their morphological similarities to L-forms this feature, together with the demonstrated lack of DNA homology between the two types (p. 47) argues against any close relationship.

NUCLEUS

Bacteria lack an organized nucleus of the type found in eukaryotes. Nevertheless, it can be seen in stained smears (Fig. 8.34) and as an area of relatively low density in electron micrographs of

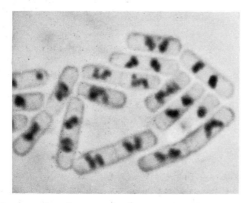

Fig. 8.34 Nuclear bodies. *Bacillus subtilis* (×3600), osmium-fixed acid hydrolysed preparation stained by a modified Feulgen stain to show chromatin bodies.

thin sections (Fig. 8.32).

Genetical evidence indicates that several bacteria at least have circular genomes and by autoradiography of the DNA, extracted by very gentle lysis from *E. coli* cells and labelled with ^{3}H-thymidine, direct evidence of this has been obtained (Fig. 8.35). Bacteria are haploid, although in rapidly growing cells nuclear division proceeds ahead of cell division (Fig. 8.34). In addition to the chromosome some bacteria possess extrachromosomal elements called plasmids (p. 41). These are non-essential to the cell and frequently code for characters such as antibiotic resistance (p. 54), colicin production (p. 109) and enterotoxin production (p. 311).

CYTOPLASMIC INCLUSIONS

All rapidly-growing bacteria contain numerous ribosomes (p. 27) in their cytoplasm. The relationship between these, the m-RNA which they are translating and the DNA which is being transcribed, can be seen in electron micrographs of the contents of gently lysed cells (Fig. 8.36a and b).

Many of the other inclusions act as reserves of energy and/or elements and their formation depends on growth conditions. Frequently such bodies consist of polymeric, osmotically inert materials which are synthesized when an element, not one of their components, becomes limiting or on the onset of other adverse environmental conditions, in the presence of an excess of an energy source.

Some of these bodies are laid down in the cytoplasm without any bounding membrane. Polyphosphate (volutin; metachromatic) granules, were used at one time in the rapid diagnosis of diphtheria (caused by *Corynebacterium diphtheriae* (Fig. 8.37)) but which are widespread in prokaryotes, are one example which lack a surrounding membrane. Some species of *Bacillus* are notable for the presence of a single, large protein crystal formed at the same time as the endospore, which is toxic for certain insects (p. 317). The crystals are not bounded by a membrane.

Polyglucoside granules in most instances are non-membrane enclosed but some deposits of this material are bounded by a membrane which may consist entirely of protein. According to their reaction with iodine solution, polyglucoside granules are often referred to as glycogen (stain red-brown) and starch (stain blue) but in only a few instances has the chemistry of the polymer been established. Other membrane enclosed deposits are poly-β-hydroxybutyrate (PHB) and sulphur. PHB granules are widely distributed among bacteria

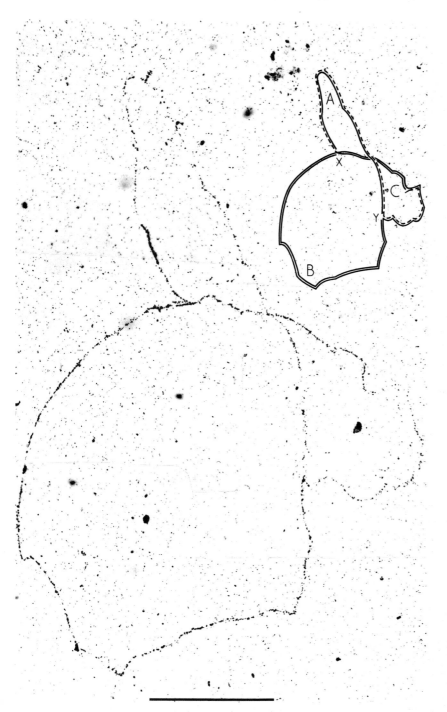

Fig. 8.35 Autoradiograph of the chromosome of *E. coli* K12 Hfr, labelled with thymidine for two generations and extracted with lysozyme. Exposure time two months. The scale shows 100 μm. Inset, the same structure is shown diagrammatically and divided into three sections (A, B and C) that arise at the two forks (X and Y). (Cairns, J. (1963) CSHSQB, **28**, 43).

(Figs. 8.20 and 38) whereas intracellular sulphur granules are found only in the purple sulphur bacteria and the colourless sulphur bacteria such as *Beggiatoa*. The green photosynthetic bacteria also deposit sulphur granules but outside the cell wall.

The photosynthetic vesicles of these organisms, unlike those of the purple photosynthetic bacteria, are delimited by a non-unit membrane. Many aquatic prokaryotes contain gas vacuoles (Fig. 8.39) whose component vesicles are bound by a membrane apparently consisting entirely of protein.

Morphogenesis and differentiation

During the life of the individual bacterium, from the moment it exists separately from the mother cell from which it is produced, until the moment when it, in turn, gives rise to a daughter cell, there is a continuous change in its morphology. This morphogenetic process in cells growing by binary fission (p. 83), under constant conditions such as are found in a chemostat (p. 92), results in the production of a population of closely similar cells. In batch cultures the size and shape of cells varies with time as the conditions change; generally rapidly growing cells in the logarithmic growth

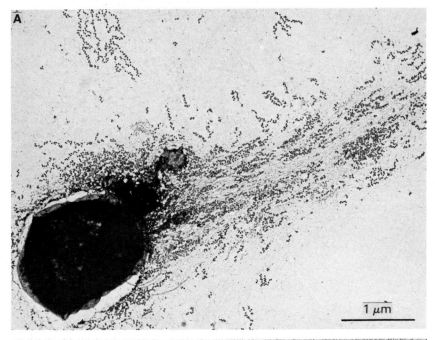

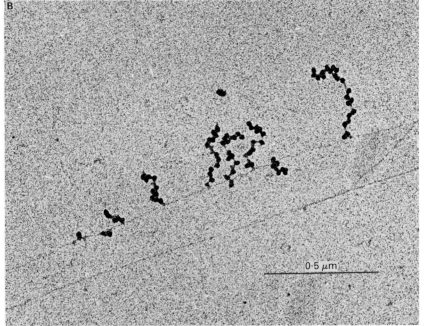

Fig. 8.36 A. Electron micrograph showing a portion of the extruded contents of an osmotically ruptured *E. coli* cell. B. Electron micrograph of an unidentified operon of the bacterium *E. coli*. The gradient of attached polyribosomes is formed by translation into protein of m-RNAs in successive stages of transcription from DNA. (Millar, O. L. (1970) *Science*, **169**, 392, Figs 1 and 3. Copyright 1970 by the American Association for the Advancement of Science.)

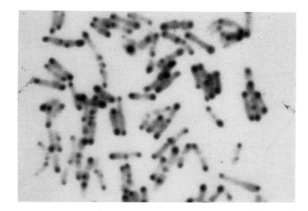

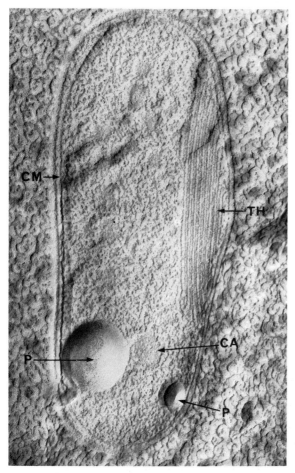

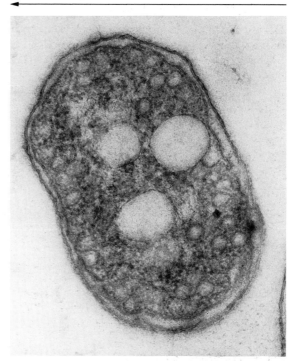

Fig. 8.39 Gas vacuoles in a thin section of *Rhodopseudomonas spheroides* (×110 000). (*Bildatlas path, Mikroorganismen, Bd.,* **111**, G. Fischer, Stuttgart)

Fig. 8.38 Inclusions in bacteria. Freeze-etching preparation of cells of *Rhodopseudomonas viridis* showing variously sized bodies of the reserve substances poly-β-hydroxybutyric acid. A thylakoid stack of lamella structures which carry the photosynthetic pigments is clearly visible. CM—cytoplasmic membrane; CA—cytoplasmic aggregate; TH—thylakoid stack; W—cell wall; P—poly-B-hydroxybutyric acid (×95 000) . (*Bildatlas path. Mikroorganismen, Bd.,* **111**, G. Fischer, Stuttgart)

phase are larger than those in the stationary phase cultures. These changes are strikingly visible in pleomorphic organisms such as *Arthrobacter* (Fig. 8.1). In the case of budding and stalked bacteria which divide asymmetrically, populations of growing cells whatever the conditions will be morphologically heterogenous.

Since the cell wall confers shape on bacteria and since the mucopeptide is the polymer responsible, it is to be expected that changes in the form of a bacterium reflect alterations to this structure. Different organisms achieve this in different ways. In *Streptococcus* virtually all the mother cell wall material is static and an entirely new area of wall is synthesized beginning at the equatorial band. In contrast, the wall material of some rod-shaped bacteria is constantly being degraded by autolysins and resynthesized and there is evidence that random insertion of new wall material occurs. Unfortunately in no instance are the chemical changes in the mucopeptide clear. In *Arthrobacter globiformis* it is possible that the mucopeptide glycan chain is shorter in coccoid forms than in rods.

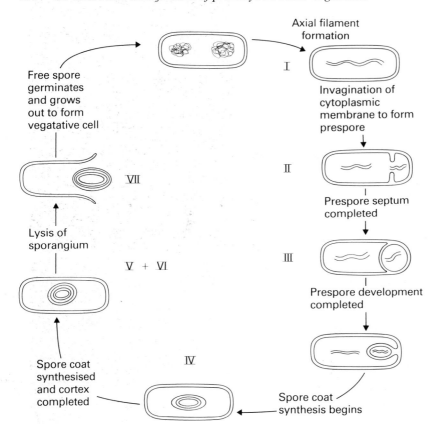

Fig. 8.40 Stages (I–VII) in the formation of an endospore.

ENDOSPORES AND SIMILAR STRUCTURES

A number of bacteria, as well as exhibiting morphological variation of vegetative cells, differentiate distinct resting cells. Endospores are formed by *Bacillus*, *Clostridium*, *Sporolactobacillus*, *Desulphotomaculum* and *Sporosarcina*, cysts by *Azotobacter*, myxospores (microcysts) by *Myxobacteria*, and conidia by Actinomycetes. With the exception of the latter, one resting cell is produced from each vegetative cell and germinates to give rise initially to a single cell; the process is therefore not one of multiplication.

Because of the attractiveness of prokaryotic resting cell formation as a relatively simple model of a differentiating system considerable effort has been made to understand the processes of formation and germination, particularly of endospores and myxospores. In no case, however, are the controlling factors completely understood. In general, resting cells are initiated when a nutritional deficiency occurs or the environment becomes otherwise adverse.

The morphological changes in endospore formation are illustrated in Fig. 8.40. The whole process typically takes about eight hours and seven different stages are conventionally recognized. With each of these there is associated particular biochemical events some of which are specifically related to spore formation. For example, spores are characterized chemically by the presence of dipicolinic acid (Fig. 8.41) and muramic acid lactam (Fig. 8.42) in the cortex which starts to accumulate in stage IV.

Endospores typically have their nuclear material

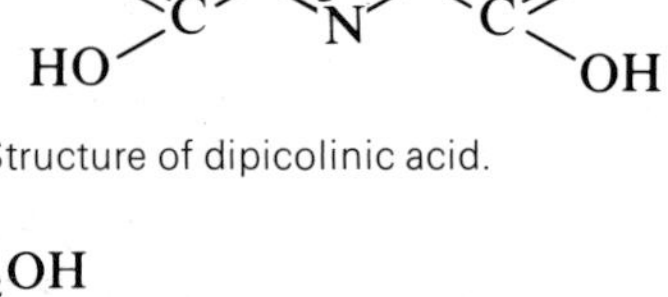

Fig. 8.41 Structure of dipicolinic acid.

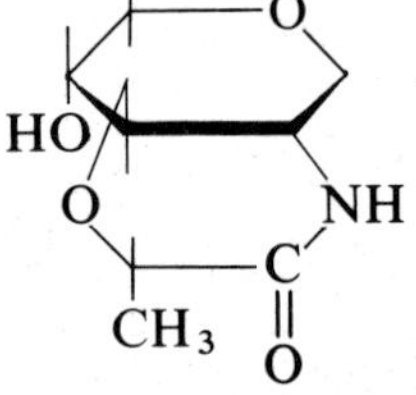

Fig. 8.42 Structure of muramic acid lactam.

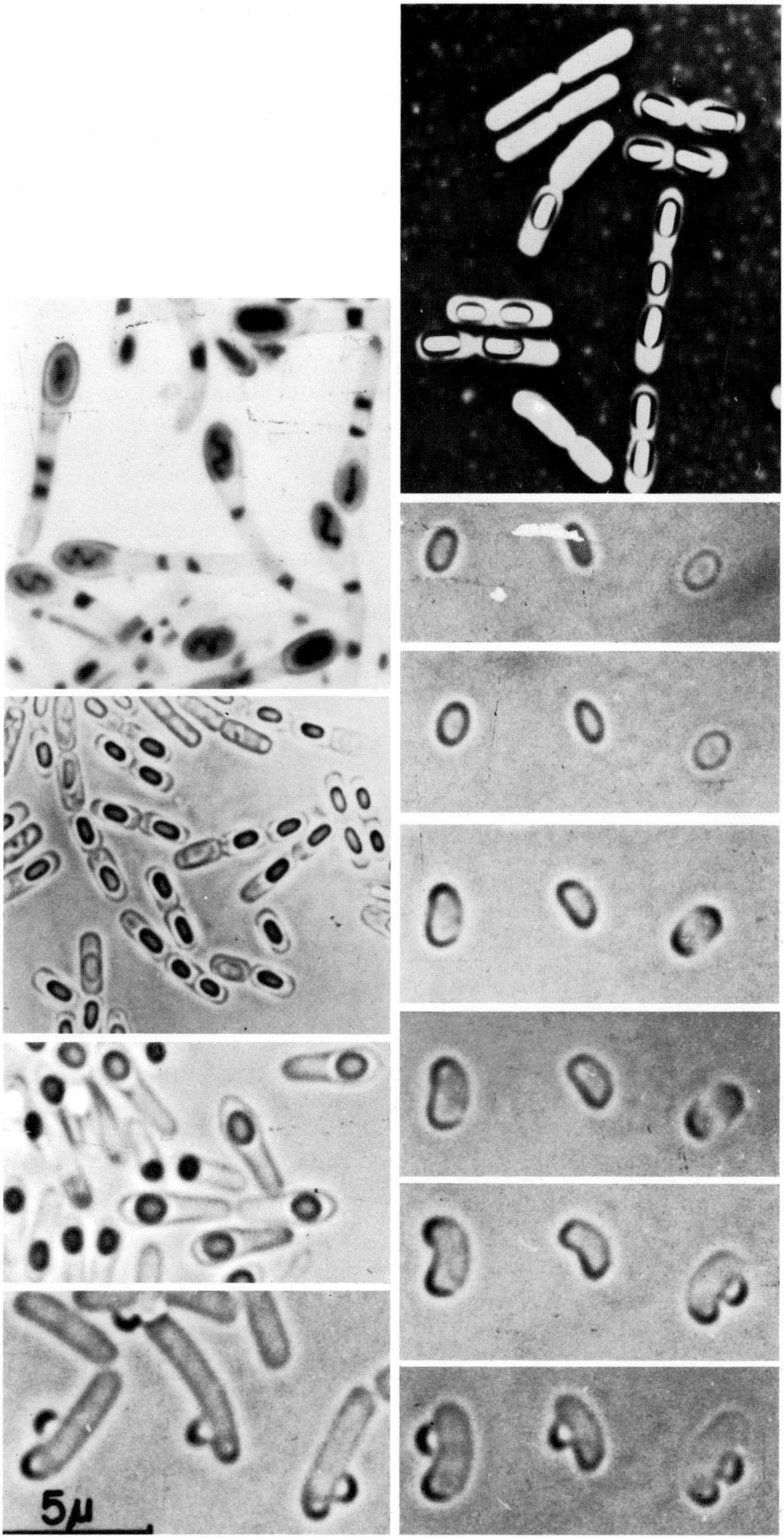

a
b
c
d
e (i)
e (ii)
e (iii)
e (iv)
e (v)
e (vi)
e (vii)

Fig. 8.43 Bacterial endospores. (by C. F. Robinow)

- a *Clostridium pectinovorum* (× 3600). Giemsa stained after acid hydrolysis, to show nuclear material arranged as a filament.
- b *Bacillus cereus* (× 3600). The preparation has been made in nigrosin, dried and mounted in air and photographed from the reverse side—a technique which emphasizes the high refractility of the spores.
- c *Bacillus cereus* (× 3600). Fixed with osmium tetroxide vapour and photographed wet between coverslip and agar surface. This organism shows the more characteristic appearance of aerobic spore-bearing organisms.
- d *Bacillus sphaericus* (× 3600). Prepared and examined as in (3). This aerobic spore bearer shows bulging of the bacillus around the spore, a feature which is unusual among aerobes and more common among anaerobes.
- e *Bacillus subtilis* (× 3600). Spore 'germination' serially photographed at the following times after inoculation on to nutrient medium: (i) 7, (ii) 22, (iii) 60, (iv) 105, (v) 165, (vi) 176, and (vii) 200 minutes.

arranged as filaments (Fig. 8.43, a); around their cytoplasm is a thin membrane which is surrounded by the cortex. This is a substantial structure containing the peptidoglycan, which is less cross-linked than that of the vegetative cell. Outside the cortex lies the stratified spore coat in which inner and outer regions may be recognized. The outermost layer is the exosporium (Fig. 8.44). The mature spore is highly refractile (Fig. 8.43, b); its position in the vegetative cell and its shape and size are characteristic of particular species. Generally *Bacillus* spores are not larger than the vegetative cell (Fig. 8.43, c and 4) whereas those of *Clostridium* frequently distend the sporangium (Fig. 8.43, a; Fig. 14.3d).

The development of a vegetative organism from an endospore is frequently referred to as germination. In fact germination is only one stage in a three-part process. First the spores must be activated by a suitable stimulus, such as exposure to heat, specific chemicals or mechanical treatment. Activation is reversible and if suitable conditions for vegetative growth do not exist, dormancy is resumed. If conditions are favourable, germination starts. This is irreversible and involves loss of spore refractility, heat resistance, liberation of organic material including calcium dipicolinate and the activation of several enzymes. Finally the developing vegetative cell grows out from the spore (Fig. 8.43, e).

The differentiation of myxospores is a more complex process since it involves the formation of a multicellular fruiting body. This occurs when a migrating colony of vegetative cells aggregates on exhaustion of nutrients. The aggregate may develop into only a simple mound of cells or into a stalk on which is subsequently formed a complex fruiting body. Within the fruiting body some cells shorten and develop into thick-walled myxospores; others autolyse and form slime which encloses the spores. The molecular mechanisms underlying these processes are imperfectly understood but some kind of cell–cell communication must be involved in their control.

Cysts of *Azotobacter* are formed by rounding off of the cell and development of a thick wall. This consists of a typical Gram-negative cell wall surrounding the central body, the intine, which is lipopolysaccharide, and the exine, which contains phospholipid (Fig. 8.45).

Conidia of actinomycetes may be formed on special hyphae (sporophores), singly, in groups of a few or in chains of indefinite length (Fig. 8.46). The

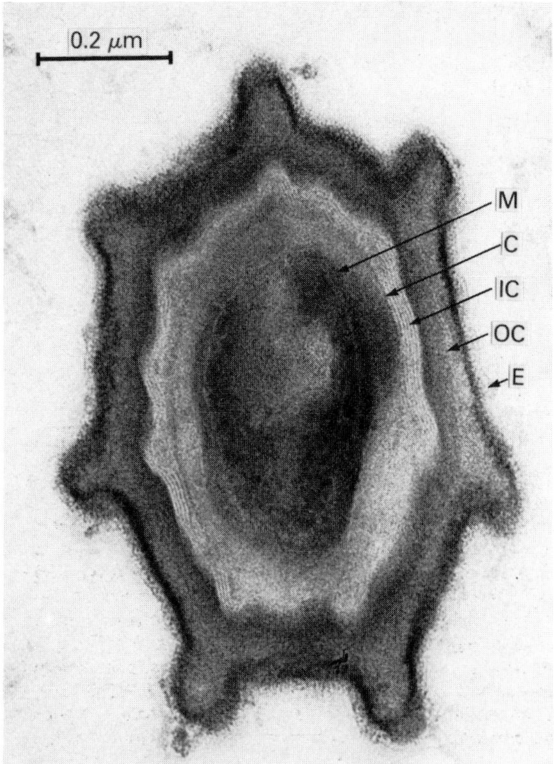

0.2 μm

M
C
IC
OC
E

Fig. 8.44 Electron micrograph of a thin section of *Bacillus circulans* endospore showing sculptured ridges in the outer spore coat (OC). IC=layered inner coat; C=cortex; M=membrane; E=exosporium. (*Bildatlas path. Mikroorganismen, Bd.*, **111**, G. Fischer, Stuttgart)

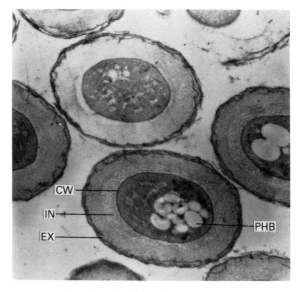

CW
IN
EX
PHB

Fig. 8.45 Cysts of *Azotobacter*. Electron micrographs of a thin section showing the cell wall (CW), exine (EX), intine (IN) and poly-β-hydroxybutyrate (PHB) granules (×23 000). (Orville Wyss).

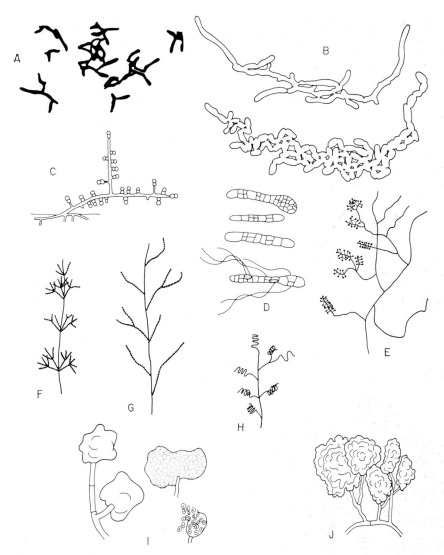

Fig. 8.46 Actinomycetes. (**A**) *Actinomyces israeli*
(× 1400). Branching elements from culture. (Drawn from
photograph by Pine, L., Howell, A. and Watson, S. J. (1960)
J. gen. Microbiol., **23**, 403.) (**B**) *Nocardia corallina*
(× 1900). A young colony already beginning to fragment;
and below, the same colony nearly nineteen hours later and
showing extensive fragmentation. (Drawn from photograph
by Brown, O. and Clark, J. B. (1966) *J. gen. Microbiol.,* **45**,
525). (**C**) *Microbispora rosea*. Schematic diagram of typical
aerial hypha. (After Lechevalier, M. P. and Lechevalier, H.
(1957) *J. gen. Microbiol.,* **17**, 104.) (**D**) *Dermatophilus
dermatonomus*. Filaments with transverse and longitudinal
septa, and one producing motile cocci. (After Thompson, R.
E. M. and Bisset, K. A. (1957) *Nature, Lond.,* **179**, 590.) (**E**)
Micromonospora sp. Fine mycelium bearing single spores.
(After Skinner, C. E., Emmons, C. W. and Tsuchiya, H. M.
(1947) *Henrici's Molds, Yeasts and Actinomycetes,* 2nd Edn.
John Wiley and Sons, Inc., N.Y.) (**F**) *Streptomyces
phaeochromogenes*. Straight or slightly wavy sporophores
arising on side branches of aerial hypha. (After Ettlinger, L.,
Corbaz, R. and Hütter, R. (1958) *Arch. Mikrobiol.,* **31**, 336.)
(**G**) *Streptomyces reticuli*. Straight branching sporophores
borne in verticils. (Source as **F**.) (**H**) *Streptomyces
violaceoniger*. Spiral spore chains. (Source as **F**.) (**I**)
Actinoplanes utahensis (× 1000). Sporangia, one showing
spores within, one dehiscing. (After Couch, J. N. (1963)
J. Elisha Mitchell Sci. Soc., **79**, 53.) (**J**) *Amorphosporangium
auranticolor* (×1000). Group of sporangia. (Source as **I**.)

resting cells have a wall consisting of an inner region
equivalent to a cell wall and an external sheath
which is hydrophobic in nature.

Myxospores, cysts and conidia all germinate by
outgrowth to form vegetative cells. The resistance
of resting cells to heat is on the whole no greater
than that of vegetative cells with the exception of
endospores, which characteristically are extremely
thermoduric: all are however more able than the
vegetative cell to stand desiccation, exposure to

ultra-violet light, hydrolytic enzymes and certain toxic chemicals. Thus they confer upon the organism producing them a survival advantage under adverse conditions.

Classification of prokaryotes

The classification of prokaryotes is still unsatisfactory. Until recently they have been divided into 'true' bacteria and the rest, a division which strongly reflects the interests and pretensions of bacteriologists rather than the true nature of the organisms. The scheme outlined in Table 8.2 is based on the latest edition (8th) of *Bergey's Manual of Determinative Bacteriology*. It is mainly arbitrary and an acknowledgement of our ignorance but allows an attempt to classify all prokaryotes based on their fine structure. The Bacteria are arranged in nineteen parts on the basis of one or more of the following principal characters: morphology, Gram stain, types of movement, cell division, sporogenesis, cell envelope (each considered in detail in this chapter), photosynthetic ability and metabolic end product (only briefly mentioned here but dealt with extensively in Chapter 2 and 5).

A brief resumé of each part is given.

PHOTOTROPHIC BACTERIA (e.g. *Rhodospirillum, Chromatium, Chlorobium*)

Predominantly aquatic organisms that have bacterio-chlorophylls and carotenoid pigments. They photosynthesize by a method different from that of blue-green bacteria and green plants (p. 36) and most, if not all, fix atmospheric nitrogen. The chief division is into: purple bacteria (Rhodospirillaeae), some of which can use sulphur as an electron donor, and the green sulphur bacteria (Chlorobiaceae). G + C ratio = 45–73 moles %.

GLIDING BACTERIA (Myxobacterales)

These organisms are characterized by their ability to glide on a solid or liquid surface and by the tough slime layer that surrounds the cells. Much interest has centred on the aggregation of individual cells to form fruiting bodies which behave as a single colonial organism. They are mostly soil organisms that produce a battery of hydrolytic enzymes capable of lysing other prokaryotic or eukaryotic cells. G + C ratio = 67–71 moles %.

SHEATHED BACTERIA (e.g. *Leptothrix, Sphaerotilus, Streptothrix*)

Bacteria grouped together because their cells are encased in gelatinous sheaths but some members have affinities with bacteria placed elsewhere, e.g. the pseudomonads. They are found in both clean and polluted waters and in activated sewage sludge. Many are able to oxidize iron and manganese salts and their sheaths may become encrusted with metal oxides. Some members produce differentiated attachment organs—holdfasts. G + C ratio = *ca* 70 moles %.

BUDDING AND/OR APPENDAGED BACTERIA (e.g. *Hyphomicrobium, Caulobacter, Gallionella*)

A collection of organisms in which the products of binary fission are usually unequal and may have specialized functions, e.g. stalks, holdfasts, swarming cells, or may resemble fungal hyphae. They are mainly soil or aquatic organisms and *Gallionella* sometimes causes problems in water works owing to its prolific growth in waters containing ferrous iron. G + C ratio = 59–67 moles %.

SPIROCHAETES (e.g. *Spirochaeta, Treponema, Leptospira*)

The main division of flexuous helical organisms is into the generally large (*ca.* 0.5–100 μm) aquatic spirochaetes and the much finer treponemes and leptospiras (*ca.* 0.1 × 15 μm). The causal organisms of syphilis (*T. pallidum*), Weil's disease (*L. icterohaemorrhagiae*), and relapsing fever (*Borrelia recurrentis*) are among the medically important members. G + C ratios: 50–66 moles % (*Spirochaeta*); 32–50 moles % (*Treponema*); 36–39 moles % (*Leptospira*).

SPIRAL AND CURVED BACTERIA (e.g. *Spirillum, Campylobacter*)

This group contains the *rigidly* helical bacteria having single or grouped polar flagella, which impel the cells with characteristic corkscrew motility. There are both free-living aquatic organisms (e.g. *Sp. anulus*) and pathogens (e.g. *Sp. minor*—a natural parasite of rats that gives rise to 'rat-bite fever' in man). *Campylobacter foetus* is one cause of venereal disease of cattle and other *Campylobacter* spp. are common in the oral flora of man. G + C ratios: 38–65 moles % (*Spirillum*); 30–35 moles % (*Campylobacter*). *Bdellovibrio*, which is often parasitic on other bacteria, is also included. G + C ratio = 43–51 moles %.

GRAM-NEGATIVE AEROBIC RODS AND COCCI (e.g. *Pseudomonas, Azotobacter, Rhizobium*)

A group that includes the pseudomonads, whose activities in the soil and in fresh and marine waters are important in the decomposition of organic matter. Some pseudomonads are pathogens of man and animals (e.g. *Ps. aeruginosa* and the glanders bacillus, *Ps. mallei*); others are plant pathogens (e.g. *Ps. solanacearum*). G + C ratio = 58–70 moles %. Also included are the nitrogen-fixing bacteria, the Azotobacteriaceae (G + C ratio = 53–70 moles %) and the Rhizobiaceae (G + C ratio = 59–66 moles %); bacteria dependent on methane or methanol as a source of energy and carbon (*Methylomonas*); and bacteria dependent for growth on high salt concentrations (*Halobacterium*). 'Genera of uncertain affiliation' tacked onto this group include the medically important genera *Brucella, Bordetella* and *Francisella*, and the extreme thermophile *Thermus*.

GRAM-NEGATIVE FACULTATIVELY ANAEROBIC RODS (Enterobacteriaceae, Vibrionaceae)

Many organisms of medical importance and their free-living relatives are included here. The Enterobacteriaceae (G + C ratio = 39–59 moles %) includes the pathogenic genera *Salmonella, Shigella* and *Yersinia* (p. 308) and the phytopathogenic Erwiniae; the Vibrionaceae includes *Vibrio cholerae* (p. 309).

GRAM-NEGATIVE ANAEROBES (Bacteroidaceae)

A group of anaerobic, non-sporing rods found in large numbers among the normal flora of the alimentary tract of man and animals, as well as in certain diseases. Many species are characterized by irregularly shaped cells. G + C ratio = 40–55 moles %. Among this group's 'genera of uncertain affiliation' is *Desulphovibrio*—a sulphur-reducing organism found in brackish waters.

GRAM-NEGATIVE COCCI AND COCCO-BACILLI (Neisseriaceae)

Most of these organisms are parasitic on man and many are pathogenic. The genera are: *Neisseria, Branhamella, Moraxella*, and *Acinetobacter*. G + C ratio = 39–52 moles %.

GRAM-NEGATIVE ANAEROBIC COCCI (Veillonellaceae)

These are parasitic bacteria found in the alimentary tract of warm-blooded animals. G + C ratio = 40–57 moles %.

GRAM-NEGATIVE CHEMOAUTOTROPHIC BACTERIA (e.g. *Nitrobacter, Thiobacillus, Siderocapsa*)

A group comprising organisms important in the nitrogen and sulphur cycles in soils and water. The Nitrobacteriaceae (G + C ratio = 51–62 moles %) contain 'nitrifying bacteria' that oxidize either nitrite (e.g. *Nitrobacter*) or ammonia (e.g. *Nitrosomonas*) to obtain energy. *Thiobacillus* and related organisms (G + C ratio = 50–68 moles %) obtain energy from oxidation of reduced sulphur compounds, whereas the Siderocapsaceae are found in iron-rich waters and oxidize ferrous or manganous compounds.

METHANE PRODUCERS (Methanobacteriaceae)

Three genera are recognized: *Methanobacterium, Methanosarcina*, and *Methanococcus*. They are found in anaerobic sediments of muddy waters and in anaerobic sewage digestors as well as in the gastrointestinal tract of animals.

GRAM-POSITIVE COCCI (Micrococcaceae, Streptococcaceae, Peptococcaceae)

This group brings together three families. The first includes *Micrococcus*—aerobic bacteria common in soils and fresh water and having oxidative metabolism (G + C ratio = 66–75 moles %), and *Staphylococcus*—facultative anaerobes commensal or pathogenic on warm-blooded animals and capable of anaerobic glycolysis (G + C ratio = 30–40 moles %). The second family includes many important pathogens of man and animals (e.g. *Streptococcus pyogenes, Strep. pneumoniae, Strep. agalactiae*) and the lactic streptococci used by the dairy industry. (G + C ratio = 33–44 moles %). The Peptococcaceae are anaerobic cocci and include members of the normal flora of man and animals which can also be secondary pathogens. (G + C ratio = 39–45 moles %).

ENDOSPORE-FORMING RODS AND COCCI (Bacillaceae)

All known endospore-forming bacteria are included here. These include the aerobic genus *Bacillus* (G + C ratio = 36–62 moles %) and the anaerobic

Table 8.2 Summary of the principal characters on which are based the divisions of BACTERIA into the nineteen parts of *Bergey's Manual of Determinative Bacteriology (Eighth edition)*. (In order to be concise some generalizations have been made to which there are exceptions.)

Part	Morphology	Gram stain	Movement	Division	Spores	Metabolism
1. Phototrophic bacteria	Spiral, spheres, ovoid	Negative	Flagella if motile	Most binary, some budding	None	Phototrophic
2. Gliding bacteria	Rods which may aggregate and differentiate to form a fruiting body and myxospores or may form microcysts	Negative	Gliding	Binary	Myxospores or microcysts	Chemoorganotrophic
3. Sheathed bacteria	Rods in chains within sheaths in which may be deposited iron or manganese	Negative	Some have single cells motile by flagella	Transverse and/or longitudinal fission	None	Chemoorganotrophic
4. Budding and/or appendaged bacteria	Cells may bear buds on appendages, be appendaged but not bud, bud, or excrete an appendage	Negative	Daughter cells may be motile by flagella	Budding or binary	None	Chemoorganotrophic
5. Spirochaetes	Helices: protoplasmic cylinder intertwined with axial fibres contained within the outer envelope	Negative	Flexion by axial fibres, rotation	Binary	None	Chemoorganotrophic
6. Spiral and curved bacteria	Rigid helically curved cells	Negative	Flagella	Binary	None	Chemoorganotrophic
7. Gram-negative aerobic rods and cocci	Straight or curved rods: a few may form branches, others are cocci, some have atypical walls lacking DAP[1] and MA[2]	Negative	Flagella	Binary	None, but some form cysts	Chemoorganotrophic
8. Gram-negative facultatively anaerobic rods	Rods	Negative	Flagella	Binary	None	Chemoorganotrophic

	Morphology	Gram reaction	Motility	Reproduction	Spores/special structures	Nutrition
9. Gram-negative anaerobic bacteria	Rods	Negative	Flagella if motile	Binary	None	Chemoorganotrophic
10. Gram-negative cocci and coccobacilli	Cocci or coccobacilli	Negative	Some show twitching motility. No flagella	Binary	None	Chemoorganotrophic
11. Gram-negative anaerobic cocci	Cocci	Negative	Non-motile	Binary	None	Chemoorganotrophic
12. Gram-negative chemolithotrophic bacteria	Rods and cocci	Negative	Flagella	Binary	None	Chemolithotrophic oxidize NH_4 or NO_2; metabolize S or deposit iron or manganese oxides
13. Methane-producing bacteria	Rods or cocci	Positive, or negative	Motile or non-motile	Binary	None	Strictly anaerobic producing methane as a result of CO_2 reduction with electron donated by hydrogen, formate, acetate etc.
14. Gram-positive cocci	Cocci	Positive	Flagella if motile	Binary	None	Chemoorganotrophic
15. Endospore-forming rods and cocci	Rods or cocci	Mostly positive	Flagella if motile	Binary	Endospores	Chemoorganotrophic
16. Gram-positive asporogenous rod-shaped bacteria	Rods	Positive	Flagella if motile	Binary	None	Chemoorganotrophic
17. Actinomycetes and related organisms	Irregular rods, filaments, branching	Positive	Flagella if motile	Binary	Conidia	Chemoorganotrophic
18. Rickettsias	Rods, cocci often	Negative	Non-motile	Binary	None	Obligate parasites
19. Mycoplasma	Highly pleomorphic: outer layer is triple layered membrane. No DAP[1] or MA[2]	Negative	Non-motile usually	Elementary body formation and release or binary	None	Chemoorganotrophic

[1] DAP—Diaminopimelic Acid.
[2] MA—Muramic acid.

genus *Clostridium* (G + C ratio = 23–43 moles %), both of which contain organisms essential to soil fertility and well-known pathogens such as *B. anthracis*, *Cl. tetani*, and *Cl. botulinum*. Other genera are: *Sporolactobacillus*, *Desulphomaculum*, and *Sporosarcina*—a sporing, motile coccus that combines features of *Micrococcus* and *Bacillus*.

GRAM-POSITIVE NON-SPORING RODS (e.g. *Lactobacillus*, *Listeria*, *Erysipelothrix*)

The main part of this group are the Lactobacillaceae—organisms associated mainly with milk and milk products (G + C ratio = 35–53 moles %). However, the pathogenic genera *Listeria* (G + C ratio = 38–56 moles %) and *Erysipelothrix* are also included. *Listeria monocytogenes* causes infections of warm-blooded animals, e.g. 'circling disease of cattle', *Erysipelothrix rhusiopathiae* causes swine erysipelas; both genera were formerly classified with the diphtheroid organisms, which they resemble.

ACTINOMYCETES AND RELATED ORGANISMS (Actinomycetales and Coryneforms)

A large group which brings together those bacteria that used to be called 'higher bacteria' owing to their possession of features reminiscent of the fungi—e.g. a tendency to a form of mycelial growth. The coryneforms (G + C ratio = 48–70 moles %) include (a) human and animal pathogens (e.g. *C. diphtheriae*, *C. pyogenes*) and parasites (e.g. *C. hofmanni*), (b) plant pathogens (e.g. *C. fascians*), and (c) non-pathogenic soil organisms, including the genera *Arthrobacter* and *Kurthia*. Among the Actinomycetales are placed the actinomycetes, streptomycetes and their relatives—mostly soil organisms, some of which are used in the production of therapeutic antibiotics, whereas others are pathogens of domestic animals. A medically-important subgroup are the mycobacteria, e.g. *M. tuberculosis*, *M. leprae*, *M. enteritidis*.

THE RICKETTSIAS (Rickettsiales and Chlamydiales (p. 115)

This is a group of obligate intracellular parasites that can be cultivated in the laboratory only by 'virological methods'. Many cause disease in man or other vertebrate or invertebrate hosts and are often transmitted by arthropod vectors (p. 308). Examples are *Rickettsia prowazekii* causing louse-borne typhus fever, *Chlamydia trachomatis* causing trachoma and lymphogranuloma venereum.

MYCOPLASMAS

These organisms lack a true cell wall and have probably been derived from a number of different origins. Some resemble wall-less forms of known bacteria but others seem to be less closely related and their relation to bacteria is unclear.

Actinomycetes

Introduction

The actinomycetes are a group of Gram-positive bacteria which produce very fine mycelium. The individual hyphae are usually not more than ca. 1 μm in diameter. The hyphae branch and possibly in all species become septate by the inward growth of rings of wall material. Within each compartment occur one or more chromatin bodies, often of irregular form. The mycelium of some species lies wholly within or in contact with the substrate, but in others there are aerial parts too.

Many actinomycetes produce spores which differ from the endospores of bacilli not only in method of formation (p. 160) but in being only mildly resistant to heat. When they germinate they form one or more germ-tubes whose walls are continuous with the spore's own walls. The combination of aerial growth and sporulation usually confers a cottony or powdery texture on the surface of the colony, while colonies which lack aerial mycelium are either glossy or matt. True bacterial endospores are however formed in one family.

Spores may be formed on special hyphae (sporophores) either singly, in groups of small numbers or in chains of indefinite length. Spore chains of indefinite length are characteristic of the genus *Streptomyces* (Fig. 8.46). The sporophore wall first becomes two-layered, and then extensions of the inner layer grow inwards to become the septa which delimit the spores. The outer layer forms a common sheath until it ruptures and frees the spores.

The sporophores in streptomycetes may sometimes be grouped to form structures resembling the coremia or pycnidia of true fungi. One family of actinomycetes (the Actinoplanaceae) is characterized by the formation of 'sporangia' (Fig. 8.46i, j). A sporangium is formed as a terminal vesicle on a hypha, and into it grow one or more filaments which then segment into chains of spores which in some species become flagellate and motile.

Most actinomycetes stain relatively easily with basic dyes and are Gram-positive. A few, principally the pathogenic mycobacteria, stain only with difficulty but once stained, resist decolorization by mineral acid. This property of acid-fastness varies between species and depends to some extent upon the conditions of growth. It is attributed to a lipid component (mycolic acid) in the morphologically intact cell, the degree of acid-fastness being proportional to the amount of mycolic acid present.

Electron microscope studies reveal that the cytoplasm contains a particularly extensive membrane system and many ribosomes. Most of the membranes are orientated randomly, but adjacent to septa they are ordered into discrete bodies which connect with the plasma membrane adjoining the developing septa. They are thought to be concerned with septum formation as in the true bacteria (p. 155).

The cell wall composition of actinomycetes resembles that of other Gram-positive bacteria. The walls are thin (*ca.* 10–20 nm) and appear to be complexes of sugars, amino-sugars, and a few amino acids. Two characteristic features are the presence of muramic acid (p. 14) and the narrow range of amino acids present. The walls contain major amounts of either 2,6-diaminopimelic acid (DAP) or lysine. The walls of actinomycetes, as of Gram-positive bacteria in general (p. 150), are conspicuously lacking in aromatic and sulphur-containing amino acids, in proline, in histidine, and in arginine.

Actinomycetes have a considerable economic importance. In nature they contribute to the mineralization of organic residues. Only a few are plant pathogens (p. 288) but a number cause debilitating or even lethal infections of animals and man (p. 306). They produce about 85% of known antibiotics (p. 362) including substances active against bacteria, fungi, protozoa, rickettsiae, larger viruses, and neoplasms.

Physiology

The germinating spore usually generates a system of filaments which grow apically and branch monopodially. In *Actinomyces* and *Nocardia* the mycelium soon fragments into pieces which grow at both ends. In *Dermatophilus* the filaments become septate both transversely and longitudinally and separate into motile coccoid spores. In other genera the substrate mycelium typically remains intact, though there are exceptions. The mycelium at first contains an optically homogeneous cytoplasm but this becomes vacuolated as it matures and fat and volutin granules may be found in the vacuoles. Anastomosis of filaments has been claimed but the evidence for this is equivocal. In *Streptomyces* and some other genera the substrate mycelium soon gives rise to aerial mycelium on which the spores are formed. The onset of sporulation in streptomycetes is related to a depletion of nutrients in the medium or to other unfavourable circumstances. The ability to form aerial mycelium implies a capacity for growth at a distance from the nutritive substrate and hence for an internal translocation of nutrients. This is a fundamentally different biological organization from that encountered in other bacteria. Growth in well-aerated submerged culture is rapid and is followed by lysis of the mycelium and even of the spores. In many streptomycetes the lytic phase is accompanied by increased liberation of antibiotics. Although some species are thermophilic and grow best at 50–56°C the majority are mesophilic. Isolates from soil and water grow best at about 25–30°C; those from warm-blooded host animals grow best at about 37°C.

The basic nutrition of the majority of saprophytic actinomycetes is simple and resembles that of a large number of other bacteria and a majority of fungi. There is a requirement for an organic source of carbon and energy, but beyond this all other nutritional needs can generally be met by inorganic salts, such as nitrate or ammonium salts for nitrogen, and salts of the necessary mineral nutrients among which phosphorus, potassium, and sulphur are needed in greatest amounts. Some actinomycetes have additional requirements for particular vitamins, or for particular amino acids. Although nitrogen fixation by free-living actinomycetes has sometimes been claimed there is still much uncertainty as to whether this really does occur. The general consensus of opinion is that it does not. Although the minimal nutrition of actinomycetes is generally simple, the total pattern of nutrition is complicated firstly by the diversity of complex macromolecular nutrients which they can digest by extracellular enzyme action; secondly by the breadth of the range of organic nutrients that can enter the metabolic pathways of the cell; and thirdly by the development in some species of considerable degrees of specificity for particular organic nutrients. Among the simpler nutrients utilized as sources of carbon and energy are glucose, maltose, lactose, cellobiose, glycerol, alcohols, and various organic acids including fatty and amino acids. The latter are also particularly suitable nitrogen sources.

A minority of actinomycetes utilize sucrose. The use of oxalic acid by some species of *Streptomyces* and *Nocardia* suggests they may have an important natural role in the detoxification of calcium oxalate in plant residues.

Most actinomycetes are aerobic organisms but anaerobes occur, principally in the family Actinomycetaceae. Anaerobic actinomycetes possess a fermentative metabolism which yields acid but not gas from a variety of carbohydrates. Even aerobic forms may produce considerable quantities of acid—largely lactic—but the acid is later consumed. The intermediary metabolism of actinomycetes is still not well known, but in several streptomycetes enzymes characteristic of the tricarboxylic acid cycle have been demonstrated. Lysine is synthesized through the α,ε-diaminopimelic acid pathway as it is in other bacteria and not through the alternative α-aminoadipic acid pathway. Actinomycetes, like other bacteria, do not contain sterols.

The smells that are characteristic of many actinomycetes are attributable to a variety of metabolites which includes acetic acid, acetaldehyde, ethanol, isobutanol, and isobutyl acetate. An earthy smelling neutral hydrocarbon oil, named geosmin, has been isolated from several species including *Streptomyces griseus*. Actinomycetes generate a great variety of substances in their environment, and these include vitamins, pigments, and antibiotics. The biosynthesis of some of these has been developed on an industrial scale (p. 361 *et seq.*).

Classification

Early attempts to classify Actinomycetes on morphological characters resulted in confusion owing to the tendency for these organisms to alter during periods of artificial culture. Lately their classification has taken into account certain chemical criteria. Of those available, the most useful are concerned with the cell wall. All cell wall preparations contain major amounts of alanine, glutamic acid, glucosamine and muramic acid, but on the basis of other major constituents nine types of wall have been recognized. Four of them correspond to the majority of aerobic actinomycetes, and comprise the *Streptomyces* type (or type I) with *L*-2, 6-diaminopimelic acid (*L*-DAP) and glycine; the *Micromonospora* type (or Type II) with *meso*-DAP, glycine and sometimes hydroxy DAP; the *Actinomadura* type (or Type III) with *meso*-DAP; and the *Nocardia* type with *meso*-DAP, arabinose and galactose. Patterns of composition of whole-cell sugars are also distinctive and correlate well with

wall types. They may be determined easily and so are useful indicators of affinity. Lipid composition is another useful taxonomic criterion. Mycobacteria, nocardiae and many corynebacteria share Type IV cell walls, but can be distinguished by their containing α-branched β-hydroxylated fatty acids of different size, viz. mycolic acids with *ca.* 80 carbon atoms, nocardomycolic acids with *ca.* 50 carbons, and corynomycolic acids with *ca.* 30 carbons.

A classification of the actinomycetes into ten families, and notes on their members follows.

ACTINOMYCETACEAE

Gram-positive, non-acidfast. Filaments may branch; fragment readily into rod-like or coccoid elements. No aerial hyphae or spores. Most are facultative anaerobes. Cell walls not of Types I–IV. Genera: *Actinomyces* (Fig. 8.46a), *Agromyces, Arachnia, Bacterionema, Bifidobacterium, Rothia.*

ACTINOPLANACEAE

Form sporangia enclosing motile or non-motile spores. Walls Type II or III. Genera: *Actinoplanes* (Fig. 8.46i), *Amorphosporangium* (Fig. 8.46j), *Ampullariella, Dactylosporangium, Planobispora, Planomonospora, Spirillospora, Streptosporangium.*

These organisms are common and widespread in soil, from which they may be isolated by baiting with pollen of *Liquidambar* and similar materials. Some species produce chains of spores in addition to sporangia. In culture most species produce brilliant, most often orange, pigments and usually lack aerial mycelium.

DERMATOPHILACEAE

Form filaments which then divide transversely and longitudinally to produce usually motile cocci. Wall Type III. Genera: *Dermatophilus* (Fig. 8.46d), *Geodermatophilus.*

FRANKIACEAE

Obligate symbionts in nitrogen-fixing non-leguminous plant nodules. No spores. Genus: *Frankia.*

A number of species are recognized on the basis of their specific associations with particular higher plants which include *Alnus, Elaeagnus, Hippophae* and *Casuarina.*

MICROMONOSPORACEAE

Mycelium extremely fine (0.2–0.6 μm diam.), profusely branched. Aerial mycelium lacking at ordinary temperatures, but abundant on hot decomposing masses of plant material; bearing heat-sensitive spores singly at apices of short lateral branches: Wall type II. Genus: *Micromonospora* (Fig. 8.46e).

MYCOBACTERIACEAE

Mycelium rudimentary or absent, no spores. Aerobic. Mycolic acids present. Wall Type IV. Genera: *Mycobacterium*, *Mycococcus*.

The members of this family range from saprophytes to obligate parasites (e.g. *M. tuberculosis*, Fig. 14.4b) and have the capacity to form rudimentary mycelium under some conditions.

NOCARDIACEAE

Substrate mycelium fragmenting into rod-like or coccoid elements. Aerobic. Contain nocardomycolic acids. Wall type IV. Genera: *Micropolyspora*, *Nocardia*, *Gordona*. Most species are soil saprophytes but a number cause diseases (nocardioses) of animals including man (Fig. 14.4a).

STREPTOMYCETACEAE

Aerobic, with non-fragmenting substrate mycelium. Aerial mycelium usually well developed, bearing chains of spores. Occasionally forming zoospores. Wall Type I. Genera: *Chainia*, *Elytrosporangium*, *Kitasatoa*, *Microellobosporia*, *Streptomyces* (Fig. 8.46f, g, h), *Streptoverticillium*.

Streptomyces species (Fig. 8.46f, g, h) are common in soil, and themselves have a characteristic earthy smell. The naming of isolates is a difficult task. The International *Streptomyces* Project is a collaborative programme of study which has produced definitive descriptions of 450 type cultures, and has facilitated identification. Criteria used in their taxonomy include colour of the aerial mycelium, ornamentation of the spores as observed by electron microscopy, straightness or coiling of spore chains, production of dark pigment in culture, and capacity to use different carbohydrates. Many saprophytic species are active producers of antibiotics which include clinically useful ones.

THERMOACTINOMYCETACEAE

Have substrate and aerial mycelium, and form heat-resistant endospores. Wall Type II. Genus: *Thermoactinomyces*.

All the endospore-forming actinomycetes are included in this monogeneric family. The endospores, unlike the spores of other actinomycetes, have a core surrounded by a cortex and multi-layered wall, and contain dipicolinic acid. They may survive even 100°C for more than 220 minutes.

THERMOMONOSPORACEAE

Forming heat-sensitive spores singly or in short chains on aerial and substrate mycelium, or on aerial mycelium alone. Substrate mycelium may fragment. Wall Type III. Genera: *Actinomadura*, *Microbispora* (Fig. 8.46c), *Microtetraspora*, *Saccharomonospora*, *Thermomonospora*. A number of members of this recently proposed family were formerly classified in the Nocardiaceae.

Blue-green bacteria (algae)

The cells of Cyanobacteria (Cyanophyte: Myxophyceae) are considerably larger than those of other prokaryotic groups and contain chlorophyll *a*, β-carotene, and the phycobilin pigments phycocyanin and phycoerythrin. Thus they rank with the algae in size and the ability to photosynthesize *via* chlorophyll *a* with the liberation of oxygen, but possess the bacterial characteristics of a prokaryotic nucleus and absence of mitochondria. The method of cell division, the copious production of mucilage and the absence of any sexual stages in the life history all draw the group closer to bacteria than to algae. They appear to be a very ancient group playing an important role in the microbiological economy.

The morphology of Cyanobacteria is basically that of unicells or aggregations of these to form colonies and filaments (Fig. 8.47). Under the microscope they show a bluish-green colour but distinct chromatophores are absent, the pigments being diffused throughout the cells. Small granules of a polysaccharide can sometimes be seen and in some species larger blackish vacuoles occur; these are the gas vacuoles containing mainly nitrogen and are present in only a small number of species. The cell walls are indistinct but often there is a prominent sheath of mucilage around the cell or filament, frequently stained brown by iron or manganese deposits and occasionally covered by crystals of carbonates. Species of Cyanobacteria are important constituents of the thermophilic flora and are also

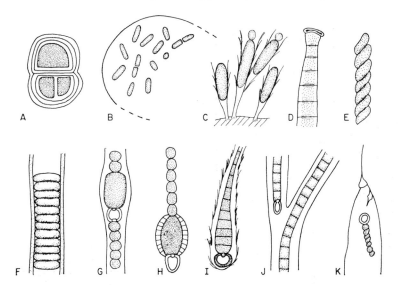

Fig. 8.47 Cyanobacteria.
(**A**) *Chroococcus*,
(**B**) *Aphanothece*,
(**C**) *Chamaesiphon*,
(**D**) *Oscillatoria*, (**E**) *Spirulina*,
(**F**) *Lyngbya*, (**G**) *Anabaena*
with heterocyst and akinete,
(**H**) *Cylindrospermum* with
basal heterocyst and akinete
above, (**I**) *Calothrix*,
(**J**) *Tolypothrix* with heterocyst
above point of branch, and
(**K**) *Richelia* in the diatom
Rhizosolenia.

found on and in the surface of rocks, tufa, etc. where they build up laminated deposits. The protoplast was thought to consist of a central region, the nucleoplasm, containing the nuclear material without any bounding membrane and an outer chroma-

toplasm in which the pigments are dispersed. The electron microscope shows in sections of Cyanophycean·cells a series of lamelae (thylakoids, Fig. 8.48) mainly in the peripheral regions of the cell but also penetrating into the centre. Although these are

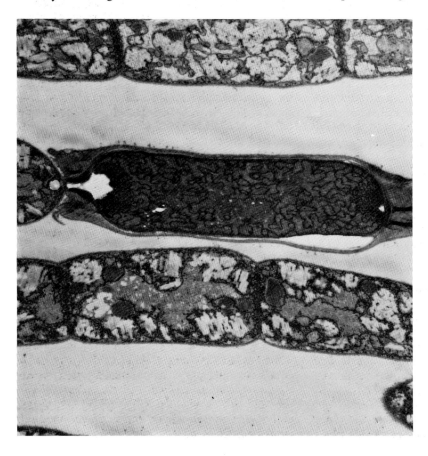

Fig. 8.48 Electron micrograph of filaments of *Aphanizomenon.* One filament with elongate heterocyst in which the thylakoids are prominent and aggregating towards the poles (×12 500 approx.). (Photograph kindly supplied by Dr R. V. Smith)

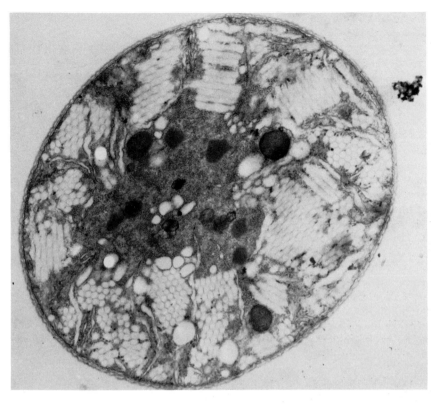

Fig. 8.49 Electron micrograph of *Microcystis* cell cut in transverse section showing the ill-defined central nuclear material and numerous gas vacuoles, each consisting of a bundle of gas cylinders. Some gas vacuoles are cut transversely, others obliquely or longitudinally. Note the very sparse thylakoids (× 30 000 approx.). (Photograph kindly supplied by Dr R. V. Smith)

not bounded by any enclosing membrane they are certainly the sites of photosynthesis. The nuclear material appears as threads and within the cell are bodies of various kinds of undetermined structure and function. Even under the light microscope the cell is seen as at least a bipartite structure (inner and outer investment) but clearly there are at least three regions in the inner investment and an outer somewhat fibrous sheath (Fig. 8.48). These are not unit membranes (p. 155). The ingrowth of cross walls is characteristic of many filamentous Cyanobacteria. Several cross walls can be initiated within one cell and the precision with which these are formed midway between older cross walls is striking.

Whilst some filamentous Cyanobacteria are relatively immobile within their mucilaginous sheaths others, e.g. *Oscillatoria*, *Spirulina*, and *Lyngbya*, glide rapidly and at the same time rotate about the longitudinal axis. Electron microscope observations have revealed rows of pores on either side of the cross walls at least in *Symploca*. The sheath material is often left behind as the filament moves and the pores in the inner investment may be involved in this gliding through the sheath. Fine pores or plasmodesmata also penetrate the cross walls whilst in the Stigonematinales, there are more

pronounced central pit-connections between the cells.

Some, but not all, planktonic Cyanobacteria are buoyant, and under calm conditions float to the surface of lakes and ponds forming so-called water blooms. The buoyancy is associated with the presence in the cells of pseudo-vacuoles or gas vacuoles. Ultrastructural studies have revealed a series of tubes (or gas cylinders) within each gas vacuole (Fig. 8.49).

Vegetative growth is the only known method of increase in cell numbers of many Cyanophytes especially of the coccoid series (Chroococcales). A small number of genera form exospores (e.g. *Chamaesiphon*, Fig. 8.47c), some filamentous genera differentiate small segments of the filaments (hormogonia) which are more actively motile and form a vegetative means of propagation. These are sometimes surrounded by a thicker wall and are then incapable of movement, the so-called hormocysts. Both types eventually germinate by outgrowth of the filament. Enlarged cells (with dense contents), sometimes with inflated walls, occur singly or in rows in *Anabaena*, *Cylindrospermum*, etc. and act as akinetes (Fig. 8.47g). These are almost certainly resting spores which germinate to form a new filament.

An apparently empty, slightly enlarged cell, the enigmatic heterocyst (Fig. 8.48), is often associated with a kinete but may also occur elsewhere in the filament. Electron microscope studies (Lang, 1968) show that the thick wall of the heterocyst is laid down between the three-layered inner investment and the outer sheath. As the wall thickens the photosynthetic lamellae become contorted and pack towards the poles of the heterocyst and the polar thickenings develop *inside* the plasma membrane. Thus the apparently empty heterocyst still contains cytoplasmic constituents, a feature observed by some early light microscopists and in line with the rare observations that under certain conditions heterocysts do germinate to form filaments. In some genera (e.g. *Tolypothrix*) branching occurs beneath the heterocyst (Fig. 8.47j).

The Cyanobacteria are widely distributed, occurring in hot sulphur springs at temperatures between 40° and 70°C, in all soils and most aquatic environments. They are particularly abundant in the tropics. They are not, very abundant in oceanic plankton (represented by *Trichodesmium* only), although the interesting symbiont *Richelia* occurs in the cells of the diatom *Rhizosolenia* (Fig. 8.47k). Symbiotic relationships are common amongst Cyanobacteria, e.g. *Anabaena* species in the fern *Azolla*, liverwort *Anthoceros*, gymnosperm *Cycas* and angiosperm *Gunnera*. Aberrant Cyanobacteria occur in the algae *Glaucocystis* and *Cyanophora*. They are common components of lichens (p. 281). Aggregation of cells and filaments in mucilaginous masses often yields macroscopic colonies which lie on lake sediments, (e.g. *Aphanothece*), float freely in the water (e.g. *Microcystis*), rest on soil (e.g. *Nostoc*) or form blackish clusters on rock faces (e.g. *Stigonema*). Colonies of *Calothrix* often form a conspicuous zone on rocks in the upper intertidal zone, in some areas associated with other blue-green bacteria which actively bore into the rock, Cyanobacteria also grow on and in mollusc shells. Other species build up deposits by precipitation of lime within the mucilage sheaths and many of the species living around thermal springs are involved in the precipitation of lime to form travertine. The black streaks down the sides of concrete buildings in wet tropical and even temperate regions are formed by a community of Cyanobacteria. Blackish patches on soils, lawns, etc. are produced by the aggregation of filamentous species, e.g. of *Anabaena*, *Phormidium*, etc. Filamentous species of *Anabaena*, *Oscillatoria*, and *Spirulina* are frequently found both floating in water (planktonic) or moving freely over the surface of underwater sediments (epipelic).

Two notable features of the metabolism of the Cyanobacteria are the ability of some species of the Nostocales to fix atmospheric nitrogen (a property which they share with certain bacteria (p. 253) and the formation by some species of very potent toxins (e.g. by *Microcystis*, *Anabaena*, and *Aphanizomenon*) which have been responsible for the death of cattle which have drunk water containing these genera.

Books and articles for reference and further study

Bacteria

ARCHIBALD, A. R. (1974). The structure, biosynthesis and function of teichoic acid. *Adv. Appl. Microbiol.*, **11**, 53–95.

ASHWORTH, J. M. and SMITH, J. E. (1973). Microbial Differentiation. *23rd Symp. Soc. gen. Microbiol.*, Cambridge University Press, Cambridge, 450 pp.

BERG, H. C. (1975). How bacteria swim. *Scient. Am.*, **233**, 36–44.

BUCHANAN, R. E. and GIBBONS, N. E. (1974). *Bergey's Manual of Determinative Bacteriology*. 8th Ed., Williams and Wilkins, Baltimore, 1268 pp.

COSTERTON, J. W. (1979). The role of electron microscopy in the elucidation of bacterial structure and function. *Ann. Rev. Microbiol.*, **32**, 459–470.

DAWES, E. A. and SENIOR P. J. (1973). The role and regulation of energy reserve polymers in micro-organisms. *Adv. Microbial. Physiol.*, **10**, 135–266.

DOETSCH, R. N. and SJOBLAD, R. D. (1980). Flagellar structure and function in Eubacteria. *Ann. Rev. Microbiol.*, **34**, 69–108.

JOHNSON, R. C. (1976). *The Biology of Parasitic Spirochaetes*. Academic Press, New York, San Francisco and London.

OTTOW, J. C. G. (1975). Ecology physiology and genetics of fimbriae and fili. *Rev. Microbiol.*, **29**, 79–108.

ROGERS, H. J., PERKINS, H. R. and WARD, J. B. (1981). *Microbial Cell Walls and Membranes*. Chapman and Hall, London and New York, 564 pp.

SHIVELY, J. M. (1974). Inclusion bodies of prokaryotes. *A. Rev. Microbiol.*, **28**, 167–187.

STANIER, R. Y., ROGERS, H. J. and WARD, J. B. (1978). Relations between Structure and Function in the Prokaryotic Cell. *28th Symp. Soc. Gen. Microbiol.* Cambridge University Press, Cambridge, 369 pp.

SMITH, P. F. (1971). *The Biology of Mycoplasmas*. Academic Press, New York and London.

SUDO, S. Z. and DWORKIN, M. (1973). Comparative biology of prokaryotic resting cells. *Adv. Microbial Physiol.* **9**, 153–224.

SUTHERLAND, I., (ed.) (1977). *Surface Carbohydrates of the Prokaryotic Cell*. Academic Press, London, 472 pp.

TROY, F. A. (1977). The chemistry and biosynthesis of selected bacterial capsular polymers. *Ann. Rev. Microbiol.*, **33**, 519–560.

Actinomycetes

KUSTER, E. (1967). The actinomycetes. In *Soil Biology*, edited by A. Burges and F. Raw, Chapter 4, pp. 111–127, Academic Press, New York and London.

LECHEVALIER, H. A. and LECHEVALIER, M. P. (1967). Biology of actinomycetes. *A. Rev. Microbiol.*, **21**, 71.

LECHEVALIER, H. A., LECHEVALIER, M. P. and GERBER, N. N. (1971). Chemical composition as a criterion in the classification of actinomycetes. *Adv. Appl. Microbiol.*, **14**, 47–72.

SYKES, G. and SKINNER, F. A., (ed.) (1973). *Actinomycetales characteristics and practical importance*, Academic Press, London and New York.

WAKEMAN, S. A. (1961). *The Actinomycetes*, Vol. 2, *Classification, Identification and Descriptions of Genera and Species*, Baillière, Tindall and Cox, Ltd., London.

WAKEMAN, S. A. (1967). *The Actinomycetes, A Summary of Current Knowledge*, Ronald Press, New York.

Cyanobacteria

DODGE, J. D. (1973). *The fine structure of algal cells*. Academic Press, London and New York, 264 pp.

DWORKIN, M. (1966). Biology of myxobacteria. *A. Rev. Microbiol.*, **20**, 75–106.

ECHLIN, P. (1966). The blue-green algae. *Scient. Am.*, **214**, 74–83.

FOGG, G. E., STEWART, W. D. P., FAY, P. and WALSBY, A. E. (1973). *The Blue-Green Algae*. Academic Press, London and New York, 420 pp.

HOLM-HANSEN, O. (1968). Ecology, physiology and biochemistry of the blue-green algae. *A. Rev. Microbiol.*, **22**, 47–70.

LANG, N. J. (1968). The fine structure of blue-green algae. *A. Rev. Microbiol.*, **20**, 75–106.

MULDER, E. G. (1964). Iron bacteria, particularly those of the *Sphaeotilus–Leptothrix* group, and industrial problems. *J. appl. Bact.*, **27**, 151–173.

STANIER, R. Y. (1942). The cytophaga group: a contribution of the biology of myxobacteria. *Bact. Rev.*, **6**, 143–196.

STARR, M. P. and SKERMAN, V. B. D. (1965). Bacterial diversity: the natural history of selected morphologically unusual bacteria. *A. Rev. Microbiol.*, **19**, 407–454.

Chapter 9

Eukaryotes

Fine structure of the eukaryotic cell

Eukaryotic cells differ from prokaryotic ones by the possession of one or more discrete nuclei each of which is enclosed by a perforated nuclear envelope. Eukaryotic cells are structurally more complex than prokaryotic ones particularly with relation to the development of cell membranes. Fungi, algae (with the exception of the blue-green algae), protozoa and slime moulds are all eukaryotic and although the cells of some of them may be less complex than those of higher plants and animals, others show considerable differentiation. Special ultrastructural features of the cells of the various groups of eukaryotic micro-organisms will be considered in the sections dealing with those groups, but first a general survey of the principal structures involved is necessary.

Cell ultrastructure

It is convenient to consider the different groups of organisms simply in terms of cells. Thus most protozoa, slime mould swarm cells, fungal and algal zoospores may be defined as unicells while the vegetative phases of most fungi and many algae are multicellular. Although a large number of organelles are common to all cells, some are found in only a few. It is, therefore, also convenient to illustrate the range of differentiation by means of a hypothetical eukaryotic cell. It must be emphasized that such a diagram (Fig. 9.1) cannot represent what is essentially a dynamic structure, or even the fixed image of an existing organism. However, with this in mind it is hoped that it will serve to summarize the morphology and spatial arrangement of some of the structures to be described in this section.

Cell walls and the plasma membrane

Cells of the thalli of algae and fungi possess a true cell wall (Fig. 9.2a, b), which is more or less rigid like that of bacteria but is of different chemical com-

position and architectural structure. Generally the wall consists of a reticulate microfibrillar skeleton or framework embedded in an amorphous matrix (Fig. 9.2b). In certain genera of green algae cellulose constitutes the microfibril layer while in most fungi the fibrils are composed of chitin, though cellulose does occur in some groups. Considerable variation is found in the wall structure of cells from different

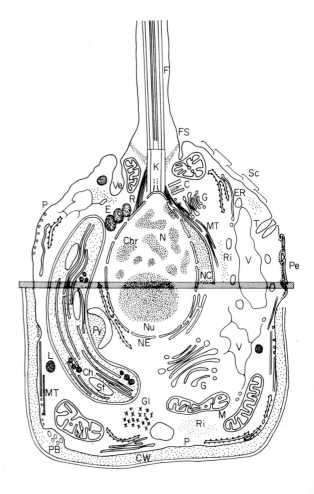

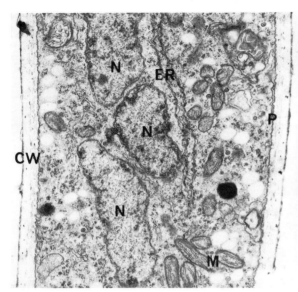

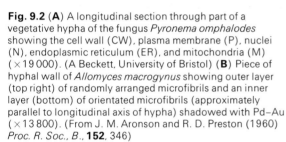

Fig. 9.2 (A) A longitudinal section through part of a vegetative hypha of the fungus *Pyronema omphalodes* showing the cell wall (CW), plasma membrane (P), nuclei (N), endoplasmic reticulum (ER), and mitochondria (M) (×19 000). (A Beckett, University of Bristol) **(B)** Piece of hyphal wall of *Allomyces macrogynus* showing outer layer (top right) of randomly arranged microfibrils and an inner layer (bottom) of orientated microfibrils (approximately parallel to longitudinal axis of hypha) shadowed with Pd–Au (×13 800). (From J. M. Aronson and R. D. Preston (1960) *Proc. R. Soc., B.,* **152**, 346)

stages in the life cycle of an organism; some fungal spore walls possess complex ornamentations of several layers, and are often pigmented; some algal groups have what might be termed a 'cell covering' composed of scales. Scales and other inorganic deposits are common in many protozoa and more typical cell walls are found enclosing the oocyst

Fig. 9.1 Diagrammatic representation of ultrastructure of eukaryotic protist cells. Upper portion includes features found in flagellated cells (flagellate protozoa, zoospores of algae and fungi, myxomycete swarm cells). Lower portion shows structures found in vegetative cells of algae and/or fungi. The organelles shown do not occur in all cells, but are included to illustrate their spatial arrangement (e.g., chloroplasts occur in algae and green flagellates but are absent from all fungi, myxomycetes, and most protozoa). Key to lettering: C=centriole, E=eyespot, F=flagellum, G=Golgi apparatus, K=kinetosome, L=lipid drop, M=mitochondria, N=nucleus, P=plasma membrane, R=rhizoplast, T=trichocyst, V=vacuole, Ch=chloroplast, Chr=chromosome, ER=endoplasmic reticulum, Fs=flagellar sheath, Gl=glycogen, MT=microtubules, NE=nuclear envelope, NC=nuclear cap, PB=paramural body, Py=pyrenoid, Pe=pellicle, Ri=ribosomes, Sc = scales, St = starch grain, Ve = vesicle.

stages of some of them. Slime layers and filamentous layers are formed at the cell surface in some protozoa but such layers are far removed from the rigid structures of fungi and algae described above.

Within the cell wall the living protoplast is bounded by a membrane, the plasma membrane or plasmalemma (Figs. 9.1; 9.2a). Many protist cells do not possess a true cell wall and are merely enclosed by the plasma membrane. Zoospores and gametes are examples of such naked cells.

The electron microscope has shown the chemically fixed plasma membrane to be a tripartite structure composed of two electron-dense layers separated by an electron-transparent layer. This tripartite structure which occurs in many biological membranes is termed a unit membrane. Microdensitometry of the plasma membrane in chemically fixed cells has shown it to possess a polarity such that one of the electron-dense layers is thicker than the other.

Detailed studies employing electron microscopy of both chemically fixed and freeze-etched material, together with a consideration of the physical and chemical properties of biological membranes, suggest that not only are various membranes constructed differently but that any one membrane may undergo structural changes with time and according to function. In view of this a rigid presentation of the ultrastructure of the plasma membrane or any other cell membrane is misleading. However, it is interesting to note that in some cases the incorporation of cytoplasmic membrane bounded vesicles into the plasma-membrane involves the fusion of the two membrane systems and suggests an interrelationship between them. Such associations occur between membranes of the Golgi system and the plasma membrane.

The reverse process whereby substances enter the cell from the exterior and pass into the cytoplasm within membrane bounded vesicles is termed pino-cytosis and is particularly common in amoebae and myxamoebae.

The plasma membrane in euglenoids is regularly folded around the underlying pellicle while in many fungal zoospores it forms a more or less smooth boundary to the cell.

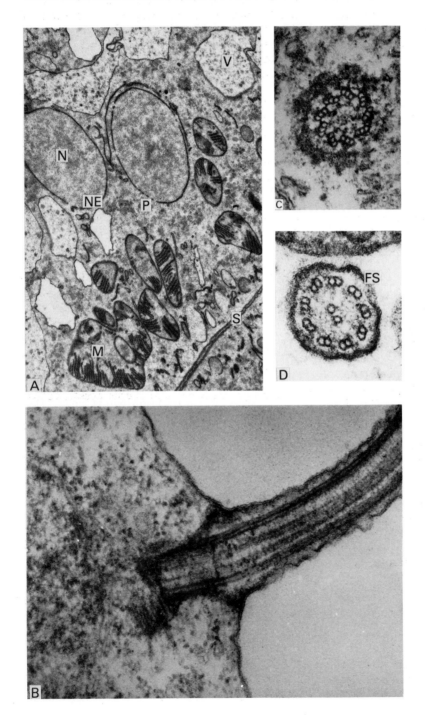

Fig. 9.3 (A) A longitudinal section through part of the zygospore of the fungus *Rhizopus sexualis* showing nuclei (N), vacuoles (V), mitochondria (M), and part of a developing septum (S). The nuclear envelope (NE), is perforated by several pores (P) ($\times 15\,000$). (M. A. Gooday, Bristol University.) **(B)** A longitudinal section through part of a zoospore of the chytrid fungus *Phlyctochytrium* showing the flagellum enclosed within the sheath which is continuous with the plasmalemma. The kinetosome, and vestigial kinetosome lie at the base of the flagellum ($\times 80\,000$). (L. W. Olson and M. S. Fuller, University of California, Berkeley) **(C)** A transverse section through a centriole of the aquatic fungus *Catenaria anguillulae*. Each of the fibres is composed of a triplet of subfibres and each triplet is radially tilted anticlockwise. This pattern is also seen in parts of the kinetosome ($\times 120\,000$). (M. S. Fuller, University of Georgia) **(D)** A transverse section through the flagellum of *Coelomomyces indicus*, a fungus parasite of mosquitoes. The flagellar sheath (FS) encloses the peripheral ring of nine double subfibres which together with the central pair constitute the axoneme ($\times 92\,000$). (A. Beckett, Bristol University)

Nuclei

THE NUCLEAR ENVELOPE

The nuclear envelope of most eukaryotic protist cells consists of two membranes. These are separated by a space termed the perinuclear space. Each membrane measures 70–90 Å (7–9 nm) thick and is perforated by a number of pores at the edges of which the two membranes are continuous (Figs. 9.1; 9.3a).

Additional membranes are found inside the nuclear envelope in some amoebae.

Freeze-etched preparations of yeast nuclei show both open and sealed pores in the envelope and it has been suggested that a regulatory mechanism may exist to control the flow of specific nuclear substances from the nucleoplasm to the cytoplasm.

In some fungal zoospores a double membrane, continuous with the outer membrane of the nuclear envelope, gives rise to an enclosed structure surrounding part of the nucleus. This structure is the nuclear cap (Fig. 9.1), within which densely packed ribosomes are found. Pores are absent from the nuclear cap membrane.

CHROMOSOMES

The electron microscope has in general been unsuccessful in contributing to our knowledge of chromosome fine structure. Following glutaraldehyde or osmium tetroxide fixation, chromosomes are usually seen as rather amorphous electron-dense structures. In some of the better preparations obtained so far, fine fibrillar components have been observed. In *Amoeba proteus* and *Pelomyxa* the nucleus contains fine threads in the form of regular helices. However, the dimensions of these structures are considerably greater than the double helix of deoxyribose nucleic acid (DNA).

Chromosomes in dinophycean algae are unique in that they consist of electron-dense banded structures. This banding is apparently due to the arrangement of the DNA fibrils. These chromosomes also differ from those of other eukaryotes by their lack of histones. In this they resemble bacterial nuclei. In the fungi, chromosomes are usually seen as irregular, dense masses.

NUCLEOLI

Nucleoli are seen in suitably prepared cells usually as compact, granular bodies consisting of electron-dense ribosomes (p. 180).

SPINDLES AND CENTRIOLES

Nuclear division in eukaryotic protist cells usually involves a spindle apparatus composed of tubular structures measuring 200–240 Å (20–24 nm) in diameter. These microtubules are often enclosed within a persistent nuclear envelope which does not break down until the telophase separation of daughter nuclei. This is the case in many fungi (p. 185), slime moulds, and certain flagellate and ciliate protozoa.

Little or no information is so far available as to what degree or how these microtubules are involved with the separation of chromosomes. In many instances where they occur, no connections with the chromosomes can be detected, while in others, some microtubules appear to terminate at or within the chromosome. Usually continuous microtubules are found passing apparently between the chromosomes and linking the polar ends of the nucleus. The mitotic nucleus of many eukaryotic cells is associated at its polar regions with an astral ray-centriole complex. Astral rays are microtubular elements often seen radiating from the cylindrical centriole. All centrioles have a system of microtubules running along their length, which, when viewed in transverse section, are seen to be arranged in a peripheral ring of nine, each composed of a triplet of tubules. Each triplet is radially tilted in a clockwise direction when viewed from the proximal end (Fig. 9.3c). The cylinder of triplets is a structure characteristic of at least part of the kinetosome at the base of flagella in motile cells (Fig. 9.3b).

Although many eukaryotic cells possess centrioles or similar structures, in some, these organelles apparently are not functionally involved in the polarization of nuclei during mitosis. Other organisms have what is termed anastral and acentric mitosis where neither astral microtubules nor centrioles are present. However, when centrioles do occur and are actively involved in mitosis, they undergo division at some stage in the mitotic cycle.

Although little is known of the mechanism involved in replication, DNA has been reported to occur in centrioles. It has been suggested that the centriole in certain protozoa, which is associated with complex fibrous structures, may be the organizing centre for protein molecules that are synthesized in the cytoplasm.

Flagella and associated organelles

In all eukaryotic protists possessing motile cell stages, flagella are composed of an axial core, enclosed within a membranous sheath continuous with the plasma membrane (Figs. 9.1; 9.3d). In most the axial core, or axoneme, consists of two sets of tubules, a peripheral ring of nine and a central pair. Transverse sections show the peripheral tubules to be doublets of tubules, and both these and the central tubules have a dense outer layer enclosing a less dense core (Fig. 9.3d).

In some protists, for example dinoflagellates and euglenoids, varying degrees of additional differentiation within the flagellum sheath have been found. Variation in ornamentation along the length of some flagella has made it possible to distinguish between the so-called whiplash flagellum (Peitschengeissel), which is devoid of appendages, and the hairy flagellum (Flimmergeissel) possessing lateral hairs. These lateral hairs have been found to be bipartite in that the distal end is very much more slender than the proximal region. Some flagellates possess both types of flagella on the one cell. In such cases the posterior flagellum is of the whiplash type, while the hairy flagellum is situated anteriorly. This arrangement is termed heterokont flagellation.

Associated with flagella and cilia, usually at or near the base, are several organelles which lie inside the cell proper. The basal body, or kinetosome (Fig. 9.3b) is a cylinder composed of a ring of nine microtubules each arranged as a triplet. However, the kinetosome does not retain this regular structure throughout its length, since a central 'cartwheel' is present usually towards the lower end of the cylinder, while in the kinetosome–flagellum transition region a complex stellate fibre pattern is often seen. In some fungal zoospores this region is marked by the gradual transition from a peripheral ring of nine doublet tubules (as in the axoneme) to nine triplet tubules, the third tubules of a triplet arising as an outgrowth of one of the doublets. Distally the kinetosome is delimited from the flagellum or cilium by a 'crosswall' of dense material, the terminal plate (Fig. 9.1).

Bands of fibrous material are closely associated with the kinetosome in many motile cells. A network of fibres linking the basal bodies is found in ciliates. This network, the kinetodesma, has been extensively studied in *Paramecium*. In many flagellates, fibres, termed rhizoplasts, link the kinetosome with the nucleus; others, known as parabasal fibres, link the kinetosome with the Golgi apparatus.

Cytoplasmic membranes

ROUGH AND SMOOTH MEMBRANES

A variety of membranes is found in most eukaryotic cells. Some of these membranes are studded with dense granules (ribosomes) similar to those found in the nucleolus, while others are smooth. These are commonly referred to as rough and smooth endoplasmic reticulum respectively. Recently the 'endomembrane concept' has been proposed for eukaryotic cells in which cytoplasmic membranes, which may be specialized locally within a cell, form part of an interrelated membrane system.

VACUOLES

Eukaryotic cells, in contrast to those of prokaryotes, possess a number of vacuoles. These are usually bounded by a single unit membrane termed the tonoplast. Vacuoles of varying shapes and sizes perform a number of functions according to cell type and state. In some fungal and algal cells they serve to localize storage products; in others they contain water. Many protozoa possess a more or less complex vacuole system, the contractile vacuole, which is sometimes connected to a system of small vesicles by tubular ducts.

Food vacuoles are formed by invagination of the plasmamembrane; the process of pinocytosis (p. 178) also results in the production of vesicles within the cytoplasm. Such processes are often associated with considerable membrane synthesis.

THE GOLGI APPARATUS

The Golgi apparatus is the term given to the complex system of smooth membranes present in many eukaryotes either in the form of dictyosomes (stacks of elongated vesicles or cisternae) or as single cisternae, from which smaller, spherical vesicles are budded. The predominant function of the Golgi system is probably secretory, but the endomembrane concept suggests that the Golgi apparatus is also an important transitional component of a subcellular pathway in membrane differentiation which links rough membranes, such as the outer membrane of the nucleus envelope and the rough endoplasmic reticulum, with smooth membranes, such as the plasma membrane and the tonoplast.

Ribosomes

Ribosomes are small electron-dense granules approximately 170–230 Å (17–23 nm) in diameter and composed of two subunits which in the functional ribosome must be joined together. Chemical analysis of isolated ribosomes has shown them to contain about 40% protein and 60% ribonucleic acid (RNA).

Ribosomes of both prokaryotes and eukaryotes are the site of protein synthesis (p. 26), and their occurrence and number may be related to the rate of synthesis within the cell.

Eukaryotic ribosomes are larger than those of prokaryotes and are situated along the surface of rough endoplasmic reticulum, the outer membranes of the nuclear envelope, inside the nucleus and free in the cytoplasm. Ribosomes are also found in chloroplasts and mitochondria where they appear to be of the smaller 'prokaryotic type'.

Mitochondria

Typical mitochondria are characterized by a double delimiting membrane and an internal differentiation in the form of flattened membranous sacs (cristae) or tubules (tubuli) (Figs. 9.3a; 9.4b).

It is within these organelles that the enzymes of the respiratory system occur. Morphologically mitochondria undergo considerable changes reflecting not only their functional state but also the metabolic state of the cell as a whole.

In many motile cells mitochondria are often intimately associated with the flagellar apparatus and in the alga *Prymnesium parvum* not only is there a relationship between the haptonema base and a mitochondrion, but also the cristae within this mitochondrion exhibit a marked polarity.

Mitochondria are apparently absent from some eukaryotes living in anaerobic environments.

Plastids, pyrenoids, and eyespots

Plastids are found in algal thallus cells and their motile stages (phytoflagellates). These plastids may be differentiated into chloroplasts, chromoplasts, or amyloplasts containing chlorophyll, carotenoids, and starch respectively. Details of their structure and distribution are given on p. 217.

Pyrenoids (Fig. 9.1) are regions of granular dense material embedded within the chloroplast or arising as an outgrowth of it, and usually associated with the deposition of reserve materials derived from photosynthates (p. 217). In many chloroplasts, discrete membrane bounded starch granules are found between the lamella, but in euglenoids the reserve material paramylon is formed outside the chloroplasts.

Following aldehyde or osmium tetroxide fixation, eyespots are seen with the electron microscope as dense osmiophilic granules (p. 217). They are considered to be associated with photoreception.

Microbodies and storage products

A variety of membrane bounded vesicles associated with ER is found in a number of eukaryotic cells. Often crystalline inclusions appear in these vesicles. Their function is obscure but storage of proteinaceous material or of enzymes has been suggested.

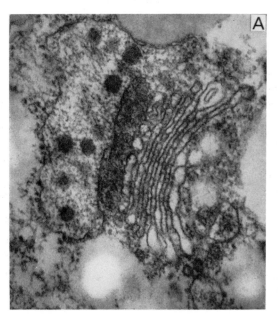

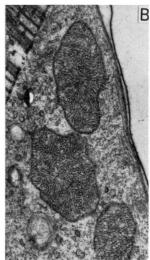

Fig. 9.4 (A) A section through part of the Golgi apparatus of *Selenidium terebellae*, an archigregarine gut parasite of marine worms. Some of the elongated vesicles are inflated at their ends forming the peripheral Golgi vesicles (×60 000). Mrs G. Dorey, Bristol University) **(B)** A section through part of the ciliate *Euplotes eurystomus* showing mitochondria with typical tubuli (×17 500). (R. Gliddon, Bristol University)

Glycogen is stored in many mature fungal cells, often in the form of dense stellate granules.

Lipid droplets are also found in many fungal cells, particularly mature spores. In osmium-fixed specimens they are areas of dense material, while after potassium permanganate fixation they are seen as electron-transparent areas bounded by what appears to be a single dense membrane. Freeze-etched preparations of yeast cells have shown these membranes to consist of fine concentric lamellae.

Cytoplasmic microtubules and miscellaneous organelles

Microtubules are found in the cytoplasm of many eukaryotic protists.

They are associated with the cell wall in many algae, while in motile cells they often occur close to the kinetosome. In motile cells of fungi belonging to the order Blastocladiales, there is apparently a close link between the kinetosome tubule pattern and the form and arrangement of microtubules which lie close to the membrane surrounding the nuclear cap.

Trichocysts are found in many eukaryotic ciliates and flagellates (p. 201). When discharged they consist of a striated fibrous shaft with a spine at the tip. Some trichocysts have a fibre pattern in their tips, similar to that found in cilia and flagella, suggesting their origin from a basal body or kinetosome.

Contractile structures or myonemes are found in some protozoa. These are usually bands of fibrous material running around the cell beneath the pellicle. Bands of contractile fibres occur in those diatoms which extrude mucilage, which is passed in streams along the outside of the cell. Active transport of this mucilage to one end of the cell by means of these contractile elements, provides the propulsive force necessary for cell movement.

In recent years the electron microscope has provided information leading to a new and complex view of the cytoplasmic organization of cells. One of the most striking features of the eukaryote is the development of membrane systems which effectively compartmentalize the cell, provide sites for enzyme location and chemical reactions, and along which active transport can occur. It is not unusual that many of these membrane bounded organelles are similarly constructed and common to a large number of eukaryotes. It is a reflection on the efficiency of the system and the limited range of molecular components involved.

The fungi

Introduction

The relatively large size of fungus cells and their possession of discrete membrane-bounded nuclei and typical mitochondria distinguishes the fungi from the bacteria. Although the yeasts and some aquatic species are normally unicellular, most fungi are filamentous. The filaments, known as hyphae, are usually much branched and collectively form the vegetative thallus or mycelium. The filamentous habit permits a greater diversity of form than that seen among bacteria. In the higher fungi the hyphae may aggregate together to form complex solid structures which in some species may reach a considerable size (e.g. the fruit-bodies of some of the wood-destroying fungi are often more than a foot in diameter). Not only do the fungi show a much wider range of form than do the bacteria, but also their life cycles are more complex.

Habitat

Fungi occupy a wide variety of habitats. Some are aquatic, some inhabiting fresh water and others the sea. The majority occupy moist situations on land although a few, notably species of *Aspergillus* and *Penicillium*, are able to withstand somewhat drier conditions.

Like the majority of the bacteria, fungi are heterotrophic (p. 69). No fungus is able to grow well in the absence of organic food substances. No fungi possess photosynthetic pigments and none is able to grow satisfactorily with carbon dioxide as the sole source of carbon. Hence they are dependent on preformed organic substances as a source of energy. Many are entirely saprophytic, feeding on non-living organic matter of diverse kinds. These include many species which are troublesome as destroyers and spoilers of man's foodstuffs and other valuable stored products (p. 354). The saprophytic abilities of others can be harnessed to our use in a variety of industrial processes (p. 361). In the presence of an external supply of sugars or other suitable organic food, many exhibit astonishing synthetic powers and produce a great variety of complex organic substances, including cell proteins and reserve foods (such as fats and complex polysaccharides), vitamins, alcohols, organic acids, enzymes, pigments, and antibiotic substances (e.g. penicillin p. 362). Many fungi obtain their organic nutrients from living host organisms. Many are active parasites of plants, including the major crop plants (p. 288), and some cause diseases of man and domestic animals (p. 304). Some parasitize insects (p. 314), others, the predacious fungi, are able to trap and destroy eelworms and soil protozoa (p. 320). Fungi are present wherever organic material occurs, if other environmental factors are favourable.

The effects of environmental factors

The general nutritional requirements of microorganisms have been considered in Chapter 4. All fungi require a relatively large amount of a carbon source and most use carbohydrates more readily than other carbon compounds. Species and often different strains of a particular species differ in the range of carbon sources that they can utilize, but in general glucose and fructose are suitable. Glucose is commonly included in artificial culture media. Fungi require nitrogen, potassium, sulphur, magnesium, and a number of micronutrient minerals (trace elements, p. 72). Many grow and sporulate well on a glucose-salts medium from which they can synthesize all the complex organic substances they require. Others require external supplies of one or more vitamins or growth substances, which they are unable to synthesize themselves. The commonest requirement is thiamin, but many fungi are deficient for biotin or other growth substances. Some can synthesize enough of such compounds to permit vegetative growth but need an external supply for the initiation of sporulation (p. 193).

Other environmental factors such as H ion concentration of the substrate, temperature, aeration, light, water content of the substrate, and humidity, may limit fungal growth and/or reproduction. The requirements of different species and of different phases in the life cycle of a given species differ widely, but in general fungi are favoured by a slightly acid reaction (in contrast to most bacteria), temperature around 20 to 25°C, good aeration, a moist substratum and a humid atmosphere.

Fine structure of the fungal cell

The cell wall

Vegetative cells and spores of nearly all fungi are surrounded by a definite cell wall. Motile zoospores and gametes lack such a wall and the vegetative thallus is naked in a group of primitive plant parasites, the Plasmodiophorales (which includes *Plasmodiophora brassicae*, the cause of the serious club root disease of cabbage and other crucifers, p. 289) and in a number of internal parasites of other fungi, for example species of *Coelomomyces*, an internal parasite of mosquitoes (p. 319) and *Erynia neoaphidis*, a parasite of the pea aphid.

The structure of the fungus cell wall is architecturally like that of the walls of algae and other green plants, and consists of a network of microfibrils (Fig. 9.2b) with matrix substances filling the spaces between the fibrils.

In most fungi the cell wall consists of 90–80% polysaccharides together with proteins, lipids, and sometimes also pigments, polyphosphates, and inorganic ions.

The microfibrillar skeleton is frequently chitin, occasionally cellulose, or, in yeasts, non-cellulosic glucans. The cementing matrix usually contains protein, lipids, and various polysaccharides. Some of the protein may be enzymic, some may be due to cytoplasmic contamination, but some must be firmly bound to the microfibrils since it is removed only by drastic treatment. The structure of the cell wall lipids differs from that of cytoplasmic lipids. The nature and proportions of the polysaccharides vary.

Fungal cell wall composition together with other information is currently used in taxonomic schemes. A tentative outline, based on admittedly incomplete data, showing correlation between cell wall composition and the main taxonomic subdivisions of the fungi has been drawn up by Bartnicki-Garcia (1968), Table 9.1. Cell wall composition is also proving a useful criterion in the classification of bacteria, particularly the actinomycetes (p. 170), and of the algae (p. 219).

The protoplast

Fungal protoplasts resemble, but in many ways are less complex than, those of animals or green plants. They are, however, definitely eukaryotic in structure and thus considerably more complex than those of bacteria (p. 155).

The plasma membrane and endoplasmic reticulum (ER) in vegetative hyphae are similar to those of other eukaryotic cells (pp. 177, 180).

Table 9.1 Cell wall composition and taxonomy of fungi (after Barnicki-Garcia (1968). *A. Rev. Microbiol.*, **22**, 87–108)

Chemical category	Taxonomic group
I Cellulose—glycogen[1]	Acrasiales
II Cellulose—glucan[1]	Oomycetes[2]
III Cellulose—chitin	Hyphochytridiomycetes
IV Chitosan—chitin	Zygomycetes[3]
V Chitin—glucan[1]	Chytridiomycetes
	Ascomycetes (except yeasts)
	Deuteromycetes (except asporogenous yeasts)
	Basidiomycetes (except mirror yeasts)
VI Mannan—glucan[1]	Saccharomycetaceae
	Cryptococcaceae
VII Mannan—chitin	Sporobolomycetaceae
	Rhodotorulaceae
VIII Polygalactosamine-galactan	Trichomycetes

[1] Incompletely characterized.
[2] Oomycetes are unique among filamentous fungi in possessing a cellulose wall.
[3] Zygomycetes differ from the filamentous higher fungi in the absence of glucan in the cell walls of vegetative hyphae; glucan has, however, been demonstrated in the spores of *Mucor rouxii* and the mannan content increases with the assumption of the yeast form by this fungus.

Local aggregations of ER occur in some fungi at sites of particularly intense metabolic activity. Ribosomes (p. 180) occur both in the cytoplasmic matrix and in association with the ER.

The tips of growing hyphae possess a zone of exclusion within which are many wall vesicles of varying size and type. Structural continuities between wall vesicle membrane and plasma membrane suggest that a process of fusion occurs between the two. This enables the vesicle contents to be incorporated into the cell wall.

Mitochondria (p. 181) have been seen in all actively growing fungal cells so far examined but are absent or deformed in some starved or anaerobically treated ones and in the protoplasts of certain internal parasitic fungi.

The mitotic apparatus varies among genera but usually involves a microtubular spindle at the poles of which lie centrioles (e.g. *Catenaria anguillulae*, Fig. 9.5a), plaque-like spindle pole bodies or SPB's (e.g. *Sordaria humana*, Fig. 9.5b) or globular spindle pole bodies (e.g. *Polystictus versicolor*). Kinetochores often occur on the chromosomes at the points where spindle microtubules terminate

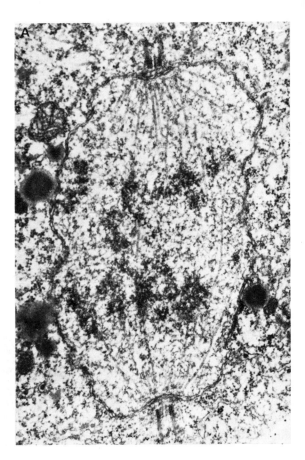

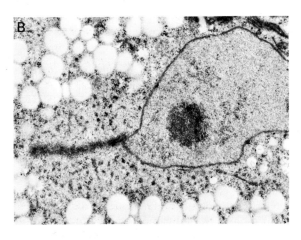

Fig. 9.5 (A) *Catenaria anguillulae.* Early anaphase nucleus. Centrioles, chromosomal fibres, continuous spindle fibres, and chromosomes are evident. The nuclear membrane is still intact (×36 000). (From Ichida, A. A. and Fuller, M. S. (1968) *Mycologia*, **60**, 141) **(B)** *Sordaria humana.* Part of an interphase nucleus following mitosis III in the ascus. The rod-like SPB is surrounded by astral microtubules. (×16 000). (A. Beckett, Bristol University).

(e.g. *Sapolegnia* and *Fusarium*). In most fungi the nuclear envelope remains intact during mitosis and meiosis but in Basidiomycetes it breaks down prior to metaphase. In *Saprolegnia* and *Thraustotheca* there is a nuclear envelope-microtubule association and the configuration of the nucleus and its associated microtubules suggests the presence of a shear force-producing interaction which may result in the sliding of the nuclear envelope along the microtubules. Such a mechanism, which may be primitive, could enable the chromosomes to separate at mitosis despite the formation in these genera of what is a comparatively small spindle.

Glycogen and lipids are the commonest reserve foods in the majority of fungi and are often abundant in fruit body cells and spores.

Some Lower Fungi are motile unicells or have motile phases in their life cycles. The flagella of these are of the complex type typical of motile eukaryotic cells (Fig. 9.3b, d) in contrast to the relatively simple ones of the bacteria.

The fungal thallus

Unicellular fungi

The simplest type of fungal thallus is that of the primitive unicellular fungi or chytrids and some allied forms. These show many affinities with the protozoa, but with only a few exceptions they possess a definite cell wall during the greater part of their life history. Among them is a number of species causing diseases of economic plants (e.g. *Synchytrium endobioticum*, the cause of black wart disease of potato) and others may occur in epidemic form on planktonic algae, but the majority are of little or no economic importance.

In contrast, the yeasts, which also normally exist in the unicellular condition, are of the greatest importance to the industrial microbiologist, chiefly owing to their ability to ferment sugars with the production of carbon dioxide and alcohol (p. 368). These organisms are almost certainly not primitive but have probably degenerated into the unicellular condition from a previous filamentous state. Even those yeasts which are usually predominantly unicellular will, under certain cultural conditions, produce chains of cells resembling the hyphae of normal fungi but less stable than those of typical filamentous species. The component cells may become elongated and the filaments are then known as 'pseudomycelia'. The life histories of the yeasts show a striking resemblance to those of certain normally filamentous species. Moreover, many species of filamentous fungi of various groups will assume a yeast-like condition under certain circumstances (p. 189).

Yeast cells are usually globose or ellipsoidal but are occasionally almost cylindrical. They are surrounded by a definite cell wall, which is thin and elastic in young cells but may become thickened and rigid in older individuals.

Under favourable conditions yeast cells divide rapidly. In most species, including those used in the fermentation industries, the cells multiply by budding. A small protuberance or bud grows out from the parent cell, enlarges to reach approximately the same size as the parent and is finally cut off and separates from it (Fig. 9.6a). Buds develop from one or more sites on the parent cells according to the species. Where the process is taking place rapidly the cells may remain partially attached to one another. Some groups of yeasts, known as the fission yeasts, show binary fission (Fig. 9.6b), resembling that of bacterial cells (p. 83). Both methods of multiplication result in similar masses or colonies of cells which are known as sprout mycelia. Some yeasts produce spores (p. 197), but these are part of the sexual cycle and are formed in a manner different from that of bacterial endospores (p. 160).

Filamentous fungi

Most fungi are filamentous and consist of a mass of branched hyphae. These may be aseptate, the vegetative thallus then consisting of a single multi-nucleate (coenocytic) cell, or regularly septate from an early stage in development.

The aseptate forms, characteristic of the lower fungi or Phycomycetes ('algal fungi'), are considered primitive. Growth in length of an individual coenocytic hypha takes place at or just behind the tip. Branches may develop from points farther back along the hypha and these in their turn grow by increase in length of the apical portions. The wall of the growing tip is thin and extensible and the cytoplasm, as in the young yeast cell, is more or less homogeneous. Farther back from the tip the walls are relatively rigid, as a result of secondary deposition of material, and the cytoplasm is granular and vacuolated. Active streaming of the cytoplasm and its contained granules along the hypha can be readily seen. Food materials are carried in this way towards the growing tip, branches, or developing reproductive organs. The nuclei are usually small, but can be demonstrated by suitable staining.

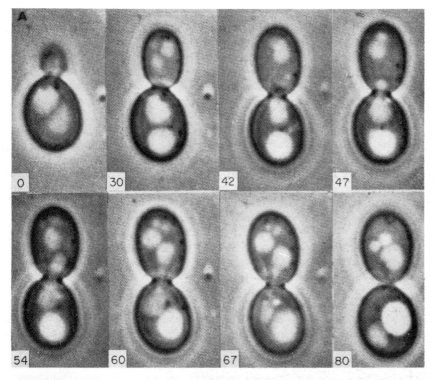

Fig. 9.6 (A) Time-lapse phase-contrast photomicrographs of living cell of a budding yeast, *Saccharomyces cerevisiae*, showing budding of cell and division of nucleus. Figures represent minutes after first photograph was taken. In first, second, and third photographs nucleus is entirely within parent cell. As bud enlarges nucleus becomes dumb-bell-shaped and extends into the bud (fourth, fifth, and sixth photographs), finally dividing, one daughter nucleus remaining in the parent cell and one in the bud. The last picture shows the bud rounded off and separated from parent (×3000). (From Robinow, C. F. (1966) *J. cell. Biol.*, **29**, 129)

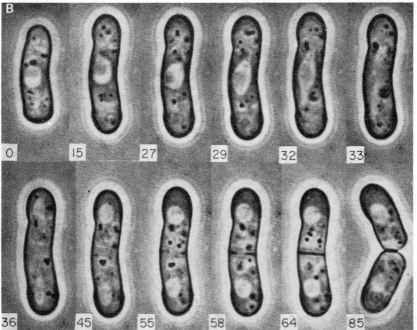

Fig. 9.6 (B) Time-lapse phase-contrast photomicrographs of a fission yeast, *Schizosaccharomyces versatilis*, showing nuclear division followed by septum formation and separation of resulting daughter cells. Figures represent minutes after first photograph was taken. First six photographs show stages in division of nucleus (light body); the next two show daughter nuclei migrated to poles of cell, the next three show septum growing inwards at middle of cell and the last shows the daughter cells separating (×1520). (From Robinow, C. F. and Bakerspigel, A. (1965) in Ainsworth, G. C. and Sussman, A. S. *The Fungi*, Vol. 1. Academic Press, New York)

Cross walls are formed in normally aseptate species in connexion with reproduction; dead or injured parts of the hypha are also cut off by septa. Old petri dish cultures of *Pythium* or *Mucor* often show young coenocytic hyphae at the edges of the plate and empty, regularly septate ones in the older central part of the colony, indicating that as the older mycelium exhausts the food supply the hyphae die and walls are formed in succession cutting off the extending dead portion.

The higher fungi are septate from an early stage. Linear growth of these, too, is limited to the extreme tip of the hypha, and the septa form behind this extending zone at a distance from the tip characteristic for the species and to some extent influenced by environmental conditions. Fungal septa do not form as a plate across the spindle of a dividing nucleus, as in higher plants, but grow inwards from the wall as a ring or annulus more or less independently of nuclear division. The wall does not usually extend right across the hypha but a central pore is left, through which there is a protoplasmic continuity from cell to cell.

In many fungi the pore is a simple hole in the septum (Fig. 9.7a) but in some of the higher basidiomycetes (p. 198) it is a more complex structure. The rim of the septum bordering the pore is swollen and a complex system of membranes develops in the vicinity of the pore (Fig. 9.7b). Such a complex system is termed a dolipore. In spite of the complexity of the adjacent membrane system, nuclei and other organelles are able to pass through the dolipore. During passage from one cell to an adjacent one nuclei usually become vermiform, regaining their original shape on reaching the neighbouring cell. A septate hypha, therefore, is not completely divided into separate compartments; protoplasmic streaming and nuclear migration can go on. The partial septa may give added rigidity to the hypha. It is from this type of septate mycelium that the complex fruit-bodies of the higher fungi are formed. A further biological advantage is that 'repairs' to a damaged hypha can be more readily accomplished by the plugging of the pore in the

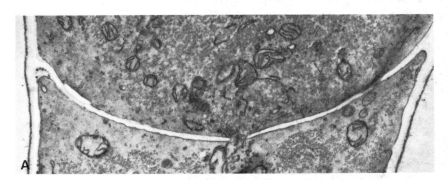

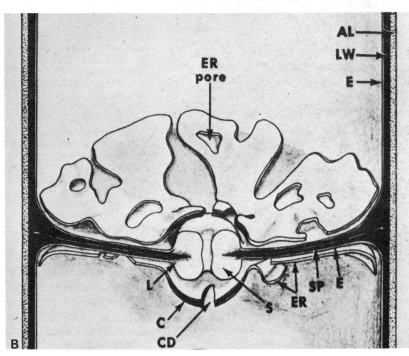

Fig. 9.7 (A) Simple pore in septum between conidiophore (lower cell) and ampulla (spore-bearing cell) of *Oedocephalum roseum*. The septum grows inwards across the young conidiophore leaving a central pore, through which organelles (including nuclei and mitochondria) pass into the developing ampulla. The photograph shows vesicles and cisternae passing through the pore (×9000). (B. Dixon, University of Bristol) **(B)** Semi-diagrammatic sketch of a mature septum of *Rhizoctonia solani*. Al=amorphous layer of lateral wall, LW=lateral wall, E=plasmalemma (ectoplast), ER=endoplasmic reticulum, SP=septal plate, composed of cross-wall lamellae, S=septal swelling, L=lamellae within the septal plates, C=septal pore cap, CD=discontinuity in the septal pore cap. (From Bracker, C. E. and Butler, E. E. (1963) *Mycologia,* **55**, 35)

septum next to the injured part than by the production of a complete new wall as is necessary when an aseptate hypha is damaged.

The cells of a septate mycelium become vacuolated as they age just as the coenocytic hyphae or yeast cells do. They may be regularly uninucleate or binucleate or may contain an indefinite number of small nuclei.

Anastomosis between hyphae of the same species takes place freely and nuclei are known to migrate across the bridging section from one hypha to another. This is of great significance in the life cycles of many fungi since nuclei of different origin may thereby be present in a single hypha (heterokaryosis, p. 65).

Many species of both coenocytic and septate

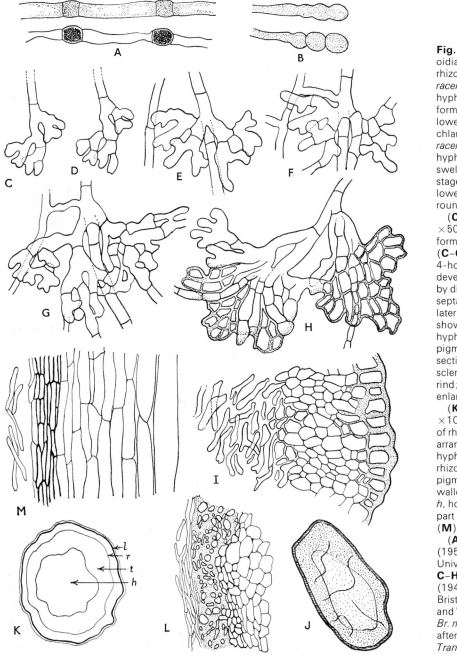

Fig. 9.8 Chlamydospores, oidia, sclerotia and rhizomorphs. (**A**) *Mucor racemosus* (×250). Upper hypha showing early stages in formation of chlamydospores; lower one with mature chlamydospores. (**B**) *Mucor racemosus* (×250). Upper hypha showing irregular swellings which are the first stages in formation of oidia; lower one showing oidia rounding off.

(**C–J**) *Botrytis cinerea* (**C–I** ×500; **J** ×20). Stages in formation of a sclerotium (**C–G**) drawn at approximately 4-hour intervals to show development of sclerotial initial by dichotomous branching and septation of the hyphae; (**H**) later stage of same initial showing coalescence of hyphae, and thickening of pigmentation of the walls; (**J**) section through mature sclerotium showing thick outer rind; (**I**) portion of this rind enlarged.

(**K–M**) *Armillaria mellea* (**K** ×10; **L** and **M** ×500). (**K**) T.S. of rhizomorph to show arrangement of zones; *l*, loose hyphae on outside of rhizomorph; *r*, thick-walled pigmented cells of rind; *t*, thin-walled cells of inner zone; *h*, hollow centre; (**L**) T.S. of part of the three zones; (**M**) L.S. of same.

(**A–B**, after Hawker, L. E. (1950) *Physiology of Fungi*. University of London Press; **C–H**, **J**, after Willetts, H. J. (1949) *Thesis*. University of Bristol; **I**, after Townsend, B. B. and Willetts, H. J. (1954) *Trans. Br. mycol. Soc.*, **37**, 213; **K–M** after Townsend, B. B. (1954) *Trans. Br. mycol. Soc.*, **37**, 232).

types tend to break down into a yeast-like mass of cells, either with ageing of the mycelium or in response to particular conditions, notably partial anaerobiosis or immersion in a strong sugar solution. The hyphae, which are normally cylindrical, become bead-like and the 'beads' round off and separate as oidia which may then start budding as in a yeast (Fig. 9.8b). Some fungal parasites of animals exhibit the phenomenon of dimorphism and are yeast-like when growing within the host and filamentous in artificial culture. Conversely some plant parasites, e.g. the grain smuts and the peach leaf curl fungus (*Taphrina deformans*), are filamentous in the host plant but produce only a yeast-like mass of cells on culture media.

Just as individual yeast cells may become full of reserve food material and develop a thick wall, in response to certain unfavourable conditions, so individual cells of filamentous hyphae or of normally thin-walled spores of some species, e.g. *Fusarium* spp., may round off and become surrounded by a thick wall. These chlamydospores (Fig. 9.8a) are probably resistant to adverse conditions and may play an important part in the survival of the fungi which form them.

Considerable differences in form may be seen between hyphae of the same thallus. Not only are the hyphae which bear the reproductive bodies often different from the vegetative ones in mode of branching, pigmentation and response to stimuli, such as light and gravity, but a mycelium may be made up of one or more types of purely vegetative hyphae. Those which penetrate the substratum are often of rhizoidal nature and quite different from the aerial hyphae (Fig. 9.9a). Special branches of distinct form may be produced, such as the haustoria, which are pushed into the host cells by many intercellular plant parasites (p. 291) or the various rings and loops with which predacious fungi trap and entangle nematodes (p. 320). Many fungi produce long runner hyphae or stolons which creep over the substrate and give rise to rhizoidal penetrating branches and aerial hyphae at intervals, as with the saprophytic *Rhizopus* or the parasitic *Gaeumannomyces graminis* (syn *Ophiobolus graminis*).

Large plectenchymatous structures

Further differentiation of fungal thalli is achieved through the aggregation of hyphae to form large structures of more or less solid nature. The septate hyphae of certain higher fungi may become closely interwoven to form a mass, the individual cells of

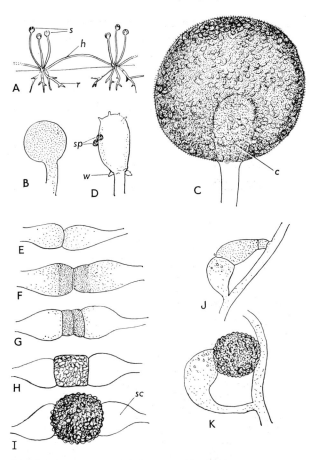

Fig. 9.9 Zygomycetes. (**A**) Growth habit of *Rhizopus* (diagrammatic, not to scale) showing sporangia (*s*), stolons or runner hyphae (*h*) and rhizoids (*r*).

(**B–D**) *Mucor plumbeus* (×500). (**B**) young, undifferentiated sporangium. (**C**) nearly mature sporangium showing columella (*c*) and partially differentiated spores. (**D**) columella of dehisced sporangium showing spines characteristic of this species, remains of sporangium wall (*w*) and two spores (*sp*) adhering to the columella.

(**E–I**) *Rhizopus sexualis* (×100). Stages in conjugation and formation of zygospore. (**E**) two equal-sized progametangia in contact and enlarged. (**F, G**) stages in differentiation of gametangia. (**H**) gametangia fused to form young zygospore which is rounding off and developing a thick sculptured wall. (**I**) nearly mature zygospore, spherical and with fully pigmented thick wall; suspensor cells (*sc*) empty.

(**J–K**) *Zygorhynchus moelleri* (×500). (**J**) young gametangia of unequal size. (**K**) nearly mature zygospore. Note large suspensor cell.

which are often inflated and globose. The resulting structure bears a superficial resemblance to the tissues of the higher plants, but is distinguished from these by its development from separate hyphae and the inability of the cells of the component filaments to divide in more than one plane. Such a pseudo-tissue or plectenchyma may, however, show differentiation into zones of different structure and function.

STROMATA

Some fungi produce a mass of plectenchyma of regular or irregular shape on or in which spores or fruit-bodies are formed, e.g. the candle-snuff fungus, *Xylaria hypoxylon*, or the ergot of rye, *Claviceps purpurea*. These are known as stromata (singular, stroma). Sometimes parts of the sub-stratum or host tissue are embedded in the stroma as with the mummified fruits of Rosaceous plants attacked by the 'brown rot' fungi (*Sclerotinia* spp).

SCLEROTIA

Other fungi produce more or less globose bodies which are termed sclerotia (Fig. 9.8c–j). These are usually differentiated into an outer protective layer of thick-walled cells and an inner mass of cells with thinner walls and containing reserve food substances, usually oily substances and glycogen. Sclerotia of species of *Rhizoctonia* and some related fungi are of looser construction and the component cells are all thick-walled and full of oily contents. Both types of sclerotium are better able to survive periods of unfavourable conditions, such as starvation or drought, than are the more delicate and relatively unprotected vegetative hyphae. Sclerotia are important in the survival of many plant parasites in the soil or on decaying plant remains.

MYCELIAL STRANDS AND RHIZOMORPHS

Hyphae running parallel to each other may become aggregated to form strands of mycelium or may form more highly organized cylindrical or strap-shaped, often branched, structures, which are known as rhizomorphs. The latter may be relatively simple in structure or may show a high degree of differentiation into an outer protective layer of thick-walled cells and an inner layer of thin-walled elements, through which food materials may be conducted. Morphologically, complex rhizo-morphs may be considered as much elongated sclerotia. They enable fungi to spread through the soil or over the surface of the host plant or other substratum while remaining in organic connection with the original mycelium. Food from the original supply or food base is carried along the strand to its growing point. The black rhizomorphs of the destructive honey agaric, *Armillaria mellea* (Fig. 9.8k,l), which destroys roots and other under-

Fig. 9.10 Ascomycetes and Basidiomycetes.

(**A–G**) *Aspergillus (Eurotium) herbariorum* (× 600). (**A**) coiled archicarp (female branch). (**B**) same at a slightly later stage with sterile hyphae growing up from base. (**C**) young fruit-body (ascocarp, cleistocarp) with archicarp still visible inside. (**D**) older fruit-body, dense mass of ascogenous hyphae visible in centre but details obscured by wall (peridium). (**E**) group of young asci. (**F**) older ascus, contents aggregating to form spores. (**G**) mature ascus containing eight ascospores.

(**H–I**) *Sordaria fimicola*. (**H** × 40) mature fruit-body (perithecium) showing short neck (beak) and ostiole (*o*). (**I** × 600) ascus containing eight black ascospores arranged in a single row (uniseriate). Thin place in wall at apex of ascus indicates pore (*p*) through which spores escape later.

(**J–L**) *Sclerotinia sclerotiorum*. (**J, K** natural size) stalked fruit-bodies (apothecia) developing from sclerotia (*s*). (**J**) cup-shaped fertile part of fruit-body still with incurved rims. (**K**) fertile part expanded and recurved. (**L** × 200) part of hymenium (fertile layer) which lines the cup, showing mature asci (*a*), young asci (*y.a.*) and sterile hairs (paraphyses, *s.h.*).

(**M–Q**) *Amanita muscaria*. (**M**) unexpanded. (**N**) expanded fruit-body (× ½). *pi*=pileus or cap, *st*=stipe or stalk, *l*=lamellae or gills, *an*=ring or annulus, *v*=volva. (**O**) diagrammatic section through gills. (**P** × 50) L.S. gill showing central part of trama (*t*) with elements running vertically, subhymenial layer (*sh*) and hymenium (*h*) consisting of basidia. (**Q**) L.S. portion of hymenium (× 500) showing basidia at all stages from the beginning of elongated to final collapse after spore discharge. *n*=nucleus, *ster*=sterigma, *sp*=spore.

ground parts of many plants, including forest trees, fruit trees, and ornamental trees and shrubs, can spread through the soil for distances up to several metres provided that they remain attached at the base to a decaying root or tree stump from which they are able to draw food. They are thus able to reach, attack and colonize plants at some distance from the original victims. This ability to travel through the soil for long distances makes this a most difficult fungus to eradicate. Complex mycelial strands are also formed by the dry-rot fungus, *Merulius lacrimans* (p. 355) and enable it to travel from one piece of timber to another over intervening areas of soil, brickwork, or metal.

COMPLEX FRUIT-BODIES

The spores of many higher fungi are borne on or in complex fruit-bodies (Fig. 9.10). Those of some of the giant puff-balls or the wood-destroying bracket fungi (p. 199) may reach a diameter of more than 30 cm. Such large structures must obviously possess a more or less rigid framework or 'skeleton' and must have some provision for the conduction of relatively large amounts of food materials to the growing tips and edges. Analysis of their structure

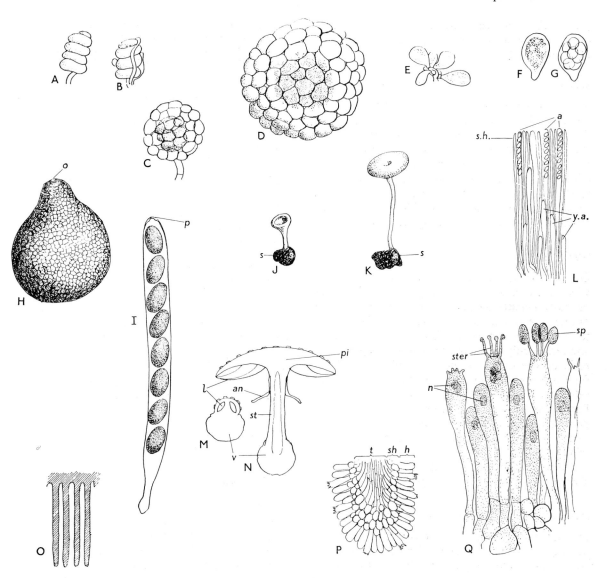

shows that they are made up of several different types of hyphae, each of which plays a special part in the organization and maintenance of the fruit-body.

Reproduction

Reproduction of fungi is usually by spores. These are most often single colourless cells, but in some species the spores may consist of two or more cells, may be pigmented, may have thick sculptured walls, or may bear appendages or be surrounded with a gelatinous layer. The spores of a particular species are remarkably uniform in size and structure, and accordingly spore characters are largely used in the identification of fungi. Spores are usually readily detachable from the parent thallus. Some are shed by special and often complicated mechanisms. Many are small and easily carried by wind or other agency and are thus dispersed over considerable distances. This ease of dispersal is a major factor in the success of fungi in colonizing suitable habitats. Under suitable conditions the spore germinates and one or more germ tubes (young hyphae) grow out (Figs. 9.11).

Physiology of reproduction

If a medium is inoculated with fungal spores or mycelium the resulting colony at first consists

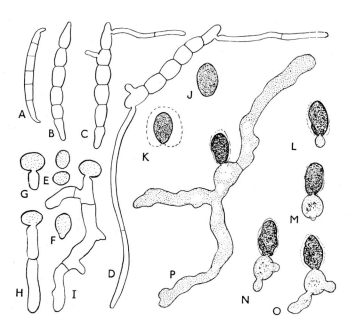

Fig. 9.11 Germination of fungus spores.

(**A–D**) *Fusarium lateritium* (× 550). (**A**) sickle-shaped, septate conidium. (**B**) same after soaking in distilled water, note swelling of cells due to uptake of water. (**C, D**) stages in germination; germ tubes emerging from any cell of the spore, elongating and developing into septate hyphae, note constriction of hyphae at point of emergence from spore.

(**E–I**) *Botrytis cinerea* (× 550). (**E**) ungerminated conidia. (**F**) conidia. (**F**) conidium swollen through uptake of water and with bulge where germ tube is about to emerge. (**G**) early stage of germination with germ tube beginning to elongate. (**H, I**) later stages of germination, germ tube septate and branching.

(**J–P**) *Sordaria fimicola* (× 550). (**J**) ascospore. (**K**) same showing gelatinous 'halo' as seen when mounted in nigrosin. (**L**) early stage in germination, contents extruding into small vesicle, spore shrinking. (**M–P**) later stages in germination, showing the original vesicle giving rise to several hyphae which then become branched and later septate.

(**A–I** after Hawker, L. E. (1950) *Physiology of Fungi*, University of London Press; **J–P** after Hawker, L. E. (1951) *Trans. Brit. mycol. Soc.*, **34**, 181).

entirely of vegetative hyphae. Some colonies remain in the vegetative condition, either because their genetic constitution is such that spore production is impossible or because of unsuitable nutritional or other factors. Most, however, will sooner or later change to the reproductive phase and produce asexual or sexual spores or both. The duration of the initial vegetative phase depends upon both internal and external factors.

INTERNAL FACTORS

The most important internal factor controlling sporulation is the genetical constitution of the fungus. Thus a single strain of a heterothallic species will not produce spores under any conditions and a heterokaryon in which the majority of nuclei are derived from a non-sporulating strain will do so only sporadically (p. 65).

Even when the genetical constitution permits sporulation, other internal factors, such as the number and distribution of nuclei or the concentration of various complex substances in the hyphae, may be unsuitable. For example, it has been suggested that certain Pyrenomycetes do not form perithecia until the concentrations of thiamin and biotin in the hyphal tips have reached a certain level. This type of internal factor is, however, almost certainly the result of a particular combination of external factors.

EXTERNAL FACTORS

The nature and concentration of the food supply is of great importance in determining the initiation, development, and maturation of reproductive structures. Usually, but not always, nutritional conditions favouring vegetative growth also favour the production of asexual spores, but those inducing sexual reproduction are often very different. In general the concentration of nutrients, and particularly of carbohydrates, optimal for sporulation, either asexual or sexual, is lower than that giving maximum dry weight. For example, the production of ascospores by the Pyrenomycete, *Sordaria fimicola* is possible only over a smaller range of concentration than that permitting vegetative growth. Although a high concentration of nutrients thus tends to prevent sporulation, spores are usually produced only on a well-nourished mycelium. A commonly used method of inducing sporulation in fungi is to transfer an established mycelium from a rich to a dilute medium.

The composition of the nutrient substrate is as important as its concentration. Many Pyrenomycetes form perithecia earlier and in greater number when the source of carbon is a di- or polysaccharide rather than a hexose sugar. Moreover, the concentration optimal for fruiting is often higher with a relatively complex carbohydrate. This has been shown to be due partly to the rate at which a particular fungus hydrolyzes the various carbohydrates, the best results being obtained when the resulting concentration of hexoses remains optimal over a relatively long period. The ease of phosphorylation of various carbohydrates has also been shown to be important.

The nature of the nitrogenous substance available and the carbon–nitrogen ratio in the medium also influence spore production. The mineral element requirements of most fungi are more exacting for spore production than for vegetative growth, and sporulation often does not begin until the concentration of minerals is significantly above that permitting minimum growth. Thus if certain species of *Fusarium* are grown without added magnesium no spores are produced, the sparse vegetative growth being supported by traces of magnesium present as an impurity in the other ingredients of the medium. Similarly the amount of copper necessary for measurable growth of *Aspergillus niger* is less than that for conidium formation. Moreover, colourless or light-coloured spores are formed sparsely at concentrations which do not permit the development of the characteristic black pigment. Finally while it has not been demonstrated that calcium is essential for vegetative growth of fungi it has been clearly established to be necessary for the formation of perithecia by *Chaetomium globosum*. Calcium is difficult to remove completely from the medium and it is likely that the requirements for vegetative growth are below, and those for fruiting are above, the amount remaining after the most stringent purification of the ingredients of the medium.

The amount of certain vitamins and other complex organic substances necessary for spore production is also greater than for vegetative growth. Many fungi, e.g. certain strains of *Sordaria fimicola*, are able to grow and even to sporulate feebly in the absence of an external supply of vitamin B_1 (thiamin), but show greatly increased fruiting when this vitamin is added to the medium. The addition of thiamin increases the rate of respiration during the early stages of colonial growth. A peak period of respiration has been shown to precede sporulation in *S. fimicola*, *Rhizopus sexualis* and some other fungi. The effect of thiamin on spore production may thus be correlated with its effects on respiration.

Physical factors such as temperature, water content of the substrate, humidity, H-ion concentration of the substrate and aeration all influence sporulation, although the mechanisms by which they do so are largely obscure. In general the range of each of these over which spore production takes place is narrower than that permitting vegetative growth.

Some fungi are strikingly influenced by light; others grow and sporulate equally well in light or in darkness.

Gravity also influences the direction of growth of the spore-bearing structures of many fungi. Many sporangiophores and conidiophores are negatively geotropic and the fruit-bodies of many higher fungi are strongly so. The sporophores of most agarics (toadstools) show a complicated response to gravity. The stipe is usually negatively geotropic; the pileus is diageotropic, expanding horizontally under the influence of gravity; the gills grow vertically downwards and are thus positively geotropic. This mechanism brings the various parts of the fruit-body into the most favourable position for spore discharge and dispersal.

The conditions favouring the initiation of sporulation are not necessarily optimal during the later stages of development and maturation. It is important that when the effects of environment on reproduction are being studied the process should be considered as a number of separate stages. For example, the production of zygospores by *Rhizopus sexualis* is inhibited by a temperature of *ca.* 10°C during the early stages, but after the gametangia have been delimited development continues at this temperature. The conditions permitting formation of perithecia of certain Pyrenomycetes do not always allow the formation of viable ascospores. Fruit-body initials of various species may be produced in quantity under conditions which permit the development of only abnormal fruit-bodies.

TYPES OF REPRODUCTION

Most fungi produce more than one kind of spore. Asexually produced spores (the so-called imperfect stage) are commonly produced in large numbers when conditions are favourable to growth and serve to spread the fungus rapidly. Later, often when conditions are no longer so favourable, either through exhaustion of food supply or from other causes such as seasonal fall in temperature, most fungi pass into the so-called perfect stage. In many species spores are then produced as the direct result of the fusion of sexual cells or branches, or after a period of secondary growth resulting from such a fusion. In a large proportion of the higher fungi, sexual fusion of gametes or of gametangia has been lost and the only trace of sex left is the fusion of nuclei at some stage during the production of the 'perfect' spores. Many fungi not only have no sex organs but are not known to produce the perfect stage under any circumstances (p. 199).

ASEXUAL REPRODUCTION

Motile asexual spores (zoospores) are found only among the more primitive aquatic or semi-aquatic groups of fungi. The zoospore has no cell wall during its motile phase. It travels by means of one or more flagella, the number and arrangement of which is constant for the group. Zoospores are formed in special sac-like cells, or zoosporangia, from which they escape either by the rupture of the sporangial wall or through special pores developing in it. After swimming for some time in water, which may be only a film of dew on a host leaf, the zoospores settle down, withdraw their flagella and encyst, that is, they become surrounded by a cell wall. Under suitable conditions they germinate, usually by putting out a germ tube which develops into a typical hypha.

In one large group of the lower fungi (the Zygomycetes, p. 196) the spores are never motile and are surrounded by a definite cell wall. In most species of this group the spores are formed in multispored sporangia but there is a tendency for the number of spores in the sporangium to be reduced in some species. In others the asexual spores are borne singly and are known as conidia.

In the higher fungi sporangia are not formed and in many groups the conidium is the typical asexual spore (Fig. 9.12). Conidia are often formed in enormous numbers and as they are usually light and readily detached from the parent hypha, or conidiophore, they are particularly efficient agents of spread of the fungus.

SEXUAL REPRODUCTION

The sexual spores of the lower fungi are the direct

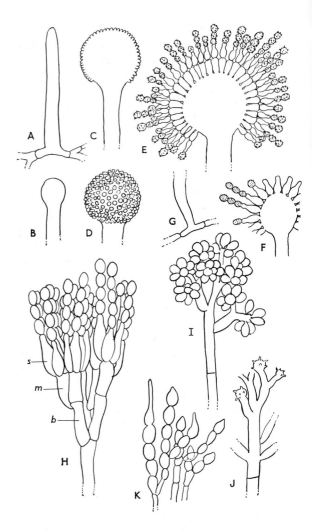

Fig. 9.12 Formation of conidia by moulds.
(**A–E**) *Aspergillus niger* (**A** ×170, **B–E** ×260). (**A**) young conidiophore growing up vertically from foot cell. (**B**) tip of slightly older conidiophore beginning to swell up to form a vesicle. (**C, D**) young vesicle with developing sterigmata (**C** in optical section, **D** in surface view). (**E**) young fruiting head (in optical section) showing secondary sterigmata bearing chains of conidia.
(**F–G**) *Aspergillus niveo-glaucus* (×300). (**F**) typical fruiting head showing single series of sterigmata (only part of head drawn). (**G**) foot cell from base of conidiophore. (**A–G**, after Thom, G. and Raper, K. B. (1945) *A Manual of the Aspergilli*, Bailliere, Tindall and Cox, London.)
(**H**) *Penicillium expansum* (×900). A typical asymmetric type of penicillus showing sterigmata (*s*), metulae (*m*) and branches (*b*). (After Raper, K. B. and Thom, G. (1949) *A Manual of the Penicillia*, Williams and Wilkins, Baltimore.)
(**I–J**) *Botrytis cinerea* (×500). (**I**) conidiophore, bearing clusters of conidia. (**J**) conidiophore showing sterigmata from which spores have been shed.
(**K**) *Sclerotinia (Monilia) fructigena* (×250). Chains of poorly differentiated conidia (drawn from a culture).

product of the fusion of gametes or gametangia. These may be isogamous (i.e. of equal size) and motile, as in most of the chytrids, motile but heterogamous (differing in size or pigmentation), as in a few chytrids and some other aquatic groups, or they may be non-motile, as in the important group of the Oomycetes (p. 196). In the Zygomycetes the characteristic sexual spore is the zygospore, produced by the conjugation of two sexual branches (gametangia) which may be equal or unequal in size according to the species. Some species form zygospores freely in monospore cultures, and are then said to be homothallic. Others form them only by conjugation of hyphae derived from two different strains of opposite mating type, termed 'plus' and 'minus' strains. These are said to be heterothallic. Heterothallism was first demonstrated in this group but has since been found in all the major groups of fungi.

In the higher fungi the sexual spores are seldom the direct result of the fusion of sexual branches. Even in the yeasts and related fungi, in which fusion of isogamous or heterogamous cells or branches takes place in a number of species, the result is not a single spore but a sporangium-like cell, the ascus, in which the ascospores are formed (p. 197). In some higher Ascomycetes a fusion of sexual branches takes place and from them a secondary mycelium grows out and finally bears the asci. In most of the higher Ascomycetes and in all the Basidiomycetes definite sexual branches are no longer formed. In both groups, however, production of the perfect stage spores is preceded by a nuclear fusion and a reduction division. These spores are haploid, germinating to give a haploid mycelium. The exact point in the life cycle at which this haploid mycelium becomes diploid differs in different species. In a few species the diploid condition arises in the spore itself.

The sexual spores of many fungi are resistant to adverse conditions, such as temporary drought or the cold of winter, either through the possession of a thick wall, as with the various resting spores, such as the oospores or zygospores of the Phycomycetes, or through being enclosed in protective fruit-bodies. Thus while the asexual spores are agents of multiplication and spread, the sexual or perfect stage spores tend to be agents of survival, although in many higher fungi they may also function as dispersal agents.

GERMINATION

The final stages of the reproductive cycle, after the formation, discharge and dispersal of spores, is spore germination (Fig. 9.11), and the initiation of a new mycelium. Germination cannot take place until the spore is mature and unless the spore wall is permeable to air and water. Some thick-walled spores, such as the oospores of Oomycetes or teleutospores of most rusts, remain dormant for long periods before germination. Most asexual spores and many ascospores and basidiospores are capable of germination as soon as they are shed from the parent hypha and, with some exceptions, retain their viability for only relatively short periods.

Even when a spore is in the right condition it germinates only if environmental factors are suitable. Many spores contain a relatively large amount of nutrients and are thus able to germinate in water or, as with the conidia of the powdery mildews, in humid air. When the reserves originally present in the spores are exhausted the germ tubes cease to grow unless they have reached an external supply of food. In the natural habitat pure water is seldom encountered, even raindrops on leaves have been shown to contain leaf exudates. Some spores with inadequate reserves of vitamins or other food substances require an external supply of these for germination.

With many external factors, such as temperature or H-ion concentration of the substrate, the range permitting germination is usually narrower than that permitting mycelial growth. Thus germination normally takes place only under conditions suitable for further growth of the germ tube and the establishment of a new mycelium.

Outline of classification
(With examples from major groups)

The lower fungi (phycomycetes)

Thallus of simple organization; unicellular (but not reproducing by budding) or vegetative phase filamentous and aseptate, or occasionally septate. Classified according to presence or absence of motile (flagellated) stages and the morphology and number (i.e. one or two) of flagella where present.

THE CHYTRIDS (CHYTRIDIALES)

Are one of a number of groups the motile cells of which possess a single posterior whiplash-type flagellum. In many species the whole of the unicellular thallus takes part in reproduction (i.e. they are holocarpic). Some species possess thread-like absorbing or attachment organs (rhizoids) which do not take part in reproduction and in others the reproductive bodies are connected by fine vegetative strands. Most species are aquatic, some are parasitic on higher plants.

The chytrids not only are simple in structure but have a simple life cycle. Parasitic species may, however, cause considerable damage to the host, e.g. *Synchytrium endobioticum*, the cause of black wart disease of potato.

THE PLASMODIOPHORALES

Are a group of parasites, mainly of higher plants, in which the vegetative stage is represented by a naked mass of protoplasm, the plasmodium, which, however, differs in origin from that of the slime moulds, and in which motile zoospores and gametes possess two anterior flagella of whiplash type and very unequal, as in the slime moulds (p. 209). The best-known example of this group is *Plasmodiophora brassicae*, the cause of club root of cruciferous plants.

THE OOMYCETES

Are fungi of aquatic or damp habitats, characterized by motile stages (absent in a few genera) bearing two flagella of approximately equal length, one of whiplash and one of tinsel type, attached anteriorly or laterally; by the cellulose nature of the microfibrils of the cell wall; and by sexual reproduction involving the passage of the contents of a male branch (antheridium) into the female organ (oogonium), where fertilization of one or more female cells (oospheres) leads to the production of usually thick-walled resting spores (oospores).

The Oomycetes include:

Saprolegniales (water moulds) Aquatic or occurring in wet soil, mostly saprophytic.

Peronosporales Semi-aquatic or terrestrial; including many parasites of higher plants, e.g. *Pythium* spp., which cause damping-off of seedlings; *Phytophthora* spp., including *P. infestans* which causes late blight of potatoes; downy mildews (*Peronospora* spp., etc.) of higher plants.

ZYGOMYCETES

Mainly terrestrial species important as soil or-

ganisms and agents of decay of organic materials (Mucorales), some parasitic on insects (Entomophthorales), others on soil protozoa (Zoopagales); motile stages lacking; asexual reproduction by sporangospores or conidia, sexual reproduction by the fusion of equal or unequal gametangia to form a zygospore.

The mycelium of the Mucorales consists of rather coarse, loosely growing hyphae (Fig. 9.9a). The colonies are at first white and become grey or brown with age or as a result of spore production. Most species (e.g. species of *Mucor, Rhizopus, Absidia, Zygorhynchus*, and *Phycomyces*) produce globose sporangia containing numerous spores and borne on special upright hyphae known as sporangiophores (Fig. 9.9b–d). In some species the sporangiophores show characteristic branching, e.g. *Syzygites*. The large multi-spored sporangia are usually produced terminally on the sporangiophore, the tip of which in most species projects into the sporangium as a globose or hemispherical columella. As the sporangium grows, cytoplasm can be seen streaming into it until the contents are noticeably denser than those of the ordinary hyphae and finally the columella is laid down. Fissures then develop in the dense cytoplasm of the sporangium cutting it up into uninucleate polygonal blocks which round off, become surrounded by a cell wall and form a dark pigment. The mature sporangium is entirely filled with these spores. The sporangial wall is usually thin and ruptures to release the spores, but in some highly specialized genera, such as the coprophilous (i.e. growing on dung) *Pilobolus*, part of the wall is thickened. Some genera, e.g. *Thamnidium*, produce smaller sporangia (sporangioles, containing only a few spores) in addition to the multispored sporangium; others produce only sporangioles and a few produce true conidia, e.g. *Cunninghamella*.

Sexual reproduction is by the fusion of two cells, the gametangia, which may be of equal or unequal size according to the species, to give a thick-walled zygote (the zygospore). In the early stages of conjugation two branches approach each other and enlarge. These are termed the progametangia. Soon after the tips of these come into contact each progametangium cuts off a terminal gametangium from the remaining part of the swollen branch, which is termed the suspensor. Most species are isogamous, e.g. *Mucor, Rhizopus* (Fig. 9.9e–i), the two gametangia being similar, but others show differences in size or behaviour, e.g. *Zygorhynchus* (Fig. 9.9j, k). In some species bristles are produced from one or both suspensor cells, e.g. *Absidia, Phycomyces*; in others the zygospore is formed in a vesicle growing out from one or both gametangia, e.g. *Piptocephalis*; in *Mortierella* the otherwise simple zygospore is protected by a sheath of sterile hyphae and in *Endogone* the conjugating branches are grouped together in a definite fruit-body in which the developing zygospores are surrounded by sterile hyphae.

The higher fungi

Thallus consisting of septate hyphae or of a sprout mycelium; asexual reproduction (imperfect stage) usually by conidia; sexual reproduction or perfect stage, when present, by characteristic spores (ascospores or basidiospores), often borne in or on complex fruit-bodies.

There are two groups; the Ascomycetes, the perfect stage spores (ascospores) of which are formed endogenously in a sac-like ascus, usually eight ascospores in each ascus; and the Basidiomycetes, the perfect spores (basidiospores) of which are borne exogenously on special cells (basidia). In addition, a third group is usually recognized, the Deuteromycetes or Fungi Imperfecti the members of which lack perfect stages. Since these are mostly forms closely resembling the asexual or imperfect stages of known Ascomycetes they are more properly treated with them. However, since many of them are of importance industrially or as parasites it is convenient to retain this 'form' group and to use the available artificial keys to identify not only those lacking a perfect stage but also others in which the stage is formed only occasionally or under unusual conditions.

ASCOMYCETES

This group was formerly subdivided according to the presence or absence of fruit-bodies (ascocarps) and to the gross morphology of these when present. It is now recognized that there is a fundamental difference between forms in which the asci have a single layered wall (unitunicate) and are formed in a layer (hymenium) within a fruit-body (Ascohymenomycetes) and those in which the asci have a two-layered wall (bitunicate) and are formed singly or occasionally in small groups in a stroma (Loculoascomycetidae). The breakdown of the matrix of hyphae surrounding the locules may produce a fruit-body resembling a hymenial form of entirely different development. Hence microscopic and developmental characters must be considered. For example:

Hemi-ascomycetidae Asci formed singly, fruit-bodies not formed.

ENDOMYCETALES (SACCHAROMYCETALES) Includes the ascogenous yeasts. Asci formed singly from conjugated or single cells, mostly saprophytic; a few parasitic on animals. The classification of the yeasts is currently under review; for the generally accepted arrangement to date see Lodder (1970).

TAPHRINALES Parasites on vascular plants.

Eu-ascomycetidae Asci unitunicate, borne on secondary ascogenous-hyphae, usually in a fruit-body.

Series A.
Plectomycetes Asci globose, evanescent, arranged irregularly within a closed spherical fruit-body (cleistocarp) or a flask-shaped one opening in an ostiole at the apex (perithecium).

PLECTASCALES (ASPERGILLALES) Fruit-body a closed, spherical cleistocarp; most saprophytic, some parasitic on plants or animals.

Most members of this group produce numerous conidia by which they are able to spread rapidly under suitable conditions. The simplest forms, such as *Byssochlamys fulva* (a cause of spoilage of canned fruits), produce asci in groups arranged irregularly on ascogenous hyphae without any peridium. Others, such as *Gymnoascus* (common on dung and other decaying organic material), surround the asci with a thin weft of hyphae, thus producing a simple fruit-body of the cleistocarp type. The cleistocarps of the more advanced genera, such as *Monascus, Aspergillus* (*Eurotium*) (Fig. 9.10a–g) and *Penicillium* are surrounded by a definite wall or peridium consisting of one or more layers of cells. The asci of all genera are evanescent and the mature fruit-body contains a mass of ascospores. Several of the dermatophytes, formerly classed as Deuteromycetes, are now known to form ascocarps of a form resembling the simple cleistocarp of the Gymnoascaceae.

MICROASCALES in which the fruit-bodies are flask-shaped includes a number of serious parasites of forest trees: *Ceratocystis ulmi* is the cause of Dutch elm disease (p. 288).

Series B.

Ascohymenomycetes Asci usually clavate or cylindrical, seldom evanescent, arranged parallel to one another in a hymenium in a cleistocarp, flask-shaped perithecium, cup-shaped apothecium or a complex derivative of the latter.

ERYSIPHALES Asci parallel in layer (hymenium) or single in a cleistocarp, asci persisting until maturity, ascospore discharge often explosive; powdery mildews, important obligate parasites of higher plants, confined to the surface of the host deriving nourishment from the epidermal cells by means of haustoria produce numerous erect chains of conidia (the *Oidium* stage) which give the powdery appearance to the infected host plant, later thick-walled fruit-bodies (cleistocarps) are formed.

Pyrenomycetes Ascocarp a perithecium (some forms now transferred to the Loculoascomycetidae).

Many are plant parasites, some saprophytic species destroy cellulose or wood. The perithecia of *Chaetomium* are more or less globose, usually covered with characteristic coiled or branched bristles and the ostiole is rudimentary or lacking. The asci are globose and evanescent but are produced in a parallel layer. *Chaetomium* has no specialized method of spore discharge and the spores ooze out through the ostiole, if present, or through a tear in the peridium. This genus thus has some features reminiscent of the Plectascales, while the initial arrangement of the asci is characteristic of the Pyrenomycetes. In more typical genera the perithecium has a beak or neck, opening by a definite ostiole, the asci are clavate (club-shaped) or cylindrical, and dehisce explosively, discharging the spores to a considerable distance. Some such as *Sordaria*, found on dung, garbage, and in soil (Fig. 9.10h–i), or *Neurospora*, the red bread mould, are of importance in industrial mycology. *Neurospora* is also of interest as the subject of much fundamental research on fungal genetics and nutrition.

Discomycetes Ascocarp an apothecium or a derivative of this, hymenium usually extensive. Many saprophytic, growing on soil or dung, some para-

sitic (including such important ones as the genus *Sclerotinia*, members of which cause brown rot of rosaceous fruits, many diseases of bulbous ornamentals, rotting of stems such as the potato haulm, etc.).

Loculoascomycetidae Asci bitunicate, formed singly or in groups in locules in a stroma; plant parasites, some of economic importance, e.g. *Venturia inaequalis*, the cause of apple scab.

Laboulbeniomycetidae (p. 319). Insect parasites, not closely related to any other Ascomycetes.

BASIDIOMYCETES

This group is divided into the Heterobasidiomycetidae, the basidia of which are septate or deeply cleft, or, if single-celled, developed from a thick-walled cell (teleutospore) and the Homobasidiomycetidae, the basidia of which are septate and usually borne in an extensive hymenium on complex fruit-bodies (basidiocarps).

Heterobasidiomycetidae Mostly specialized parasites of higher plants.

USTILAGINALES or smuts

UREDINALES or rusts

Homobasidiomycetidae Basidia one-celled, usually in complex fruit-bodies.

Hymenomycetes Hymenium exposed before maturity, mushrooms, toadstools, bracket fungi, elf clubs, etc. (Fig. 9.10m–q).

Agaricus campestris (the common field mushroom) is typical of the most advanced order of the Hymenomycetes, the Agaricales (agarics, commonly known as mushrooms and toadstools). The fruit-bodies begin as small globose masses of hyphae borne on branched mycelial strands. These develop into the familiar mushroom with a stalk or stipe which bears an expanded cap or pileus, from the under side of which thin plates or lamellae (the gills) hang down vertically. These are covered by the hymenium of basidia each bearing four basidio-spores (most cultivated mushrooms are forms of *A. bisporus* and have two-spored basidia). The gills are protected until they are nearly mature by a thin membrane consisting of several layers of hyphae, which later ruptures along the edge of the pileus as the latter expands, and remains attached to the stipe as a ring or annulus. The gills are at first white and become pink, purple and finally almost black as the spores ripen and become pigmented.

Other agarics differ in spore colour, absence of a ring, and in a few species, including the death cap (*Amanita phalloides*, p. 341), in the presence of an enveloping universal veil (volva) through which the growing fruit-body finally bursts leaving a cup-shaped structure at its base and often a few scraps of the torn volva adhering to the pileus.

Another order of the Hymenomycetes, the Poly-porales, contains many economically important wood-destroying fungi. The fruit-bodies of mem-bers of the family Polyporaceae are more or less bracket-shaped and the hymenium lines numerous pores on the under side of the fructification. These pores may be merely shallow depressions, as in the dry-rot fungus, *Merulius lacrymans* (which is con-sequently sometimes placed in a separate family, Meruliaceae), or, more commonly, they are long and tubular, as in species of *Polyporus* or the perennial woody fruit-bodies of *Fomes* and *Gan-oderma*.

Gasteromycetes Hymenium enclosed until ma-turity in complex fruit-body; saprophytes or mycor-rhizal fungi. (Puff-balls, earthstars, bird's nest fungi, stinkhorns, etc.).

DEUTEROMYCETES (FUNGI IMPERFECTI)

Perfect stage unknown or rare (Fig. 9.12).

HYPHOMYCETES (MONILIALES)

Conidia borne di-rectly on the mycelium or on specialized co-nidiophores which are usually entirely free or may be found in tufts or pulvinate masses; saprophytes and parasites on plants and animals some of which are of major economic importance.

This group is a heterogeneous collection of species, the only characters common to all being the production of conidia and the absence or extreme rarity of the perfect stage. In the absence of the latter the subdivision of the group is based on arbitrary conidial characters. Although modern systematists have attempted to devise a more natu-ral arrangement, the classical one, devised mainly by Saccardo, is still the most generally used for *identification* of species.

In this system the Hyphomycetes are divided into four families as follows:

(i) *Moniliaceae.* Conidiophores, if present, dis-tinct from one another, or spores borne scat-tered over undifferentiated hyphae, spores and conidiophores bright coloured or colourless, not dark;

(ii) *Dematiaceae.* Similar, but spores or co-nidiophores, or both, dark coloured;

(iii) *Stilbaceae.* Conidiophores interwoven to form a short cylindrical or spine-like coremium (synnema);

(iv) *Tuberculariaceae.* Spores produced in fascicles in dense patches, sporodochia.

There is much overlapping between these sub-families. Thus *Aspergillus niger* which has black spores is obviously closely related to other species of *Aspergillus* with white or brightly coloured ones, but if the system is strictly followed should be placed in the Dematiaceae. Some species of *Penicillium*, which normally produce distinct conidiophores, produce coremia under certain environmental con-ditions. Despite these and other similar difficulties, this arrangement remains the most workable one available for the rapid identification of these fungi. Current studies of these fungi may well produce a more natural classification. Further subdivision of these families is based on spore characters.

SPHAEROPSIDALES

Conidia borne in globose or flask-shaped fruit-bodies (pycnidia) or in cavities within host or stroma; plant parasites.

MELANCONIALES

Conidia and conidiophores in cushion-like masses (acervuli) developing below surface of host and bursting through at maturity; plant parasites.

Some species not known to produce spores of any type are classed as MYCELIA STERILIA and include a number of plant parasites.

Protozoa

Introduction: form, life cycles, and nutrition

Protozoa are eukaryote Protista which have the combination of locomotion and heterotrophic nutrition that is characteristic of animals. The Protozoa are thus simple animals, but, since the evolutionary divergence that led to 'higher' animals and plants probably involved flagellated eukaryotes, among which there is still abundant evidence of adaptive evolution, it is not surprising that some plant-like organisms show clear evidence of relationship with genuine animals. A complete account of the flagellate Protozoa would therefore necessarily include a number of autotrophic organisms that can also justifiably be classified as algae (p. 215).

Since the different types of Protozoa show such diverse features a general description of a 'typical' protozoan cannot be devised. The characteristics and diversity of Protozoa are indicated in the account which follows.

Many species within the group have the characteristics of single cells, but many have two or more nuclei enclosed within a single membrane and many more have a colonial organization of interconnected individuals. The size of individuals ranges from a few μm to several millimetres. Protozoa frequently contain a range of specialized and complex organelles and may have complex life cycles in which the individual passes through a number of specialized stages. The organization and life of the protozoan individual is different in character from that of a cell of a multicellular organism, and many protozoologists prefer not to refer to an individual protozoan as a cell. Multicellular animals are assumed to have evolved from protozoan ancestors in the distant past, but there is a lack of evidence as to the forms of Protozoa which may have been the starting points.

The major subgroups of Protozoa are based on their organelles of locomotion and characteristics of their life cycles. Two of the groups are entirely parasitic, with limited powers of locomotion and complex life cycles; they have spores of characteristic types. In the Cnidospora the complex spores contain one or several amoeboid sporoplasms (the infective individuals) and possess one or several polar filaments which are ejected before emergence of the sporoplasms from the spore. The spores of the Sporozoa contain infective sporozoites of characteristic shape; they are simpler and do not have polar filaments. Members of the Ciliophora (ciliate Protozoa) are characterized by the possession of numbers of cilia (organelles of unilateral beat), and (normally) two types of nucleus. The flagellate Protozoa (Mastigophora), which have only one type of nucleus and move by means of flagella (organelles with an undulatory beat), are now grouped with the amoeboid Protozoa (Sarcodina), whose organization depends on pseudopodia for locomotion and food capture. The occurrence in some organisms of both flagella and pseudopodia, either simultaneously or at different parts of the life cycle, led to the taxonomic linking of the flagellates and amoebae to form the Sarcomastigophora.

The nutrition of some of the pigmented flagellates may be exclusively photo-autotrophic, but the majority of Protozoa are heterotrophs, dependent upon organic molecules as sources of carbon and nitrogen as well as for energy. Many protozoans (phagotrophs) eat other organisms or large fragments of these by phagocytosis, taking the organic material into food vacuoles for digestion. Such vacuoles may be formed at a permanent 'cell mouth' or cytostome in ciliates and some flagellates, or at many parts of the body surface in amoeboid organisms and some others. Numerous Protozoa (osmotrophs) feed saprozoically, absorbing soluble organic materials, with or without pinocytosis; in many species the relative importance of this method of feeding has not been evaluated, but in some forms, including many parasites, it is probably the only mode of nutrition. Since the Protozoa include both autotrophic and heterotrophic forms, it is not surprising that dietary requirements range from a simple selection of inorganic salts to a complex mixture containing many amino acids, vitamins and other growth factors.

The waste products of metabolic activities of Protozoa are believed to escape by diffusion, although in some organisms crystalline deposits may accumulate, internally or in skeletal structures. The contractile vacuole is believed to serve an osmoregulatory rather than an excretory function.

Studies of protozoan ultrastructure indicate that the basic form of a species can be related to a characteristic cyto-architecture which is primarily fibrous, but which also includes membrane elements and secreted structures. The structure and formation of this architecture has been fully studied in relatively few species, but the organization of some skeletal structures and of the ciliate and flagellate pellicles shows abundant evidence of the utilization of micro-tubular fibres, bundles of thin filaments and membranous vesicles, both as components of the complex structures and as morphogenetic organizing systems.

In most protozoan species the ability to reproduce is retained by all individuals although they may have a complex body organization. Usually reproduction involves binary fission of an organism to form two more or less equal daughters, so that no parent remains. Such reproduction involves not only the replication of the nucleus, but also the production of new organelles; in many cases existing organelles may be broken down and a new set produced in each of the daughters. Fission of an organism to produce a larger number of daughters is not uncommon; in the best-known example, called schizogony, the nucleus of an organism divides many times in succession, and then each nucleus separates from the main mass with a small amount of cytoplasm and many daughter organisms are formed.

Some sort of sexual process occurs in all major groups of Protozoa, although there are species in several groups which are not known to have any form of sexual activity, and there are also forms with modified sexual phases which have lost the full genetic potentialities of meiosis and cross-fertilization. Some flagellates and all sporozoans are haploid and show a meiotic division of the zygote nucleus, while other flagellates, the ciliates, and some sarcodines are diploid with meiosis in the formation of the gametic pronuclei. A life cycle involving intermediary meiosis and an alternation of a haploid generation with a diploid one (comparable in development with that of most cryptogamic land plants) is shown by several species of foraminiferan sarcodines and may be normal throughout that group. The occurrence of sexual phases may be restricted to certain parts of the life cycle, as in foraminiferans and sporozoans, or may involve 'unspecialized' individuals. Examples are known in both flagellates and ciliates where the likelihood of fertilization between clones (rather than within clones) is increased by mating-type systems, in which only individuals of different mating-types show sexual fusion. These mating types are analogous to the cross-pollination devices of some higher plants. They bear no direct connection with sexuality, since, in ciliates at least, each individual is monoecious and produces gametes of both types.

In some Protozoa the life cycles are regular and include a consistent sequence involving divisions, sexual fusion, cyst formation, etc.; such regularity is characteristic of parasitic forms. The life cycles of others are more dependent upon environmental changes of a physico-chemical or biological nature for the determination of cyst formation or characteristic types of reproduction, e.g. resistant cysts are commonly produced by soil Protozoa which are subjected to periods of drought, or predatory Protozoa may encyst when the supply of food organisms is exhausted. The alternation of a motile larva and sessile adult might be regarded as a simple cycle, but such organisms may also show sexual and cyst phases of irregular frequency.

Protozoa may occupy any trophic level in the food chains of natural communities where free water is present. Autotrophic flagellates are abundant in the sea and in fresh waters as well as in symbiotic associations with animals at various levels of organization, including other Protozoa. Herbivorous members of the sarcodine and ciliate groups form important items in the diet of many animals including carnivorous ciliates and sarcodines. Of particular importance in the economy of many communities, in damp terrestrial habitats as well as in truly aquatic ones, are the saprozoic and bacteria-feeding Protozoa which make use of the substances and organisms involved in the final 'decomposition' level of food chains and themselves form the food of organisms at a higher trophic level, thereby recirculating organic matter.

The biology of the main groups of Protozoa

Sarcomastigophora

This subphylum includes flagellates, sarcodines and the taxonomically enigmatic opalinids. They are forms moving by flagella or pseudopodia, or both, and have only one type of nucleus. The flagellates and opalinids can further be distinguished from ciliates by the form of binary fission, which in the flagellated organism is basically longitudinal, producing two mirror-image daughters, while in ciliates anterior and posterior daughters are produced by a transverse division running across the kineties (p. 207).

Many of the simpler flagellates possess chlorophyll and other pigments, or are heterotrophic derivatives of the autotrophic forms. These flagellates are classed by protozoologists as Phytomastigophorea, many of which are described in the section on Algae (p. 215). Many pigmented flagellates feed osmotrophically or phagotrophically and at the same time use photosynthesis, e.g. *Ochromonas* (Fig. 9.13a). Colourless heterotrophic species like *Chilomonas* (Fig. 9.13b) and *Peranema* (Fig. 9.13c) occur in most of the major groups of algal flagellates. The majority of the plant flagellates have few flagella associated with simple root structures and the body surface is often strengthened with cellulose or proteinaceous pellicular thickenings or scales.

By contrast, members of the Class Zoomastigophorea frequently have large numbers of flagella whose bases are associated with complex fibrous organelles, and the body surface is seldom stiffened with anything more than subpellicular microtubules. This group contains mainly parasitic forms with fairly complex structure as well as a small number of simple heterotrophic forms for which no relationship with any group of the Phytomastigophorea has yet been established. The latter category includes the collar flagellates (Fig. 9.13d), with a single flagellum surrounded by a collar of microvilli (found elsewhere only in sponges), and bicosoecid flagellates (Fig. 9.13e). Some simple flagellates show both pseudopodia and flagella either simultaneously, e.g. *Mastigamoeba* (Fig. 9.13g), *Actinomonas* (Fig. 9.13h), and *Dimorpha* (Fig. 9.13i), or at different stages of the life cycle, e.g. *Naegleria* (Fig. 9.13f).

The most important flagellate parasites are *Trypanosoma* and its relatives, which possess a DNA-rich mitochondrial organelle called the kinetoplast near the flagellar base. There are many

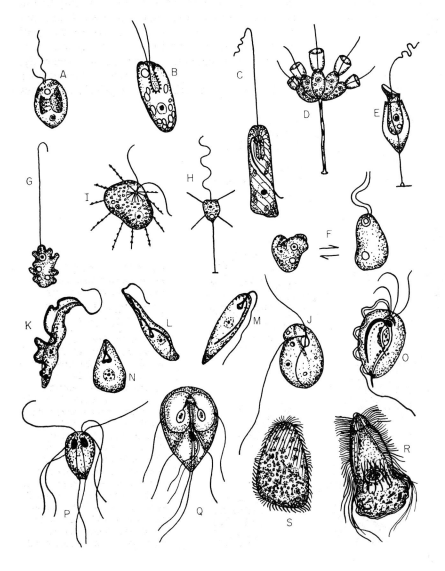

Fig. 9.13 Flagellate and opalinid Protozoa. (**A**) *Ochromonas* sp. 10 μm. (**B**) *Chilomonas paramaecium* 25 μm. (**C**) *Peranema trichophorum* 50 μm. (**D**) *Codonosiga botrytis* 15 μm. (**E**) *Bicoeca* sp. 12 μm. (**F**) *Naegleria gruberi* 20 μm. (i) amoeboid and (ii) flagellate phases. (**G**) *Mastigamoeba* sp. 15 μm. (**H**) *Actinomonas* sp. 10 μm. (**I**) *Dimorpha mutans* 20 μm. (**J**) *Bodo saltans* 10 μm. (**K**) *Trypanosoma brucei* 25 μm. (**L**) crithidial form of trypanosome 25 μm. (**M**) leptomonad form of trypanosome 25 μm. (**N**) leishmanial form of trypanosome 15 μm. (**O**) *Trichomonas* sp. 20 μm. (**P**) *Hexamita intestinalis* 10 μm. (**Q**) *Giardia muris* 10 μm. (**R**) *Trichonympha campanula* 200 μm. (**S**) *Opalina ranarum* 500 μm. Drawn by Margaret Attwood from live material and a range of published illustrations; normally several sources of information have been used for each figure. The average length of each organism is given in the legend.

species of *Trypanosoma* which spend part of their life cycle in the circulatory systems of vertebrate hosts and part of the cycle in blood-sucking invertebrates, such as insects and leeches; some are pathogenic species causing diseases in man and domestic animals, and others are said to be non-pathogenic. At different stages of the life cycle trypanosomes may possess different body forms, related particularly to the origin of the anteriorly directed flagellum (Fig. 9.13k, l, m). *Trypanosoma rhodesiense* and *T. gambiense*, which cause sleeping sickness in man in Africa, are transmitted by tsetse flies; parasites sucked from the blood plasma of an infected man multiply in the stomach of the insect, change form, migrate to the salivary glands and may be injected into the next man bitten by the insect. Species of the related genus *Leishmania* cause kala azar in man and other diseases of mammals in which the flagellate enters the white blood corpuscles.

The more complex zooflagellates are sometimes referred to collectively as metamonads. Species of the polymastigote *Trichomonas* (Fig. 9.13o) occur as parasites of man and domestic animals as well as in other situations. Some other forms have a double array of organelles, flagella and nuclei, e.g. *Hexamita* (Fig. 9.13p)—common in frogs, and *Giardia* (Fig. 9.13q)—common in mice. Hypermastigote species with numerous flagella like *Trichonympha* (Fig. 9.13r) form part of the abundant symbiont fauna of the termite gut (p. 000); they are xylophagous, engulfing wood through a naked posterior surface.

The opalinid 'flagellates' are saprozoic parasites which are found in the intestine of cold-blooded vertebrates, principally anuran amphibians. In the best-known genus, *Opalina* (Fig. 9.13s), the body is extremely flat, carries a dense covering of ciliary organelles in oblique longitudinal rows and contains a large number of similar nuclei. Binary fission is typically longitudinal, along the rows of cilia, and sexual reproduction involves the fusion of two small multiflagellate gametes. In two related genera there are only two similar nuclei and a less flattened body.

Sarcodines do not possess surface specializations of the type characteristic of the pellicle of ciliates

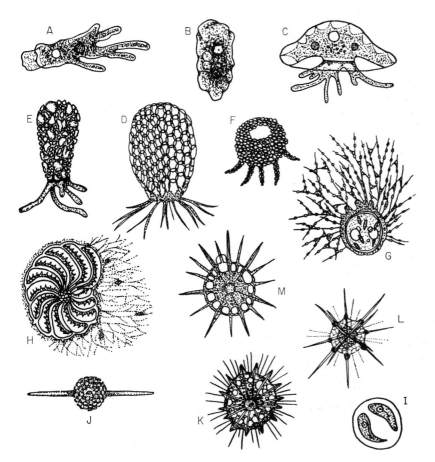

Fig. 9.14 Amoeboid Protozoa. (**A**) *Amoeba proteus* 500 μm. (**B**) *Entamoeba histolytica* 20 μm. (**C**) *Arcella vulgaris* 100 μm. (**D**) *Euglypha* sp. 150 μm. (**E**) *Difflugia* sp. 200 μm. (**F**) *Centropyxis aculeata* 130 μm. (**G**) *Gromia* sp. 150 μm. (**H**) *Elphidium crispum* 1 mm. (**I**) *Babesia bigemina*, two trophozoites within an erythrocyte, 5 μm. (**J**) Internal siliceous skeleton of a radiolarian 750 μm. (**K**) *Aulacantha* sp. 1·5 mm. (**L**) *Acanthometron* sp. 500 μm. (**M**) *Actinophrys sol* 50 μm. Drawn by Margaret Attwood from live material and a range of published illustrations; normally several sources of information have been used for each figure. The average length of each organism is given in the legend.

and flagellates; their surface flexibility permits the production of cytoplasmic extensions called pseudopodia, which are used in locomotion and in phagocytic feeding. The form of the pseudopodia and the characteristics of any shell, test, or skeleton are features used in classification.

The plasticity of body form is most evident in such naked forms as *Amoeba* (Fig. 9.14a), where the pseudopodia are broadly lobed. Contractile activity in the cytoplasm causes cytoplasmic flow, pseudopod formation and circulation of body organelles, so that the latter do not have a fixed position. These lobose amoebae range in size from a few microns to several millimetres in length, and may have one to many nuclei. They include parasites like *Entamoeba histolytica* (Fig. 9.14b) which causes amoebic dysentery in man and is transmitted as a cyst. Many protozoologists believe that the slime moulds, particularly the acrasian forms, are related to these amoebae (p. 214). Amoebae with lobed pseudopodia may secrete a chitinous test; this may remain chitinous (*Arcella*, Fig. 9.14c), or may incorporate siliceous plates (*Euglypha*, Fig. 9.14d), sand grains (*Difflugia*, Fig. 9.14e) or other debris (*Centropyxis*, Fig. 9.14f).

Fine branching pseudopodia occur in foraminiferans (Fig. 9.14h); in typical forms like *Elphidium* the reticulate pseudopodial system is granular and the cell body is enclosed in a chitinous test of one or many chambers whose walls are usually perforated and may be thickened with foreign material or a calcareous deposit. These forms are mostly marine and have been abundant since the early palaeozoic; since they form good fossils they are valuable to geologists in identifying rock strata.

In the Actinopodea the spherical body is surrounded by slender pseudopods (axopodia) supported by some form of axial structure. The planktonic marine Radiolaria (e.g. Fig. 9.14j) are known by their internal silica skeletons of very diverse form; many of them contain symbiotic algae. In the related Acantharia (Fig. 9.14k, l, m) the radial skeletal spines are impregnated with strontium sulphate. The Heliozoa (Fig. 9.14m) are predominantly freshwater animals in which the pseudopodia are supported by protein fibres, but the body may have siliceous spines or scales.

Sporozoa

All members of this subphylum are parasites, often with complex life cycles (Fig. 9.15a–e). In the main class, the Telosporea, the infective stage is a ver-

Fig. 9.15 Sporozoa and Cnidospora.

(**A**) Diagrams illustrating the life cycle of a monocystid gregarine. A sporozoite which hatches from the sporocyst (**i**) in the gut of an earthworm migrates and enters a cell in the seminal vesicle of the host (**ii**). Having outgrown the host cell the trophozoite lives free in the seminal fluid (**iii** and **iv**) and eventually pairs with another mature trophozoite (**v**). Within the gametocyst (**vi**) each gametocyte produces many gametes; fertilization occurs to form zygotes (**vii**), each of which develops into a thick-walled sporocyst containing several sporozoites (**viii**) which may later be released from the gametocyst (**ix**).

(**B**) Diagrams illustrating the life cycle of an eimeriid coccidian. A sporozoite which has left the sporocyst emerges from the oocyst (**i**) and enters a cell of the host intestine (**ii**), grows (**iii**), and divides (**iv** and **v**). The merozoites released at the breakdown of the host cell enter other intestine cells and may repeat the growth and schizogony stages (**iii**, **iv**, and **v**) or may become gametocytes (**vi**) from which flagellated microgametes (**vii**) or stationary macrogametes (**viii**) are formed. The zygote encysts (**ix**) and within this oocyst meiosis takes place (**x**); a number of sporocysts develops (**xi**), each of which may contain several sporozoites (**xii**).

(**C**) Diagrams illustrating the life cycle of *Plasmodium*. A sporozoite which has been injected into the blood of the vertebrate by the mosquito first enters an endothelial cell in the liver (**i**) where it grows (**ii**) and undergoes schizogony (**iii**) to release many merozoites which enter red blood corpuscles (**iv**) within which schizogony again occurs (**v**). Merozoites released by erythrocytes may undergo repeated schizogony (**iv** and **v**) or may develop into gametocytes (**vi** and **vii**). If these gametocytes enter the stomach of a mosquito they mature to form long microgametes (**viii**) or rounded macrogametes which are fertilized by microgametes (**ix**). The zygote (**x**) migrates between the stomach wall cells of the insect (**xi**) where it grows (**xii**) and divides to form many spindle-shaped sporozoites (**xiii**); these migrate within the insect to the salivary glands (**xiv**), whence they may be injected into a vertebrate and start a new cycle (**i**).

(**D**) *Toxoplasma gondii* 8 μm.

(**E**) *Sarcocystis tenella* 10 μm.

(**F**) Spore of *Haplosporidium* sp. 10 μm.

(**G**) Spore of *Myxobolus pfiefferi* 12 μm.

(**H**) Spore of *Triactinomyxon ignotum* 30 μm.

(**I**) Spore of *Nosema* sp. 5 μm.

Drawn by Margaret Attwood from live material and a range of published illustrations; normally several sources of information have been used for each figure. The average length of each organism is given in the legend.

miform sporozoite which may be injected naked into a host by an insect vector, or, more usually, may hatch in the host intestine from spores which are eaten by the host. The sporozoites migrate to specific host cells and feed saprozoically or sometimes phagocytically within the cell. At this stage parasites of the subclass Gregarinia leave the host cell and continue to grow in an organ cavity of the host body. In this case reproduction occurs only at spore formation (sporogony).

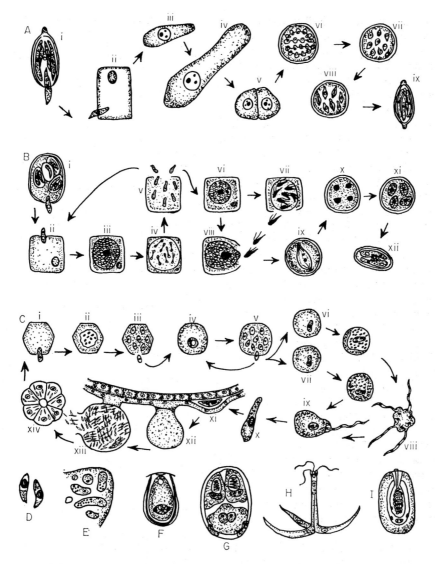

In the subclass Coccidia the trophozoites do not grow very large since they remain intracellular (Figs. 9.15b). After growth they undergo schizogony to produce many new infective individuals (merozoites) which enter other cells, so that a very heavy infection can be built up by repeated schizogonies. Trophozoites ultimately develop into gametocytes, some of which give rise to single macrogametes and others to several microgametes, which may be flagellate. Fusion of gametes in pairs produces zygotes, each of which develops into a few or many sporozoites, either naked or enclosed in a spore. Naked spores are characteristic of the Haemosporina (e.g. *Plasmodium*), and in the Eimeriina (e.g. *Eimeria*) the sporozoites are enclosed in a walled spore (or sporocyst).

Members of the genus *Eimeria* (Fig. 9.15b) cause disease (coccidiosis) in domestic mammals and birds as well as in other animals. The whole of the life cycle in these forms takes place in a single host individual, usually in the gut wall. Malaria, which is probably the most important human disease caused by a pathogenic protozoan, is caused by species of *Plasmodium* which are transmitted from man to man by *Anopheles* mosquitoes (Fig. 9.15c). Many other species of *Plasmodium* occur in other terrestrial vertebrates.

Toxoplasma (Fig. 9.15d) and *Sarcocystis* (Fig. 9.15e) are two sporozoan forms until recently known only as trophic stages and resistant cysts. It has now been shown that they resemble *Isospora* (a coccidian that occurs occasionally in man) and possess sexual reproduction and spore stages of the coccidian type, although the resistant cysts can apparently also be infective.

The piroplasms, e.g. *Babesia* (Fig. 9.14i) also inhabit the erythrocytes of vertebrates, but do not form spores. They are transmitted by ticks, and the cattle disease Texas red-water fever was the first disease whose life cycle was shown to involve an arthropod vector.

Cnidospora

Parasites of this subphylum have complex spores (Fig. 9.15f–i) surrounded by a single membrane or a two- or three-valved structure. They mostly contain one to six polar filaments, usually coiled in polar capsules, and one to many sporoplasms. The more complex spores found in the class Myxosporidea are formed from several cells during sporulation, and there is some doubt that these organisms are really protozoans. They occur in lower vertebrates and many types of invertebrates. In the host intestine, where the polar filaments are said to be extruded and anchor the spore, the amoeboid sporoplasms hatch from the spore and migrate through the gut epithelium to the organs character-istic of the species. Within the tissues the parasites grow, become multinucleate and split up into multinucleate masses, in each of which one or two spores develop.

The simpler spores of members of the Micro-sporidea are of unicellular origin. They occur in animals of most phyla, but are most commonly found in insects and fishes, where they multiply within host cells and cause hypertrophy of host tissues. The best-known microsporidian diseases are caused by species of *Nosema* (Fig. 9.15i) which parasitize intestinal cells of honey bees and other insects. The Haplosporea (e.g. Fig. 9.15f) are para-sites of invertebrates and fish and are probably related to microsporidians although they lack the polar filament.

Ciliophora

The most distinctive features of typical members of the Ciliophora are the possession of two types of nucleus and many cilia which are used in character-istic ways in movement and in feeding.

The two types of nucleus found in ciliates are the smaller diploid micronucleus and the larger poly-ploid macronucleus. At least one of each is normally present in every individual, but in some species there may be several or many of either or both types. In a

Fig. 9.16 Ciliated Protozoa. (**A**) *Tetrahymena pyriformis* 50 μm. (**B**) The arrangement of ciliary bases and kinetodesmata in the region of the buccal cavity of *Tetrahymena*. Down one side of the mouth is a double row of kinetosomes; only the outer of these rows has ciliary shafts which form the undulating membrane; at the other side of the mouth are three groups of basal bodies which mark the bases of the membranelles. Somatic kinetosomes lie alongside the kinetodesmal fibres. (**C**) The arrangement of fibres around the kinetosomes of *Tetrahymena*. Each basal body gives rise to a striated kinetodesmal fibre which runs forward, and to two groups of microtubular fibrils, one of which runs to one side and the other runs backwards; a third group of microtubular fibrils lies above the kinetodesmata, but does not make contact with the kinetosomes. (**D**) A longitudinal section through the base of a cilium of *Tetrahymena*. Beneath the cell membrane is an alveolar space in the pellicle and below this lie the fibres associated with the kinetosome, including the striated kinetodesmal fibres and the transverse and longitudinal microtubular fibrils shown in (**C**); two mitochondria lie nearby. (**E**) *Loxophyllum helus* 200 μm. (**F**) *Didinium nasutum* 150 μm. (**G**) *Nassula aurea* 200 μm. (**H**) *Colpoda cucullus* 75 μm. (**I**) *Paramecium caudatum* 250 μm. (**J**) *Cyclidium glaucoma* 25 μm. (**K**) *Spirochona gemmipara* 100 μm. (**L**) *Condylostoma arenarium* 500 μm. (**M**) *Stentor coeruleus* 1 mm. (**N**) *Spirostomum ambiguum* 2 mm. (**O**) *Halteria grandinella* 30 μm. (**P**) *Epidinium ecaudatum* 120 μm. (**Q**) *Oxytricha* sp. 120 μm. (**R**) *Euplotes patella* 100 μm. (**S**) *Aspidisca* sp. 30 μm. (**T**) *Tintinnopsis campanula* 100 μm. (**U**) *Vorticella* sp., expanded and partially contracted, body length 100 μm. (**V**) *Trichodina pediculus* 60 μm. (**W**) *Podophrya collini* 50 μm. Drawn by Margaret Attwood from live material and a range of published illustrations; normally several sources of information have been used for each figure. The average length of each organism is given in the legend.

few well-known species, e.g. *Tetrahymena, Did-inium*, there are strains which lack micronuclei. The macronucleus has a somatic function, providing for the routine synthetic activities of the cell, while the micronucleus is not concerned with these activities. Both types divide during binary fission of a ciliate, and nuclei of both types pass to each daughter. However, during the sexual process of conjugation, only the micronucleus provides genetic continuity.

Conjugation involves the exchange of genetic material (haploid products of meiotic division of the micronuclei) between two ciliates which come to-gether with a temporary, local, cytoplasmic fusion; a zygote nucleus is normally formed in both ciliates during the process. During conjugation the old macronucleus of each conjugant disintegrates and one of the products of the first mitotic division of the micronuclear syncaryon develops to form a new polyploid macronucleus by repeated replication of the nuclear material. A form of self-fertilization, called autogamy, may occur in which the two pronuclei fuse with each other within the single individual.

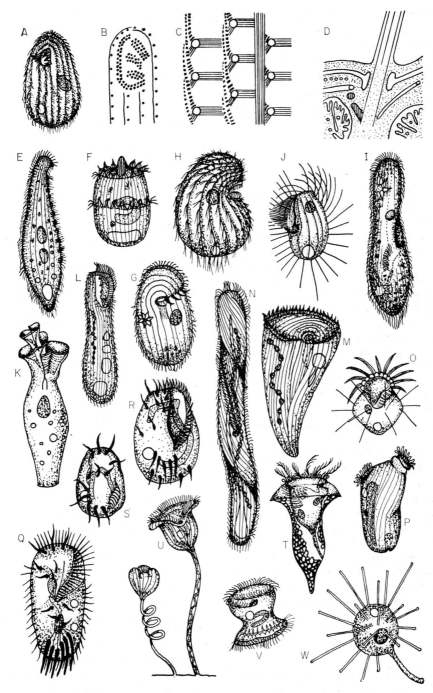

The cilia are arranged in longitudinal meridional rows called kineties (Fig. 9.16a), which are made up of rows of pellicular units called kinetids. Each of these consists of a cilium arising from a basal body (kinetosome), to which various horizontally arranged fibres are attached (Fig. 9.16c) and around which is a pair of surface alveoli, so that the pellicle is mostly comprised of three membrane layers underlain by a lattice of fibres (Fig. 9.19d). These fibre systems probably assist to maintain the shape and rigidity of the body, and the alignment of the kinetids; there is no evidence that they co-ordinate the activity of the cilia. Fibrous systems with an active contractile function occur in the stalk of

Vorticella (Fig. 9.16u) and in the pellicle of *Stentor* (Fig. 9.16m) and *Spirostomum* (Fig. 9.16n). The pellicle may contain arrays of trichocysts which can explosively discharge a thread which may have toxic properties.

The ciliature is often specialized for feeding or locomotion by the aggregation of the cilia into bands or compound structures, as in the mouth of *Tetrahymena* (Fig. 9.16b). In many cases there is a long adoral zone of membranelles (AZM) (e.g. Fig. 9.16r), while the undulating membrane may be extended in a number of forms (e.g. Fig. 9.16l). Compound locomotor cilia called cirri are present on some forms (e.g. Fig. 9.16r).

The classification of ciliates reflects the distribution and specialization of their ciliary organelles. The holotrich ciliates have characteristically a fairly uniform covering of cilia, but show a progressive development of mouth cilia (Fig. 9.16a–j). They mostly feed on bacteria and small organisms, but the simpler forms like *Loxophyllum*, *Didinium* and *Nassula* eat organisms larger than bacteria whole; the chonotrichs (Fig. 9.16k) are epizoic on crustaceans. The spirotrich ciliates (Fig. 9.16l) have a long AZM and show the trend for reduction from the complete ciliation of *Condylostoma*, *Stentor* and *Spirostomum* to the specialized cirri of the hypotrichs *Oxytricha*, *Euplotes* and *Aspidisca*. The peritrichs *Vorticella* and *Trichodina* (Fig. 9.16v) appear to form a different evolutionary line in which the undulating membrane has been extensively developed. The suctorian ciliates (e.g. Fig. 9.16w) lack cilia in the adult stage when they feed through tubular tentacles, but resemble mouthless holotrich ciliates in the non-feeding larval stage which is formed from the adult by budding.

Outline of the classification of the Protozoa

Phylum PROTOZOA

Subphylum **Sarcomastigophora**. With flagella or pseudopodia or both, a single type of nucleus, and usually without spore production. Sexuality involves syngamy.

Superclass **Mastigophora**. The flagellates.

Class **Phytomastigophorea**. The pigmented flagellates and their relatives, e.g. *Ochromonas*, *Euglena*, *Peranema*, *Chilomonas*, *Noctiluca*.

Class **Zoomastigophorea**. The animal flagellates, e.g. *Codonosiga*, *Bicosoeca*, *Bodo*, *Trypanosoma*, *Trichomonas*, *Trichonympha*.

Superclass **Opalinata**. With numerous ciliary organelles and 2 to many nuclei, parasitic.

Class **Opalinidea**. e.g. *Opalina*.

Superclass **Sarcodina**. Pseudopodial forms, body at least partly naked, but shell or skeleton often present.

Class **Rhizopodea**. With lobed, filose or reticular pseudopodia, e.g. *Amoeba*, *Arcella*, *Gromia*, *Elphidium*, *Dictyostelium*.*

Class **Actinopodea**. Spherical floating forms with axopodia or slender pseudopodia or both, often with skeleton or spicules, e.g. *Aulacantha*, *Sphaerozoum*, *Acanthometron*, *Actinophrys*.

Subphylum **Sporozoa**. Usually with simple spores containing sporozoites, parasitic.

Class **Telosporea**. With spores and characteristic sporozoites, e.g. *Monocystis*, *Eimeria*, *Plasmodium*, *Toxoplasma*, *Sarcocystis*.

Class **Piroplasmea**. Small piriform or amoeboid parasites in vertebrate erythrocytes, transmitted by ticks, e.g. *Babesia*.

Subphylum **Cnidospora**. Spores with polar filaments and amoeboid sporoplasms, parasitic.

Class **Myxosporidea**. Spore of multicellular origin, e.g. *Myxobolus*, *Triactinomyxon*.

Class **Microsporidea**. Spore of unicellular origin, single tubular polar filament, e.g. *Nosema*, *Haplosporidium*.

Subphylum **Ciliophora**. With cilia and two types of nucleus. Sexuality involves conjugation.

Class **Ciliatea**

Subclass HOLOTRICHIA. Body ciliature usually simple and uniform, mouth ciliature simple, e.g. *Loxophyllum*, *Didinium*, *Nassula*, *Colpoda*, *Paramecium*, *Tetrahymena*, *Pleuronema*.

Subclass PERITRICHIA. Body cilia absent in mature form, oral cilia in rows winding to mouth, often attached to substrate by stalk or basal disk, e.g. *Vorticella*, *Carchesium*, *Trichodina*.

Subclass SUCTORIA. No cilia in mature stage which feeds by tentacles, larva ciliated, e.g. *Podophrya*, *Dendrocometes*.

Subclass SPIROTRICHIA. Mouth cilia composed of many membranelles, forming a long adoral zone, bodily cilia usually reduced or compounded to form cirri, e.g. *Blepharisma*, *Stentor*, *Spirostomum*, *Halteria*, *Euplotes*, *Aspidisca*, *Epidinium*.

*(See also p. 210).

The phylum Gymnomyxa (the slime moulds)

Introduction

The mycetozoans and their associates comprise the true slime moulds (myxomycetes), the protostelids, the dictyostelid and acrasid cellular slime moulds, the labyrinthulas and the thraustochytrids. The plasmodiophorids, organisms with unequally laterally biflagellate zoospores and forming multinucleate protoplasts within host cells, are sometimes included in the phylum, but their relationships with other members are obscure. They are treated amongst the true fungi on p. 196).

The Gymnomyxa are a heterogeneous assemblage of organisms requiring organic nutrients. All at one stage in the life cycle consist of single cells which are motile and trophic. Most produce stationary spore-bearing fruit bodies of varied degrees of complexity. This combination of animal-like and plant-like phases in the one life cycle is a feature which characterizes the mycetozoans. The labyrinthulas and thraustochytrids (the 'associates' of the mycetozoans) do not form fruit bodies and probably are only distantly related to the rest of the group.

The various sorts of *cellular* slime mould share the characteristic that their fruit bodies result from the aggregation of previously independent individual cells. Though this phenomenon of aggregation of individuals to yield what appears to be a single multicellular organism is so strange, it nevertheless appears to have arisen independently in several different groups of organisms. Consequently its manifestation does not of itself imply phylogenetic affinity. It is found not only in the cellular slime moulds discussed here but also in a recently discovered but unnamed ciliate and in the prokaryotic myxobacteria (p. 164).

In some classifications the mycetozoans and associates are treated as a major taxon within the Fungi. In others they are treated as an independent phylum, the Gymnomyxa, within the Kingdom Protista. In the recent classification of L. S. Olive (1975) the mycetozoans comprise a single subphylum (the Mycetozoa) with two classes. One of these, the Eumycetozoa, or true mycetozoa, accommodates the subclasses Protostelia, Dictyostelia and Myxogastria, while the other, the Acrasea, contains but a single group. The associates of the mycetozoans, viz. the labyrinthulas and thraustochytrids, constitute a third subphylum—the Labyrinthulina.

Protostelia

The protostelids are widespread and common in soil, dung and plant remains from which they may be isolated and grown in two-membered culture with an appropriate bacterium, yeast or mould to serve as food.

The spore liberates a single protoplast which proceeds to multiply. The protoplasts of different species have from none to several flagella which may or may not be of equal size (Fig. 9.17a). Their pseudopodia are characteristically long and slender. Cell and nuclear divisions are not always closely associated, and protoplasts with several nuclei may appear. In some species multinucleate plasmodia arise, but they lack the rhythmic shuttle streaming of cytoplasm which occurs in the plasmodia of many true slime moulds.

At sporulation, individual protoplasts behave as pre-spore cells, while plasmodia, if present, segment into such cells. Each pre-spore cell develops a sheath and then secretes a non-cellular, probably cellulosic, narrow tubular stalk upon which the protoplast is raised (Fig. 9.17b). The protoplast then yields one to several spores which in different species may be retained, shed or even forcibly expelled.

One protostelid, *Ceratiomyxa fruticulosa*, forms unusually massive structures at sporulation (Fig. 9.17c). Its plasmodium, accompanied by mucus, emerges from rotting wood and produces a minute cushion-like body from which arise digitate processes that may branch and sometimes anastomose. On their surfaces the plasmodium takes on a network appearance and becomes a mosaic of cells as it segments into pre-spore cells, each of which proceeds to form a single spore on a slender stalk. *Ceratiomyxa* is the only protostelid in which sexual reproduction has been discovered. In the trophic phase similar flagellate cells fuse. Meiosis occurs in the pre-spore cells or in the spores. Each of the latter on germinating releases a protoplast which often cleaves into eight flagellate cells. This organism was formerly classified in the true slime moulds (Myxomycetes) as the sole representative of the subclass Exosporeae. With the discovery and description since 1960 of sixteen microscopic protostelids the natural affinities of *Ceratiomyxa* have become better understood.

The Protostelia are thought to be primitive Eumycetozoa evolved from simple algae and later

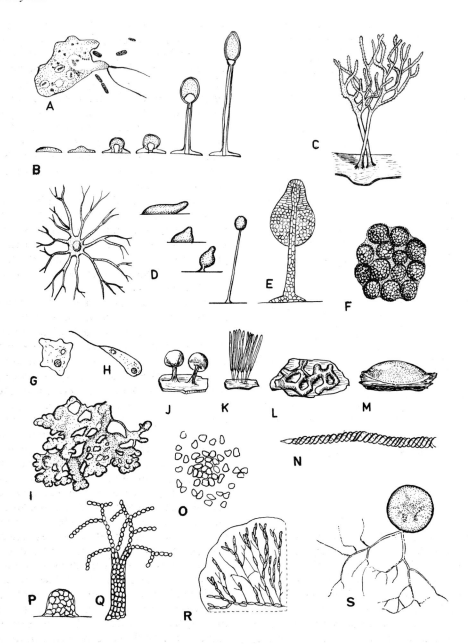

giving rise to the myxomycetes and dictyostelids. Evidence of an algal ancestry is shown by one species, the amoeboflagellate cells of which bear scales resembling those of chrysophycean flagellates.

Dictyostelia (Dictyostelid cellular slime moulds)

The general life cycle follows the sequence: germination of the spore to yield an amoeba, multiplication of amoebae, their aggregation, migration of

the entire aggregate, and finally transformation of the aggregate into a sorocarp containing spores (Fig. 9.17d). There is no true plasmodial stage, nor obligatory sexual phase, though evidence of sexual processes is accruing. There are two families. In the Acytosteliidae the sorocarp consists of a head of spores borne on a very slender (1–2 μm wide) hollow stalk while in the Dictyosteliidae the stalk is broader and packed with empty cells. All of the aggregating amoebae in the Acytosteliidae eventually become spores while in Dictyosteliidae some become stalk

Fig. 9.17 Gymnomyxa. (**A–C**) Protostelia. (**A**) *Cavostelium bisporum* (× 1300). Flagellate cell ingesting bacteria. (**B**) *Nematostelium ovatum* (× 350). Stages in sporogenesis from prespore cell to mature sporocarp. (**C**) *Ceratiomyxa fruiticulosa* (× 7). Habit sketch of fructification. (**D–F**) Dictyostelia. (**D**) *Dictyostelium*. Aggregation, migrating grex and two stages in culmination leading to formation of the stalked sorocarp. (**E**) *Dictyostelium discoideum*. Diagrammatic median longitudinal section of a developing sorocarp, drawn at approximately 4 × the magnification of the similar stage in (**D**), showing formation of the stalk as a tube, into the distal end of which prestalk cells enter. They then become vacuolated, secrete walls and die. (**F**) *Dictyostelium mucoroides*. A cluster of macrocysts. (**G–N**) Myxogastria. (**G, H**) *Didymium difforme*. Amoeba (× 550) and flagellated swarm cell (× 400) respectively. (**I**) *Badhamia utricularis* (× 4). Part of a plasmodium, showing reticulate posterior regions and lobed margins. (**J**) *Physarum nutans* (× 8). Sporangia. (**K**) *Stemonitis splendens* (× 1) Group of sporangia. (**L**) *Cienkowskia reticulata* (× 4). Plasmodiocarp. (**M**) *Reticularia lycoperdon* (× 0.7). Aethalium (**N**) *Trichia subfusca* (× 500). Spirally ornamented thread from capillitium. (**O–Q**) Acrasea. *Acrasis rosea*. (approximately (× 120). (**O**) Aggregation of amoebae. (**P**) Early sorocarp. (**Q**) Mature sorocarp. (**R, S**) Labyrinthulina. (**R**) *Labyrinthula* sp. (approximately × 150). Colony with spindle cells on ectoplasmic net. (**S**) *Thraustochytrium motivum* (× 350). Developing sporangium with rhizoid-like ectoplasmic net. (**B, E, O, P, Q** and **R** after Olive, L. S. (1975) *The Mycetozoans*, Academic Press, New York, San Francisco and London, **A** and **S** drawn from photographs in same source, **G, H, J, K, L, M** and **N** after Lister, A. (1925) *A Monograph of the Mycetozoa* (3rd Edn., by G. Lister), British Museum, London, **C** after Macbride, T. H. and Martin, G. W. (1934) *The Myxomycetes*, Macmillan, New York; **D** and **F** after Bonner, J. T. (1967) *The Cellular Slime Moulds*, 2nd Edn., Princeton University Press, New Jersey).

---

cells. Sorocarps of malnourished *Acytostelium* bear only one or two spores and are then indistinguishable from protostelids. The Dictyosteliidae is the larger of the two families and contains three genera, of which *Dictyostelium* has about twenty known species, some with virtually worldwide distribution. Soil and vegetable debris are their main habitats. For their isolation a diluted suspension of soil may be mixed with a suspension of bacteria, such as *Escherichia coli*, which serve as food, and poured over a dilute hay infusion agar buffered with phosphate to pH 6. Clones of amoebae should appear in three to four days.

Each dictyostelid spore on germination in an aqueous medium liberates an amoeba. These are small, uninucleate, and non-flagellate, have filose and sometimes lobose pseudopodia, and live in water films in which they feed by ingesting bacteria. They are usually haploid but diploid clones have been obtained. Their nuclei divide by mitoses in which the nuclear membranes persist and intranuclear spindles form. If nourished properly the amoebae divide every few hours. They can be grown in two-membered cultures in association with a wide variety of bacteria, and selected strains of *Dictyostelium* have been grown axenically on complex liquid media. Amoebae of a *Polysphondylium* species have even been cultured in a defined liquid medium. Food vacuoles are abundant and contractile vacuoles continually function. Sometimes individual amoebae develop thin protective walls and become microcysts.

When eventually the food supply is depleted the amoebae enter a pre-aggregation phase. They stop multiplying, their food vacuoles progressively disappear, the amoebae shrink, and their staining properties alter. Henceforth their development is sustained by endogenous reserves. After a pause they begin to aggregate.

Details of subsequent behaviour vary with different species but have been well studied in members of the Dictyosteliidae. The amoebae elongate and orient themselves radially about a number of centres towards which they begin to migrate in rhythmic surges. The converging amoebae form streams in which the cells adhere end-to-end. Aggregation involves chemotaxis governed by a substance named *acrasin* that is produced by the amoebae themselves. For several dictyostelids acrasin is cyclic 3′,5′-adenosine monophosphate (cyclic AMP). Chemotaxis requires that a cell in some way perceives a gradient in concentration of the chemotactic principle. There is, however, no single gradient of acrasin concentration associated with an aggregation pattern of amoebae. Instead a particular cell or group of cells produces acrasin briefly. The pulse diffuses outwards and induces nearby cells not only to orient themselves so that they are moving up the gradient but also to produce a further pulse of acrasin. By repetition of this process a wave of orientation and acrasin production is propagated centrifugally. This appears to be the cause of the rhythmic convergence of the aggregating amoebae. Successive pulses of acrasin production are initiated at the centre as the cells there recover from the preceding pulse. Cyclic AMP produce by the amoebae is accompanied by the enzyme phosphodiesterase which hydrolyses it to 5′-AMP. The simultaneous production of an inactivator may be biologically advantageous, for it increases the concentration gradient and also prevents the development of a high and uniform background concentration in confined environments. Besides cyclic AMP, folic acid and related compounds strongly attract amoebae of several dictyostelids including some for which cyclic AMP is not the natural acrasin. Both cyclic AMP and folic acid are produced by bacteria, so may help feeding amoebae to locate their prey. It is possible that folic acid or related substances are the acrasins of certain dictyostelids.

The nature of the cells that initiate aggregation centres has aroused controversy. It has been suggested that there is a small fixed proportion of special initiator cells in a population, but current opinion favours the view that there are simply some cells or groups of cells that happen to raise the local acrasin concentration. It has been shown that amoebae of different genotypes may participate in the formation of a single aggregate, and eventually, of a single sorocarp. This phenomenon has been termed heterocytosis.

As long as cells that are aggregating remain separate they appear to be guided only by chemotaxis, but once they begin to meet, a process known as 'contact following' can operate. Meeting cells cohere in short chains which lengthen and join up with others, and eventually continuous centripetal streams form.

The aggregate of cells eventually acquires a fairly definite columnar shape, topples on to its side, and proceeds to migrate over the surface of its substrate as if it were a single multicellular organism. This allows the organisms in the natural environment to move from the wet conditions ideal for growth to the dry condition best for sporulation. The phase of migration may however be eliminated in dry conditions or if the aggregate of cells is small. The migratory stage has been termed the 'grex'. The term pseudoplasmodium, though widely used, has little to commend it. In some species, for example *Dictyostelium mucoroides* and species of *Polysphondylium*, the grex as it migrates continually secretes a prostrate cellular stalk which is left behind, but in other species the grex forms no stalk and may then be termed a 'slug'. The slug of *Dictyostelium discoideum* ranges from about a fifth to five millimetres long and consists of a few hundred to hundreds of thousands of cells. Its size depends on the number of amoebae entering the aggregation and the density of amoebae in the culture. The slug has a distinct polarity and becomes increasingly differentiated inside as it migrates. Its anterior third comprises cells which differ histochemically, serologically, and in size from the rest. These anterior cells will eventually form the stalk of the sorocarp, while the posterior ones will form the spores. Those amoebae that arrived first at the aggregation centre

becomes the anterior pre-stalk cells; those that arrive later become the posterior pre-spore cells. In *Dictyostelium discoideum* those that are the very last to arrive eventually form the basal disc that is characteristic of the sorocarp of this particular species. Up to a certain stage the differentiation of cells within a slug is reversible, for slugs cut in half will give rise to normal sorocarps containing all three types of cell, i.e. stalk, spore, and basal disc. Further, cells removed from aggregates, slugs, or immature sorocarps will multiply and produce clones of cells able eventually to form normal sorocarps. The differentiation of these cells is therefore only phenotypic.

The grex migrates at about 0.25–2.00 mm/hour by a slow gliding movement, all the while leaving a collapsed slime sheath behind. The slime sheath, which contains cellulose fibrils, is probably secreted by all the cells of the grex. It is itself stationary, the individual amoebae moving forward through it in a coordinated way. The direction of movement of the grex is influenced by illumination and extremely small gradients of temperature, but appears to be indifferent to acrasin concentration. The positive phototaxis depends on the optical properties of the grex and on the presence of a haem-protein photoreceptor termed phototaxin. Light focussed internally on the distal side of the grex causes the cells there to surge ahead, so causing the grex to turn towards the light source. As would be predicted, general illumination increases the velocity of the grex as a whole. Single amoebae, however, do not respond to illumination.

Eventually the grex undergoes the process known as culmination by which the erect sorocarp is produced. Culmination is hastened by a degree of desiccation, and involves extensive and well ordered movements of the constituent cells and their differentiation into specialized types. In *Dictyostelium discoideum* culmination proceeds as follows (Fig. 9.17d). The grex halts, becomes rounded and its tip rises to become a papilla on top of the cell mass. A stalk tube initial then appears beneath the papilla as a more or less cylindrical, hyaline membrane and the cells that are within this tube become large and vacuolated. More pre-stalk cells migrate into the open upper end of the stalk tube which is growing in length by the deposition of material by the pre-stalk cells as they ascend its outside prior to entering its open end. The lengthening of the stalk is accompanied by its downward extension through the supporting mass of cells until its first-formed basal part contacts the substrate. Thereafter further extension elevates the growing end of the stalk (Fig.

9.17e). The pre-spore cells of the grex follow the pre-stalk cells in their ascent of the outside of the stalk, develop cellulose walls, and become a lemon-shaped or spherical mass of spores embedded in mucilage. In *Polysphondylium* rings of cells are successively pinched off from the hind end of the grex as it ascends the erect portion of the stalk, and each forms a whorl of small sorocarps, each essentially like that of *Dictyostelium*.

Non-sexual genetic exchanges can occur between amoebae, probably by means of temporary protoplasmic bridges, but sexual reproduction, though not unequivocally demonstrated, probably occurs through structures known as macrocysts (Fig. 9.17f). These form in some species as an alternative to sorocarp formation. Under suitable conditions a roughly spherical aggregate of amoebae becomes surrounded by a layered wall. Within this macrocyst, a larger-than-normal amoeba, suspected of being the product of cell and nuclear fusions, progressively ingests the other amoebae and eventually fills the macrocyst. Ultrastructural evidence suggests meiosis then occurs. Subsequent cleavage eventually yields normal-sized amoebae which escape as the walls of the germinating macrocyst break down. Some strains, especially of *Dictyostelium mucoroides*, are homothallic in respect of macrocyst formation, but others, including most strains of *D. discoideum*, are heterothallic. Only when appropriate clones are paired do macrocysts form. There is thus much to suggest sexual reproduction occurs, but it remains to be proven that macrocysts are structures in which genetic recombination takes place.

Myxogastria (Myxomycetes or true slime moulds)

Myxomycetes have a distinctive life cycle. The spore liberates one or more uninucleate protoplasts which feed, multiply, and eventually may fuse in pairs. The resulting zygotes develop into plasmodia—slime-ensheathed multinucleate masses of cytoplasm which feed and creep about. Eventually the plasmodia transform into fruit-bodies in which cellulose or chitin or both have been reported. Most of the 400–500 species are probably cosmopolitan. The plasmodia are often seen on damp decaying twigs and logs.

The unicellular, usually uninucleate, haploid spores are enclosed in ornamented walls. Some spores under favourable conditions germinate after only 20 minutes, others take up to 1–2 weeks, while in some species germination has proved very difficult to obtain. On germination, up to 8 naked cells generated by mitosis and cleavage, escape through a crack or irregular pore in the spore coat. These cells may be amoebae or anteriorly flagellated swarm cells and in many species the one can change into the other (Fig. 9.17g, h). Swarm cells are usually unequally biflagellate, or at least are potentially so. Two basal bodies are consistently present. The shorter of the flagella may be visible only after special fixation. Neither flagellum bears mastigonemes. Flagellum-like pseudopodia also are sometimes present. The cells feed by ingesting solid particles such as bacteria, and can multiply rapidly, their nuclei dividing by extranuclear mitoses; the nuclear membranes break down in late prophase and a spindle with polar centrioles forms. Contractile vacuoles are active in swarm cells. Under unfavourable conditions the protoplasts may encyst.

Fusion between pairs of haploid cells is normally a prerequisite of plasmodium production. A few species may be homothallic, but in most species plasmodia arise only when appropriate clones of cells are paired. In some heterothallic myxomycetes multiple alleles govern mating type so that either mating type in one race of a species may fuse with either type of another race. Cell fusion is closely followed by fusion of nuclei. Thereafter nuclear divisions take place without cell division so that a multinucleate diploid coenocytic plasmodium forms. Nuclear divisions in plasmodia are synchronized. The mitoses are intranuclear; the nuclear membranes persist and the intranuclear spindles lack polar centrioles.

There are several types of plasmodia. The simplest, the protoplasmodium, is a minute sheet of undifferentiated granular protoplasm with barely detectable cytoplasmic streaming. When eventually it fruits it forms a single sporangium. The aphanoplasmodium is a rather inconspicuous, very flat sheet of transparent, non-granular protoplasm which virtually lacks a sheath of gelled ectoplasm but exhibits streaming. The phaneroplasmodium is a conspicuous, thick, fan-shaped mass of cytoplasm organized into veins of gelled ectoplasm within which there is a shuttle flow of granular endoplasm on a massive scale (Fig. 9.17i). The flow follows a network of veins at speeds as great as 1 mm/sec. Protoplasmic streaming and locomotion of the plasmodium as a whole are intimately linked processes, but their mechanisms are not yet understood. Two contractile proteins, myxomyosin and myosin B, and possibly others also, are implicated in streaming. One hypothesis for the mechanism is that ATP-induced contraction of contractile protein complexes in the ectoplasm forces the fluid endoplasm through the veins. Microfibrils seen in plasmodia might represent such contractile elements.

Plasmodia from the field are often cluttered with relics of ingested food particles, but there is evidence that plasmodia may be nourished by soluble foods also. They may show chemotactic responses to certain food and other materials. They are sometimes coloured, often in shades of yellow, orange, or buff, but the chemical nature of the pigments is unknown. Sometimes large plasmodia on meeting fuse together completely if there is genetical identity at particular loci on their diploid genomes, but some fusions are followed by lethal reactions. Plasmodia sometimes cut off large numbers of amoebae and swarm cells of unknown ploidy, perhaps as a means of reproduction.

Vegetative plasmodia eventually fruit. Starvation of the plasmodium appears to be necessary to induce fruiting, and those which are pigmented also need exposure to light. The plasmodium usually differentiates into small clumps which become sporangia that may remain sessile or develop stalks. Sporangia are usually a millimetre or so high, and may be borne directly on the substrate or on a membranous hypothallus (Fig. 9.17j, k). Nuclear divisions occur within the sporangia, and the formation of cell walls gives rise to large numbers of spores. In some species the fructification resembles a prostrate strand of the plasmodium, the 'plasmodiocarp' (Fig. 9.17l). In others, the plasmodia coalesce as they develop and form an 'aethalium' which may be very large (e.g. up to 30 cm across) (Fig. 9.17m). Most sporangia at maturity have a double wall, the outer one tough and sometimes impregnated with calcium carbonate, the inner thin and transparent. Within the sporangium there may be a system of threads, the capillitium, which consists of irregular tubes or filaments, and sometimes contains calcium carbonate. In some genera the threads are beautifully and regularly ornamented (Fig. 9.17n). The capillitium usually arises by the deposition of material in or on a system of vacuoles which develops within the sporangial protoplasm. Meiosis occurs within the sporangium. Abundant evidence, much of it fine-structural, suggests it occurs within the delimited spore while it matures, three of the four haploid nuclei produced then degenerating.

Fewer than a sixth of the known species of myxomycetes have been induced to complete their life cycles in culture. Usually myxomycetes are cultured in association with food organisms (e.g.

bacteria or yeasts), but axenic culture of both amoebae and plasmodia is possible. Plasmodia of a very few species have been cultured axenically on *fully defined* media, but successful cultivation of the amoebae of the haplophase on synthetic media has been achieved only recently. Suitable defined media contain the growth factors haematin, thiamine and biotin.

Acrasea (Acrasid cellular slime moulds)

The members of this class were formerly classified in the same taxon as the dictyostelids because their sorocarps, although of simpler structure, are produced from amoebae which aggregate (Fig. 9.17o). They are now classified separately, being thought to have evolved independently. Resemblances in sorocarp formation are believed to be the results of convergent evolution. Acrasids have distinctive amoebae which have eruptive lobose pseudopodia and which move relatively swiftly by backward wavelike movements of hyaloplasmic bulges, in contrast to the amoebae of dictyostelids which produce a succession of lobose pseudopodia at the advancing front of the amoeba (p. 211). Moreover, the acrasid sorocarp lacks the cellulosic stalk-tube characteristics of dictyostelids.

Known acrasids are rather diverse in structure and life cycle, and are possibly polyphyletic. In *Acrasis rosea*, the best-known species, each spore liberates a single uninucleate carotenoid-containing amoeba which feeds and multiplies. Some amoebae encyst, to become microcysts. Others begin to aggregate by an undiscovered mechanism which does not involve the formation of convergent streams of cells as in dictyostelids (Fig. 9.17o). The bright orange slime-ensheathed grex thus formed does not migrate (Fig. 9.17p), but proceeds to form a sorocarp by the elevation of a mass of amoebae on a stalk made up of encysted amoebae. The latter remain alive, unlike the stalk cells of dictyostelids. The stalk may be one to several cells thick. The head composed of amoebae then proceeds to rearrange itself in the form of one to several simple or branched chains of cells which become spores (Fig. 9.17q). Ring-shaped scars mark the attachment of each spore to its neighbour in the chain. Other acrasids differ from *Acrasis* in such respects as having biflagellate trophic cells, or sorocarps without stalk cells. Sexuality in the group is unknown.

Labyrinthulina (Labyrinthulans and Thraustochytrids)

Representative members of the two families, Labyrinthulidae and Thraustochytridae, which constitute this predominantly marine subphylum look very different but share the key features that their cells produce an ectoplasmic net which originates at specialized morphologically characteristic organelles termed sagenogenetosomes. In the labyrinthulans the net takes the form of a system of 'slimeways' through which the spindle-shaped multiplying cells of the organism move swiftly by an obscure mechanism (Fig. 9.17r). In the thraustochytrids, formerly classified as Oomycetes among the true fungi because of their unequally biflagellate zoospores, the ectoplasmic net generated by a single vegetative cell constitutes a rhizoid-like system which serves in the anchorage and nutrition of the organism and also furnishes its limited capacity for movement (Fig. 9.17s). In both families the net proliferates over the cells of animals, plants, and fungi, often dissolving pores in their walls and digesting their contents. The organisms take up food by absorption, not by ingestion. Organisms in both groups have been cultured axenically.

In labyrinthulans, vegetative cells within the ectoplasmic net eventually cluster and give rise to rounded cells which probably function as sporangia. There is evidence that meiosis occurs within each sporangium and that unequally biflagellate zoospores are released. The location of nuclear fusion in the life cycle is uncertain but may follow fusion of spindle-shaped cells. *Labyrinthula* species are implicated in the etiology of 'wasting disease' of eel grass (*Zostera marina*) but are probably not its sole cause.

In thraustochytrids, the thallus has a laminated wall which is composed of overlapping circular scales, wholly unlike the wall of chytridiaceous true fungi (q.v.) which thraustochytrids superficially resemble. The cell becomes an epibiotic sporangium which liberates zoospores. Sexuality is unknown in these organisms. Curiously, the first herpes-like virus to be found in a non-vertebrate organism has been found within a species of *Thraustochytrium*. This may prove to be a convenient experimental system for the study of such viruses.

The algae

Introduction

The algae and the Cyanobacteria (p. 171) are the major carbon-assimilating micro-organisms and since seventy per cent of the global surface is water the importance of algal photosynthesis cannot be under-estimated. It is probable that at least as much carbon is fixed by algae as by the land flora, since although most algae are microscopic, the rate of turnover of the population is great and continuous throughout the year. The processes involved in this fixation of carbon are essentially similar to those of higher plants and the same pigment, chlorophyll *a*, is present in all algae. They possess a wide range of accessory pigments some of which are involved in photosynthesis and although starch is the reserve product of some algal groups (as with higher plants), various other polysaccharides, fats, etc., are the main reserve substances of others. Cellulose, often in a microfibrillar form is present in walls of some, but other carbohydrates, e.g. mannose, xylose, sulphated polysaccharides, alginic acid, fucoidin, laminarin, etc., occur in walls of other species. Certain groups deposit silica or calcium carbonate in a pectin or mucilaginous base or as intricate scales.

Habitat

Algae require moisture for the movement of their reproductive cells but many can exist for long periods in habitats subject to intense desiccation. They are common on trees as epiphytes, especially in moist tropical regions; on bare rock surfaces, especially when moistened by sea-spray, and on soils. In all fresh-water and marine situations algae thrive down to depths to which photosynthetically usable light penetrates. Between the land and any body of water there is an intermediate zone where the soil (or rock) algal flora intermingles with that of the underwater soil. On permanently submerged silt or rock surfaces characteristic algal floras occur; on the former mainly motile and on the latter mainly attached forms. Similar but not identical species coat the larger aquatic plants and also occur on many animals, particularly molluscs and hydroids. The boundary between land and sea has its very characteristic algal flora, visually dominated by the large brown and red algae which are richly coated with microscopic species. In open water, species of many algal groups are maintained in circulation by the turbulence of the water; this community extends over the whole ocean surface as a thin layer of

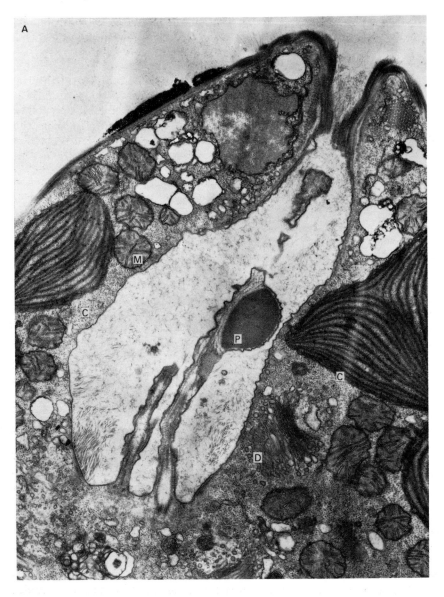

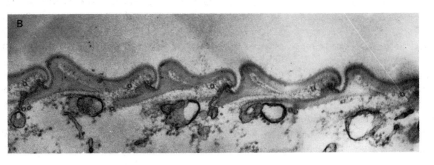

Fig. 9.18 (A) *Lepocinclis ovum* v. *butschlii*—a euglenoid flagellate closely related to *Euglena*. Electron micrograph of cell apex showing the anterior invagination with the two flagella arising from the base. The larger flagellum which extends out of the invagination, though not seen to do so in this section, bears the swollen photoreceptor (p). The pellicular strips can be seen entering the opening of the invagination but ceasing there. Chloroplasts (C), mitochondria (M), and a dictyosome (D) are also visible (×15 000). (Electron micrograph kindly supplied by G. F. Leedale, University of Leeds) **(B)** Transverse section of the periphery of a *Euglena* cell showing several pellicular strips, with muciferous bodies lying beneath each strip together with various microtubules. (Electron micrograph by G. F. Leedale, University of Leeds)

carbon-fixing species. A further marine algal community exists in and on the under surface of sea ice and is of considerable importance in polar waters.

Fine structure of algal cells

The main features of algal cells resemble those of other eukaryotic protists (p. 176), but some organelles show characteristic features and others, such as chloroplasts, are found among micro-organisms only in algae.

Plastids

Both pigmented (chloroplasts, chromoplasts) and non-pigmented (amyloplasts) plastids occur in algal cells, amyloplasts so far being known only among siphonaceous Chlorophyta. All eukaryotic algal plastids are membrane bound and composed of a granular matrix through which is dispersed a varying number of elongated flattened vesicles, termed thylakoids (Fig. 9.18a). In disrupted dinoflagellate chromatophores they are seen to be circular.

The plastids may also contain pyrenoids, starch grains, osmiophilic granules and pigment-containing granules grouped to form the stigma (eyespot). The bounding membrane is usually a double structure corresponding to two closely appressed unit membranes (*sensu* Robertson) and in some algae it is continuous with elements of the endoplasmic reticulum. The ER may form additional layers around the plastid, as in Haptophyceae where there is a double endoplasmic reticulum membrane outside the double plastid one. In dinoflagellates, euglenoids, and xanthophyta the plastid membrane is distinctly three-layered. The thylakoids are usually stacked into lamellae which run along part of or the whole length of the plastids. Material of the matrix does not penetrate the flattened thylakoids. In the red algae the thylakoids run singly through the matrix (cf. the situation in the Cyanobacteria, formerly known as blue-green algae, p. 171, which differ, however, in the absence of a bounding membrane) whilst in all other algal divisions they are grouped, e.g. two thylakoids in the Cryptophyta, two to six in Euglenophyta and Chlorophyta, three in dinoflagellates, diatoms, Chrysophyta, and Phaeophyta.

Highly characteristic 'girdle' lamellae encircle the inner lamellae in Phaeophyta, Chrysophyta, and Xanthophyta and some diatoms; in the latter they are distinctive features even of the male gametes. Most algae lack the characteristic stacks of disk-like vesicles, known as grana, found in the chloroplasts of Angiosperms, but invagination and folding of the thylakoids result in 'pseudograna' in *Carteria* and perhaps even true grana in some *Acetabularia* species.

Colourless strains of some *Euglena* species can be induced by growth at high temperatures; chloroplast replication is inhibited whilst cell division is not. Cell division then produces some cells containing only a single chloroplast; each of these on further division yields one without plastids. These cells cannot then be induced to form chloroplasts; further evidence for the theory that chloroplasts cannot arise *de novo* but only from existing chloroplasts.

Fine arrays of microtubules obliquely striate in longitudinal section have been found in the chloroplasts, vegetative cells, zoospores and eggs of *Oedogonium*, but not in other groups of green algae.

Pyrenoids, either immersed in the plastid or as outgrowths from it (e.g. in some dinoflagellates and Phaeophyta), are areas of fine-grained dense material without bounding membranes except in some diatoms. In all algal groups the thylakoids are displaced by the aggregation of the pyrenoid matrix, but one or two thylakoids of the lamellar system may continue into and sometimes through the matrix. In some Chlorophyta the normally flattened thylakoids reduce to tubules as they pass into the pyrenoid and in cross section these tubules contain peripheral rod-like structures, which may be the reduced inner thylakoids of the lamellae. Starch is formed usually as discrete plates around the pyrenoid, though in the diatoms which do not form starch no reserve product is visibly associated with this structure. Light microscopy suggests that the projecting pyrenoids are unconnected with starch formation, but the electron microscope reveals that some of them have a thin layer of some reserve substance surrounding the matrix, e.g. in Dinoflagellates, where in the species with stalked pyrenoids the plastid membrane also covers the pyrenoid, polysaccharide material is deposited outside the membrane. Similarly in the euglenoid algae the paramylon granules lie outside the chloroplasts. The starch grains in *Scenedesmus* are cut off from the pyrenoid by intrusion of lamellae, the pyrenoid disappearing completely at cell division and arising *de novo* in the daughter chloroplasts.

Eyespots are composed of collections of osmiophilic granules in which carotenoid material is dissolved. When they occur within the plastid, they form either a single or multiple row(s) of granules located between thylakoids usually adjacent to the outer chloroplast membrane. The eyespots in the gametes of fucoid Phaeophyta appear as a series of chambers containing the pigment rather than as separate granules and it is noteworthy that in these organisms the hind flagellum is swollen as it passes over the eyespot. Although not actually attached to the gamete in this position it is pressed close against the cell membrane. The eyespot in euglenoid flagellates is free in the cytoplasm usually around the base of the anterior invagination and consists of clusters of osmiophilic granules and some microtubular elements (as in some Chlorophyta). The eyespot granules are clustered in groups of 2–3 surrounded by a membrane of unit membrane dimensions, but the eyespot as a whole has no bounding membrane. Osmiophilic granules are common in the chloroplasts of many algae including euglenoid genera and may be channelled between the thylakoids towards the eyespot when this is present.

Colourless amyloplasts occur in some siphonaceous Chlorophyta and resemble chloroplasts but contain a single large starch grain and a small concentric mass of lamellae. This small concentric aggregation of lamellae also occurs in the chloroplasts of species containing amyloplasts but not in those siphonaceous species without amyloplasts.

Mitochondria

Though basically similar to those of other plants, algal mitochondria show subtle variations between groups, particularly in the form of the cristae. In *Micromonas pusilla* there is only a single mitochondrion, but in most algal cells they are numerous, often precisely located within the cell, e.g. around the nucleus and between the lobes of the chloroplast (*Chlamydomonas eugamatos*), between the plasmalemma and the chloroplast (other *Chlamydomonas* spp., *Carteria* and colonial species). They are often closely associated with flagella bases (kinetosomes) and in the Prasinophycean genus *Heteromastix* a single mitochondrion lies pressed against the inside of the 'C' shaped chloroplast.

Dictyosomes

Dictyosomes (Golgi bodies) are very distinctive components of algal cells and are often precisely located. They consist of stacks of elongate vesicles dilated at the periphery. In some colonial Chlorophyta they surround the nucleus and are contained within a series of 'amplexi' formed of endoplasmic reticulum continuous with the outer layer of the nuclear membrane. Associated with the dictyosomes are vesicles budded off from the cisternae; in the Haptophyceae and Prasinophyceae the characteristic scales are formed within these vesicles (p. 224). In the dinoflagellate *Gonyaulax* the dictyosomes form a 'spherical shell' closely associated with the concave side of the 'C' shaped nucleus and within this area vesicles, presumably formed from the dictyosomes, give rise to the trichocysts (p. 182). The dictyosomes in *Chlorella* pair prior to cell division to form corresponding 'mirror' images between which the new partition membrane is formed.

In the diatoms special membrane-limited vesicles appear during cell division and within these the new silica valves are deposited. The surrounding membrane has been termed the silicalemma and is distinct from the plasmelemma.

Vacuoles

Contractile vacuoles appear as empty membrane-bounded spaces often with subsidiary cisternal elements around them and fusing to form the main vacuole. In dinoflagellates more complex vacuoles (pusules) are associated one with each of the flagellar canals. These are in clusters of approximately 40 vesicles, each with a double membrane and opening by means of a narrow pore, either into a central vesicle and then into the flagellar canal or individually into the canal. These pusules are permanent vacuoles, unlike the temporary contractile vacuoles which enlarge, discharge and enlarge again.

Miscellaneous organelles

Structures comparable to the lomasomes of the fungi have been seen in the algal genera *Chara* and *Nitella* of the Charophyta and in *Platymonas* of the Prasinophyceae but not in other groups.

Muciferous bodies are often located beneath the plasma membrane (Fig. 9.18b). They may discharge through it or into microtubules.

Trichocysts of the dinoflagellates are complex organelles which eject fine, banded, probably proteinaceous threads from the cell. The threads are extruded through thin areas in the wall after the mature trichocyst has migrated from the central dictyosome area. The trichocyst has a single membrane, thickened by hoops or spirals of material, and a central square core of amorphous material. Along the upper end of the core small backwardly directed tubules occur, and attached to the apex is a group of twisted fibrils ending in even finer ones which just touch the closing membrane. Discharge is accompanied by a change in hydration of the core which takes place in a matter of milliseconds. Trichocysts occur also in some other algal groups.

Cell walls

Wall structure in the algae is extremely variable, though usually based on polysaccharides within or on which other substances are deposited. Mucilages are common but these may have structural microfibrillar elements within them, e.g. in colonial volvocalean genera. In many Chlorophyta the microfibrillar material is cellulose and in the classic examples *Cladophora* and *Chaetomorpha* it is deposited in layers of microfibrils running at different angles to the long axis of the cells. Shorter microfibrillar units of xylose or mannose form the wall material in some of the siphonaceous green algae.

The Haptophyceae and Prasinophyceae have a 'wall' layer composed of scales formed in the vesicles arising from the dictyosomes. In the former group they contain mannose, and have a fibrous structure with an inner face with radiating fibres and an outer one of randomly arranged fibres, the two faces being two sides of a flattened disk. In the Coccolithophorids, the calcium carbonate is deposited upon the scales. Simple scales without the calcified part (coccolith) are also found on the body of the cell. Thus in the coccolithophorids at least two scale types can be demonstrated on any one cell and in the other section of the group containing the genus *Chrysochromulina* several precisely arranged purely organic scale types can be found on any cell (Fig. 9.20i). In the Prasinophyceae it is common to find two scale types on the flagella and three types over the body of the cell (e.g. in *Pyramimonas*). In *Platymonas*, fine scale-like structures, also produced in the dictyosome vesicles are liberated and coalesce outside the cell to form the theca. The dinoflagellates have an 'armoured' cell wall consisting of precisely arranged plates joined together by sutures. These plates at least in some species are sandwiched between outer and inner membranes and the latter connect the edge of the plates with the underlying plasmalemma.

The range of vegetative form

The range of vegetative morphology includes (1) unicells which may be non-motile (*Chlorella*), motile by means of flagella (*Chlamydomonas, Euglena, Ceratium*), motile by extrusion of mucilage or other means (desmids and diatoms). (2) aggregations of cells into colonies—motile (*Volvox,*

Synura), non-motile (*Pediastrum*), (3) filaments— unbranched (*Ulothrix*), branched (*Ectocarpus*), (4) complex thalli based on filamentous construction (Rhodophyta and some Phaeophyta), (5) 'parenchymatous', i.e. without distinct filamentous structure in the mature stages (*Ulva, Laminaria, Fucus*), (6) coenocytic, simple unicellular (*Protosiphon*), branched unicellular (*Bryopsis*) or divided into multinucleate segments (*Cladophora*).

Life-cycles of algae

The simplest life cycle is shown by *Fucus*, where the spermatozoids and eggs are the only haploid phase in the life history. In other algae there are varying degrees of isomorphic/heteromorphic development of gametophytes and sporophytes culminating in the most complex of all plant life histories in the Rhodophyta. The formation of gametangia lacking a sterile wall layer around the fertile tissue is common in all algae, distinguishing them from higher plants. The sexual process may involve simple fusion of nucleated cytoplasmic masses or of flagellated gametes and non-flagellated egg cells. Accessory reproduction of either or both gametophyte and sporophyte may be achieved either by non-motile spores or by motile zoospores.

The groups of algae

The classical concept of algae as consisting of four basic groups distinguished on the chemical basis of pigmentation into the blue-green (Cyanophyta), green (Chlorophyta), brown (Phaeophyta), and red (Rhodophyta), was an over-simplification. The blue-green algae are now considered to be distinct from the rest and are treated as prokaryotes, the Cyanobacteria (p. 171). Studies during the last decade and the increasing use of electron microscopy has led to a clearer segregation of eukaryotic algae into a series of divisions (phyla) each of which requires consideration to give a balanced view of the group.

Chlorophyte series

Four major groups are included. With chlorophyll α they also possess chlorophyll β and small amounts of carotenoid pigment. Starch or starch-like polymers are the common reserve product.

EUGLENOPHYTA (Fig. 9.19a–d)

Pigmented genera of the phylum (Fig. 9.19a–c), are

abundant, particularly in organically polluted fresh waters. Although photosynthetic they require organic molecules in the form of vitamins (that is, they are photoauxotrophic), e.g. *Euglena gracilis* requires vitamin B_{12} complex and has been used in biological assay. Many green species are also facultative heterotrophs, that is, they can grow in the dark when supplied with organic compounds. Most genera can be cultured in a bi-phasic medium with soil at the base of the culture vessel and liquid above. The interior invagination (Fig. 9.18a) is not a gullet and solid particles cannot be ingested through it; the complex contractile vacuole system discharges into this invagination. Within it the two very unequal flagella rise; the shorter does not extend to the outside and is smooth (i.e. does not have fine hairs or 'flimmer') but the larger may be as long as or longer than the cell, has a single row of hairs and near the base a swelling which is the photo-receptor. No other algal or protozoan group has hairs in a single row. The so-called 'pellicle' of euglenoids is pliable and stretchable allowing the cell to change its shape; it consists of a series of strips of proteinaceous material wound around the cell like a number of bandages which fuse in pairs at the base and apex. The intricate means by which these 'pellicular strips' engage one another is shown in Fig. 9.18b, together with the associated muciferous bodies, microtubules and external wart-like granules which are present in some species. This skeletal system is really an endoskeleton since the cell membrane is external to it. The chromatophores vary greatly in shape and size but the paramylon granules (a β-1:3 linked glucan) are a most characteristic diagnostic feature as also in the large red eyespot which is free in the cytoplasm. Some species contain endophytic bacteria which occur in the cytoplasm and/or within the nucleus. One green genus, *Colacium* (Fig. 9.19d) is colonial, attaching itself by secretion of a mucilage stalk at the apical end; another, *Euglenamorpha* has three emergent flagella each with basal flagella swellings.

CHLOROPHYTA (Fig. 9.19i–u)

This is the largest and most complex group of green algae. The variously shaped chloroplasts possess pyrenoids around which starch (an α-1, 4 glucan) is deposited, though granules of starch may also be free in the cell. Eyespots when present are embedded in the chloroplast. Contractile vacuoles open directly to the exterior and the two (sometimes four) flagella arise on either side of an apical papilla; they are equal in length and although devoid of 'flimmer'

(p. 180) many have a felt-like coating. The cell wall is frequently composed of cellulose and is often microfibrillar. Sexual reproduction is widespread and when motile gametes are involved these are usually biflagellate though all stages through to oogamy are found. Chlorophyta are most widespread in fresh waters; Bourelly (1966) describes 520 genera from fresh water. One or two groups, however, are predominantly marine, e.g. the Bryopsidophyceae.

The simplest cells are merely spheres of cellulosic wall enclosing the cytoplasm, which contains a single chloroplast and a nucleus, e.g. the ubiquitous *Chlorella* (Fig. 9.19k) which grows in laboratory glassware, condensers, transparent tubing, etc. Such a cell divides into four similar small daughter cells which then enlarge. In the same group as *Chlorella* are a large number of other single-celled algae and also genera which form colonies when the products remain together, e.g. *Scenedesmus* (Fig. 9.19l). Some form motile zoospores but all are nonmotile in the vegetative phase. Some, e.g. *Trebouxia*, are the phycobiont of many lichens (p. 281). Unicells or groups of cells embedded in mucilage and often provided with channels through the mucilage, easily mistaken for flagella, comprise another subgroup of which *Tetraspora* is one of the commoner genera. The vegetative cells often retain the contractile vacuoles and eyespots, derived from the motile stage. Of greater complexity are the motile cells of the *Chlamydomonas* type (Fig. 9.19i) which have all the characteristics of a *Chlorella*-like cell plus contractile vacuoles, an eyespot and two flagella (emerging from pores in the cellulose wall). Not all motile genera possess a true cell wall, others resemble *Trachelomonas* in having a distinct theca around the cell. Aggregation of chlamydomonad-like cells into mucilaginous colonies (e.g. *Gonium*, *Volvox*) and into non-mucilaginous aggregates (e.g. *Spondylomorum*) parallel the colonial habit of the coccoid forms.

The desmids are specialized coccoid cells with a highly sculptured, poroid wall usually consisting of two halves joined by a bevelled edge at an isthmus, e.g. *Cosmarium* (Fig. 9.19t) and *Staurastrum*. Mucilage is extruded through the pores and frequently forms a thick capsule around the cell. Bacteria are commonly found lodged in the channels between the blocks of mucilage which form the capsule. Some genera form along filaments owing to the possession of special spines or mucilage pads which prevent the separation of the daughter cells during division. There is a second smaller series in which the wall is composed of a single piece; here again

Fig. 9.19 Chlorophyte genera. (**A**) *Euglena.* (**B**) *Phacus.* (**C**) *Trachelomonas.* (**D**) *Colacium.* (**E**) The apex of a *Euglena* showing the pellicular strips curving into the anterior invagination. (**F**) Diagrammatic cross section of the pellicular strips as seen in electron microscope section. Showing the articulation of adjacent strips, mucilaginous warts (w), mucilage organs (mo), and microtubules (m). (**G**) *Pterosperma.* (**H**) *Pyramimonas.* (**I**) *Chlamydomonas.* (**J**) *Gonium.* (**K**) *Chlorella.* Cell and formation of four autospores. (**L**) *Scenedesmus.* (**M**) Fragment of *Pediastrum* colony. (**N**) *Ulothrix.* (**O**) *Oedogonium,* one cell of unbranched filament with cap cells. (**P**) Base of *Bulbochaete.* (**Q**) *Enteromorpha* (left) and *Ulva* (right). (**R**) *Spirogyra.* (**S**) *Cladophora.* (**T**) *Cosmarium,* cell and two stages in cell division. (**U**) *Acetabularia.*

some genera produce short loose filaments. The desmids are exclusively fresh-water species of world-wide distribution (most frequently in acid waters) and are often found in mucilaginous clumps on wet rock faces and on peaty soils. Many are cosmopolitan but a few genera have restricted geographic distributions; arctic and tropical species can be distinguished.

The remaining groups of the Chlorophyta are all filamentous or siphonaceous. Filaments may be unbranched, e.g. *Ulothrix, Spirogyra, Oedogonium* (Fig. 9.19n, r, o) or branched, e.g. *Chaetophora, Bulbochaete* (Fig. 9.19p). In some the initial fila-mentous morphology is expanded into a pseudo-parenchymatous state, e.g. *Ulva* and *Enteromorpha*

(Fig. 9.19q). Many have a basal cell modified as an attaching organ others have adopted a creeping habit and are firmly attached to the substrate by nucilage. They play a very important part in epiphytic associations and are themselves hosts to numerous other algae, protozoa, bacteria, and fungi. Some are particularly adapted to an aerial environment, e.g. *Apatococcus* and *Desmococcus* (formerly known as *Pleurococcus, Protococcus,* etc.) which grow on tree bark and develop a branched filamentous form only in moist situations or in culture. A further group, of which *Trentepohlia* is the commonest, are also terrestrial, growing on rock and tree surfaces and yet others, e.g. *Cephaleuros,* are serious leaf parasites of commercial crops including tea and coffee. This section is also exceptional amongst the Chlorophyta in forming oil rather than starch and also copious carotenoid pigments which give an orange coloration to the thallus.

The siphonaceous genera are mainly marine in distribution. Some have cross walls delimiting multi-nucleate segments of the thallus, e.g. *Cladophora* (Fig. 9.14s); others have branching siphons which aggregate to form complex leaf and rhizome-like structures of considerable size. Many are calcified and form subtidal turf-like growths. Amongst them is the uninucleate *Acetabularia* (Fig. 9.19u), a valuable tool in biochemical and morphogenetic research. It consists of a cap, a stalk, and a rhizoid anchoring it to the substratum. Although the organism is several centimetres long, it has only a single nucleus which is situated in the rhizoid. This cap is of complex structure, and its form is a species characteristic. If the cap is amputated, it regenerates, this occurring even in the absence of the nucleus. Enucleate individuals survive for several months. It is possible by amputating the cap and replacing the nucleus by one transferred from another species, to produce a system in which the formation of a cap is jointly controlled by the nucleus of one species and the cytoplasm of another. The form of the regenerated cap at first resembles that of the species from which the cytoplasm was derived but finally becomes identical with that of the species from which the nucleus was obtained, suggesting that factors controlling differentiation of the cap are produced by the nucleus and persist for some time in the cytoplasm after removal of the original nucleus, thus permitting regeneration of enucleate individuals.

Asexual reproduction is common in the Chlorophyta. In the filamentous Conjugatophyceae convolutions of the cross walls result in fragmentation of the filaments. The desmids split apart at the isthmus and two new semi-cells are formed (Fig. 9.19t): thus every desmid cell has one old and one new semi-cell. Thick-walled spores (akinetes) are formed in some filamentous species, e.g. *Cladophora*. Production of bi-quadri-flagellate zoospores, usually by cleavage of the cytoplasm within the cell wall to form four or more cells each of which acquires two isokont flagella, is by far the commonest mode of asexual reproduction. The zoospores escape either after dissolution of the parent cell, e.g. *Chlamydomonas*, or through special pores, e.g. *Cladophora*. In *Oedogonium* they possess a ring of paired flagella. Attachment is usually at the flagella pole in those which germinate to form an attached cell or filament. Sometimes the products of division are non-motile, e.g. in *Chlorella*.

Sexual reproduction in many species is by the fusion of motile gametes produced in a similar manner to the zoospores and often morphologically indistinguishable from these. In a few species the gametes are unequal in size and in some the process is oogamous. Many groups of the Chlorophyta range from isogamy to oogamy whilst others, e.g. the Oedogoniales, are completely oogamous. The zygote usually forms a thick-walled resting cell. In the Conjugatophyceae the gametes are formed from all or part of the cytoplasm of the cell and are non-flagellate masses which are transferred either into a conjugation tube between the two filaments or completely into the filaments of opposite strain. Homothallism and heterothallism both occur.

Both isomorphic and heteromorphic alterations of haploid and diploid phases occur in the Chlorophyta. Some of the heteromorphic genera have an alternation between a filamentous or thalloid phase and a unicellular (*Codiolum*) stage which is often endophytic in other algae.

PRASINOPHYCEAE

The members of this group occur in both freshwater and marine habitats and the non-motile spheres of *Halosphaera* and *Pterosperma* are frequent in marine plankton where they float to the surface. Formerly many of the species were included as anomalous forms of the Chlorophyta. Many genera are motile and most have four rather thick flagella arising from an apical pit around which the cell is extended into lobes (e.g. *Pyramimonas*, Fig. 9.19h). The cell surface and also that of the flagella is covered by layers of scales and hairs, demonstrated only with the aid of the electron microscope. The scales are of organic material and in some species it has been shown that they form within vesicles in the cells. The pyrenoid is often large and in some genera penetrated by fingers of the chloroplast or nuclear membranes or both. Whilst some species are motile cells in the vegetative stage, others, e.g. *Halosphaera* and *Pterosperma* (Fig. 9.19g), are coccoid and yet others are dendroid (e.g. *Prasinocladus*), the typical motile cells being produced only during asexual reproduction.

CHAROPHYTA

This is a distinctive group, forming macroscopic radically organized plants. These grow from apical cells which cut off cells at the base, each one of which divides again to give (i) an upper nodal cell, which then gives rise to whorls of branches, and (ii) a lower inter-nodal cell, which enlarges greatly but is incapable of branching. They are anchored by a rhizoidal system and have complex oogonia and antheridia. Since the large internodal cells may grow several inches long they have proved suitable for many biochemical/biophysical studies. They are common fresh-water/brackish plants growing anchored in the sediment.

Chromophyte series (Fig. 9.20a–s)

This is a heterogeneous collection in which the ratio of chlorophylls to carotenoid pigments is such that in most groups the plastids are brown coloured. Chlorophyll *b* is absent and carbohydrate reserve products tend to accumulate outside the chromatophore; in many instances oil globules are prominent in the cells. There are many modes of flagellation but two apical flagella occur only in a relatively small number of genera. In some they are accompanied by a further organelle, the haptonema. There are at least seven distinct groups in this series.

CRYPTOPHYTA

Only the flagellate genus *Cryptomonas* (Fig. 9.20a) is at all common but hardly a body of fresh water can be sampled without encountering this genus; it is found also in marine habitats. *Cryptomonas* is readily recognizable from the lateral invagination at the upper end out of which the two somewhat unequal flagella arise. The invagination is often lined with rows of trichocysts. Pigmentation is varied but olive/green to brown is common, although blue and red forms occur.

PYRROPHYTA (DINOPHYTA)

This phylum is important in the microbiology of the oceans (p. 263). The flagellates (Dinoflagellates) are often extremely abundant in the surface layers of the sea and contribute a considerable amount to the primary production within the phytoplankton; many species occur in fresh water. Many have a distinct 'armour' of interlocking plates (hence the prefix 'dino'). Equally abundant are the unpigmented genera, best considered as protozoa of the class Mastigophora (p. 201). Non-motile stages of pigmented dinoflagellates are common symbionts (the zooxanthellae of earlier authors) in corals, clams, sea anemones, etc. (p. 276). Recent

work suggests that even the apparently 'unarmoured' forms have thin plates embedded in the outer layer of the cell. Two basic morphological types exist. One has a theca in the form of two watchglass-like halves sealed together around the edge and with the two flagella arising from adjacent pores and extending out from the cell (e.g. *Exuviaella*, Fig. 9.20h). The other has a series of plates fused together to form an epicone, a medium furrow and a hypocone in which there is a longitudinal furrow (e.g. *Peridinium*). The two unlike flagella emerge through pores where the two furrows join; one runs around and within the median furrow and the other runs down the longitudinal furrow out into the surrounding medium. The plates are precisely orientated and joined together by sutures within which some slight growth of the cell occurs. The cells are often flattened and some genera have the angles developed into long spines (e.g. *Ceratium*, Fig. 9.20g). The chromatophores are usually brown disks with stalked pyrenoids (cf. the Phaeophyta). The nucleus is often large and the chromosomes stainable at all times. Many dinoflagellate cells have trichocysts beneath the cell membrane. These develop from the Golgi vesicles and eject banded proteinaceous threads. Highly characteristic saclike structures, the so-called 'pusules', are associated with the flagella pores. There are often smaller associated sacs opening by constricted tubes into the central sac. Their function is not yet known although they may be involved in either uptake or excretion.

Cell division is by a splitting of the cell along an oblique plane between the plates followed by the growth of the new halves. Sometimes the cells remain together after division to form colonies. Sexual reproduction has rarely been recorded but fusion between anisogametes has been authenticated in *Ceratium*: this is followed by encystment of the zygote and germination after several months with the production of motile cells; meiosis occurs during formation of these motile cells.

Cysts which have been called 'hystrichospheres' by palaeontologists are now known to be formed by dinoflagellates and can be found in marine plankton. On germination they form typical dinoflagellate cells. Some marine dinoflagellates produce a toxin which is poisonous to fish and man but not to shell-fish which feed on the dinoflagellates. Normally the concentration of toxin in the shell-fish is negligible but during times of intense growth of the dinoflagellates the toxin becomes concentrated in lethal quantities and sale of the shell-fish has to be prohibited. This is normally only a tropical or subtropical occurrence but recently cases of non-lethal poisoning occurred in the north of England. The reasons for the massive growths of the dinoflagellates are not known. They are sometimes so intense that the sea is coloured by them—the so-called 'red tides'.

Some dinoflagellates are able to emit flashes of light, i.e. they are bioluminescent, as are certain bacteria and fungi. This bioluminescence has been shown to be under an endogenous 'biological clock' control system.

CHRYSOPHYTA

Many genera of this phylum are flagellate with considerable range in type of flagellation. Coccoid, tetrasporal and filamentous genera also occur, but it is not known whether some of these are stages in the life history of flagellate forms. The simplest cells have a single flagellum and discoid brown chromatophores. Several apparently uni-flagellate species, e.g. *Chromulina* (Fig. 9.20b), have been shown to have a second flagella base within the cell, often associated with an eyespot. In some, fine organic scales cover the cell membrane. In the larger flagellate *Mallomonas* (Fig. 9.20d) the body is covered by relatively massive siliceous scales (Fig. 9.21f) each provided with a long siliceous bristle. Colonial flagellate forms also occur, e.g. *Synura*, which has two slightly unequal flagella and numerous siliceous scales coating each cell (Figs. 9.20e, 9.21e). In other genera the cell is enclosed in a theca, e.g. *Chrysococcus* (Fig. 9.20c), which is comparable to *Trachelomonas* in that a spherical theca encloses the cell leaving only a pore for the single flagellum. In *Dinobryon* the thecae are vase-shaped, composed of microfibrillar material and joined together to form colonies (Fig. 9.20f).

Many Chrysophyte flagellates form characteristic siliceous cysts. These are endogenous structures formed within the cytoplasm of the cell, which slowly move into the cyst which is than 'plugged'.

Sexual reproduction by an isogamous fusion has been reported in some species.

HAPTOPHYTA

This group of algae is probably of great importance in the seas. The whole group of flagellates formerly known as the Coccolithineae has been transferred to this group. These are biflagellate planktonic flagellates which secrete intricately formed calcareous scales on to the outside of the cell. In some genera these have now been seen to form an outer layer superimposed on fine organic scales with characteristic fibrous markings. The calcareous scales (coccoliths) occur abundantly in some chalk deposits indicating that this group is an ancient one and an important constitutent of marine plankton over a long period of geological time. Non-motile stages also occur as filaments on rocks around coasts. Another group of biflagellate cells, e.g. *Chrysochromulina*, forms intricate organic scales in layers on the outer membrane (Fig. 9.20i). There is, in addition, a further organelle, the haptonema, adjacent to the two equal flagella. It is a long, often coiled, tube of similar dimension to the flagella but constructed differently with three concentric membranes surrounding a group of seven or eight microtubules. It is sometimes swollen at the end and acts as an attachment organ. The cells have two or more brown chromatophores.

XANTHOPHYTA (formerly the Heterokontae)

This is the apparent anomaly in the 'chromophyte' series, since there is not sufficient carotenoid pigment in the chromatophores to obscure the green colour given by the chlorophylls. In other features they are a distinct group with many characters which ally them with the Chrysophyta–Bacillariophyta groupings. Motile vegetative cells and gametes are 'heterokont', i.e. they have one long and one short flagellum, the former having a coating of fine hairs. The chromatophores are almost always parietal and disk-like; oil, but never starch, is stored in the cells, and the cell walls tend to have a greater proportion of pectin than cellulose. In some genera the cell wall is composed of striate H-pieces which fit closely together (e.g. *Tribonema*) but their actual nature is revealed only by chemical treatment. Apart from the 'heterokont' flagellation, no single character distinguishes a species of the Xanthophyta from one of the Chlorophyta. Cells with discoid plastids and/or without starch are rare in the Chlorophyta and should be regarded as possible heterokont algae.

Flagellate species are rare. *Ophiocytium* (Fig. 9.20j) is perhaps the commonest coccoid species. The unbranched filaments of *Tribonema* are abundant in fresh waters; branching filamentous genera occur also on soils. *Vaucheria*, the commonest genus, is siphonaceous and oogamous with male gametes of typical heterokont form. Asexual reproduction is by means of a large multi-nucleate swarmer. A pair of smooth slightly unequal flagella is associated with each nucleus.

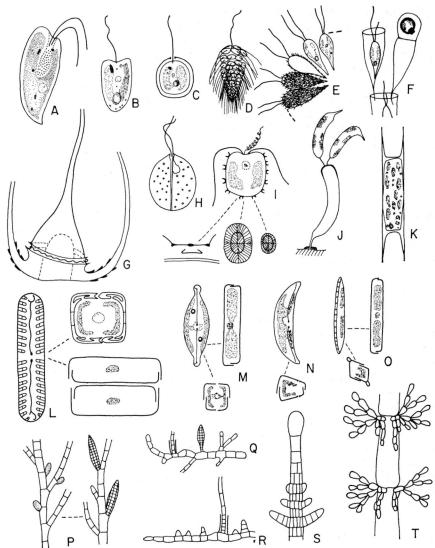

Fig. 9.20 Chromophyte genera. (**A**) *Cryptomonas*. (**B**) *Chromulina*. (**C**) *Chrysococcus*. (**D**) *Mallomonas*. (**E**) Fragment of *Synura* colony, two lower cells showing scales. (**F**) Fragment of *Dinobryon* theca, on right with resting spore. (**G**) *Ceratium*. (**H**) *Exuvaiella*. (**I**) *Chrysochromulina* cell and scales of two sizes and in section on the cell (below). (**J**) *Ophiocytium*. (**K**) One cell of a *Tribonema* filament. (**L**) *Pinnularia*, valve view of a cell, cross section and cell division. (**M**) *Caloneis*, valve view, girdle view, and cross section. (**N**) *Cymbella*, valve view and cross section. (**O**) *Nitzschia*, valve view, girdle view, and cross section. (**P**) *Ectocarpus*, unilocular plant (left), plurilocular plant (right). (**Q**) *Streblonema*. (**R**) *Myrionema*. (**S**) Apex of *Sphacelaria*. Rhodophyte genus. (**T**) Fragment of branching thallus of *Batrachospermum* showing axial cells and branches.

BACILLARIOPHYTA

Few phyla of algae are as well circumscribed as the diatoms (bacillariophyta). The cardinal feature of the group is the possession of a siliceous skeleton composed of discrete segments which fit together to form basically a 'pill-box' or 'date box' structure sometimes modified by curvature of one or more of the three planes (Fig. 9.20l–o; 9.21a–d). Within this hyaline silica case the diatom has the usual organelles (a centrally suspended nucleus, oil globules and two or more brown chromatophores, to which are attached deposits of the polysaccharide chrysose). A large number of genera are motile but without visible means. The problem of motility has not yet

been satisfactorily solved, but it is likely that systems of fissures and pores in the silica skeleton known as the raphe system are involved. The cell membrane lies internal to the silica wall which may thus be considered as an exoskeleton. In some genera there is conclusive evidence that there is also a fine organic 'skin' outside the silica. The siliceous parts consist of upper and lower lids (epitheca and hypotheca, which in 'centric' forms are shaped like two halves of a Petri dish). These thecae have intricate pores, often radiating out from the central point but changing form at the rim which itself may be supplied with a series of spines and tubes (Fig. 9.21b). The two thecae do not overlap one another but are joined by two or more 'hoops', known as

Fig. 9.21 (**A**, **C**, and **E**) Transmission electron micrographs. (**B**, **D**, and **F**) Scanning electron micrographs. (**A**) A centric diatom *Stephanodiscus* in oblique girdle view (upper left) and valve view (lower right). (**B**) Another species of *Stephanodiscus*. Complete cell appearing like a petri-dish with a girdle band overlapping top and bottom valve. Note the concentric indentations on the valve face. (**C**) A *Navicula* species showing the raphe system and pores through the silica wall. (**D**) Another *Navicula* species showing the boat-like nature of the cell. (**E**) A field of scales from *Synura* spp. and *Mallomonas* showing the perforate base plates, various ornamentation, and spines on these. (**F**) A single scale of a *Mallomonas* species showing the raised V-shaped central ridge, the overturned rim and the protruberance from which the spine arises. (Magnifications: (**A**) × 3200, (**B**) × 2848, (**C**) × 3200, (**D**) × 2060, (**E**) × 3600, (**F**) × 4000, F. E. Round, Bristol University)

girdle bands, which overlap one another and also the edges of the two thecae. These 'hoops' may be plain bands of silica or they may be poroid. The mode of formation of these parts and their spatial relationship is clarified by electron microscopy (Fig. 9.21a–d). At cell division the nucleus (all diatoms are diploid in the vegetative phase) divides and the daughter nuclei move to lie one adjacent to the hypotheca and one adjacent to the epitheca. Two new thecae are laid down in membrane-limited vesicles between the two nuclei. Since these are formed internally they are slightly smaller than the parent thecae. The fate of the old girdle bands and the mode of formation of new bands is not yet clear. When the new thecae are fully formed the two cells

separate (Fig. 9.20l). At each division the diatom gets smaller; the very large centric forms can slowly reduce in size from around 400 μm in diameter down to 50 μm. It is not known with certainty whether all diatoms undergo this reduction in size; some may have thecae which are flexible enough to allow the new thecae to expand to the same size as the old. Diatoms are similar to desmids in that each cell has one old and one new half. Reduction in cell size is reversed only by the occurrence of sexual reproduction. This has been adequately studied in only a small number of species. In some 'centric' forms it is oogamous, cells being converted into single oogania or into numerous spermatozoids each with a single apical flagellum. In *Lithodesmium* the flagellum is unusual in that the two central microtubules are missing. The zygote (auxospore) enlarges and becomes surrounded by a siliceous wall (perizonium). On germination new thecae and girdle bands are all formed endogenously within the auxospore. In some genera the auxospore is produced by enlargement of the cytoplasm and parting of the thecae without any sexual process and in others particularly in 'pennate' diatoms (see below), it is formed after conjugation between pairs of diatoms and the fusion of amoeboid masses of cytoplasm (cf. desmids).

The box-like structure of the 'centric' diatoms is repeated in the 'pennate' forms but here the cells are elongated so that viewed from above (the valve view) they are cigar-shaped and from the side (girdle view) oblong (e.g. *Pinnularia*, Fig. 9.20l). Both planes can be curved or contorted in various ways (Fig. 9.20n–o). In addition the 'pennate' diatoms tend to have the rows of pores arranged in lines on either side of the centre line leaving a clear central area which may be completely solid (a pseudoraphe), or it may have two raphe fissures running from central nodules to apical ones. The fissures and associated nodules, although not obvious with the light microscope are in fact complicated systems of pores in the silica wall; the whole complex is termed the raphe system and is involved in cell movement. Various theories have been proposed involving extrusion of substances through the fissures/pores, flow of cytoplasm or water along the fissures, undulating membranes in the fissures and even cilia projecting through them. The latter is certainly not true and motility is probably achieved by a combination of processes involving extrusion of substances and adhesive/surface tension properties of the cell.

Colony formation occurs in both groups, the daughter cells formed at cell division remaining attached, assisted by the interlocking of spines and/or combined with extrusion of mucilage. Others secrete mucilage from special pores to form attachment disks or stalks.

Diatoms are the most widespread of all algae, common on the surface of soils and wet rocks, in all freshwater, brackish and marine habitats and on the undersurfaces and in interstitial spaces of the sea ice around the poles. Since the silica incorporated in the skeletons is relatively insoluble, under natural conditions the dead cells which sediment out of lakes and over the ocean beds build up considerable deposits. Some of these when raised up on to the land are used commercially, i.e. the so-called diatomite (kieselgühr) often composed of almost pure silica and used in industrial filtration processes. Examination of cores taken from lakes or ocean basins reveals details of earlier diatom floras and yields information on past changes in the environment. Since silica is absorbed in a soluble form from the water in which the diatoms live, the concentration of this substance in the water often changes radically during the growth cycles and does not accumulate in the water.

PHAEOPHYTA

The 'brown algae' are mostly macroscopic structures outside the realm of microbiology. There are, however, a considerable number of small filamentous and encrusting forms. Microscopic stages in the life cycles of larger forms (e.g. the Laminariales) require microbiological techniques for their study. The brown chromatophores have club-shaped pyrenoids. Around the nucleus there are often clusters of fucosan vesicles containing a tannin-like substance. The cell walls are characterized by the occurrence of the polysaccharides, laminarin and alginic acid in addition to cellulose and fucoidin. The sugar-alcohol mannitol is frequently found in brown algal cells.

The simplest morphological types are branched filaments often growing by means of intercalary meristems, e.g. *Ectocarpus* (Fig. 9.20p). In species of *Sphacelaria* (Fig. 9.20s) there is a precise segmentation at the apex to give alternating 'nodal' and 'internodal' cells, only the former giving rise to branches. Some species grow as creeping filaments, e.g. *Steblonema*, others form upright branch systems from the basal creeping filaments, e.g. *Myrionema*. Some grow between the tissues of other macroscopic algae.

The motile cells are always pear-shaped with two unequal flagella inserted in a lateral position, one pointing forwards and bearing hair, the other curving backwards and attached to the body in the region of the eyespot. The motile cells may act as zoospores or as gametes (isogametes or the male gametes of the oogamous genera).

The Phaeophyta are almost entirely marine in distribution and occur mainly in the coastal region, the microscopic forms being attached to other algae or growing on rock surfaces.

Rhodophyte series (Fig. 9.20t) (The 'red algae')

This the most complex group of the algae falls outside the 'chlorophyte' and 'chromophyte' series but in some respects shows similarities to the

Cyanophyta. Red algae are abundant as macroscopic growths in coastal waters but many, including almost all those which occur on soils or in fresh water, are microscopic.

The plastids contain the phycobilin pigments, phycoerythrin and phycocyanin, the former in sufficient quantity to impart a red coloration to the algae. A variety of starch (Floridean starch) is formed and the cell walls contain, in addition to pectin and cellulose, gel-like, sulphate polysaccharides (one of which is agar, a sulphated galactose polymer).

Only a few unicellular red algae are known (e.g. *Porphyridium*, which is a soil alga) the remainder are filamentous (e.g. *Batrachospermum*, Fig. 9.20t), though the filamentous nature tends to be obscured in the larger red seaweeds. As with Phaeophyta there is a large number of creeping and upright branching filamentous forms. These often occur as red crusts on rocks in the intertidal and subtidal zone, or as minute red filamentous pustules on rocks or plants. Some are also endophytic in other algae and a few genera have become completely parasitic.

The succession of phases in the life history and the changes associated with the post-fertilization developments of the female organ or carpogonium are more complex than in any other plant group. There are no known flagellate stages and the male cell is a minute single-celled structure (spermatium) budded off the filaments in the male gametangia, produced in large numbers on the male plants and transported by water currents to the female plants, where they adhere to the long-drawn-out trichogyne of the carpogonium. After fusion of the nuclei in the carpogonium the diploid nucleus is transferred by various means to other cells in the thallus and eventually diploid carpospores are formed which on germination grow into a diploid plant usually identical with the sexual plants. Reduction division takes place only on the diploid (commonly known as the 'tetraspore plant') leading to the production of four spores (tetraspores), two of which germinate into male plants and two into female.

The microbiological activity of the numerous fine filamentous forms growing at solid/water interfaces in the intertidal and subtidal zone is as yet relatively unknown. Several produce antibiotic compounds, many precipitate calcium/magnesium carbonate to form rock-like strata, and many play an important part in the economy of coral reefs.

Books and articles for reference and further study

Fine structure of the eukaryotic cell

BECKETT, A., HEATH, I. B. and MCLAUGHLIN, D. J. (1974). *An Atlas of Fungal Ultrastructure*. Longman, London, 221 pp.

DODGE, J. D. (1973). *The Fine Structure of Algal Cells*. Academic Press, London and New York, 261 pp.

GRELL, K. G. (1973). *Protozoology*. Springer Verlag, Berlin, Heidelberg and New York, 554 pp.

HEATH, I. B. (1976). Ultrastructure of Freshwater Phycomycetes. In *Recent Advances in Aquatic Mycology*, edited by E. B. Gareth Jones, Elek Science, London, pp. 603–650.

PERKINS, F. O. (1976). Fine Structure of Lower Marine and Estuarine Fungi. *Ibid*. pp. 279–312.

PICKETT-HEAPS, J. D. (1975). *Green Algae: Structure, Reproduction and Evolution in Selected Genera*. Sinauer Associates, Inc., Sunderland, Mass., 606 pp.

SLEIGH, M. (1973). *The Biology of Protozoa*. Edward Arnold, 315 pp.

The fungi

AINSWORTH, G. C. (1975). *Ainsworth & Bisby's Dictionary of the Fungi*. 6th Edn., Commonwealth Mycological Institute, Kew, Surrey, 663 pp.

AINSWORTH, G. C. and SUSSMAN, A. S. (eds.) (1965). *The Fungi*. Vol. I *The Fungal Cell*. (1966), Vol. II *The Fungal Organism* (1968) Vol. III *The Fungal Population* (1973); with F. K. SPARROW, Vol. IV A Taxonomic Review with Keys: *Ascomycetes and Fungi Imperfecti*. Vol. IV B. *Basidiomycetes and Lower Fungi*. Academic Press, New York.

ALEXOPOULOS, C. J. (1962). *Introductory Mycology*. 2nd Edn., Wiley, New York, 163 pp.

BURNETT, J. H. (1976). *Fundamentals of Mycology*. 2nd Edn., Edward Arnold, London, 688 pp.

BULLER, A. H. R. (1909–1950). *Researches on Fungi*. Vol. I–VI, Longmans Green, London; Vol. VII, University of Toronto Press, Toronto.

HAWKER, L. E. (1974). *Fungi: An Introduction*. 2nd Edn., Hutchinson University Library, London, 216 pp.

INGOLD, C. T. (1971). *Fungal Spores: their Liberation and Dispersal*. Clarendon Press, Oxford, 302 pp.

MADELIN, M. F. (ed.) (1966). *The Fungus Spore*. Colston Papers No. 18, Butterworth, London, 338 pp.

LANGE, M. and HORA, F. B. (1963). *Collins Guide to Mushrooms and Toadstools*. Collins, London, 257 pp.

LODDER, J. (ed.) (1970). *The Yeasts: a Taxonomic Study*. 2nd Edn., North Holland Publishing Co., Amsterdam and London, 1385 pp.

SMITH, G. (1969). *An Introduction to Industrial Mycology*. 6th Edn., Edward Arnold, London, 400 pp.

ROSE, A. H. and HARRISON, J. S. (eds.) (1969). *The Yeasts*, Vol. I, *Biology of Yeasts*; (1970) Vol. II. *Physiology and Biochemistry of Yeasts*; (1970) Vol. III. *Yeast Technology*. Academic Press, London and New York.

SMITH, J. E. and BERRY, D. R. (1974). *An Introduction to Biochemistry of Fungal Development*. Academic Press, London and New York, 326 pp.

SMITH, J. E. and BERRY, D. R. (eds.) (1975). *The Filamentous Fungi*. Vol. I. *Industrial Mycology*; (1976) Vol. II. *Biosynthesis and Metabolism*. Edward Arnold, London.

WEBER, D. J. and HESS, W. M. (eds.) (1976). *The Fungal Spore: Form and Function*. John Wiley & Sons, New York, London, Sydney and Toronto. 895 pp.

WEBSTER, J. (1970). *Introduction to Fungi*. Cambridge University Press, Cambridge, 424 pp.

THE PROTOZOA

BAKER, J. R. (1969). *Parasitic Protozoa*. Hutchinson, London, 176 pp.

CHEN, T. T. (ed.) (1967). *Research in Protozoology* (in four volumes). Pergamon Press, Oxford.

CORLESS, J. O. (1973). *The Ciliated Protozoa*. Pergamon Press, Oxford, 310 pp.

GRASSÉ, P. P. (ed.) (1932–1953). *Traité de Zoologie*. Vol. I. *Protozoaires*, Part I. (1071 pp) and Part II. (1160 pp). Masson, Paris.

GRELL, K. G. (1973). *Protozoology*. Springer-Verlag, Heidelberg.

HONIGBERG, B. M. and committee (1964). A revised classification of the phylum Protozoa. *J. Protozoal.*, **11**, 7–20.

JAHN, T. L. and JAHN, F. F. (1949). *How to Know the Protozoa*. Brown, Dubuque. Iowa, 234 pp. (for identification).

KUDO, R. R. (1966). *Protozoology*. 5th Edn., Thomas, Springfield, Illinois, 1174 pp.

MACKINNON, D. L. and HAWES, R. S. J. (1961). *An Introduction to the Study of Protozoa*. Oxford University Press, London, 506 pp.

PITELKA, D. R. (1963). *Electron-microscopic Structure of Protozoa*. Pergamon Press, Oxford, 269 pp.

SLEIGH, M. A. (1973). *The Biology of Protozoa*. Edward Arnold, London 315 pp.

SLEIGH, M. A. (1974). *Cilia and Flagella*. Academic Press, London, 500 pp.

WARD, H. B. and WHIPPLE, G. C. (1959). *Fresh-Water Biology*. 2nd Edn., edited by W. T. Edmondson, Wiley, New York, 1248 pp. (Chapters 6, 8, 9 and 10, for identification of Protozoa).

Gymnomyxa (The slime moulds)

ALEXOPOULOS, C. J. (1963). The Myxomycetes II. *Bot. Rev.*, **29**, 1–78.

ALEXOPOULOS, C. J. (1966). Morphogenesis in the Myxomycetes. In *The Fungi*, edited by G. C. Ainsworth and A. S. Sussman, Chapter 8, 211–234. Academic Press, New York and London.

ALLEN, R. D. and KAMIYA, N. (eds.) (1964). *Primitive Motile Systems in Cell Biology*. Academic Press, New York and London, 642 pp.

ASHWORTH, J. M. and DEE, J. (1975). *The Biology of Slime Moulds*. Institute of Biology's Studies in Biology, No. 56, Edward Arnold, London, 68 pp.

BONNER, J. T. (1963). Epigenetic development in the cellular slime moulds. *Symp. Soc. exp. Biol.*, **17**, 341–358.

BONNER, J. T. (1965). Physiology of development in the cellular slime moulds (Acrasiales). In *Encyclopedia of Plant Physiology*, 15, Pt I, edited by W. Ruhland,

612–640, Springer-Verlag Berlin, Heidelberg and New York.

BONNER, J. T. (1967). *The Cellular Slime Moulds*. 2nd Edn., Princeton University Press, 205 pp.

CAVENDISH, J. C. and RAPER, K. B. (1965). The Acrasieae in Nature; I, II and III. *Amer. J. Bot.*, **52**, 294–296, 297–302, 302–308.

GRAY, W. D. and ALEXOPOULOS, C. J. (1968). *Biology of the Myxomycetes*. Ronald Press Co., New York, 288 pp.

GREGG, J. H. (1966). Organization and Synthesis in the Cellular Slime Moulds. In *The Fungi*. II, edited by G. C. Ainsworth and A. S. Sussman, pp. 235–289. Academic Press, New York and London.

LISTER, A. A. (1925). *A Monograph of the Mycetozoa*. (3rd Edn. by G. Lister). British Museum, London, 294 pp and 220 plates.

MACBRIDE, T. H. and MARTIN, G. W. (1934). *The Myxomycetes*. The Macmillan Co., New York.

MARTIN, G. W. and ALEXOPOULOS, C. J. (1969). *The Myxomycetes*. The University of Iowa Press, Iowa City, 561 pp.

OLIVE, L. S. (1975). *The Mycetozoans*. Academic Press, New York.

SHAFFER, B. M. (1962). The Acrasina. In Abercrombie, M. and Brachet, J. (eds.), *Advances in Morphogenesis*, **2**, 109–182. Academic Press, New York and London.

SHAFFER, B. M. (1963). The Acrasina, *Ibid.*, **3**, 301–322.

VON STOSCH, H. A. (1965). Wachstums- und Entwicklungsphysiologie der Myxomyceten. In Ruhland, W. (ed.), *Encyclopedia of Plant Physiology*, **15**, Pt I, 641–679. Springer-Verlag, Berlin, Heidelberg and New York.

YOUNG, E. L. (1943). Studies on *Labyrinthula*: the etiologic agent of the wasting disease of eel-grass. *Amer. J. Bot.*, **30**, 586–593.

The algae

BONEY, A. D. (1966). *A Biology of Marine Algae*. Hutchinson Educational Press, London, 216 pp.

BOURELLY, P. (1966–1968). *Les Algues d'eau douce. Initiation a la systematique*. Tome I *Les algues vertes*, 511 pp. Tome 2. *Les algues jaunes et brunes*, 438 pp. N. Boubée et Cie., Paris.

DAWSON, E. W. (1966). *Marine Botany, An Introduction*. Holt, Rhinehart & Winston, Inc., New York, 377 pp.

DODGE, J. D. as on p. 000.

FRITSCH, F. E. (1935, 1945). *The Structure and Reproduction of the Algae*. Vol. I. 791 pp., Vol. II, 939 pp., Cambridge University Press, Cambridge.

FOTT, B. (1959). *Algenkunde*. Fischer, Jena, 482 pp.

KYLIN, H. (1956). *Die Gattungen der Rhodophyceen*. Lund, Sweden, 673 pp.

LEEDALE, G. F. (1967). *Euglenoid Flagellates*. Prentice Hall, Englewood Cliffs, New Jersey.

ROUND, F. E. (1965). *The Biology of the Algae*. Edward Arnold, London, 269 pp.

SMITH, G. M. (1950). *The Fresh-water Algae of the United States*. McGraw-Hill Co., New York, 719 pp.

SMITH, G. M. (1951). *Manual of Phycology*. Chronica Botanica Company, Waltham, Mass. 375 pp.

Chapter 10

Principles of classification and numerical taxonomy

Some definitions

The terms of taxonomy have been subjected to almost as many variant definitions as those of mathematical statistics—a science with which it has many points of contact.

In this chapter *classification* means 'the act of arranging a number of objects (of any sort) into groups (or *taxa*) in relation to attributes possessed by those objects'. The word classification is also applied to the result of any such arrangement. *Taxonomy* (Greek táxis—arrangement; nómos—law) is concerned, *inter alia*, with definition of the aims of classification, the design of rules by which arrangements may be achieved, and with the evaluation of the end results. In brief, taxonomy may be described as the scientific study of classifications.

In biological classifications the primary objects (animals, plants, bacteria) are usually arranged in groups which are themselves members of larger groups (and so on) in such a way that any item or any group appears as a member of only one larger grouping, i.e. the groups are non-overlapping. This method of classification is the familiar *hierarchical* system which can be conveniently represented by a 'family tree' or *dendrogram*.

For ease of reference, the units at each level (taxonomic rank) of a hierarchical system are given distinctive names or labels—a branch of taxonomy known as *nomenclature*. Every student of biology will be familiar with the system of nomenclature normally used for living organisms, which derives from that used by the great eighteenth-century taxonomist Linnaeus (Carl von Linné). In this system the basic unit (the *species*) is given two names—one denoting its membership of a taxon at the rank that we label *genus* (generic name), followed by a second denoting the particular species (specific name). These names are written in a latinized form and constitute a so-called *latinized binomial* (e.g. *Staphylococcus aureus, Aspergillus niger*). Taxa of higher rank (families, orders, etc.) are given single latinized names with characteristic endings (e.g. Enterobacteriaceae, family; Eubacteriales, order). The naming of newly discovered organisms or of newly proposed taxa of higher ranks is governed by rigid rules standardized by international agreement.

So familiar is this system that it is perhaps worth emphasizing that it is by no means the only possible one. The simplest system would be merely to label the different types of organism with some sort of catalogue number which referred to a listed description. A much more useful approach might be similar to one proposed for the naming of viruses, viz., the virus is given a group name (probably latinized) which is followed by a descriptive formula akin to that used by botanists in floral diagrams (or to the antigenic formula of a *Salmonella* species, p. 331). The later method is, in fact, reminiscent of that often used by Linnaeus, who sometimes followed his latinized generic name with up to a dozen descriptive 'specific' epithets.

Ideally, the coining of new names is contrived to convey as much information as possible about the organism or taxon. Unfortunately, both the restriction to latinized binomials and, often, the rules of precedence make this aim difficult to achieve.

Having very briefly stated the areas of taxonomy with which classification and nomenclature are concerned, we must consider the meaning of *identification*. In essence, this simply involves the comparison of an 'unknown' object (e.g. a newly isolated bacterium, a collected plant or animal) with all similar objects that are already known. If the 'unknown' object matches up with a 'known' one we say that the former has been *identified*; if not, it may be considered to be a 'new' species, variety, or strain and, when adequately described, is added to the list of 'known' objects. In practice this act of comparison is normally carried out not between two actual objects but between the 'unknown' *isolate* and a recorded description of previously discovered bacteria, plants, animals, etc. The inadequacy of recorded descriptions of many microbial 'species' can sometimes make accurate identification very difficult, if not impossible.

It is not always appreciated that neither identification (as defined above) nor nomenclature need necessarily be connected with classification. In astronomy the identification of a particular star is made by matching the observed celestial coordinates with those recorded in star catalogues for stars previously described. Nomenclature is either a matter of a number in the star catalogue or of a name deriving from its membership of the constellation that suggested shapes to the eyes of early astronomers, e.g. α-centauri. In contrast, the classification of stars is based upon grouping together stars of similar physical nature (e.g. red dwarf, nova, blue giant). It is interesting to note that this sort of classification results in an arrangement in which stars of the same state of development are generally grouped together, i.e. it becomes an *evolutionary* classification—a point which will be taken up in the next section.

The aims of classification

If the unsuspecting student of biology is asked 'Why do we classify living organisms?' the chances are high that his reply will contain a reference to 'showing evolutionary relations between the organisms'. So strong has been the influence of evolutionary criteria on taxonomy during the post-Darwinian period that production of such *phyletic* classifications is often thought to be the sole aim of the taxonomist. It is therefore necessary to consider whether other possible aims are valid and, indeed, whether any other approach might lead to classifications of greater value than the purely phyletic.

To do so we must first make the distinction between *special* (or *artificial*) classifications and natural classifications. A special classification is one made for a single, defined purpose; it assists in finding the answer to a specific question. A well-known example is the classification of enteric bacteria according to the biochemical differential tests, as used by the water bacteriologists (p. 274). The purpose of this classification is to group together those organisms which may indicate recent faecal pollution of a water supply and to separate them from other similar bacteria which do not have this significance. When a bacterial isolate is identified as falling within a particular group of this system an answer to the question of possible faecal pollution is obtained. A further example is the system of classification, used by medical bacteriologists, which places great weight on the pathogenicity of an organism in separating it from otherwise very similar bacteria, e.g. the anthrax bacillus from 'anthracoid' bacilli such as *Bacillus cereus*; the diphtheria bacillus from other 'diphtheroid' Corynebacteria. The question answered here is whether a fresh isolate is likely to cause disease—a question of paramount importance to the medical bacteriologist.

Such classifications are perfectly valid and perform an important function, but they make no pretence to be natural systems. In special classifications an organism may be separated from its fellows by differing in a single key attribute (e.g. toxigenicity) whereas the residue may be grouped under a common taxonomic title (e.g. species name) and yet differ among themselves in several attributes.

What then constitutes a *natural* classification? To answer this question we must first examine what guided the taxonomists during the pre-Darwinian period. We find that the taxonomic logic of this era can be traced back to the ideas of Aristotle, in particular to his Logical Division Theory, which governed the ideas of Linnaeus and held sway up to the beginning of the present century. The basic notion was that organisms (or any other items) should be classified according to their *essential nature*, i.e. according to 'what they really are'. This idea is linked to the Aristotelian notion of the *species infimae*—the ultimate unit of classification which became the basis of the Linnaean *species*. The *species infimae* was rather analogous to the atom of classical chemistry: it was the smallest unit into which more complex groupings could be broken down by repeated division into components. A classification based on such principles would be *sensu stricto* 'natural' but it is easily applied only to the classification of items which are clearly defined, e.g. geometrical shapes, where the essential nature is contained within the definition.

When attempts were made to apply this logic to the classification of living organisms taxonomists were faced with two connected difficulties which were impossible to overcome. The first and most fundamental of these is that Aristotle's principle is one of *deductive* logic and yet taxonomists tried to apply it to situations where only *induction* is possible. We cannot deduce that dogs are different from cats, we can only recognize that they differ on the basis of our observations (because we do not know the essential nature of 'dog' or 'cat'). The second difficulty is that of biological variation which makes the decision of which attributes are more 'essential' than others even more likely to be arbitrary. After the publication of Darwin's works on the origin of species, the earlier approach to classification was replaced by one that was thought to be at least equally 'natural', viz., the phyletic system. Once the doctrine of evolution had been accepted it seemed reasonable to argue that organisms of similar 'essential nature' would have shared common lines of descent. The great advantage to taxonomists of the phyletic approach was that speculation about which attributes reflected most accurately the essential natures of organisms was replaced by decisions based on more tangible evidence such as fossil records.

Even so, difficulties still remain. To mention only three: (1) fossil records are seldom adequate; (2) biological variation (both phenotypical and genetical) still poses the problem of the taxonomic level at which organisms are to be separated from each other; (3) the *homology* of various structures or other attributes is often in doubt. This point, which is a consequence of the lack of evolutionary evidence, refers to the identity of attributes which appear to be similar but which may in fact have evolved along quite different lines of descent (*convergent evolution*), e.g. the similar structural form of the truffles (Ascomycetes) and false truffles (Gasteromycetes), or may be due to quite different processes. Moreover, as Sneath and Sokal (1973) point out, taxonomists have often been uncertain about the meaning of *homology*, and circular arguments can frequently be found in the literature. If we consider micro-organisms in particular with regard to these three points we find that (1) fossil evidence is virtually non-existent; (2) biological variation is great—owing both to the ease with which mutations may be expressed (especially in laboratory environments) and to extreme phenotypic adaptability, e.g. the possession of inducible enzymes and other 'switchable' control systems; (3) the homology of attributes is frequently unknown; to mention one example, we can divide all bacteria into Grampositive or Gram-negative organisms but there may be more than one mechanism by which cells avoid decolorization.

The problem of convergent evolution and homology raises a question of fundamental importance to the formulation of the aims of natural classifications. Darwin believed that evolutionary convergence was never so great as to obscure ancestral resemblances and that natural taxa were therefore always phyletic. With regard to higher plants and animals, this may well be so; the Biblical grouping of the bat with birds and the whale with fish is clearly a 'special' classification based on only superficial resemblances. The lack of fossil evidence makes it much more difficult, if not impossible, to decide whether apparently similar micro-organisms have evolved from a common ancestral organism or whether convergence, due perhaps to the selective pressures of sharing a similar habitat, has been responsible.

For the sake of illustration, let us suppose that we have two bacterial strains that share a large number of *what appear to be* similar attributes. Let us further suppose that we also know that the lines of evolution of these strains converged from very different origins. Would the objective of a natural classification be best achieved by grouping these strains together on the basis of their mutual overall similarity (*phenetic* classification) or by separating them so as to reflect their different origins (*phyletic* classification)? An argument for the phyletic approach might be that this best reflects the 'essential natures' of the two strains, to which the counter argument might be that, because of convergent evolution, their 'essential natures' have become similar.

Such arguments could be continued indefinitely, were it not possible to approach the question from a different angle, i.e. by asking 'what is the *practical value of a natural classification*?' The answer to this question is that, to be of maximum use, a natural classification should have good predictive value (information content). This means that if we are told that an organism is classified in Group 'X' of a hypothetical taxonomic scheme, then, from our knowledge of the general properties of Group 'X', we should be able to predict that the organism in question *certainly* has features, a, b, c, d; *probably* has features p, q, r; and *may* have features x, y, z. When approached from this viewpoint it is possible to describe a natural classification as one which gives the greatest information to the 'general' user. In contrast, a special or artificial classification yields particular information to the specialized user. If we accept this distinction we may now resolve the hypothetical question posed above, since it is clear that the phenetic classification would allow the most *general* predictive properties, whereas the phyletic system would offer information that is primarily of use to the student of evolution, i.e. it is a special classification.

It is possible to see a resemblance between the grouping of organisms on the basis of phenetic classification and the use of statistical parameters in characterizing sets of data. For example, if we are told that a certain set of data is normally distributed about a mean value of 10.0 with Standard Deviation of 0.2, we may predict that in roughly 95 % of cases a single datum, randomly selected from the set, would have a value between 9.6 and 10.4, although we should be unable to predict the *exact* value of the datum. This is similar to saying that organisms placed in hypothetical Group 'X' of a phenetic classification would be expected to vary about an 'average' or modal form only between defined limits, although we should be unable to predict the exact complete description of a single organism selected randomly from Group 'X'. Indeed, in the systems of Numerical Taxonomy, to be outlined later, it is possible to describe these limits in numerical terms. Again, if the range of bacterial variation were so great that between each 'typical' or modal strain there was an almost continuous gradation of 'intermediate' strains (and some authors have suggested that this may indeed be so) a phenetic classification would still have practical use in much the same way that a histogram may allow us to group—and so handle—what is in fact a continuous spectrum of data.

So useful is this method of classification that we commonly apply it in everyday speech. For example, if we are told that Major Smith is a 'military' type and Dr Jones is a 'donnish' type of person we build up certain mental pictures of these individuals. When we actually meet Major Smith and Dr Jones we are not surprised that they differ in details from our preconceived mental pictures but we should be amazed if they do not share a large number of attributes with them.

Monothetic and polythetic classifications—the concept of weight

From what has been written above, it is obvious that the best phenetic classification is one built on comparisons based upon as many attributes as possible. Organisms which share a large number of attributes would cluster together to form a 'natural' group and such groups would separate from each other at 'points of rarity', i.e. at combinations of attributes which never, or very rarely, occur. If 'points of rarity' are absent it means that a continuous spectrum of 'intermediate' types of organism exists and the classification is then arbitrary (but could still be useful, as explained above). A phenetic classification based on the frequency of shared attributes (overall similarity) is termed *polythetic*.

The student of biology will certainly have encountered what is termed *monothetic* classification, since this principle is much used in the construction of artificial dichotomous keys for identification of both higher organisms and micro-organisms. The essence of such a system is that certain key characters are selected, the possession of which automatically places the organism to be identified into a group of organisms that *uniquely* possess these characters.

We see immediately that once a key character is selected it assumes great *weight* (importance) in determining the classificatory position of an unknown organism and we should therefore enquire whether we are justified in giving some characters more weight than others. It is obvious that, in principle, the use of key characters could nullify the aims of 'natural' classification outlined above. For example, if a new strain of micro-organism were discovered that differed in a single key character from those already classified together in a group, and yet had a large number of characters in common with that group, we should be forced to place the strain in a separate group according to the monothetic system, whereas it would obviously join the existing group in a polythetic system.

It is easy to justify the use of certain key characters in artificial classifications, since they may reflect the very criteria that were used in setting up the classification. For example, a special classification based on the criterion of *pathogenicity* would justifiably separate *Corynebacterium diphtheriae* from closely related 'diphtheroid' bacilli on the sole key character, toxigenicity, which thus takes an over-riding weight. In the case of natural (phenetic) classifications the justification of weighting certain characters is less easy. One possible justification is in cases where we know that certain characters are homologous whereas we are unsure about others. Here we may logically argue that greater weight should be given to the homologous characters in deciding the classification. A second possibility is to argue that more weight should be given to those characters that are strongly correlated with others, a single one of these could then be used as a key character; e.g. a Gram-positive reaction in bacteria usually shows correlation with cell-wall structure, penicillin-sensitivity, sensitivity to basic dyes, etc. Two things follow from this example. First, that the same weighting would be obtained by giving *equal* weight to each of the individual correlated characters, which would then act in concert in influencing the classificatory position. Second, that if eventually found that all of these correlated characters stemmed from a single genetical feature then their weight would disappear since all would be expressions of the same thing; a paradox that has been pointed out by Sneath (1957a). However, we are usually in doubt about the homology of apparently similar characters in micro-organisms, nor do we at present know the precise genetical reasons for observed correlations between characters. What is more, even if it were decided that some characters should bear greater weight, we should be faced with the equally difficult question: how much greater? There is, therefore, an increasing trend in microbial taxonomy towards the idea that, in our position of ignorance, the best natural classification is one based upon comparison of micro-organisms with respect to as many characters as possible, *each character being given equal weight* in contributing to the grouping and separation of different organisms (i.e. a polythetic system). Once such a classification has been made it is *then* possible to search for key characters which may be of use in a method of *identification*. It is, however, still unlikely that single key characters could be used as in the familiar dichotomous system, rather a set of such characters would have to be examined together in order to narrow down the possible classificatory location of an unknown organism (see, for example, Cowan and Steel, 1974).

The idea of phenetic classification based on characters of equal weight is not new. Indeed this approach stems from one used by the eighteenth-century French botanist, Michel Adanson, and it is now usual to apply the term *Adansonian* to such classifications.

Numerical taxonomy

Microbiologists, especially bacteriologists, have long felt that the systems of classification in general use need revision, largely owing to the obviously arbitrary nature of many of the groupings (taxa) and to evident confusion between natural and special classification.

The most widely used general classification of bacteria is embodied in *Bergey's Manual of Determinative Bacteriology* and is the system followed in this book. In previous editions we commented on the then current 7th edition of *Bergey's Manual*, pointing out that it was a mixture of phenetic classification and a quasi-evolutionary approach. The amount of information on which phenetic comparisons were based varied greatly—and often inevitably—among the groups. Although the classification was arranged in a Linnaean hierarchy of taxa, it was obvious that what constituted a *species* differed strikingly among the various 'genera' (e.g. the serotypes of *Salmonella* were given specific rank, while those of the pneumococcus were types of the

single species *Diplococcus pneumoniae*). Furthermore, the unequal weighting of features caused the classification of some organisms in groups with which they had very little overall similarity (e.g. *Corynebacterium pyogenes*).

One disturbing property of such a system is that if a group of organisms is re-examined across a set of characters completely different from those used in making the original classification, it is possible that the classification will need radical alteration in order to accommodate the new information. This instability, which is obviously undesirable in a natural classification, is far less likely to be a feature of classifications built on Adansonian principles (see later).

During the past twenty-five years a large number of publications have appeared concerned with what has become known as *Numerical Taxonomy*. This branch of the subject owes much to the availability of high-speed digital computers, and interest in its application to bacterial classification was stimulated by the pioneer work of P. H. A. Sneath. Normally the term Numerical Taxonomy is applied to systems of classification which are basically Adansonian but in which the degree of similarity of organisms is assessed in quantitative, rather than merely qualitative, terms. There are many advantages in having some numerical estimate of the degree of phenetic similarity or difference between a pair of organisms, of which the most obvious is that it can provide a rational basis for deciding the levels of taxonomic rank. It was soon shown by Sneath and others that there was at least as much difference between pairs of species of certain established genera as there was between pairs of genera.

The techniques of numerical taxonomy

A detailed discussion of these techniques will not be attempted here, and the interested reader is referred to the publications listed at the end of this chapter. What follows is an outline intended to illustrate the general principles.

There are several distinct approaches to Numerical Taxonomy, but all start by:

1 Collecting the organisms, or groups of organisms, to be compared, which are now known as *Operational Taxonomic Units* (OTUs).
2 Observing these OTUs for presence or absence (or quantity) of a large set of *characters*.
3 Drawing up a table of OTUs versus characters, i.e. of *character states*.

A character is usually defined as an attribute

about which a single statement can be made, e.g. 'present' or 'absent' or some quantitative measurement. It is important to give careful thought to what constitutes a single character before drawing up the OTU × character table. Some attributes are obviously not proper characters, e.g. the number of the OTU in the collection. Other apparent characters may not be permissible because they are redundant, i.e. are expressions of an already listed character. For instance, if an OTU ferments both glucose and sucrose with the formation of acid and gas this may generate three distinct characters, viz., acid from glucose; gas from glucose; sucrose fermented. It is improper to score 'gas from sucrose' as a separate character if we know that the fermentation of sucrose involves an initial hydrolysis to glucose, which is subsequently fermented to acid and gas.

Furthermore, it is essential to the principle of Numerical Taxonomy that each of the OTUs should be examined across the complete set of characters, so that true comparisons may be made. Care must be taken, however, not to make comparisons that are illogical. Suppose that one OTU ferments glucose to acid and gas, whereas a second OTU does not ferment glucose at all. In the case of the first OTU we may score a positive character for each of the attributes: acid from glucose; gas production. However, with regard to the second OTU we may score a negative character for lack of production of acid from glucose, but it is now illogical to score a result for 'gas production' since this depends on the prior formation of formic acid which we have already noted as absent. We therefore score 'No Comparison' (NC) for gas production by OTU number 2, which means that this character cannot be used when comparing the similarity of OTU number 2 with any other OTU.

Further questions are prompted by practical considerations, such as:

1 Since observation of characters is necessarily carried out under the artificial conditions of the laboratory, can we make a true comparison of micro-organisms which might behave differently in their natural environment?

2 If we have among the OTUs some organisms that can carry out certain reactions at one temperature of incubation but not at a higher in comparison with organisms that can carry out the same reaction only at the higher temperature, should we then use different temperatures of incubation in order to correctly characterize the different OTUs?

The answer to the first question is that a comparison of micro-organisms under laboratory conditions (1) is the 'best we can do' and (2) according to our practical definition of a 'natural' classification, is satisfactory because other investigators will be observing the micro-organisms under similar conditions. The answer to the second question is more difficult. If there are many temperature-sensitive reactions, we may bias the comparison of OTUs, compared under standard conditions, towards an emphasis of dissimilarity when the temperature sensitivity may be due to only a few underlying causes (which we do not know).

In our position of ignorance of the complete genetical and biochemical bases of observed characters it is generally considered best to compare OTUs over a *rigidly standardized* set of tests. Although it is almost inevitable that certain of these conditions will introduce bias when measuring the degree of similarity between pairs of OTUs, this course of action is adopted for two chief reasons: (1) practical expediency; (2) if a wide range of characters are observed the bias should be 'diluted out'. Of course, tests should not be used to generate characters when it is *known* that bias is inherent in the test condition. For example, we may adjust the sensitivity of a test for urease production so that it is read as positive only with those Enterobacteriaceae that we call *Proteus* spp. To use this test in a phenetic comparison of Enterobacteriaceae would obviously introduce bias, since we have prejudged the issue by distinguishing certain species as urease-positive beforehand. Such a test, however, would be perfectly valid if applied to unrelated organisms.

The kinds of morphological, structural, and metabolic attributes commonly used as classificatory characters are typified by those which appear elsewhere in this book in descriptions of the various micro-organisms. More recently attention has turned to other potentially valuable sources of characters. These include (1) cell-wall chemistry, (2) electrophoretic studies on esterases and other soluble proteins, (3) infra-red adsorption spectra, (4) DNA base composition, and (5) gas chromatography of cell pyrolysis products (see references at end of chapter).

From this discussion it is clear that comparisons of OTUs based on a large number of characters are likely to be more accurate (free from bias) than comparisons based on only a few characters (we have already discussed the bias that can be caused by the use of a single key character). *How many characters should we observe?* A guide to the answer may be obtained from elementary probability theory, which tells us that we are most likely to succeed in distinguishing different organisms when the number of characters is of the same order as the number of OTUs, and that we should have limited confidence in an S-type Similarity Coefficient (see below) calculated on the basis of fewer than 50 characters.

A special difficulty may exist when an attempt is made to compare organisms that have very different growth-rates under standardized conditions (e.g. pathogenic and saprophytic Mycobacteria). Here there is clearly the possibility of bias due to comparison of characters that depend on metabolic rate when similarities are calculated after an incubation period that is suboptimal for the slower-growing strains. We may either incubate all strains so that the reactions of the slowest grower are realized—when difficulties may arise due, for example, to alkaline reversion in carbohydrate fermentation tests with the fast-growing strains, or we may have recourse to special methods of calculation that attempt to separate effects due to *Vigour* (growth-rate) from that due to *Pattern* (Sneath, 1968).

Methods of comparing similarities

After an OTU × character table has been compiled, all possible pairs of OTUs are compared and their similarities computed. There are three basic methods by which measures of similarity may be computed. These are:

1 Correlation coefficients.
2 Measures of taxonomic distance.
3 Similarity coefficients (S).

The first two methods have the advantage that characters which are expressed as quantitative data may be more or less directly incorporated into the calculations of similarity.

The correlation coefficients are closely related to the commonly used statistic *r*, which expresses the degree of correlation between two sets of bivariate data and can vary from +1 (absolute correlation), through 0 (no correlation at all), to −1 (absolute negative correlation). Thus two organisms that were absolutely identical in all characters studied would generate a coefficient of +1, two organisms that were absolutely opposite in every character (if this were possible) would generate a coefficient of −1, whereas a coefficient of 0 would indicate no correlation of the characters of the first organism with those of the second.

Measures of taxonomic distance attempt to plot the relative positions of the OTUs in multi-dimensional space (one dimension for each character studied) in such a way that if two OTUs were identical their *mean taxonomic distance* would be 0 whereas if they were absolutely dissimilar their *mean taxonomic distance* would be +1.

However, it is the similarity coefficients (S) that have been used in most attempts at numerical classification of micro-organisms, owing to their ease of computation and handling in subsequent stages of the process. Although many different forms of these coefficients have been described, all measure the overall similarity between a pair of OTUs by the *proportion* of features in which they match. They range, therefore, from 0 (no matching characters) to +1 (all observed characters match). Such coefficients also have the advantage of being 'common-sense' measures, in contrast to some of the highly complicated taxonomic measures in use today.

Their most obvious disadvantage is that the majority can be applied only to *quantal* (i.e. yes/no) character states, and *quantitative* observations must therefore be 'coded', i.e. broken down, into a number of discrete quantal characters. This immediately introduces an element of arbitrariness into the calculations and generates many technical problems, some of which were outlined in previous editions of this book. Nevertheless, similarity coefficients continue to be widely used and generally produce acceptable and self-consistent classifications.

During the past decade, Numerical Taxonomy has excited the interest of a growing number of mathematicians, some of whom have applied themselves to studying the intuitively satisfying, but mathematically difficult, methods that use taxonomic distance, while others have investigated coefficients based on information theory or probability theory.

Cluster analysis

After numerical estimates of the degree of similarity between all possible pairs of OTUs have been generated, the next step is to form the groups (or *clusters*) which are the basis of the final classification—an operation known as *cluster analysis*. Although some Numerical Taxonomic methods have their own peculiar techniques, most of the coefficients referred to above may be clustered in three ways:

1 by Single Linkage.

2 by Average Linkage.
3 by Total Linkage.

Suppose that at a certain level of similarity (say, $S = 0.8$) there exist two distinct clusters, *A* and *B*. When we examine these clusters at a lower level of similarity (say, $S = 0.6$) we may find a tendency for *A* and *B* to fuse into a single cluster. For fusion by Total Linkage, each OTU in *A* must have $S \geq 0.6$ with every OTU in *B*. For fusion by Single Linkage, it is sufficient that one OTU in *A* has $S \geq 0.6$ with a single OTU in *B*, so forging the 'single link' between the two clusters.

Total Linkage will obviously yield clusters that are homogeneous owing to the high internal similarity imposed by the rules of formation. However, these rules are so rigorous that they frequently produce only a few dense clusters which are unsatisfactory for making a classification, and Total Linkage has been little used by Microbiologists.

The most popular method has undoubtedly been Single Linkage. As in the choice of a suitable measure of similarity, here again the ease of operation of rules for clustering has weighed heavily in favour of Single Linkage. This method is not only easy to understand but also leads to straightforward computer programming. Furthermore, the results obtained are often satisfying in spite of the inherent possibility of *chaining*. This is the production of heterogeneous clusters by the forging of a chain of single links, which permits considerable dissimilarity between OTUs at opposite ends of the chain, as is illustrated in the diagram below:

Character	OTU		
	A	B	C
1	1	1	0
2	1	1	0
3	1	1	0
4	1	0	1
5	1	0	1
6	1	0	1

Here, the similarity between *A* and *B* is the same as that between *A* and *C* ($S = 0.5$) but, clearly, there is no similarity at all between *B* and *C* ($S = 0$).

Clustering by Average Linkage avoids the chaining problem by permitting clusters to fuse only when the *average* similarity between them is sufficiently high. This is probably the most satisfactory of the types of clustering method described here, but it is much more difficult to write computer programmes for Average Linkage than for the other two methods and Average Linkage can have serious mathematical snags.

When all clusters have been formed at the various levels of similarity, it is usual to represent the result in the form of a *dendrogram*, which resembles the diagrams of familiar hierarchical classifications but differs in having numerical estimates of similarity rather than qualitative criteria to determine its branching points, thus:

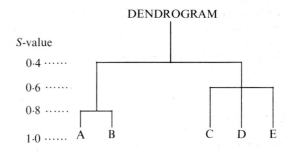

DENDROGRAM

Anyone who constructs a dendrogram quickly realizes that it is only a diagram and not a spatial representation of the degree of relatedness of the various OTUs. A true spatial representation requires multi-dimensional space in order to avoid distortion—a difficulty that often crops up in studies using taxonomic distances.

The stability of numerical (Adansonian) classifications

It was stated above that an Adansonian system of phenetic classification based on a large number of character comparisons is likely to be stable or self-consistent. This intuitive notion was earlier lent support by a statistical argument called the *matches asymptote hypothesis*, which supposed that there was some true measure of similarity (e.g. *S*-value) between a given pair of OTUs that would be known only if an 'infinitely' large number of characters could be observed. In any actual comparison of OTUs the calculated value of *S* was regarded as an estimate made on a 'random sample' of possible characters. If this were true then classical statistical theory would predict that the sample estimates of *S* cluster round the true value in the same way that sample means cluster round the true (population) mean. Therefore, any new estimate of similarity would tend to reinforce rather than disrupt the previously calculated values.

It is now clear that the matches hypothesis is unlikely to be true. One reason is that arbitrary factors (e.g. coding) enter into the calculations, thus influencing the actual numerical values of coefficients. Another reason is that not all characters *can* be independent and the concept of a random sample becomes strained.

Nevertheless, it is equally clear that *in practice* these numerical classifications do show a high degree of stability due to what has been called *inertia* (Sneath and Sokal, 1973). That is, once a taxonomic measure has been made on a large selection of characters, it becomes progressively less sensitive to alteration by newly added characters, thus making for the highly desirable property of stability.

Phyletic classification

Although it was remarked above that a purely phyletic classification was a special classification, it is certainly true that attempting to plot the course of evolution excites the interest of many Micro-biologists.

One apparently promising approach is to determine the amino acid sequences in homologous proteins occurring in all the organisms studied and then to construct possible schemes whereby the genes determining one form of the protein could have mutated so as to determine another form, and so on, to build up a complete evolutionary tree. The dependence of this approach on the availability of computers is evident when one realizes that a tree connecting only ten mutant proteins can be constructed in more than 34 million ways.

What may be less obvious is that many of the Numerical Taxonomic methods developed primarily for phenetic classification can readily be adapted to produce and test the possible phyletic schemes.

The present position

Even were the logic of Numerical Taxonomy generally acceptable to all Microbiologists—which it is not—there would still remain formidable problems for those attempting to apply its principles to the classification of micro-organisms. Apart from theoretical difficulties, some of which have been outlined above, a strict application of the logic would entail examining examples of all existing microbes across a full set of characters. Even with the availability of high-speed computers, the task is clearly impossible to tackle in one continuous investigation.

Two less-demanding approaches are possible. First, one may rely on published data for making wide-ranging classifications. This is unlikely to be very successful, owing to the incompleteness of most published descriptions (e.g. tests not done; results described as 'variable'). The result of an early attempt to derive a numerical classification in this way is shown in Fig. 10.1, which shows some (not unexpected) deviations from established classifications.

The second approach is to select examples from an 'established' group of organisms and its near relatives, and to take a more limited 'bite' at determining their numerical relations. This approach has been much used and many studies have now been published on diverse groups of bacteria and on yeasts. Fewer Numerical Taxonomic studies have been made on the filamentous fungi, algae, and protozoa, possibly because their existing classifications are felt to be more satisfactory than the bacterial classifications.

The results of these studies have had a large impact on bacterial classification, and this has been recognized in the scheme adopted by the current (8th) edition of *Bergey's Manual*. Whereas among higher organisms the *species*, at least, is based on the reality of the interbreeding unit, among the mainly asexually reproducing bacteria even this taxonomic rank is arbitrary. However, the editors of the *Manual* take the view that a reasonable case can be made for recognizing both *species* and *genus* among bacteria and for retaining the latinized binomial species name. With few exceptions, all other hierarchical taxonomic structure and terminology has been abandoned and replaced by a division of bacteria into 19 major groups based on some easily recognizable property (e.g. Phototrophic Bacteria; Gliding Bacteria; Sheathed Bacteria; etc. p. 166).

This striking departure from tradition is a reflection of the realization of our ignorance of the degree of relatedness among larger groupings of bacteria as well as being a reminder of the *Manual's* chief use, i.e. as an aid to *identification*. Within the groups, the current classification has often incorporated information gained from those Numerical Taxonomic studies that were available.

As both theoretical and practical studies continue to be made, we move closer to the possibility envisaged by A. J. Cain (Symposium, 1962)—that just *because* it is unencumbered with fossil evidence and a sexual definition of species, and because many characters can easily be studied, bacterial classification may become the most firmly based of all taxonomies.

Classification of eukaryotic micro-organisms

The range of form shown by eukaryotic micro-

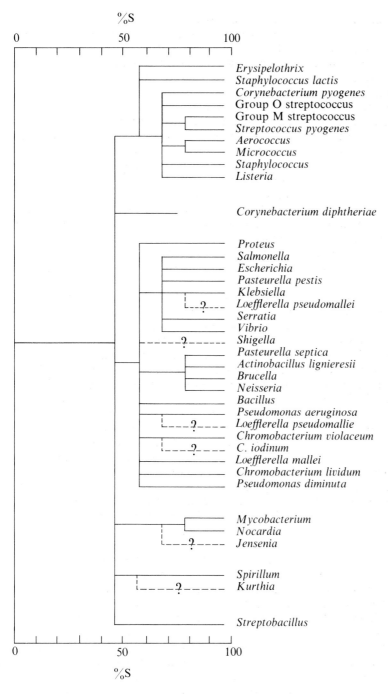

Fig. 10.1 Taxonomic dendrogram of a wide range of bacteria. (From Sneath and Cowan (1958) *J. gen. Microbiol.*, **19**, 551–565; also cited in Symposium (1962) p. 34, cited in references)

organisms is much greater and their life cycles are more complex than those of prokaryotes. Hence classification based on structure and development is often satisfactory and has the advantage that the worker in the field such as the plant pathologist is able to identify organisms without recourse to long and complicated laboratory tests. Nevertheless,

ultrastructural and chemical characters are being increasingly used in the classification of difficult groups. Algae have traditionally been subdivided according to the nature of their pigments and with increasing knowledge of these, modern classification of algae has evolved (p. 219). The species of yeasts, many of which are of similar form although not closely related are necessarily distinguished partly by chemical tests (p. 197). Even the field mycologist studying agarics makes increasing use of colour reactions when certain chemicals are applied to a cut surface. More sophisticated methods are being used, particularly in the taxonomy of fungi. These include the comparison of cell wall composition (Bartnicki-Garcia, 1968; Chapter 9, Table 9.1), serology (p. 326), and the analysis by various techniques of nucleic acids and proteins.

Books and articles for reference and further study

General

AINSWORTH, G. C. and SNEATH, P. H. A. (eds.) (1962). *12th Symposium of the Society for General Microbiology*. Cambridge University Press, 483 pp.

BARGHOORN, E. S. (1971). The oldest fossils. *Scient. Am.,* **224**, No. 5, 30–44.

BREED, R. S., MURRAY, E. G. D. and SMITH, N. R. (eds.) (1957), *Bergey's Manual of Determinative Bacteriology*, 7th Edn., Bailliere, Tindall and Cox, London, 1094 pp.

BUCHANAN, R. E. and GIBBONS, N. E. (eds.) (1975). *Bergey's Manual of Determinative Bacteriology*, 8th Edn., Williams and Wilkins, Baltimore, 1268 pp.

CAMPBELL, I. (1974). Methods of Numerical Taxonomy for various genera of yeasts. *Ad. appl. Micro.,* **17**, 135–156.

COWAN, S. T. and STEEL, K. J. (1974). *Manual for the Identification of Medical Bacteria*. Revised by S. T. Cowan. 2nd Edn., Cambridge University Press, 238 pp.

CARLILE, M. J. and SKEHEL, J. J. (eds.) (1974). *Evolution in the Microbial World*. 24th Symposium of the Society for General Microbiology. Cambridge University Press, 430 pp.

DAYHOFF, M. O. (1969). Computer analysis of protein evolution. *Scient. Am.,* **211**, No. 1, 86–95.

DICKERSON, R. E. (1972). The structure and history of an ancient protein. *Scient. Am.,* **226**, No. 4, 58–72.

JARDINE, N. and SIBSON, R. (1971). *Mathematical Taxonomy*. Wiley, London, 286 pp.

SCHOPE, J. W. (1975). The age of microscopic life. *Endeavour,* **34**, 51–58.

SNEATH, P. H. A. (1957a). Some thoughts on bacterial classification. *J. gen. Microbiol.,* **17**, 184–200.

SNEATH, P. H. A. (1957b). The application of computers to taxonomy. *J. gen. Microbiol.,* **17**, 201–226.

SNEATH, P. H. A. (1968). Vigour and pattern in taxonomy. *J. gen. Microbiol.,* **54**, 1–11.

SNEATH, P. H. A. and SOKAL, R. R. (1973). *Numerical Taxonomy*. Freeman, San Francisco, 573 pp.

SOKAL, R. R. (1966). Numerical taxonomy. *Scient. Am.,* **215**, No. 6, 106–116.

SOKAL, R. R. and SNEATH, P. H. A. (1963). *Principles of Numerical Taxonomy*. Freeman, San Francisco, 360 pp.

Part III

The activities of micro-organisms in their environment

Introduction

This third section (Chapters 11–16) considers the relation between micro-organisms and their environment, including both natural habitats and man-made ones. The habitat of a living organism is usually complex and seldom constant. Many environmental factors are continually changing and, moreover, micro-organisms are in competition with others of many different species for food, oxygen and living space. The metabolic products of one organism may either stimulate or inhibit growth of another. Interactions between the components of a mixed population may thus be highly complex.

Vigorously growing saprophytes, able to utilize dead organic matter, rapidly colonize suitable habitats. Less generally favourable sites are occupied by species which may be unable to compete with more vigorous organisms under optimal conditions for growth but which can survive through their ability to resist sub-optimal or even unfavourable conditions. The occupation of a particular habitat does not necessarily mean that it is the most favourable one for the occupying species. Thus many successful parasites actually grow better in pure culture on artificial media, but cannot survive in nature outside the host owing to competition with saprophytic organisms. Some soil fungi normally grow fairly deep in the soil through their ability to tolerate conditions of poor aeration, but are unable to survive the intense competition in the better aerated surface layers. Some habitats are rendered more suitable for the growth of micro-organisms by the action of pioneer species, which break down complex materials, thus making food available for a wider range of micro-organisms. Thus a definite succession of organisms occurs on such a substrate and the pioneer species are often overcome and excluded by the later growth of less specialized ones.

Both as saprophytes and as parasites, micro-organisms are of importance to man. Saprophytes attack stored products of various kinds and may cause serious economic loss (Chapter 16). On the other hand their activities in the soil, notably the breakdown of complex organic materials, is essential to soil fertility (Chapter 11) and thus to all life. Some micro-organisms are employed in industry to bring about desirable chemical syntheses or degradation (Chapter 16). Some habitats (e.g. air, Chapter 11) do not support propagation of micro-organisms but are nevertheless not sterile being contaminated from the environment; others, like water (Chapter 11), carry a microbial flora in proportion to the level of nutrients present. Food is a natural source of nutrients for micro-organisms and their use in obtaining fermented foods and problems associated with food processing are considered (Chapter 15). Parasites cause many serious diseases of man, domestic animals and crop plants (Chapters 13 and 14). Control of these organisms is both difficult and expensive, the study of the morphology, physiology, ecology—the natural history—of these organisms and of the host's responses to attack is essential to the improvement of existing control methods and the introduction of new ones.

Chapter 11

Microbiology of soil, air and water

Microbiology of soil

Introduction

Early work in soil microbiology was almost entirely confined to the study of bacteria. The pioneer work of Beijerinck and Winogradsky in the late nineteenth century on bacteria involved in the transformation of sulphur and nitrogen compounds provided explanations for some of the chemical changes which were known to take place in the soil and to be related to soil fertility. These studies were the forerunners of many concerned with the soil bacteria and the processes they mediated, but it was not until about thirty years later that comparable interest was shown in the fungal flora and its activities. The microfauna was neglected for even longer and only recently has interest in the soil algae increased.

Soil consists of mineral particles and organic residues which are, to a greater or lesser extent, wet. The interstices of these materials are usually occupied by a gaseous mixture but when the soil is waterlogged they become full of water.

The mineral particles are derived from the parent rock by weathering, which includes physical processes such as the action of heat, rain, running water, ice, and blown sand. According to the nature of the parent rock, chemical processes such as oxidation and solution by weak acids (arising from carbon dioxide dissolved in water) also play a part. The size of particles ranges from that of clay to gravel (see Table 11.1) and the whole range of sizes may be found in one soil, or perhaps as a result of sorting by water, particles of only one size may be found.

Table 11.1 Size of soil particles (mm)

Clay	0.002
Silt	0.002–0.02
Fine sand	0.02–0.20
Coarse sand	0.20–2.00
Gravel	2.00 +

Organic residues from plant and animal sources are added to the minerals. The amount varies greatly from one situation to another as does the state of decay of the residue when it reaches the soil. This diverse organic matter is incorporated into the soil by the action of soil organisms and, if it is not first oxidized, is converted to humus, the dark, amorphous colloidal material (p. 249) which plays such an important part in soil processes. The result of these physical and chemical processes is not a homogeneous mixture of mineral and organic particles, considerable variation in soil constitution occurs at points separated by very short distances. The mineral particles are bound together by organic material to form soil crumbs. These aggregates are extremely important for soil fertility since they increase the pore space essential for the good drainage and aeration necessary for high levels of microbial activity. The volume occupied by the various soil constituents is illustrated for one type of soil in Table 11.2.

Table 11.2 Approximate percentage of the total volume occupied by various components of a fertile soil

Mineral material	51%
Pore space	40%
Organic material	9%
Organisms	1%

The main source of soil water is precipitation. After heavy rain the pores will be full, but in a well-structured soil much of the water will drain away under the influence of gravity leaving water only in the capillary spaces and bound to inanimate soil constituents. The gravitational and capillary water is available to soil organisms and plays an important part in solute transport, whereas the bound water is not available and does not function in solute transport.

The soil atmosphere differs from the air above the

surface in that it usually contains more carbon dioxide and less oxygen. When gaseous exchange is hindered by permanent waterlogging, anaerobic conditions develop and not only carbon dioxide but also methane, hydrogen, and hydrogen sulphide may accumulate.

In addition to the point-to-point variation in soil constitution at a particular level it can readily be seen with the naked eye that the soil overlying the parent rock exists in distinct layers or horizons. A pit dug in the soil will reveal a profile of these horizons. The system of nomenclature is illustrated in Fig. 11.1.

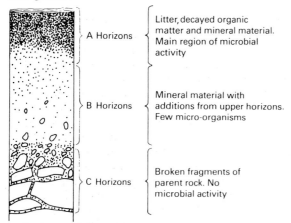

	A Horizons	Litter, decayed organic matter and mineral material. Main region of microbial activity
	B Horizons	Mineral material with additions from upper horizons. Few micro-organisms
	C Horizons	Broken fragments of parent rock. No microbial activity

Fig. 11.1 Soil profile.

The chemical nature of a soil affects the type and size of its microbial population. The soil also acts as a buffer, reducing the range of physical conditions as compared with those found on and above its surface. Its effectiveness as a buffer will depend on its chemical composition and physical make-up. The clay fractions are particularly important because of their relatively large surface area and their ability to adsorb nutrients, enzymes and micro-organisms. It is increasingly being recognized that microbial activity at surfaces, not only in soil, is of great consequence. The more fertile soils on the whole ensure a more constant supply of water and nutrients than do infertile ones. However, even in fertile soils adjacent habitats separated by only a few micrometres may be extremely favourable for growth and reproduction of micro-organisms on the one hand and completely inhospitable on the other.

The soil population and its assessment

The complexity of the soil environment, the diversity of its population and the minute size of the habitats makes sampling and assessment of the microbial population very difficult. Many small samples are usually collected and combined and then subsamples taken for the assessment. There are three main methods: cultural studies, direct examination and activity measurements.

For culturing, the soil is usually diluted and aliquots plated out onto agar or placed in tubes of liquid media. All media are, to some extent, *selective* but some are specially designed to discourage organisms other than those required. For example, a combination of rose bengal and streptomycin in an agar medium reduces the growth of bacteria and also restricts the rate of growth and spread of some fast-growing fungi, thus allowing the development of slower growing types. *Elective* media, while encouraging the required organism, do not positively deter the growth of others. *Enrichment* cultures are used to obtain a large popluation of organisms carrying out a particular transformation. For example, if cellulose is added to a soil sample which is then incubated, cellulose-decomposing organisms will usually be present in large numbers after a suitable period and can then be isolated by selective or elective methods more easily than from non-enriched soil. Cultural methods do not allow a complete enumeration of soil micro-organisms, mainly because of the problems with media, but they do usually permit identification of the isolates. They are not very satisfactory for fungi since the importance of spores is greatly over emphasized by the dilution plates.

Direct microscopic examination gives an idea of the distribution of the organisms in relation to the soil particles and the numbers recorded are frequently one or two orders of magnitude higher than dilution plates. Direct examination is time consuming and it is difficult to locate and count organisms in opaque soils, though various staining techniques, including fluorescence microscopy, assist in this.

The activity of micro-organisms may be estimated by measuring the respiration rate of soil samples or by assaying the levels of various key metabolites, (e.g. ATP) or the activity of common enzymes (e.g. phosphatase or dehydrogenase). No information on the species of organisms is obtained by these methods.

In general as many methods as possible should be used to study a soil and even then the numbers, biomass or activity levels obtained are only rather rough measures of the microbial populations. However in Table 11.3 some estimate of the numbers in the various main categories of micro-organism is given together with a value for the

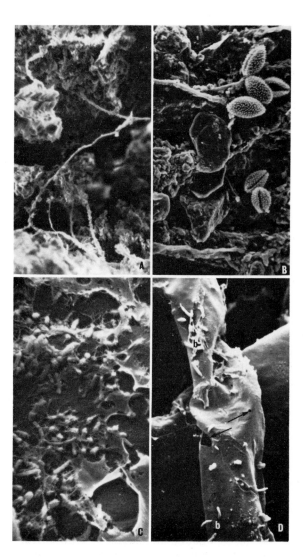

Table 11.3 Numbers and biomass of some organisms in the top 15 cm of agricultural soil. Data taken from a number of sources

	Number of organisms per gram	Biomass g/m²
Bacteria	9.8×10^7	160
Actinomycetes	2.0×10^6	160
Fungi	1.2×10^5	200
Algae	2.5×10^4	32
Protozoa	3.0×10^4	38
Nematodes	1.5	12
Earthworms	0.001	80

biomass of each of the groups. The biomass figure is possibly more meaningful than numbers alone because it takes into account the mass of organisms and hence gives a more realistic indication of the activity of a particular group of organisms in the soil though they may or may not be active.

It can be seen that although bacteria are more numerous than any other group of organisms the biomass of the fungi is larger than that of the non-filamentous bacteria. Also the combined biomass of protozoa, nematodes, and earthworms is not much different from that of the more important groups of the soil microflora. If the biomass of all soil invertebrates is taken into account the total would be about a quarter or a half of that of the microflora. Often in the past the importance of animals in the soil processes has been underestimated and this biomass value probably gives a truer indication of their significance.

Bacteria (excluding Actinomycetes)

Most bacterial types could probably be isolated from the soil if the search were diligent enough and if the right techniques were used; however, this does not mean that all the bacteria isolated are soil inhabitants. This description can be applied only to relatively few species which are fairly well categorized. These exhibit almost the whole range of bacterial morphology, and physiologically they range from aerobes to obligate anaerobes, from heterotrophs to autotrophs, and from saprophytes to mutualistic and parasitic symbionts.

Members of the genera *Pseudomonas*, *Achromobacter*, and *Bacillus* will be found in most aerobic soils; where conditions are anaerobic *Clostridium* spp. will occur. The numbers of such organisms will often increase dramatically when suitable organic substrates are added to the soil. Similarly the addition of suitable inorganic substrates will cause

Fig. 11.2 Scanning electron microscope pictures of micro-organisms in soil. (**A**) Hyphae probably of an Actinomycete crossing the air gap between soil crumbs (× 950). (**B**) Fungal hyphae and spores on a root surface. Notice the small clay particles which are also lying on the root (× 1100). (From Campbell, R. and Rovira, A. D. (1974). The study of the rhizosphere by scanning electron microscopy. *Soil Biol. Biochem.*, **5**, 747–752) (**C**) Bacteria on a root surface: each bacterium is surrounded by mucilage which has a fibrous appearance and there is also mucilage of plant origin which has dried during preparation into smooth sheets, e.g., at lower right (× 2300). (From Rovira, A. D. and Campbell, R. (1974) Scanning electron microscopy of micro-organisms on the roots of wheat. *Microbial Ecology*, **1**, 15–23) (**D**) A lysed fungal hypha colonized by bacteria (*b*); the wall of the hypha has several holes in it (arrowed) (× 2000). (From Rovira, A. D. and Campbell, R. (1975) A Scanning electron microscope study of interactions between micro-organisms and *Gaeumannomyces graminis* on wheat roots. *Microbial Ecology*, **2**, 177–185).

a multiplication of chemoautotrophic organisms such as the nitrifying bacteria *Nitrosomonas* and *Nitrobacter* and the sulpher oxidizers of the genus *Thiobacillus*. Organisms whose numbers increase in this way as a result of special soil conditions may be regarded as showing a zymogenous (fermentative) response.

The bacteria generally found to be the most numerous, and often comprising a majority of the population of the soil, are the coryneforms. This genus is variable not only in its morphology but also its Gram-staining, and contains the most important members of the autochthonous or indigenous population. In any one situation their numbers remain relatively constant and are not affected by soil amendment. *Agrobacterium* species are also members of this group of organisms.

The size of the bacterial population depends not only on the nutrients available but on other environmental factors too. Temperature, moisture, and indirectly gas content of the soil atmosphere and pH will vary with the prevailing climatic and seasonal conditions and with depth in the soil. These variations will introduce variations in the soil population.

The spatial distribution of bacteria in the soil is complex. Generally there is a decrease in numbers with depth (Table 11.4) which is a reflection of decreasing organic matter content of the soil. There are, however, instances where it is known that the maximum bacterial population does not correspond with the level containing most organic matter, possibly because the conditions favour the development of one of the other groups of soil organisms rather than bacteria. Superimposed on the general distribution are local variations due to a number of factors. For example, plant roots cause an increase in the population in the soil adjacent to them (p. 256); large populations develop in association with big pieces of organic matter such as lengths of fungal mycelium and plant remains.

Bacteria not associated with residues of this kind exist in the soil in small colonies, 60 per cent of which contain 2–6 cells per colony, 20 per cent contain 7 or more cells per colony, and others occur as individuals. These colonies are often associated with the film of colloidal material which coats the soil mineral particles. Within a soil crumb it is found that the population is largest near the surface and falls to its lowest level at or near the centre.

Actinomycetes

Most actinomycetes are soil inhabitants. The characteristic odour of soil after it is ploughed or wetted by rain is largely due to these organisms, many of which produce strong earthy smells in pure culture. The numbers of actinomycetes in soils as revealed by the dilution plate method are second only to the numbers of bacteria, but there is no accurate way of estimating their abundance in the soil. The large count on dilution plates suggests that actinomycetes are present there very largely as spores. Actinomycetes are worldwide in their distribution. Their numbers generally increase with the warmth of the climate and decrease with depth in the soil. The difficulty with which streptomycete spores are wetted may enable them to be dispersed in the air–water interfaces in the upper layers of the soil rather than to sink to lower regions. Like most other bacteria their growth is favoured by alkaline conditions. Cultivation practices which improve soil aeration often lead to increased numbers of streptomycetes which are the commonest of the soil-inhabiting actinomycetes. In acid and waterlogged soils they are relatively scarce.

The important and unique role of actinomycetes in the soil is attributable to their ability to flourish after faster growing micro-organisms have transformed organic residues into a dark homogenous mass, rich in lignin, certain hemicelluloses, and proteins, which is no longer favourable for the

Table 11.4 Numbers of micro-organisms of the major groups, determined by the dilution plate method, and the amount of humus present in various horizons. From Starc, A. (1941). *Arch. Mikrobiol.*, **12**, 329.

Horizon	Humus %	Depth cm	Organisms/gram of soil $\times 10^3$				
			Aerobic bacteria	Actino-mycetes	Anaerobic bacteria	Fungi	Algae
A_1	3.00	3–8	7,800	2,080	1,950	119	25
A_2	1.28	20–25	1,804	245	379	50	5
A_2–B_1	0.91	35–40	472	49	98	14	0.5
B_1	0.37	65–75	10	0.5	1	6	0.1
B_2	0.41	135–145	1	–	0.4	3	–

continued growth of these other organisms. Actinomycetes compete effectively with the latter only through their ability to break down residual nutrients remaining in the soil.

In view of the large numbers of soil actinomycetes that produce antibiotics under laboratory conditions one might expect these substances to be of importance in nature. However, the formation of antibiotics by actinomycetes actually living in the soil is hard to demonstrate, and occurs, if at all, to only a slight extent owing to the unsuitable nutritional conditions. Further, any antibiotic effect would be limited to the vicinity of the colony responsible because antibiotics are inactivated by adsorption on to such surfaces as clay minerals and are destroyed by non-biological and biological reactions.

If compost is allowed to become heated by microbial action, the surviving microflora which continues the decomposition contains many thermophilic actinomycetes. The inhalation of dust from mouldy hay which has spontaneously heated during its maturation may produce an allergic response known as 'Farmer's Lung' in persons who have become hypersensitive to antigens in the dust. Spores of thermophilic actinomycetes, notably *Thermoactinomyces vulgaris* and *Micropolyspora faeni* are the principal sources of this antigen. Because these spores are small they can penetrate deeply into the respiratory tract, in the peripheral parts of which the allergic symptoms develop.

Fungi

As with bacteria the problem exists of distinguishing true soil inhabitants from organisms present by chance in the sample examined, but experience has shown that certain organisms occur constantly and others frequently. Isolation experiments suggested that in general *Mucor*, *Penicillium*, *Trichoderma* and *Aspergillus* predominate and that *Rhizopus*, *Zygorhynchus*, *Fusarium*, *Cephalosporium*, *Cladosporium*, and *Verticillium* occur commonly, but all of these grow quickly and sporulate copiously and are therefore favoured by the dilution plate method (p. 244). Direct examination of the soil by special methods (p. 244) shows that basidiomycetes (p. 198) are numerous and that dark sterile hyphae, probably of members of the Dematiaceae (p. 199), are common. Representatives of these last two types are seldom isolated from soils by the dilution plate technique. The biomass of fungi in cultivated soils often exceeds that of any other group of organisms and in acidic environments they are usually numeri-

cally dominant too. Addition of organic matter to a soil stimulates the fungal flora in the same way as it does the zymogenous bacterial population. The heavily sporing soil fungi are regarded mainly as zymogenous; Basidiomycotina form the major part of the autochthonous fungal flora.

Water also affects fungi in the same way as it does bacteria. A minimum level is required for activity, although since more fungi than bacteria have structures resistant to desiccation, the fungal population of arid environments is greater than that of bacteria. Waterlogging and the consequent reduction of oxygen concentration inhibits most filamentous fungi as the majority are strict aerobes, hence this group is virtually absent from permanently saturated soils.

Fungal mycelium is interwoven among the soil particles and it binds these together, thus improving the texture of clay soils. Mycelium is also intimately associated with larger organic particles which it frequently penetrates. Hyphae may become associated together in strands containing only a few threads or, particularly in Basidiomycotina, into complex rhizomorphs which are commonly found in the surface layers of forest soils. Generally, as can be seen in Table 11.4, the major part of the fungal flora occurs in the upper soil horizons where there is most organic matter, although the relative decrease in numbers with depth is not as marked as with other groups of organisms not excluding the anaerobic bacteria.

Algae and Cyanobacteria

Recent investigations suggest that soil algae are more widespread and abundant than was formerly supposed since, even in regions as dry as the Negev desert in Israel, rich algal floras have been recorded both on the surface and also underneath stones and even within cracks in the stones. If the sand is sufficiently translucent a layer of algae may occur in the slightly moister region under the upper millimeter of sand. This flora consists mainly of Cyanobacteria (e.g. *Schizothrix* and *Microcoleus*) and coccoid Chlorophyta. The algal biomass on surface soil must be great since it is frequently visible to the naked eye. Certain algae have been found at considerable depths in undisturbed soil but it is unlikely that they are present in an active state since although some species are capable of heterotrophic growth in the dark, utilizing carbohydrate substrates, it is impossible that they could compete effectively with heterotrophic soil bacteria and fungi under these conditions. Thus relatively large algal populations far below the surface of the soil must be the result of wash-down.

Algal species living in soils are notoriously difficult to identify; many are smaller celled versions of aquatic species. They include flagellate, coccoid, or filamentous species. The coccoid species reproduce via motile stages (e.g. *Chlorococcum*) whilst almost all the diatoms occurring are motile. Motility or multiplication via motile cells is an obvious advantage in sites which are periodically wet. Attached species, however, also occur in suitable situations. Insufficient is known of soil algae to list the commonest forms but *Nostoc*, *Cylindrospermum*, and *Anabaena* of the Cyanobacteria, *Chlorella* and *Chlorococcum* of the Chlorophyta, together with certain diatoms (e.g. *Hantzschia* and *Navicula*) are frequently isolated from soil samples. Cyanobacteria are common constituents of neutral, alkaline and saline soils; species and genus differences also occur in other groups depending on the soil pH and nutrient status. The growth of soil algae and Cyanobacteria undoubtedly affects the surface soil. Depletion of some nutrients has been demonstrated and the addition of organic matter is sometimes considerable, e.g. in algal crusts on desert soils and on the surface of peats. Several of the Cyanobacteria occurring on soils fix atmospheric nitrogen.

The soil fauna

The soil fauna contains numerous protozoa and representatives of metazoa, most of the latter falling outside the scope of this book. However, their activities in some soil processes are of such moment that they cannot be excluded on grounds of size alone. The smallest species, protozoa and nematodes, are widely distributed in the soil. The former generally occur in the water film surrounding soil particles and the latter usually in the soil of the upper horizons among plant roots. The larger animals are on the whole much more mobile yet particular types are usually found in characteristic environments, e.g. under stones or large pieces of debris on the surface, some in the surface soil and others in deeper layers, but most occur at or near the surface. Some animals such as earthworms are exceptional in that they frequently move from horizon to horizon and hence play a very important part in the transport of materials not only within the soil but from the surface to the soil proper. Prior to this mixture of organic residues with the soil a large number of animals have taken part in the primary attack of the debris.

On very wet or very dry soils in which few animals exist, surface litter tends to accumulate and under these conditions fungi will be the principal agents of decomposition. Where animals are absent from a soil, for example because of the use of toxic pesticides, it is significant that little of the humus so important in producing a well-structured fertile soil occurs.

Transformations in the soil

The soil population is responsible for transformations which are important for its continued fertility and which ensure the removal of natural litter from the surface of the earth. Many of the transformations are cyclic systems comprising sequential reactions of compounds containing a common element such as carbon, nitrogen, sulphur, etc. However, it is rare to find even the majority of the steps in a cycle occurring in a particular situation at any one time. Some stages in these systems result in an amelioration of the soil, whereas others decrease its fertility, thus both the encouragement and control of these stages is agriculturally desirable. In some instances stages in a cycle are mediated only by organisms which are extremely sensitive to environmental conditions, others take place as a result of the action of a wide variety of micro-organisms and proceed under a wide range of conditions.

Carbon cycle

Quantitatively the most important series of transformations in which soil organisms are concerned is that involving carbon compounds. Organic residues added to the soil contain about 50 per cent carbon which is eventually converted to carbon dioxide as a result of the action of the soil population. The microflora is responsible for the evolution of 95 per cent of the gas produced and animals for only 5 per cent. Carbon dioxide is fixed mainly by green plants, though heterotrophs also use small quantities of this gas, and the rate of the process is such that it has been estimated that the atmospheric supply would be exhausted in twenty-five to thirty years if it were not replenished by oxidation. A much simplified representation of the cycle is given in Fig. 11.3.

DECOMPOSITION OF ORGANIC RESIDUES

The main types of carbon compounds added to the soil are plant and animal remains in which the

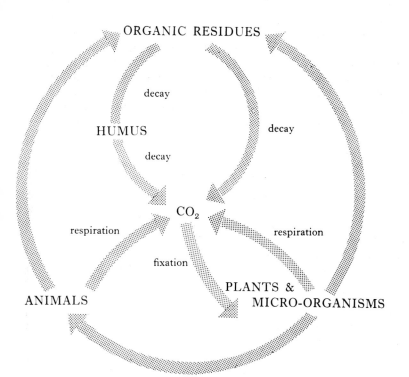

Fig. 11.3 The carbon cycle.

carbon is chiefly included in high molecular weight compounds. In plants the main compounds of this type are cellulose, hemicelluloses, and lignins, together with smaller quantities of fats, waxes and oils, proteins, and nucleic acids. In micro-organisms and animals a similar range of high molecular weight compounds exists. However, the main structural components are different. In invertebrates and fungi the polymer chitin occurs widely and in bacteria mucopeptide is found (p. 14).

On the surface of the soil, residues are attacked by bacteria, fungi, and animals (p. 248). The predominant microbial types involved and the rate of the attack depend on the chemical nature of the residue, the environmental conditions and the nature of the underlying soil. For example, nitrogenous materials are generally degraded more rapidly than residues with high C:N ratios. Of the environmental conditions, moisture is particularly important and in dry conditions the attack will be slow. Temperature also affects the rate of degradation and the composition of the attacking flora. The underlying soil determines to a large extent the population available for the degradation. Within the soil the conditions are more favourable for the microflora, in particular the bacteria, and the rate of degradation is increased.

At every stage in this process, the initial attack on the high molecular weight polymers is by enzymes liberated into the soil or acting at the outside surface of the producing cell. Thus environmental factors may alter the rate of decay by their effect both on the microflora and on the enzymes outside the cell.

Most of the carbon contained in organic residues is oxidized to carbon dioxide but some is incorporated in microbial tissues which themselves will eventually be decomposed. The remainder is incorporated into humus. This is the dark-coloured amorphous organic material which is so important for soil fertility as is evidenced by the infertility of soils from which it is absent. Its formation is not understood, although it may be that soil animals and micro-organisms are important in the process. Similarly its chemistry is obscure. It is not a single chemical substance and its composition varies but it is essentially polyphenolic in nature and contains some amino acids and amino sugars.

DEGRADATION OF PURE SUBSTRATES

In view of the complexity of the processes involved in the breakdown of complex natural residues much of our understanding of the microbiology of the degradation of carbon-containing polymers has come from the study of the decay of pure substrates such as cellulose and chitin.

Cellulose decomposition has been studied by

burying strips of cellophane supported on glass coverslips. In general the course of the decomposition in approximately neutral soils is similar. Fungal hyphae colonize the surface rapidly and hydrolyse cellulose; genera include *Botryotrichum*, *Chaetomium*, *Humicola*, *Stachybotrys*, and *Stysanus*. After a few weeks the mycelium dies off and attack by bacteria, sometimes simultaneously with animals, occurs. The animals evidently are important in this stage of the degradation since in their absence cellophane persists for long periods. Cultivation causes an overall increase in the number of cellulose decomposing micro-organisms but the increase in bacterial numbers is much greater than that of the fungi. In particular *Cytophaga* and *Sporocytophaga* may be increased three- or fourfold.

The availability of nitrogen is a factor limiting the rate of breakdown of any organic residue but is particularly important with respect to pure cellulose which contains none. The average C:N ratio of microbial cells is about 10:1, thus when material with a higher proportion of nitrogen than this is added to a soil, nitrogen is not limiting, indeed release of inorganic nitrogen will occur. Conversely residues with higher C:N ratios than this are decayed at a rate related to the exogenous nitrogen supply, other things being equal.

In contrast to the decomposition of cellulose the degradation of chitin in neutral and slightly alkaline soils is mainly achieved by bacteria and in well-aerated soils predominantly by actinomycetes. In waterlogged anaerobic soils both cellulose and chitin are attacked by *Clostridium* species.

Cellulose and chitin account for much of the carbon added to soil yet it is impossible to predict fully the composition of the microflora involved in the various stages of their degradation even when added to the soil as purified substances. Much remains to be learnt about the decomposition of more complex natural residues.

DECOMPOSITION OF HUMUS

The amount of humus in a mature soil varies little from year to year provided it is not depleted as a result of cultural practices. Humus is relatively resistant to degradation either because it is intrinsically stable or because it is adsorbed on to clay particles, but whatever the reason, only a small fraction of the total is broken down in any one year. As with the degradation of plant and animal residues added to the soil, the rate of breakdown is influenced by environmental factors such as aeration, temperature, pH, and in addition the amount of organic matter in the soil; when the amount is large the rate of humus decomposition is high. This is probably because the zymogenous flora attacks the humus fraction in addition to the added organic residues. Normally, however, humus degradation is achieved mainly by the autochthonous flora.

METHANE FORMATION AND OXIDATION

In permanently waterlogged anaerobic soils, *Clostridium* species may convert organic matter to organic acids, carbon dioxide, and hydrogen which are in turn converted by certain strictly anaerobic bacteria to methane. This process is of no significance in agricultural soils which are only temporarily waterlogged but a combination of poor drainage and high organic matter content may very rarely produce conditions such that methane is produced and liberated at the surface as gas. In waterlogged anaerobic paddy field soils where methane generation occurs, a methane oxidizing flora exists at the surface and converts most of the gas to carbon dioxide.

Nitrogen cycle

The activity of some of the organisms involved in the nitrogen cycle results in nitrogen being made available to plants in an assimilable form. These organisms therefore act in an analogous way to those involved in the carbon cycle which make carbon dioxide available. An important difference between the two systems is that nitrogen turnover determines productivity in most agricultural situations, hence the activity of those soil organisms metabolizing nitrogen compounds in such a way as to remove nitrogen assimilable by higher plants from the soil, may be very important with respect to loss of soil fertility.

There are two conspicuous differences between the carbon and nitrogen cycles in terrestrial environments. Firstly, carbon is made available to plants in gaseous form as carbon dioxide, whereas nitrogen is utilized by plants mainly in the form of ammonium or nitrate ions. Secondly, most of the micro-organisms involved in the carbon cycle in terrestrial environments are heterotrophic (Table 4.1), whereas many of those taking part in the nitrogen cycle are autotrophic.

Living organisms contain about 10–15 percent nitrogen and the annual turnover of this element has been estimated to be between 10^9 tons and 10^{10} tons per year. The cycling of nitrogen is mainly

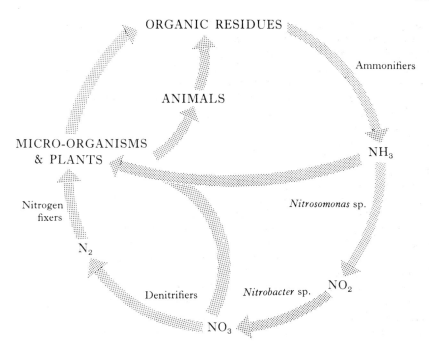

ORGANIC RESIDUES

Ammonifiers

ANIMALS

NH_3

MICRO-ORGANISMS
& PLANTS

Nitrosomonas sp.

Nitrogen
fixers

N_2

Nitrobacter sp.

NO_2

Denitrifiers

NO_3

Fig. 11.4 The nitrogen cycle.

achieved by biological processes (Fig. 11.4) but certain chemical reactions mediated by lightning and ultra-violet light can cause fixation of atmospheric nitrogen. In addition, nitrogen compounds in solution can be converted by chemical processes under certain soil conditions.

DECOMPOSITION OF ORGANIC NITROGEN-CONTAINING
COMPOUNDS

Almost all nitrogen in soil is in the form of organic compounds but it has been found that although both available ammonium and nitrate ions may under certain circumstances be present their levels fluctuate rapidly and they usually account for less than 2 per cent of the total soil nitrogen. Three main types of organic molecule contain much of the soil nitrogen. Between 20 per cent and 50 per cent is combined in amino acids largely in humus, 5–10 per cent in amino sugars and only about 1 per cent in purine and pyrimidine bases. Thus about half or more of the total bound nitrogen in the soil is not accounted for. Some of this may be in the form of non-exchangeable ammonium ions bound to clay particles, but the majority is combined in compounds whose chemistry is not understood.

The three types of compound mentioned above occur in organisms mainly as polymers but may be added to the soil as monomers as a consequence of autolytic processes. More usually the residue incorporated into the soil will be of high molecular weight and will initially be degraded by extracellular enzymes or enzymes liberated into the soil by cell lysis. A wide variety of soil bacteria are capable of hydrolysing protein. *Bacillus* and *Pseudomonas* species particularly are responsible for the breakdown of pure protein added to the soil. These and other bacteria such as *Arthrobacter* are probably responsible for the degradation in alkaline soils of proteinaceous material from natural sources too. But in acidic soils fungi are predominantly responsible. The fate of the amino acids produced depends on their type and the soil conditions. Some, such as glutamic acid, may be utilized directly, most will be deaminated with the formation of ammonia and the corresponding acid, but a few, such as threonine, lysine, and methionine, will be attacked only very slowly. Aminopolysaccharides are broken down by a much more restricted flora than are proteins but in the average agricultural soil, these macromolecules never occur at high levels. They are far more important in marine sediments which are often anaerobic and where breakdown is mediated mostly by species of *Clostridium*. These polymers are usually hydrolysed to the constituent amino sugar and then deaminated. Nucleic acids are broken down by a relatively complex series of processes. Firstly, the polymer is degraded to form the constituent nucleotides. Several fungi, such as

Aspergillus and *Penicillium*, play a part in this as do the bacteria, *Clostridium*, *Bacillus*, and *Achromobacter*. The nucleotides are then dephosphorylated and then the residual base–sugar complex is broken down. The nitrogen in the base is finally liberated in the form of urea, ammonia, or μ-alanine, according to the particular base and the soil conditions, as a result mainly of bacterial action. Thus the main product of the breakdown of organic residues is ammonia. The fate of this depends on the C:N ratio of the soil. When carbohydrate is abundant the ammonia will be used immediately, but when the C:N ratio has fallen to 12:1, it will be oxidized by nitrifying bacteria or utilized by other organisms.

NITRIFICATION

The oxidation of ammonia to nitrite and nitrate is achieved chiefly by organisms of two strictly autotrophic genera: *Nitrosomonas* and *Nitrobacter*.

Nitrosomonas is a small, Gram-negative rod with polar flagella. It is strictly aerobic and mediates the overall reaction:

$$2\,NH_4^+ + 3\,O_2 \rightarrow 2\,NO_2^- + 4\,H^+ + 2\,H_2O$$

The intermediates in this oxidation are not yet confirmed, but one is probably hydroxylamine.

Nitrobacter is similar to *Nitrosomonas* but is slightly smaller and converts nitrite to nitrate. The process can be represented by the equation:

$$2\,NO_2^- + O_2 \rightarrow 2\,NO_3^-$$

Compared with the transformation involved in the production of ammonia these processes are very sensitive to environmental conditions and are carried out effectively only in neutral or slightly alkaline soils. At soil pH values below 6, ammonium ions usually accumulate and in very alkaline soils high concentrations of nitrite may be found. This is due to the sensitivity of *Nitrobacter* to free ammonia and it is generally true that this organism is more sensitive to adverse conditions than is *Nitrosomonas*. Extremes of temperature and drought may also cause accumulation of nitrite. Nitrification is also suppressed by addition of carbohydrate to the soil. This is not due to the inability of these autotrophic organisms to tolerate organic matter but to competition for ammonium ions with the large heterotrophic population involved in carbohydrate degradation. Both *Nitrosomonas* and *Nitrobacter* are strict aerobes, consequently waterlogging of the soil and concomitant

reduction in gaseous exchange renders them inactive.

The supposed importance of nitrification with relation to soil fertility rests upon the fact that nitrates, as opposed to ammonium salts, are often preferentially absorbed by plant roots. Nitrification takes place in most cultivated soils but in many forest, orchard, and grassland soils the nitrifying bacteria are inactive because of soil or climatic conditions. The plants in these situations must therefore assimilate ammonium nitrogen. Ammonium ions have the advantage that they are chemically stable in acidic conditions and as they may be bound to negatively charged clay particles, are not leached out of the soil. Nitrate and nitrite, in contrast, are easily leached out and nitrite is also converted in acid soils to gaseous nitrogen and nitrous oxide (see below). There are, however, some disadvantages associated with the nitrogen in the soil remaining as ammonium ions. Only a relatively small proportion of the ammonium ions bound to clay particles is exchangeable and available for assimilation. Above certain levels they may be toxic to plants and in very alkaline soils they may be volatilized. Notwithstanding these problems, it has been suggested that nitrification may be an agriculturally undesirable process and a pyridine derivative suppressing *Nitrosomonas* has been marketed. The value of such control is not yet certain but it is clear that attempts to control nitrification must be directed at *Nitrosomonas* and not *Nitrobacter*, since it is high levels of nitrite which are phytotoxic.

DENITRIFICATION

The process of denitrification involves the reduction of nitrate or nitrite to molecular nitrogen or to nitrous oxide both gaseous products released into the atmosphere causing a reduction in the level of soil nitrogen. It does not include assimilation of these ions by plants or micro-organisms or loss by leaching. This distinction must be made clear since the initial stages of the denitrification and assimilation pathways are probably the same.

$$HNO_3 \rightarrow HNO_2 \begin{cases} \rightarrow NH_3 & \text{Assimilation} \\ \rightarrow \begin{array}{l} N_2O \\ N_2 \end{array} & \text{Denitrification} \end{cases}$$

Nitrate competes as an alternative electron acceptor to oxygen under anaerobic conditions and improved aeration will reduce denitrification although even in a very well-aerated soil anaerobic micro-environments provide suitable conditions for denitrification. The soil organic matter provides the energy source for the heterotrophic denitrifying bacteria but in soil with a reaction lower than pH 5.0–5.5 little or no nitrate loss occurs. However, in acid soils nitrite is chemically unstable and is converted to nitrogen and nitrous oxide.

In spite of the enormous population potentially capable of causing denitrification this activity is apparently confined to bacteria, particularly the genera *Pseudomonas*, *Achromobacter*, and *Bacillus*. Fungi and actinomycetes are probably not involved, though they may be more important in acid forest soils.

NON-SYMBIOTIC NITROGEN FIXATION

A wide range of prokaryotes is capable of non-symbiotic nitrogen fixation in the laboratory. The bacteria include the aerobes *Azotobacter*, *Beijerinckia*. *Pseudomonas*, and *Nocardia*, the facultative anaerobes *Klebsiella*, *Bacillus polymyxa*, and *Rhodospirillum*, and the anaerobic *Clostridium pasteurianum*, *Desulphovibrio*, *Methanobacterium*, *Chromatium*, and *Chlorobium*. Many Cyanobacteria can also fix nitrogen, particularly those with heterocysts though these structures are not essential. Although *Azotobacter* is an extremely active nitrogen fixer in vitro, the efficiency of its nitrogen fixation is low; to fix 5–20 pounds of nitrogen per acre these organisms would require and entirely utilize 1000 pounds (454 kg) of carbohydrate. It is therefore improbable that *Azotobacter* makes any great contribution to soil fertility by nitrogen fixation in temperate agricultural soils. Other free-living organisms capable of fixing nitrogen are probably unimportant in most soils, although there is evidence that in some specialized environments some do contribute to soil fertility. In some very fertile Egyptian soils *Azotobacter* cells are very numerous and may have a significant role. In tropical paddy fields an extensive population of blue-green bacteria develops and it is thought that the nitrogen fixed by these organisms accounts for the ability to take successive crops of rice from these fields without any nitrogenous supplementation of the soil. Also in one sub-arctic soil there is some indication that nitrogen fixation by a variety of non-symbiotic organisms may be significant.

SYMBIOTIC NITROGEN FIXATION

Several genera of bacteria form symbiotic associations with the roots of plants to form root nodules which fix nitrogen. The most important symbiosis is between *Rhizobium* and Leguminosae, (Chapter 12). In contrast to the free-living genera the amount of nitrogen fixed is considerable and economically important.

MECHANISM OF NITROGEN FIXATION

The biochemistry of nitrogen fixation is similar in all organisms that have been studied. The enzyme, nitrogenase, is very sensitive to oxygen, whether it is extracted from aerobes or anaerobes. Nitrogen gas is bound to molybdenum and iron atoms in the main enzyme protein, though a second protein containing just iron is required. Magnesium and ATP are also closely associated with active nitrogenase. The nitrogen is reduced, in two-electron steps, while enzyme bound with ferrodoxin as a co-factor. Ferredoxin is a non-haem iron protein with the very low redox potential (-0.42) needed to reduce nitrogen, and other co-factors are also known (e.g. flavodoxin from *Azotobacter*). The source of ATP and reducing power does not seem to be important: ATP usually comes from oxidative phosphorylation, or from pyruvate and inorganic phosphate, though photosynthetic organisms may use ATP from cyclic phosphorylation of photosystem 1.

Ammonia is the first product of the reaction, being released as the last of the three bonds between the nitrogen atoms is broken.

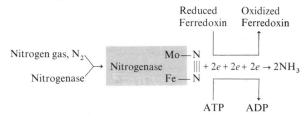

Sulphur transformations

Many of the organisms taking part in the sulphur cycle (Fig. 11.5) are not active in the soil, e.g. the photosynthetic bacteria involved in the oxidation of sulphides, which are important in natural waters.

In most agricultural soils sulphur is present in sufficient concentrations to meet the requirements of the soil population and crop plants. This supply derives from the parent rock, from organic residues and also, in agricultural areas adjacent to industrial regions, from volatile compounds.

Most soil sulphur is combined in organic compounds, though higher plants normally use sulphate

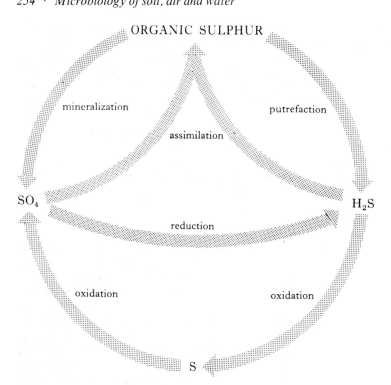

ORGANIC SULPHUR

mineralization

putrefaction

assimilation

SO₄ ... H₂S

reduction

oxidation ... oxidation

S

Fig. 11.5 The sulphur cycle.

ions as their sulphur supply although they can also assimilate sulphur-containing amino acids. Micro-organisms are, as a group, more versatile and can utilize most of the sulphur compounds and are also instrumental in making sulphur available to higher plants in a utilizable form.

DECOMPOSITION OF ORGANIC SULPHUR-CONTAINING COMPOUNDS

If cystine or cysteine are added to a well-aerated soil, sulphate is rapidly formed by a route not involving hydrogen sulphide. However, when these amino acids are incorporated in a protein, at least some hydrogen sulphide will be produced even in an aerobic soil. In waterlogged anaerobic soil hydrogen sulphide will be the major product, forming black metal sulphides in the presence of iron and manganese particularly.

Environmental factors affect the mineralization of sulphur in the same way as they affect that of nitrogen and carbon, but their effects are less well understood. However, the C:S ratio of a residue apparently has to be less than about 200 before sulphur is made available to higher plants.

OXIDATION OF SULPHUR COMPOUNDS

Sulphur may exist in many oxidation states, and

micro-organisms, particularly bacteria, can carry out conversions between them depending on the aeration of the microhabitat. *Thiobacillus* spp. are chemosynthetic aerobes which can oxidize sulphides, elemental sulphur, sulphite, etc. to sulphate and which are present in cultivated soils in small numbers. It is uncertain whether they play a part in the oxidation of hydrogen sulphide since oxidation occurs spontaneously in the presence of oxygen. *Th. thioparus* and *Th. thiooxidans* oxidize other sulphur compounds and may cause a reduction in pH because of the sulphate produced. Sulphur metabolism is complex but it usually involves adenosine–5′–phosphosulphate (APS). Sulphur oxidations are also performed under anaerobic conditions by the photosynthetic sulphur bacteria which cleave H_2S instead of H_2O, which eukaryotes and some other prokaryotes use in photolysis: the photoautotrophs are not important in soil. Various filamentous bacteria such as *Beggiatoa* and *Thiothrix* may live on the surface of muds rich in hydrogen sulphide and they oxidize it to elemental sulphur, though they probably live mostly heterotrophically.

REDUCTION OF SULPHATE

The organisms chiefly responsible for reducing sulphate in the soil are members of the genus *Desulfovibrio* which use sulphate rather than oxygen as a terminal electron acceptor. These organisms are widespread in nature and can be found growing heterotrophically, using carbohydrates, organic acids, and alcohols as electron donors. In agricultural soils they do not play a very important part, since the aeration conditions are such that they are not significantly active. However, in waterlogged paddy field soils where anaerobic conditions prevail, they are responsible for the generation of hydrogen sulphide which may damage the roots of rice plants growing there. Apart from this, sulphate-reducing organisms are economically important in a number of ways. They are responsible for causing extensive corrosion to underground iron pipes by the removal, and use as an electron donor, of hydrogen which normally forms a protective layer around these pipes in the absence of air. Further corrosion then occurs. It is also thought that they may play a part in liberating oil from oil-bearing shales. In the past they probably took part in the formation of sulphur deposits by producing hydrogen sulphide which was then oxidized by micro-organisms or by purely chemical processes.

ASSIMILATION

Hydrogen sulphide in addition to being oxidized may be assimilated as such by some microorganisms, but sulphate is the form more usually taken up by these organisms and by higher plants. Within the cell sulphur is mainly in a reduced form and the reduction process involves APS and 3′–phosphoadenosine–5′–phosphosulphate.

Transformations involving phosphorus

Phosphorus is needed in substantial quantities by living organisms for nucleic acid synthesis and in lesser amounts for phospholipids and high energy phosphorus compounds such as ATP. Soils are frequently deficient in this element and the supply from plant and animal residues is often supplemented with a phosphate fertilizer. The transformations of phosphorus do not, like those of sulphur and nitrogen, involve changes in oxidation states by bacteria.

Organic residues added to the soil are degraded by the heterotrophic microflora, the rate of the process depending on the environmental conditions, as do the rates of sulphur and nitrogen mineralization. In fact, the conditions for rapid ammonification are the same as those for speedy phosphate release.

Substantial amounts remain bound in organic material. The phosphorus content of the added residue determines whether phosphate is made available to crop plants or whether it is incorporated into microbial tissue. The level above which phosphate is made available is about 0.2 per cent. Phosphate is also made available to plants as a result of the liberation by micro-organisms of organic acids which dissolve insoluble inorganic phosphate compounds in the soil. Associations between fungi and plant roots (mycorrhiza) have been particularly studied in this respect (Chapter 12, p. 283).

Other transformations

The only other major nutrient required by microorganisms is potassium which is taken up and used mainly as K^+. Its availability is probably largely determined by the chemical nature of the soil and by physical processes such as adsorption. Microorganisms immobilize it but are not known to have any great effect on its availability. Trace elements (e.g. Zn, Cu, Co, etc.) are released by microbes during decay and their uptake by higher plants can be affected by bacteria growing on the root surface (e.g. Mo uptake).

The oxidation of iron, and possibly manganese, is used as an energy source by chemoautotrophs, and both are released and immobilized by heterotrophs. The fungal flora appears to be less sensitive to iron and manganese availability than are the bacteria. *Bacillus* and *Arthrobacter* have a particular high requirement for iron. In soils of normal pH, transformation of manganese is the result of biological activity. Only at pH values in excess of 8 is this element converted to the insoluble manganic form, usually manganic oxide (MnO_2), by chemical means. Below pH 5.5 manganese exists predominantly as the divalent manganous form, which is the form mainly utilized by plants, though manganese oxides can also act as a source of assimilable manganese. As a result of the activity of the manganese-oxidizing flora soils may become deficient in available forms of this element and a reduction in the crop yield may occur. Bacteria oxidizing manganese are common only in the upper layers of heavy soil but are more widely distributed in sandy soil—a reflection of the aeration of the two soil-types. Certain heterotrophic bacteria (*Psuedomonas, Corynebacterium*, and *Flavobacterium*) can oxidize manganese as can two budding bacteria: *Metallogenium* is widely distributed in soil but apparently only develops in association with a fungus; *Pedomicrobium*, which resembles *Hypho-*

microbium (p. 164), may oxidize not only manganese but also iron. Reduction of manganese compounds to a form readily utilizable by plants can result from the production of acid and/or by a reduction of the redox potential of the soil as a result of the activity of the microflora. However, the importance of bacteria in making manganese available to plants is uncertain since the return of aerobic and neutral pH conditions results in its reoxidation.

In contrast to the role of the soil microflora in manganese transformation the part played by bacteria in the conversion of iron compounds in neutral and alkaline soils is unimportant. Some heterotrophic bacteria may cause deposition of iron from organic complexes by using the organic compound as an energy source and it is possible, but not certain, that iron deposited in this way may play a part in the formation of the iron pans which occur in some soils. Other heterotrophic bacteria can also reduce ferric to ferrous iron under anaerobic conditions. The chemoautotrophs, such as *Thiobacillus ferrooxidans*, and heterotrophs including the filamentous forms such as *Sphaerotilus* and the unicellular *Gallionella* are active in water rather than soil.

Interactions and relations of soil organisms

Competition and antagonism

The varied soil micro-organisms interact with one another in a number of ways. Because of the density of the population, interaction between soil organisms is often competitive. The outcome of competition between two organisms may be determined because one is better adapted than the other to the prevailing environmental conditions. Alternatively, or additionally, one may produce an antagonistic substance which inhibits or kills the other. Changes in the environment, either caused by the micro-organisms themselves (e.g. the use of nutrients) or imposed from outside (e.g. seasonal effects), result in different microbes becoming dominant members of the soil population as conditions alter. Such a sequence of dominant organisms is called a *succession* and is exemplified by the colonization of cellulose described on p. 250.

In a few instances the activities of the dominant microflora change the environment in such a way that succession is impossible. The formation of peat bogs is one example of this. The digestion of the plant material under anaerobic conditions results in the production of considerable quantities of acid which so lowers the pH that biological processes virtually cease.

The part played by antibiotics, and agents such as bacteriocins and phage, in the interactions between soil micro-organisms is in doubt. Many organisms producing antibiotics in culture can be isolated, but antibiotics cannot be demonstrated in significant concentrations in soil. The antibiotics may operate over very small distances in microhabitats so that they are not detectable in the bulk of the soil or they may be adsorbed onto clays and colloidal organic matter. The dormancy of many of the soil micro-organisms (zymogenous flora) is probably due to low nutrient levels, though antibiotics may also play a part. All members of the soil population compete to some extent though only some of them are involved in symbiotic relationships such as mutualism (Chapter 12), predation and pathogenesis (Chapter 14). Soil-borne plant pathogens must compete with soil saprophytes or at least survive as spores until they can recolonize their host. It is at this stage that they are most susceptible to unfavourable environmental influences and control measures, either biological or chemical, are often applied at this time.

The rhizosphere

In the region of the soil under the influence of plant roots, the rhizosphere, many of the types of interaction mentioned above occur. It is this region of the soil through which any soil organism influencing the plant via its root must pass or transmit its effect and through which all plant nutrients pass. It has therefore been extensively studied.

The extent of this zone depends on the activity of the plant and the type and state of the soil, but generally the influence of the root extends for only up to one or two millimetres. Its effect on the soil population is often expressed as the R/S ratio, that is, the number of organisms in rhizosphere soil as compared with the number in the same soil beyond the influence of the root. R/S values of over 100 are sometimes obtained (Table 11.5) and even higher values are known for some bacteria. Bacterial response to the rhizosphere conditions is greater than that of other groups, and sometimes where a high bacterial response is recorded the R/S ratio for protozoa is also large. Fungal numbers, estimated by the dilution plate count, increase only slightly in the rhizosphere although more than do those of algae which are usually unaffected since the algae live mainly on the soil surface.

The nutritional requirements of the rhizosphere flora and that of the soil flora in general are different. There is the greatest increase in the

Table 11.5 Ratios of the numbers of micro-organisms in the rhizosphere of wheat to those in control soil. From Rouatt, J. J., Katznelson, H. and Payne, T. M. B. (1960). *Proc. soil Sci. Soc. Amer.*, **24**, 271–273.

Micro-organisms	Nos./g dry weight		Significance of the difference	R:S ratio
	Rhizosphere soil	Control soil		
Bacteria	1200×10^7	5×10^7	**	240.0
Actinomycetes	46×10^6	7×10^6	**	6.6
Fungi	12×10^5	1×10^5	**	12.0
Protozoa	2.4×10^3	1×10^3	**	2.4
Algae	5×10^3	27×10^3	*	0.2

** = Significant at the 1% probability level.
 * = Significant at the 5% probability level.

number of bacteria requiring one or more amino acids.

The types favoured are Gram-negative rods belonging to the genera *Pseudomonas*, *Achromobacter*, and *Agrobacterium* and probably other members of several functional groups (e.g. the ammonifiers, denitrifiers, aerobic cellulose decomposers and the nitrifying bacteria *Nitrosomonas* and *Nitrobacter*, although it has been recorded that the latter group is inhibited as are the anaerobic cellulose decomposers and nitrogen fixers). *Bacillus* species decline somewhat in number although there is a qualitative change in the population of this group. The fungal population also changes qualitatively, though as with the bacteria the species recorded depend on the isolation technique used.

The plant influences the rhizosphere flora mainly by sloughing off dead cells from the growing root cap and older parts of the root and by exudation of organic compounds from the young part of the root. These compounds are released in substantial quantities and wide variety; for example, 21 amino acids have been found. In addition the root affects the microbial environment by raising the carbon dioxide level, decreasing the oxygen concentration and taking up nutrient ions and water.

The microflora influences the plant root in a number of ways. The possibility of a specific stimulation of *Rhizobium* by legume roots is considered on p. 285. There are also more general effects on root branching and on root-hair production. The physiology of the root may also be affected by the secretion of polypeptide membrane-active antibiotics which stimulate further leakage of plant cell contents into the soil. The micro-organisms also affect the availability of certain inorganic nutrients, particularly phosphate (p. 255), compete with the plant for nutrients and water, and affect the root by producing carbon dioxide and taking up oxygen.

The increase in size of the microbial population demonstrates effectively that as a whole the microflora benefits from the presence of plant roots in the soil. It is more difficult to assess the significance of the rhizosphere organisms for the plant in most instances. However, there is evidence that plants grown in sterile soil grow less well than those in soil inoculated with micro-organisms. Also, such plants are more susceptible to infection by re-introduced soil-borne plant pathogens.

Pesticides and micro-organisms

It has become obvious since biocides have been used on a large scale that many of these substances added to the soil remain in it for long periods without being degraded. While persistence may be advantageous from the point of view of pest control, the effects of these compounds, which if very resistant may accumulate significantly, on other organisms and ultimately on man give cause for serious concern. Most of the modern biocides are biodegradable.

One approach to the problem of resistance to degradation (i.e. recalcitrance) is to investigate the susceptibility of different compounds to decay and to study the effect that modification of their chemical structure has on the rate of the process. With variously substituted phenols and benzoates, it has been established that the type, position and number of the substituents affect the rate of degradation of the compound. For example, chlorophenols which are fungicidal and chlorobenzoates which are herbicidal are both more resistant than the corresponding unsubstituted phenols and benzoates. Para-substituted chlorines are more resistant than meta or ortho substitutions.

Another approach is to study the degradation of non-persistent pesticides, to discover the pathways and to characterize the enzymes involved in their breakdown. As a result of this approach a considerable amount is known about the decay of herbicides. Even the persistent compounds like DDT may be broken down eventually but the degradation may not be complete and the products (e.g. DDD) may be just as toxic and as resistant as the original chemical. It is fortunate that pesticides appear to have little if any lasting effect on the soil microflora. Even soil where populations have been severely depleted as a result of soil fungicide and fumigant treatment are eventually recolonized. Some of the more resistant insecticides such as aldrin and dieldrin are removed from the soil by volatilization, but this process is slow. The damage that biocide residues do must be balanced against the increased food production and disease control which their use permits, though careful monitoring of their effects on the environment is needed.

Microbiology of air

Introduction

The atmosphere of the earth contains many minute particles of solid matter, a large proportion of which are of biological origin. Most viable airborne particles are spores which are to some extent suited for survival in such an environment. Not only spores of fungi, myxomycetes, bryophytes, and pteridophytes, but also pollen grains, moss gemmae, propagules of lichens, cells of algae, vegetative cells and spores of bacteria, cysts of protozoa, and virus particles may occur in the air and constitute the 'air spora'. The largest of these particles are the pollens which may range up to 200 μm in diameter, though most are 20–50 μm. Most airborne fungal spores are between 3 and 30 μm in diameter, most bacterial cells about 1 μm in diameter, and viruses 0.1 μm. The interpretation of the spora in a given air mass requires an understanding of the probable movements that brought it there. It is therefore necessary to consider first some physical aspects of the atmosphere.

Air is in constant motion, the kinetic energy of this circulation deriving primarily from radiation from the sun. The amount of radiant energy received by the earth is balanced over a period of time by the amount radiated back into space, but in the course of the year the region between 40°N and 40°S receives more energy than it loses by radiation while the regions nearer to the poles lose more than they receive. The net transport of heat from the equatorial belt towards the poles is effected chiefly by winds, warm air flowing towards the poles and cold air towards the equator. The vertical transport of energy eventually to be radiated into space is augmented locally by convection currents, especially over land in intense sunshine. Because of the drag of the ground the various major currents of air generate a layer of turbulence which extends upwards for perhaps 500–1000 metres. However, in the immediate vicinity of the ground and other surfaces the air is virtually still or flows in an orderly fashion. This constitutes the laminar boundary layer which may be several metres thick on clear nights when the wind is light, but around a blade of grass on a windy day may be only about a millimetre in thickness.

Estimation of micro-organisms in the air

Study of the particles in the air requires first that they be trapped. The air spora may be trapped by filtration, centrifugation, electrostatic deposition, or by accelerating the air to a high speed so that airborne particles collide with suitably placed sticky surfaces or fluids. No single method of trapping is universally satisfactory because of the diversity of particle sizes. Small particles will tend to follow flow-lines of air as it travels around intercepting surfaces. Only if particles are large enough or travelling fast enough will they strike the surface. If the surface itself is large then only extremely large or fast-moving particles will be caught. Most techniques used nowadays sample the spores by drawing air at known rates through a suitable trap. Two widely used aspirator traps are the cascade impactor, in which sampled air passes through successive jets at different velocities so that particles of different size ranges are trapped at different stages, and the Hirst spore trap in which the particles impinge on a slowly moving target so that the numbers of particles captured from the air at different times can be recorded.

Methods used to collect samples from air in buildings are similar in principle (p. 261). It is possible to use animals directly as a means of assessing airborne contamination with animal pathogens but quantitative studies are possible only if a single organism can give rise to an initial lesion. The level of contamination of the air in a confined cubicle from a patient suffering from pulmonary tuberculosis has been determined by this means.

After the constituents of the air spora have been trapped by mechanical means it is generally necessary to identify them. Trapped particles may be identified visually under the microscope if they are sufficiently large and distinctive. Pollen, various spores, lichen propagules, algae, and protozoa may be identified in this way with more or less precision. A total count of bacteria may be obtained by direct inspection but their small size makes this particularly difficult. They can seldom be identified unless they possess specific staining reactions (as, for example, does *Mycobacterium tuberculosis*).

If components of the trapped spora are too small or too uncharacteristic to be identified or countable by visual means it is generally necessary to culture them on artificial media or appropriate host organisms or tissues. The use of fluorescent antibody techniques or specific bacteriophages can allow colonies to be identified at very early stages. Serological tests can also be applied. Viable counts may be obtained by inoculating culture media with known volumes or dilutions of a fluid used to scrub the organisms from the air. More generally, bacterial counts are made on 'settle plates' of solid culture media exposed directly to air, time being allowed for particles to settle. This technique can be made more sensitive and quantitative by the use of the 'slit-sampler' which draws a known volume of air through a narrow slit immediately above a rotating plate of nutrient medium, the plate making one complete rotation during the exposure.

A number of factors can greatly influence the counts. These include the choice of nutrient media and the temperature of incubation. No one medium will support growth of all the trapped organisms. The medium used is often selected to favour particular organisms, such as microbial pathogens. In any method involving the cultivation of trapped cells it is also necessary to recognize that the very process of trapping may kill some of them.

Airborne viruses and bacteriophages have been recovered from air in suitable fluids such as 10 per cent skim milk. Animal viruses are inoculated into living animal tissues including embryonated eggs and tissue cultures (p. 124). Bacteriophages are mixed with a suitable host culture in soft agar and poured on to the surface of a nutrient agar plate (p. 127).

Origin, distribution and movement of the air spora

Spores remain suspended in the air for as long as their fall speeds are less than the speeds of frequently recurring upward air currents. The terminal velocity of a falling particle in air is proportional to the square of its radius, and for a body $20\,\mu m$ in diameter it is about 1 cm/sec. Thus for most fungal spores it is 0.05–2.0 cm/sec and for most pollens 1–10 cm/sec. Convection and turbulence can generate upward air speeds considerably in excess of these velocities, but spores must first of all enter regions where this sort of movement is frequently encountered before they begin to rise. Most of the air spora derives from the surface of vegetation or vegetable debris above ground level. The laminar boundary layer which adjoins these sources is too tranquil for shed spores to remain long in suspension. However, many organisms have structures or mechanisms which introduce spores more or less directly into the turbulent zone. Under many circumstances it is necessary that the spore be raised only a few millimetres above the parent surface in order to do this. The elevation of spores of many fungi on erect conidiophores places them in situations where their chances of being swept up by gusts of wind are greater. Numerous organisms expel their spores forcibly for distances ranging from a few to many millimetres by active mechanisms in which the propulsive force is generated within the organism. Modifications towards this end are found among ferns, fungi, and bryophytes. It has recently been suggested that violent liberation of conidia of many fungi is an electrostatic mechanism, the acquisition of like charges leading to repulsion between the conidium and conidiophore. Viruses, bacteria, algae, and protozoa have no special mechanisms of take-off. Animal hosts by sneezing and coughing may assist the launching of some bacteria and viruses, but this is biologically significant only in restricted spaces (p. 261). Adventitious physical disturbances are probably the agents mainly responsible for the take-off of these various organisms.

Populations of fungal spores near ground level sometimes display a diurnal periodicity. Spores of shadow yeasts are most abundant in the air spora before dawn, spores of *Phytophthora infestans* late in the morning, and spores of *Cladosporium*, *Alternaria*, and *Ustilago* in the early afternoon. The major cause of this periodicity is probably diurnal periodicity of spore release. Release of the spores of many fungi is influenced by environmental factors among which light is often important. Also some release mechanisms depend on the activity of turgid cells, so it is likely that only when the environment is moist will release occur. It is for this reason that after rain there is often the appearance near ground level of a distinctive 'wet-air spora', rich in basidiospores, ballistospores of shadow yeasts, and ascospores. This replaces rather than augments the 'dry-air spora', which consists of pollen grains and the dry spores of *Cladosporium*, powdery mildews, *Alternaria*, smuts, and rusts, and which is largely washed out of the air by the same fall of rain that generates the wet-air spora.

Once a spore has taken off it will rise or fall in the air according to the relative influences of turbulence and gravity. Spores liberated in stable air soon settle to earth, but upward movement may be swift in unstable conditions. Upward air speeds of 30 cm/sec are common near the ground, while well above ground these may exceed 600 cm/sec. The vertical profile of spore concentration in many parts of the world generally shows a decrease in concentration with height. Spores may even be undetectable at heights of 2–6 kilometres. Nevertheless, sampling of very large volumes of air with balloon-borne equipment at much greater heights has yielded small numbers of viable cells of moulds, yeasts, and bacteria. Theory predicts that concentration will decrease logarithmically with height and actual observation sometimes show this relationship. Bacteria and fungal spores are the most numerous members of the air spores near ground level where an average concentration of spores in country air in summer is $10000/m^3$, but over short periods the concentration may greatly exceed this.

Two aspects need to be distinguished in considering the horizontal distribution of spores in the air: firstly the behaviour of a cloud of spores arising from a single restricted source and secondly the background concentration of spores from a myriad distant sources. The concentration of spores in the atmosphere decreases rapidly with distance from a source and with time from the moment of release. More than 90 per cent of spores from near-ground sources are often deposited within 100 metres. The progressive dilution of the spore cloud by upward migration of the remainder leads to spores from the local source becoming undetectable at near-ground level within a short distance. The spore cloud may be envisaged spreading downwind as a widening, gently ascending plume.

Despite the numbers of spores entering the atmosphere very few are ubiquitous and catches made at sea show rapidly decreasing numbers of micro-organisms in the air as distance from land increases. Few organisms other than marine bacteria are caught in mid-ocean.

Spores which continue their ascent may eventually rise into regions of high wind where they may remain suspended for several weeks. Under these conditions they are exposed to low temperatures, desiccation, and intense and harmful solar radiation before they eventually descend. Survival of long journeys of this sort is unlikely, though very low concentrations of viable fungal and bacterial cells have sometimes been detected at heights in the range 18–27 km.

The airborne flight of a spore draws to a close when various factors tend to accelerate the spore's downward velocity. Rain is an important factor in this process. Inside rain clouds the majority of particles in excess of $0.2 \mu m$ diameter become the nuclei of cloud droplets which aggregate and fall as rain. Falling rain drops range up to about 5 mm in diameter. Whether spores are captured by falling drops depends on whether the spore has sufficient inertia to resist displacement by air lying ahead of the falling drops. Spores which are larger than $2 \mu m$ diameter, as are most eukaryotic spores, are readily intercepted by raindrops. Raindrops 2 mm in diameter are reported to have the greatest collecting efficiency and the efficiency increases with the size of the spore. Another mechanism of deposition is sedimentation in association with boundary layer exchange; spores from a cloud of particles overhead diffuse into the boundary layer of air in which settling is mainly gravitational. Deposition of spores is also achieved by their impaction against solid objects. Impaction can be efficient for relatively large spores ($> 10 \mu m$ diameter) encountering small sticky objects at high speeds but is not efficient for small particles or for large obstacles.

Some consequences of the existence of an air spora

Airborne dispersal of spores out of doors has considerable biological and economic importance. Numerous plant diseases are caused by air-borne fungi (p. 293). The rapid spread of some diseases may be the result of many successive acts of dispersal and infection rather than long individual journeys, but long-distance dispersal is very important in the epidemiology of some. For example, uredospores of the wheat rust *Puccinia graminis* do not survive the cold winters in the northern parts of the north American wheat belt, nor the hot summers in the southern part. Spring-sown wheat in the northern USA and Canada is showered by rust uredospores blown northwards from rusted autumn-sown wheat in Mexico and Texas. Sometimes the spores travel large distances in a single hop, while at other times a succession of shorter hops is punctuated by phases of infection and multiplication. A reverse air movement later in the season carries spores south to infect the next autumn-sown wheat crop in the south. Migration of uredospores of cereal rusts have been reported also in India, Russia, and elsewhere.

Because most human and animal pathogens spread from host to host directly, being unable to multiply in the non-living environment, outdoor air is not a serious source of infection. The principal exceptions to this generalization are fungus diseases such as histoplasmosis and coccidioidomycosis, both of which are caused by organisms able to multiply in the soil. Many sorts of airborne biological particles, particularly pollen grains and spores of certain fungi, are, however, important causes of respiratory allergies in man (p. 332).

The microbiology of air inside buildings

Many organisms normally found in the outside atmosphere will also be found inside buildings where they are introduced by air currents but the main sources of contamination arise from within buildings from animal, human or vegetable sources. Since most airborne micro-organisms have no special structures to facilitate their dispersal into the atmosphere, they are dependent on physical disturbance for their 'take-off'. The microbial flora of the air within a building at any time will depend therefore both on the numbers and variety of organisms carried on the occupants and vegetable materials present and on the mechanical movements of animal or human activity within the enclosed space. Apart from movement of occupants, airflow in a building varies widely according to its shape, size and furnishings, and the design and manner of operation of heating and ventilation systems. Airflow is particularly pronounced near a concentrated heat source. Unless a room is adequately ventilated to dilute the concentrations of micro-organisms or otherwise treated (p. 337), the number in the airborne flora increase with time.

Dust as a vehicle of airborne contamination may arise from textiles (especially bedding, handkerchiefs, and clothing contaminated from contact with man) and desquamated skin scales and hairs which are being continuously shed. In buildings housing animals, the air pollution is particularly high owing to the presence of hay, straw or other fodder, bedding and dried excreta, and contamination from the animals' coats. The denser particles settle rapidly but particles of 1 μm or less in diameter remain suspended permanently. The dry-sweeping of floors, the dusting of objects, shaking of cloths, making of beds, movement of people, and draughts can break up the original substrates into finely divided particles or disturb settled dust and cause it to become airborne.

The other important group of micro-organisms found in the air flora comprises those ejected from the respiratory tract in droplets of moisture. During talking, coughing and sneezing air is forced, under considerable pressure, through the nose and the constricted apertures between the teeth. Organisms from the upper respiratory tract are thus ejected into the air. Fluid picked up by this expelled air is broken down into droplets. Sneezing is the most vigorous of these mechanisms and a single sneeze may generate as many as one million droplets less than 0.1 μm in diameter and thousands of larger drops, mainly from the saliva at the front of the mouth. Not all expelled droplets become airborne; the larger ones exceeding 100–200 μm in diameter fall to the ground before evaporation but the smaller ones rapidly evaporate down to their non-volatile residuum and remain suspended as 'droplet-nuclei'.

The survival of airborne micro-organisms will depend on many factors. Spores which can tolerate dehydration will survive longer than vegetative cells and the length of survival of all micro-organisms is increased in humid atmospheres away from the bactericidal rays of sunlight which penetrate glass windows to only a limited extent (p. 337). The presence of organic material, as from saliva, skin, etc. gives added protection to many organisms.

The microbial content of the air inside buildings may include viruses, pathogenic and non-pathogenic bacteria, and fungi. Endospores of the genera *Bacillus* and *Clostridium*, especially *Cl. perfringens*, are commonly found in occupied rooms, hospital wards, and even operating theatres. *Rhodotorula* and other yeasts, and spores of species of *Aspergillus*, *Penicillium*, *Mucor* and other moulds are commonly present. These may contaminate food and moist, perishable organic materials, such as leather, and by inhalation may cause respiratory infections of man and animals and allergic reactions, such as asthma.

Many viruses, including those of influenza, the common cold, virus pneumonia, measles, German measles, and smallpox are dispersed by air and may produce infection when inhaled. Airborne bacterial pathogens include the organisms causing scarlet fever and tonsillitis (*Streptococcus pyogenes*), tuberculosis (*Mycobacterium tuberculosis*), diphtheria (*Corynebacterium diphtheriae*), whooping cough (*Bordetella pertussis*) and Q-fever (*Rickettsia burnetii*). All but the last of these have been demonstrated in the dust of fever hospital wards. Various occupational diseases result from the inhalation of contaminated dust associated with particular occupations. Thus 'woolsorter's disease' was caused by the inhalation of dust particles contaminated with spores of the anthrax bacillus from wool imported from parts of the world where anthrax is common; this is now controlled by the prior treatment of the wool with formalin. The development of intensive methods of husbandry of calves, pigs and poultry has resulted in an increase in the incidence of respiratory diseases and special care is necessary to maintain an adequate airflow through the buildings to limit this animal health hazard.

Certain fungi with airborne spores are responsible for pulmonary disease. These may be derived from sources outside buildings, such as mouldy grain, straw, damp vegetation, and compost heaps, some of which may be brought inside farm buildings as bedding or fodder. Outbreaks of aspergillosis in pigs, for instance, have been traced to consignments of bedding straw contaminated with *Aspergillus fumigatus*, and penguins, when exposed to the stress of transport, frequently become infected from contaminated bedding and develop aspergillosis when kept in captivity. The spores of *A. fumigatus* have been found in 80 per cent or more of samples of dust examined from city houses. This fungus can produce one or other of several forms of disease in man. It is able to grow in the mucoid secretions of bronchi, without invading the tissues, producing a hypersensitive state (p. 000), the patient demonstrating an allergic response either to the fungus already in his respiratory system or when spores of the same species are inhaled later. Aspergilloma is an X-ray-detectable solid lesion resulting from a saprophytic colonization by *A. fumigatus* of an old lung cavity such as may be caused by a healed tuberculous cavity; the lesion can be surgically removed. In the invasive type of aspergillosis the fungus invaded lung or other tissues, often secondarily to any grave systemic disease, the fungus probably contributing to death of the patient.

Spores of other fungi, including *Cladosporium herbarum*, cause allergies in man. The serious disease of agricultural workers known as 'farmer's lung' is, however, caused by the development of hypersensitivity to spores of thermophilic actinomycetes (p. 171) derived from mouldy hay and not to spores of *Cladosporium* and other fungi even though these are also present in the same hay. Ultimately the inhalation of only a limited number of actinomycete spores may cause severe pulmonary symptoms.

Until recently many microbial agents responsible for ripening of cheese gained entry into the curd from air of the dairy. Hence certain localities became famous for particular types of cheeses owing to the presence of specific organisms (e.g. species of *Penicillium* giving their characteristic flavour to gorgonzola, stilton, roquefort or camembert cheese). Nowadays cheese is largely produced on a factory scale and the ripening processes are initiated by the introduction of pure cultures as 'starters' (p. 349). Bacteriophages, acting against bacterial starters, may be also present in the air of cheese factories where they present an important problem.

Microbiology of water

The aquatic environment

Seventy per cent of the global surface is covered by water in which micro-organisms occupy almost every niche and are intimately linked with the biological processes of decay and production. Rain, snow, or hail remove large numbers of micro-organisms from the air. Hence water reaching the earth by precipitation is not sterile. Over most of the earth's surface rain-water collects micro-organisms from the surfaces of plants, buildings, or soil on which it falls. Large numbers are acquired from the soil and pass out with the drainage water into streams and rivers and thence into fresh-water lakes and eventually into the ocean or inland seas. Water from natural springs or artesian wells is usually relatively free of organisms owing to the filtering effect of natural percolation through the surface rocks. The number and type of micro-organisms in surface water varies according to the source of the water, its organic and inorganic content and with geographical, biological, and climatic factors.

The organisms inhabiting large volumes of water are insulated from the extremes of climate found on land and so are rarely frozen or exposed to harmful high temperature, desiccation, or sudden changes in concentration or chemical composition of the water. Organisms inhabiting the soil water or small volumes of water such as pools, artificial ponds or ditches, are exposed to greater extremes and are liable to periods of desiccation. Only those species able to form resting stages are likely to survive in such habitats. In running water the effect of the current is important.

Most naturally occurring water contains nutrients and other substances in solution, together with colloidal and particulate matter. Pure water probably never occurs in nature but the concentration of dissolved substances varies from negligible amounts in some upland waters to progressively higher concentrations in fresh water which has collected by drainage of agricultural and industrial areas, brackish waters of estuaries, the saline waters of the oceans and the extremely saline waters of some inland lakes in regions of excessive evaporation, e.g. the Dead Sea. The most concentrated of these is still a relatively dilute solution, but the range is sufficient to exert a selective effect. Some organisms inhabiting tidal estuaries are able to tolerate a wide range of salinity but few, if any, extend over the whole range. In fresh water the concentration of nutrients does not determine the species present but influences the 'productivity' or number of individuals developing. Productive waters are termed eutrophic, unproductive waters, oligotrophic.

Aquatic micro-organisms

The algae and Cyanobacteria: primary producers of organic matter

The photosynthetic algae are the main producers of organic matter in aquatic habitats although they occupy the water column only down to the point where photosynthetically available light penetrates.

This may be to the bottom of a shallow lake or to the sea-bed near the coast but is only a relatively thin 'veneer' of water in the open ocean.

Factors affecting the growth of algae and methods of study are similar in both marine and freshwater habitats. The organisms in the two environments are, however, almost entirely distinct, few if any species being common to both. The same genus may occur, but not the same species, e.g. *Asterionella formosa* (Fig. 11.6, E1) is a common freshwater form and *A. japonica* (Fig. 11.6, F1) is the marine counterpart. There are two major spheres in the aquatic environment; the benthic, which encompasses a vast range of habitats all associated with solid/liquid interfaces and therefore usually with the bottom, e.g. the surfaces of rocks, silt, plants, animals, etc. (Fig. 11.4), and the planktonic, which encompasses all the niches within the water mass itself and within which the organisms must either float or swim.

The algal flora tends to be quite discrete in each habitat. In the benthos two distinct life forms are found, one attached and non-motile, the other unattached and capable of horizontal and vertical movements. The attached species (Fig. 11.6a, b) grow on rock or stone surfaces (epilithic) either as single cells or in groups, and include many diatoms (e.g. *Synedra*, *Meridion*, *Licmophora*), the blue-green *Chamaesiphon*, creeping filaments or semi-circular mucilage colonies (e.g. *Hildenbrandia*, *Rivularia*, *Chaetophora*), or filamentous outgrowths attached by basal holdfast cells (e.g. *Cladophora*, *Sphacelaria* and many small Rhodophytes). Many of these are firmly 'glued' to the substrate, some even penetrating it (e.g. the lime-boring species of Cyanobacteria). Some produce mucilage in which they are embedded whilst others precipitate calcium carbonate around themselves and build up nodular growths some of which become fused with the rock. 'Beach rock' is formed in a similar manner in many tropical regions by the cementing of sand grains and calcium carbonate amongst the algal cells. Very much smaller algal cells are attached amongst the coating of bacteria on both fresh-water and marine sand grains (epipsammic flora, Fig. 11.6a, b); the majority belong to the Bacillariophyta, though some Cyanophyta and Chlorophyta are also reported. A dense flora develops in some situations; the cells are usually attached by mucilage pads or short mucilage stalks and tend to be so small that they are protected by the slight concavities of the grains. A similar attached flora occurs on the undersurface of sea ice and even penetrates into the interstitial water.

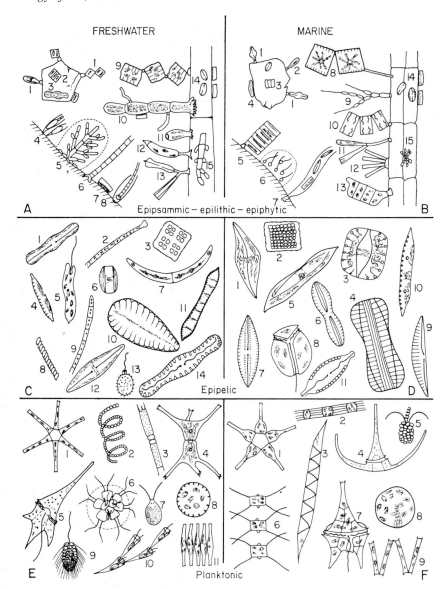

Fig. 11.6 Examples of the algae occurring in the various freshwater and marine associations. (**A, B**) Freshwater and marine epipsammic, epilithic, and epiphytic.

(**A**) 1. *Opephora*. 2. *Fragilaria*. 3. *Achnanthes* on sand grains. 4. *Chamaesiphon*. 5. *Chaetophora*. 6. *Ulothrix*. 7. *Cocconeis*. 8. *Calothrix* on rock. 9. *Tabellaria*. 10. *Oedogonium* (also with *Achnanthes*, *Cocconeis*, and *Eunotia*). 11. *Characium*. 12. *Ophiocytium*. 13. *Gomphonema*. 14. *Cocconeis*. 15. *Aphanochaete* on plant material.

(**B**) 1. *Raphoneis*. 2. *Opephora*. 3. *Fragilaria*. 4. *Amphora* on sand grains. 5. *Fragilaria*. 6. *Rivularia*. 7. *Navicula* (in mucilage tube) on rock. 8. *Striatella*. 9. *Acrochaetium*. 10. *Grammatophora*. 11. *Licmophora*. 12. *Synedra*. 13. *Isthmea*. 14. *Cocconeis*. 15. Filamentous Rhodophyte on plant material.

(**C**) Freshwater—epipelic. 1. *Caloneis*. 2. *Oscillatoria*. 3. *Merismopedia*. 4. *Nitzschia*. 5. *Euglena*. 6. *Amphora*.

7. *Closterium*. 8. *Spirulina*. 9. *Phormidium*. 10. *Surirelia*. 11. *Cymatopleura*. 12. *Navicula*. 13. *Trachelomonas*. 14. *Pinnularia*.

(**D**) Marine—epipelic. 1. *Gyrosigma*. 2. Holopedia. 3. *Campylodiscus*. 4. *Amphiprora*. 5. *Tropidoneis*. 6. *Diploneis*. 7. *Navicula*. 8. *Amphidinium*. 9. *Amphora*. 10. *Nitzschia*. 11. *Mastogloia*.

(**E**) Plankton—freshwater. 1. *Asterioenall*. 2. *Anabaena*. 3. *Melosira*. 4. *Staurastrum*. 5. *Ceratium*. 6. *Pandorina*. 7. *Chlamydomonas*. 8. *Cyclotella*. 9. *Mallomonas*. 10. *Dinobryon*. 11. *Fragilaria*.

(**F**) Plankton—marine. 1. *Asterionella*. 2. *Sceletonema*. 3. *Rhizosolenia*. 4. *Ceratium*. 5. A coccolithophorid. 6. *Chaetoceros*. 7. *Peridinium*. 8. *Coscinodiscus*. 9. *Thalassionema*.

N.B. The various genera are drawn diagrammatically and are *not* to scale.

The second large group of attached species occurs on larger plants (epiphytic, Fig. 11.6a, b) and on animals (epizoic). Many species occur in only one or the other habitat. The same morphological types are present as in the lithophilic associations but there is a greater degree of host specificity, e.g. on some large marine algae a single diatom species such as *Isthmia* or *Licmophora*, is dominant. This is often the situation also when small Crustacea bear epizooic algal populations. On many fresh-water plants a seasonal succession of epiphytes occurs. The relationship between the growth of the host, the seasonally changing environments, and the epiphytes is not yet clear. The metabolic activities of the host affect the epiphyte populations by the excretion of nutrients and metabolic products. Owing to the intensive concentration of the epiphytic biomass at a convenient surface it forms one of the most important 'grazing grounds' for protozoa, etc. and this flora is grazed often almost exclusively by the vegetarian fish, particularly of tropical waters. Amongst the primarily attached forms of the true epiphyton occur numerous other unattached organisms; algae such as desmids, coccoid and flagellate green algae, motile diatoms, etc. collectively termed metaphyton. It is also extremely rich in bacteria, fungi, and protozoa.

Wherever sediments are deposited in water, shallow enough to receive adequate radiation, a rich motile algal flora grows (epipelic, Fig. 11.6c, d). For life in this habitat the species must be able to maintain themselves on the surface, either because they are motile (e.g. flagellates, diatoms, and Cyanobacteria), or by forming flocculent masses which sediment more slowly than do the inorganic particles. These algae frequently form growths obvious to the naked eye and some mat together and rise to the surface buoyed up by the bubbles of oxygen formed during photosynthesis. This habitat abounds in species of diatoms which have raphe systems on the two valves and many of these display an innate diurnal rhythm of motility. The motility rhythm is even more pronounced where this association occurs on intertidal sediments. In this latter habitat the algae (diatoms and *Euglena* species) move beneath the silt at sunset and move up on to the surface during the early morning. Tidal cover during the daylight period induces a downward movement *before* the tide actually reaches the algae; this rhythm is one which can be demonstrated under constant laboratory conditions, showing that it is under cellular control and is not impressed by the environment.

The same diatom genera are often present on both fresh-water and marine sediments but the species are different, e.g. *Caloneis, Diploneis, Navicula, Pinnularia, Amphora, Cymatopleura, Campylodiscus, Nitzschia*, and *Surrirella* are all common, but whereas *Diploneis* and *Amphora* are represented in the fresh-water epipelon by four or five species there are up to 100 or more in the marine epipelon. Cyanobacteria (e.g. *Aphanothece, Merismopedia, Anabaena, Oscillatoria*) and desmids (e.g. *Closterium*) are more common in fresh water than in the sea, where the other components tend to be dinoflagellates and occasional Euglenoids.

The plankton does not support such a variety of life forms since the habitat is isotropic. One small variant is formed by the algae, bacteria, and protozoa which live at the water/air interface (neustonic). The only common motile forms are flagellates some of which are capable of vertical migration especially under relatively calm conditions. The other components tend to sink slowly and are kept in suspension only by water movement, e.g. the bulk of the diatoms (*Asterionella, Fragillaria, Cyclotella, Stephanodiscus*, and *Melosira* in fresh water; *Asterionella, Chaetoceros, Rhizosolenia, Bacteriastrium, Guinardia, Leptocylindrus*, and colonial *Nitzchia* species in the marine habitat). A few species have buoyancy mechanisms, e.g. gas vacuoles in Cyanobacteria (*Microcystis, Anabaena, Aphanizomenon*), oil globules (*Botryococcus*) and unknown mechanisms in forms such as *Halosphaera*. Even the forms without apparent buoyancy have some subtle mechanism which aids their maintenance in the plankton since death leads to rapid sedimentation. Some species by an unknown mechanism maintain themselves at given depths, e.g. some dinoflagellates are 'shade plants' only found in the subsurface ocean water, others are 'sun plants' and predominate at the surface. All these are large forms easily collected by drawing fine silk nets through the water. However, between these large particles and passing through most collecting nets are a whole assemblage of minute organisms (less than 5 μm in diameter) often collectively called nannoplankton with the smaller species sometimes separated off as μ-plankton. These are algal and bacterial components and they may be very numerous, although in fresh water their total biomass is probably less than that of the larger species. In the oceans many of these forms are flagellates (e.g. *Chrysochromulina*) and many are extremely delicate and difficult to collect and preserve. Even less is known of their distribution and biology than for the larger forms, since techniques involving direct observations are not available and plating techniques are a selective procedure. In nature the species comprising the plankton are probably never randomly dispersed either horizontally or vertically. Swarms of species occur in a complex interaction between growth rates, grazing, sexual reproduction, nutrient depletion, etc. In large lakes and the oceans even more extensive patches occur in the various current systems. Grazing by animals is not as general as commonly supposed whilst epiphytism of one alga on another, of bacteria and protozoa on algae, or parasitic fungi and protozoa on and in algae are all common.

Three interrelated major parameters affect the whole aquatic microbiological system, viz., the radiation flux reaching the sea or lake surface, the thermal properties of the water, and the nutrient flux within the water. Primary fixation of carbon by algae is dependent on the radiation flux and this varies from a very high figure at O° latitude, where two minima and two maxima occur every twelve months, to the single maximum towards the poles where, for a short summer period, the daily flux exceeds that of the tropics but where at all other times light is absent or severely limiting to algal growth. Most algae grow intermittently as do annual land plants and the commencement of growth may be related to the light climate. Radiation is absorbed and its spectral composition modified as it penetrates water, thus even in pure water (which does not exist in nature) some 2 per cent of blue light and 4 per cent of the yellow is absorbed in one metre. Since particulate matter, including the suspended algae, bacteria, etc. reduces the amount transmitted, these figures are approached only in the purest oceanic water whilst in coastal water 40 per cent or more may be absorbed in the first metre. The effect of this light absorption can be readily shown if the algae living on the sediments from the shore are studied down a depth profile. These can be estimated either as cell numbers, cell volume, or by estimation of some cell

component (e.g. chlorophyll α). In a relatively clear lake (e.g. Lake Windermere) this population has virtually disappeared between 6 and 8 m, but in a very small lake with turbid water (e.g. Abbots Pond near Bristol) a comparable population exists down to only 2 m. Measurement of a planktonic population through all depths and hence light limitation of growth is better determined by inoculating plankton into stoppered bottles which can be hung at each depth. 'Growth' may be measured over short intervals by measuring photosynthesis (Fig. 11.7) either by the change in oxygen content in the bottles or by uptake of ^{14}C supplied in a soluble form. However, as Fig. 11.7 shows, the highly illuminated surface water often has a lower rate of carbon fixation than has the subsurface; this is related to inhibition of carbon fixation by high light intensity. In general the red end of the spectrum is absorbed rapidly and the blue penetrates to the greatest depths. Thus algae at various depths are exposed to light of different wavelengths as well as to different amounts.

The second major effect of radiation entering the water is its heating effect, which if the water were pure and undisturbed would show an exponential fall with depth. However, owing to cooling of the surface by evaporation and movement via wind-induced currents, a very characteristic pattern de-

Fig. 11.7 A diagrammatic section from the shore out into deep water in either a freshwater lake or in the sea. The distribution of the algal associations is shown and three graphs illustrating the decreasing illumination (I) with depth (surface illumination 100 per cent) ; winter temperature in degrees centigrade (Θ) (dotted), summer temperature (solid line), position of the thermocline (T), photosynthesis (P), without surface inhibition (solid line) and with inhibition (broken line) ; maximum photosynthesis represented as 100 per cent. On the temperature diagram the plankton is shown distributed down the complete profile during winter and confined to the water above the thermocline in summer. HW—high water, LW—low water.

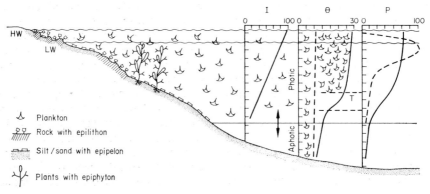

velops for all except very shallow waters. The radiation heats the surface waters and this is transmitted to lower depths by wind-induced currents. When heating becomes excessive in early spring the resultant reduction in density of the surface heated water makes it difficult to mix this water into the denser cold layer below (rather like pouring water on to syrup and blowing on the surface to mix the two liquids). The layer of warm water then rests all summer on the cold lower layer and the region of rapid temperature drop between the two is known as the thermocline (Fig. 11.7); the water mass is said to be thermally stratified and active circulation is then confined to the upper layer. In the autumn, surface cooling and increased wind stress eventually break down the stratification, sometimes within 24 hours, and the water then circulates to a greater depth, which may be to the bottom in many lakes or through 100 m or more in the open ocean. Thus during the most unfavourable time of the year the algae are circulating throughout a greater depth and therefore spend a longer period in the region where photosynthesis is impossible. In the summer, they are trapped in the warm surface water which is also well illuminated (Fig. 11.7).

This stratification has two major effects on the nutrient supply to the micro-organisms, firstly absorption of nutrients is excessive in the surface water above the thermocline and nutrient concentrations may thus be reduced to limiting amounts. Replenishment is slow, since transport across the thermocline is hampered by relative lack of water movements. Secondly, in the lower water, oxygen is depleted by respiration of all the inhabitants, both micro- and macro-organisms, and is neither replaced by downward movement, since the thermocline is a barrier, nor by photosynthesis, since it is below the limit of light penetration. Various degrees of de-oxygenation result, accompanied by a steady build-up of nutrients in this region. In lakes these tend to diffuse out of the sediments when the surface changes from an oxidized to a reduced state. Thus the surface few metres of fresh-water lakes and the upper 100 m or so of the ocean may be rich in organisms but depleted of nutrients whilst below this depth is a rich reservoir of nutrients unavailable to the surface crop until mixed by some agency which breaks down the thermocline. In the sea this nutrient-rich water upwells in some coastal areas (e.g. off the Peruvian coast) and this results in tremendous algal productivity. The breakdown of thermal stratification in fresh water allows the deep nutrient-rich water to mix with the surface water, but this usually

occurs in autumn by which time light intensity is often too low to allow much growth before the spring.

The factors affecting the growth of aquatic micro-organisms in nature have been studied for only a few species. Nutrient depletion has been shown to be responsible for the decline of the growth rate of such diatoms as *Asterionella* and *Stephanodiscus*. Diatoms have an absolute requirement for silica and this substance is supplied only slowly, via inflows or by re-cycling from the lower waters, and is easily exhausted. This effect is pronounced on organisms in the plankton but less obvious amongst the benthic forms which are living close to a supply of silica in the sediments or possibly diffusing from some plant stems. Other algae which require carbon and not silica for their cell walls are unlikely to be limited by lack of wall material since carbon is continuously supplied via solution of carbon dioxide and the interaction of pH, carbonates, and bicarbonates. Instead nutrients such as nitrogen and phosphorus become limiting to many algae especially in midsummer when supply from the lower waters is limited. There is other evidence, however, to show that some algae, such as *Dinobryon*, grow in lakes when the concentration of the major ions has been reduced and this is coupled with further experimental evidence that high phosphate concentrations are deleterious.

The onset of growth of vernal species is not related to a build-up of nutrients, since these have been in high concentration throughout the winter, but is triggered by increase in radiation flux. Temperature is still very low and although it does affect the growth rate it is not the prime factor. Other species (e.g. *Oscillatoria rubescens* and *Melosira italica*) behave as winter or cold-water shade plants. Both these species tend to disappear at least from the surface waters of lakes in spring. *M. italica* has been shown to remain alive on the sediments for several years and its sudden appearance in the plankton in the autumn is due to mixing of this material upwards by autumn gales. It is a heavy species and if ice cover occurs it sinks owing to the reduction of wind-induced turbulence. In spring, as the temperature increases and stratification begins, it sinks out of the surface water.

Since all species in the aquatic environment have patterns of seasonal growth it is obvious that the problems are complex, and, in addition, there are well authenticated examples of biological interactions which influence the primary fluctuations correlated with the cycle of physico-chemical variables. Indirect effects of the production of antibiotics have been demonstrated in culture, but although these are undoubtedly important, it is extremely difficult in the natural habitat to separate effects due to antibiotics from those of other factors. Much more drastic, however, are the effects of parasitism, e.g. of the chytrid *Rhizophidium* (p. 196) on planktonic diatoms. Such parasites occur on almost all fresh water planktonic species and recently a whole series of minute animal parasites have also been found in or on planktonic algae. So far few have been detected amongst benthic organisms but this may be merely due to lack of detailed study. Recently algal viruses have been reported (p. 131) and these too may have a profound effect on algal populations.

An imperfectly understood aspect of the chemical composition of the environment is its effect on the species composition of all the algal associations. No two water masses are alike in their species complement but certain broad features can be discerned particularly where there is a single over-riding factor (e.g. saline lakes tend to be dominated by certain Cyanobacteria, such as *Spirulina*; very acid, low-calcium lakes by desmids and Chrysophyceae; organically rich waters by Euglenoids). However, almost all algae can be cultured in defined media containing the common elements plus the vitamins thiamin, cyanocobalamin and biotin, all of which are generally present in natural waters. Although an obvious simplification, it is clear that certain species are abundant in low-calcium/acidic waters (e.g. the diatoms, *Frustulia*, *Stenopterobia*, *Actinella*) whilst others occur only in higher calcium/alkaline waters (e.g. *Gyrosigma*, *Cymatopleura*, *Epithemia*). In the marine environment, however, there is no such easy distinction, since the waters are all alkaline and salinity varies only slightly, at least in the open oceans which have a similar flora throughout the world, though ecological races occur related to temperature and salinity. Brackish waters of varying degrees are an exception and have characteristic floras. Further study may also reveal differences in the benthic flora.

Heterotrophic micro-organisms

Heterotrophic micro-organisms are usually present in natural waters in direct proportion to the amount of organic matter available as food. Bacteria, aquatic fungi, and protozoa are associated with the various communities of algae described above. By their activities they break down organic compounds and thus return plant nutrients to the water.

Bacteria are found wherever sufficient organic matter is present and vary in number with the amount of such food material available. Bacteria are numerous around submerged vegetation and in and just above the mud layer at the bottom of both fresh water and the sea. In the absence of algae, competition for oxygen is intense on the bottom sediments and oxygen becomes limiting for aerobic species. Anaerobic bacteria and a few fungi with an unusually low oxygen requirement then predominate.

The number of bacteria present in freshwater plankton is directly correlated with the amount of organic matter present and hence is greatest in eutrophic waters. In waters relatively unpolluted by intestinal organisms, both pigmented and non-pigmented bacteria occur, including species of *Pseudomonas*, *Chromobacterium*, *Achromobacter*, *Flavobacterium*, and *Micrococcus*. Organisms washed into water from soil are mainly spore-bearing bacilli, including species of *Bacillus* (aerobic) and *Clostridium* (anaerobic), but also include members of the coli–aerogenes group (Eschericheae) normally found on plants and decaying vegetation. Nitrifying bacteria and species of *Streptomyces* are frequently present. In water polluted with animal or human excreta many organisms characteristic of the intestinal canal survive for considerable lengths of time and include *Escherichia coli*, *Streptococcus faecalis*, *Proteus vulgaris*, and *Clostridium perfringens*. Occasionally intestinal pathogens, such as *Salmonella typhi*, and, in certain parts of the world, *Vibrio cholerae* and other water-borne pathogens may be found (p. 333).

Bacteria, including species of *Pseudomonas*, *Vibrio*, *Flavobacterium*, and *Achromobacter*, are present in the surface layers of the sea but their numbers differ widely with conditions. They are most numerous where the organic content of the water is high, that is near coasts, particularly where land drainage occurs, in harbours and in areas of highly productive plankton. Various counts show decreases from more than 1 000 000 per ml in harbours and around 400 000 per ml near the shore to as low as 10–250 per ml a few miles offshore. Among the seaweeds of the Sargasso Sea the count rises to more than 1000 per ml. A large proportion of marine bacteria are pigmented and thus protected from the sun's rays. Some, such as the purple surphur bacteria, may be sufficiently numerous to impart a red tinge to the water. Many are phosphorescent. Many yeasts have been recorded and strains of *Streptomyces*, *Micromonospora* and *Actinoplanes* have been demonstrated in fresh and salt water. The activity of actinomycetes in reservoirs may impart unpleasant odours to water supplies. Marine bacteria tend to be smaller than freshwater species, but sulphur bacteria and the spirochaetes are exceptions. All are salt-tolerant (halophilic) and most are relatively intolerant of high temperatures. Marine bacteria play similar parts in the nitrogen, sulphur, phosphorus, and carbon cycles in the sea to those played by other species in the soil (p. 248). Some bacteria may be troublesome by eroding metal (e.g. ships' plates) immersed in sea water.

Where oxygen is available, as in moving water, or in epiphytic algal communities, a wide range of aquatic fungi is found. Reference has already been made (p. 268) to the chytrids parasitic on algae. Other species are parasites on other aquatic fungi or on protozoa. Many are saprophytic on organic matter. Stems of such fresh-water aquatic plants as *Phragmites* and species of *Scirpus* are parasitized by aquatic Ascomycetes, including many Discomycetes. A few brown seaweeds are parasitized by species of *Mycosphaerella*, some Pyrenomycetes and some Fungi Imperfecti. Pyrenomycetes (e.g. species of *Ceriosporopsis*) also attack and erode wood submerged in the sea. Decaying leaves of deciduous trees in well-aerated streams and lakes support a characteristic flora of aquatic Hyphomycetes. These are taxonomically unrelated forms and although most of them resemble one another in the production of conidia with projecting arms or spines or consisting of a curved or branched row of cells, these superficially similar spores arise by quite different processes. Their shape, however, is well adapted to their habitat and they readily become entangled in submerged leaves which they colonize after germination. Many of the aquatic Ascomycetes have ascospores of a similar shape to the conidia of the aquatic Hyphomycetes. *Saprolegnia* (p. 196) and related species, and *Mono blepharis* (found on submerged sticks) are also limited to well-aerated water. These produce motile spores (zoospores) and resting spores (oospores) which are non-motile and relatively thick-walled.

Where oxygen is scarce, as in deposits of leaves or other organic matter at the bottom of stagnant ponds or where excessive growth of bacteria has depleted the supply, the fungus flora is quite different from that of similar substrata in well-aerated waters. Certain Hyphomycetes, such as species of *Clathrosphaerina* and *Helicodendron*, which have distinctive coiled spores may be present as hyphae but usually produce their conidia only when removed from the water and placed in humid air. They have been described as 'aero-aquatic'

fungi. Members of the Blastocladiales may also be present. It has been shown that *Blastocladia pringsheimiana* is almost anaerobic and that it produces its resting spores only under conditions of high carbon dioxide concentration.

Most water moulds of the family Saprolegniaceae are less common in strongly eutrophic waters than in rather less productive ones, but the related *Leptomitus lacteus* (the so-called sewage fungus), increases with the organic content of the water and is particularly common in water polluted by sewage or organic industrial effluents, where it may become a serious nuisance by actual mechanical blocking of channels and by exhaustion of the available oxygen supply.

Aquatic protozoa are common in both fresh and salt waters. Planktonic protozoa (ciliates, flagellates, and the Heliozoa or 'sun animalcules' which possess radiating strands of protoplasm) and other animals, particularly the grazing species, also fluctuate more or less with the phytoplankton. They are numerous in marine plankton, and include flagellates, foraminifers (members of which sink and build up thick calcareous deposits on the ocean floor, e.g. the globigerina ooze), radiolaria (which similarly contribute to siliceous deposits, e.g. radiolarian ooze) and ciliates. Other small animals and larval forms are also present. Grazing of the phytoplankton by these microscopic animals and by the young fishes may be a factor in the fluctuation of the algae and even be a contributory cause of the frequent disappearance of phytoplankton under conditions of good nutrition.

Micro-organisms and the problems of water pollution, sewage disposal, and water supply

Pollution of natural waters

The increasing use of fertilizers and the disposal of wastes in streams is resulting in rapidly increasing nutrient content (eutrophication) in many natural water bodies (e.g. Lake Erie), and enormous algal crops result, followed by their decay giving massive pollution of shores. In spite of considerable study few algicides have proved effective and non-toxic to other organisms. Sodium pentachlorophenate has been used but it is toxic to fish. Quinone derivatives, e.g. 2,3-dichloronaphthoquinone (dichlone), are more effective and apparently less harmful. Similar problems of excessive algal growth arise in industrial plants, e.g. in power-station cooling towers, and ponds. Here the above chemicals, quaternary ammonium compounds (p. 103) and tetrachlorobenzoquinone (chloronil) have been successfully used. The potent fish-killing alga *Pyrmnesium parvum* has been controlled in Israeli fish ponds by treatment with ammonium sulphate. The degree of pollution of both fresh and marine waters can be determined by studying the algal growth. Benthic algae are very sensitive indicators and in the most extreme polluted zones only algae such as *Oscillatoria chlorina* and *Spirulina jenneri* occur and in the slightly less polluted zones the diatoms *Nitzschia palea* and *Gamphonema parvulum* appear. Pollution often reduces the diversity of the algal flora but within limits increases enormously the growth of the resistant species, owing to the removal of competitors and predators. This is particularly noticeable in marine habitats where removal of molluscs, etc. results in intense growth of diatoms and allows the sporelings of green algae such as Enteromorpha to become established. A further deleterious effect of algae is their growth on all objects placed in water or which conduct water for industrial processes. Ships are particularly prone to colonization by algae and in spite of the use of algicidal paints there is still no completely satisfactory solution to this expensive problem.

The recent growth in number of nuclear power stations, from which a large amount of excess heat has to be dissipated, is giving cause for concern over 'heat pollution' of rivers and estuaries. If the waters of these are used for cooling not only is there direct danger to fish, but interference with the numbers and types of micro-organisms may disrupt the food chain or favour harmful species.

Sewage disposal

In highly industrialized countries such as Britain where large communities have developed, the disposal of industrial and domestic waste presents an enormous problem. The simplest means of sewage disposal, still practised by coastal and estuarine towns, is to run the raw sewage into the sea or nearby river. The larger inland communities have been forced, by lack of convenient dumping grounds, to develop sewage purification systems whose end-products are mainly pure water and some harmless solids.

One of the important means of observing the quality of a sewage is by the measurement of its biochemical oxygen demand (BOD). This is defined as the weight of dissolved oxygen (usually in mg) required by a definite volume of liquid (usually 1 litre) during five days' incubation at 65°F. Nowadays, 20°C is recognized as the international standard temperature for incubation. The aim of purifying sewage is to reduce the BOD so that if effluent waters are run into a stream or river, the indigenous flora and fauna will not die from lack of oxygen.

Sewage consists mainly of water containing organic and inorganic dissolved and suspended substances along with many micro-organisms. When sewage is received at a processing plant, there is a preliminary screening to remove solid matter. Grit and stones are allowed to settle out from a slow-flowing stream of sewage. There is a regular removal of putrefactive sludge to prevent the onset of anaerobic digestion. The remaining material which is removed after varying periods of sedimentation (usually 15 hours in Britain) consists of liquid with suspended flocs of organic matter. This may be treated in one of several ways.

TREATMENT OF WATER EFFLUENT

Activated sludge treatment   This process is used at large sewage works and consists of running the sewage into tanks where it is aerated either by compressed air or stirring. This vigorous agitation continues from 5 to 15 hours, and during this time flocculation of the organic matter takes place. The effluent is subsequently run into settling tanks, where the sediment is recycled back to provide an inoculum for the incoming sewage, whilst the water is sufficiently pure to be discharged into a river. When the sewage contains an excess of certain substances, e.g. carbohydrates in brewery effluent, it is necessary to add nitrogen to keep the balance correct.

The chemical and microbiological changes that take place in activated sludge are not fully understood. The dominant floc-forming bacterium is *Zooglea ramigera*. Nitrifying bacteria are also present, and while these do not compete with the heterotrophs for nutrients, there is considerable competition for oxygen. Fungi do not become well established, and if *Geotrichum* species appear it usually indicates an upset in the balance of the process. Protozoa are present both as the free-swimming forms and the stalked species, the latter being dominant. Their presence is most important because they feed on bacteria, and effluent from this process lacking protozoa is often cloudy.

The activated sludge process will remove 85–95 per cent of BOD and the same percentage of suspended materials. Thus, it enables the solids to be spread on the land without danger to health, and water to be returned to the rivers without causing pollution.

Biological filters These consist of a circular tank, above ground to a height of about 4 ft, packed with inert material such as coke that acts as a filter bed. Over the top of the bed, four sparge arms slowly rotate distributing effluent evenly over the whole surface. The micro-organisms become stratified within the filter bed, different types growing at a level where the oxygen concentration and composition of the effluent suits their particular requirements best. In the upper layers such heterotrophs as *Zooglea ramigera* are to be found and as the organic matter diminishes in the lower layers chemolithotrophs (Table 4.1) such as *Nitrobacter* and *Nitrosomonas* grow. In addition, *Fusarium* and *Geotrichum* grow on the surface and moulds are more numerous than in the activated sludge process. Protozoa abound, the motile species grow in the upper layers whilst the stalked kinds grow lower down the beds. A macrofauna of small animals graze upon the solid matter that gets caught in the interstices of the bed, thus preventing the plant from becoming blocked. Reduction of BOD is about the same as that of the activated sludge process, whereas the removal of suspended solids is slightly less, varying from 70 to 90 per cent.

Synthetic detergents have been added to sewage in recent years. These are surface-active agents that cause considerable foaming and also reduce oxygen transfer by about 20 per cent in the activated sludge process. By altering a substituent group in the benzene ring, biological degradation can take place through this group. Furthermore, the straight-chain compounds are more readily attacked than branched chain ones. The tendency now is to produce detergents containing a higher proportion of straight-chain compounds, and some countries now have legislation prohibiting the manufacture of detergents resistant to biological degradation.

Oxidation ponds Settled sewage is run into ponds and held for about 30 days. The action of bacteria produces carbon dioxide and ammonia. The presence of these substances encourages the growth of algae which releases oxygen during photosynthesis. These conditions permit aerobic bacteria to grow which decompose the organic matter further, leaving a minimum of residue.

Land treatment Settled sewage is run on to arable land where accumulated organic matter is removed by microbiological oxidation. This system is not used today as a major method for dealing with effluent, but is usually applied to the cleaning of effluents from other processes such as activated sludge.

TREATMENT OF SOLIDS Sludge of this type is derived from the initial screening of raw sewage, from the settled deposits of activated sludge or as humus sludge from percolating filters. These are combined and placed in a 'digester'. This is the part of the sewage that may contain pathogens and it is, therefore essential that this material is rendered innocuous before disposal. The digester, in reality, is nothing more than a large-scale septic tank whose contents are stirred and heated. The most common temperature range is 30–35°C which is applied for about one month. Thermophilic digestion at a temperature of 50°C is sometimes practised, but is less usual than the mesophilic digestion which is more easily controlled. During digestion, methane, with a calorific value of about 700–750 BThU/cu ft, is evolved and in a large sewage works this is used as a source of heat and power. Organisms producing methane belong to several genera and include: *Methanobacterium formicum*, *Methanobacillus omelianskii*, *Methanosarcina barkerii*. They utilize a variety of substrates including alcohols, fatty acids, CO_2 and H_2 to produce methane and may be highly substrate specific. An example of a methane producing reaction by *Methanobacterium suboxydans* is illustrated below.

$$2\,CH_3CH_2CH_2COOH + 2\,H_2O + \overset{*}{C}O_2 \rightarrow$$
Butyric acid

$$\overset{*}{C}H_4 + 4\,CH_3COOH$$
Methane   Acetic acid

Experiments using labelled carbon (*) have shown that the CO_2 carbon becomes the methane carbon. Other changes taking place during digestion are the liquefaction of solids and the breakdown of carbohydrates, fats, and proteins. The digestion process can give rise to very obnoxious odours since indole, skatole, and compounds containing mercaptans are produced.

When sludge digestion is complete, the excess liquid is returned to the digester and the residual solid material is dried by vacuum filtration, by heat or in under-drained beds. The remaining dry solid can then be disposed of by incineration or applied to the land as manure.

The disposal of trade wastes presents many problems. Some industrial liquors such as the effluents from flax retting, from factories dealing with sugar beet or milk, or from slaughterhouses or breweries, can be mixed with sewage or treated in similar disposal plants. Other trade wastes may contain poisonous organic material (e.g. effluents from factories processing leather, cellulose, textiles, or glue) or poisonous inorganic substances (e.g. effluents from coke ovens, which contain inorganic cyanides, from metal works, where acids and metallic compounds will be present, or from lead or copper mines). These cannot be dealt with by ordinary methods of sewage disposal without preliminary treatment to remove the toxic substances. A different type of pollution is that from china-clay mines where a fine deposit of clay covers the bottom of streams and is continually replenished, thus preventing growth of most micro-organisms. This is only a local effect and is not a danger to public health.

PURIFICATION OF WATER SUPPLIES

Storage Where water is stored in a reservoir the amount of suspended food material normally tends to decrease, both by sedimentation and as a result of the activities of micro-organisms. This in turn leads to a reduction in the number of micro-organisms and in particular of pathogenic bacteria which are unable to survive in competition with saprophytic species.

Populations of planktonic and benthic algae similar to those in natural water occur in water supply reservoirs. These algae are controlled by factors similar to those in nature and the same principles apply with the added complication of control by man, e.g. thermal stratification may be prevented by inflow design, pumping, etc. The algae are here normally beneficial since they absorb certain nutrients, supply oxygen to the water, have some anti-bacterial effect, assist filtration, etc. but these activities can readily become deleterious when the algal growth becomes too great. Water reservoirs receiving acid drainage water with low nutrient concentrations rarely give trouble, but the same water impounded in lowland reservoirs or river water containing large amounts of dissolved chemicals can produce enormous algal crops (the so-called 'water blooms'). These may decay and produce very unpleasant by-products, causing taints or odours. If the water is being drawn off during such a growth, beds of sand through which the water is filtered in the treatment plants become clogged within a few hours and have to be drained and the surface layer of sand removed. Algae, such as *Cladophora*, may grow actually on the filters and then must be removed. Control of the algae in the reservoirs may be achieved by adding small amounts of copper sulphate to the water, but this does not always have the desired effect since the removal of one alga is often followed by the growth of another and as is often the case after copper sulphate treatment this may be a smaller and even more troublesome species. Reservoir management, including the installation of jets at the inlets to cause the water to circulate, switching to alternative reservoirs, drawing off the water at the most suitable level from a stratified reservoir, and the regulation of the storage period to give the maximum reduction of bacteria with the minimum increase of algae, is important in overcoming the problem of excessive algal growth.

Filtration Even if no additional pollution occurs during the process, storage of water in open lakes and reservoirs is insufficient to render it safe for drinking, hence artificial methods of purification must also be employed. The water is first piped to a filter plant which may be a *slow sand filter* suitable for reasonably pure waters and for use where adequate space is available for the large filters needed, or a *rapid sand filter*, suitable for turbid waters or where insufficient land is available.

SLOW SAND FILTERS The natural process of filtration through rocks is imitated by allowing the waters to percolate slowly through clean sand. Purification does not, however, depend only on mechanical straining of the water by the sand but is achieved mainly by the activity of the micro-organisms present in the system. After filtration has proceeded for several days a slimy gelatinous mass composed of bacteria, protozoa, and algae (the 'Schmutzdecke') accumulates, particularly in the upper layers of the sand. This layer brings about biological oxidations and reductions, slowly closes up the pores of the filter thus rendering it more effective by slowing down the rate of filtration, and provides predatory protozoa which feed on bacteria. The net result is that the water emerging from the filter is chemically and biologically purer than before filtration. Eventually the slime layer becomes too thick and slows filtration too much. The filter bed is then cleaned, relaid with fresh sand and filtration is renewed. A newly prepared filter bed is of low efficiency and the water from it cannot be used until a new slime layer has developed.

RAPID SAND FILTERS differ from the slow filters in

having a smaller filter area and in the rapidity and entirely mechanical nature of the process. The organic matter in the water is first precipitated by chemicals such as ammonium aluminium sulphate, and the sticky, bulky precipitate is allowed to sediment in large settling basins. Bacteria as well as any colouring matter in the water are readily adsorbed on to the precipitate and are thus carried down with it. The clean supernatant fluid is finally run over a sand filter to remove any precipitate still in suspension and the purified water is collected at the bottom of the filter.

BACTERIOLOGICAL STANDARDS OF WATER SUPPLIES

Bacterial pollution of water may originate either from individuals with clinical symptoms of disease or from symptomless carriers of enteric (i.e. intestinal) pathogens such as typhoid bacilli.

Pathogens are difficult to detect in water for a number of reasons. They may be present only sporadically owing to intermittent excretion and dilution and, unless the water is being continuously polluted from an infected source, the pathogens usually disappear before the disease is recognized in someone infected by drinking the contaminated water. The isolation of pathogens necessitates the concentration of large volumes of the suspected water and the use of selective media, techniques which are complex and unsuitable for routine examinations. In contrast, harmless members of the normal faecal flora are constantly being excreted and may persist much longer in polluted water. Bacteriological tests to determine the suitability of water for drinking are designed therefore to detect the presence in the water of organisms of the normal flora of the gut. The detection of these organisms is relatively simple but their presence is indicative of faecal contamination which renders the water potentially dangerous. Simple routine tests for faecal pollution, described below are employed at regular intervals to screen supplies of water used for drinking purposes.

Viable counts Counts are made from dilutions of the water sample on a standard medium after incubation at 22°C and 37°C. Most saprophytic bacteria grow at 22°C and all organisms able to grow on the standard medium at this temperature will be included in this count. Since the number of saprophytes present is likely to be proportional to the amount of organic matter available for their nutrition, this count gives a rough indication of the relative amount of organic material present in the water. The majority of organisms growing at 37°C will be parasites or potential parasites of man and animals, derived from soil, excreta, or sewage. A high count at 37°C, relative to the count at 22°C, almost always indicates pollution with animal or human excreta. In unpolluted water the ratio of the count at 22°C to that at 37°C is usually greater than 10:1; in polluted water the ratio is much lower and may be 1:1.

Demonstration of organisms of faecal origin The water is further examined for the presence of specific bacteria of known intestinal origin. In Great Britain, *E. coli* (*Bacterium coli* faecal type I), frequently present in human and animal excreta, is used as the indicator strain in official tests. Many coliform bacteria are found on vegetation and in soil but *E. coli* is found only transiently outside the alimentary canal and its presence in water is thus almost certain evidence of recent faecal pollution. The demonstration of even a few *E. coli* in water is sufficient to condemn it as unfit for human consumption even though no pathogens have been found.

Two tests are used to detect *E. coli*: (1) the presumptive coliform count and (2) the identification of isolated strains by differential biochemical tests.

PRESUMPTIVE COLIFORM COUNT This count is performed in a selective nutrient fluid medium to which bile salts, lactose, and a pH indicator are added. The bile salts inhibit the growth of most non-intestinal organisms but not that of *E. coli* and allied organisms. The presence of coliform bacilli under these conditions is demonstrated by the production of acid and gas by fermentation of lactose. For statistical reasons a range of volumes and dilutions of the water under test is added to appropriate volumes of the bile–lactose medium. The greater the number of *E. coli* present and the larger the sample of water examined, the higher will be the chance that *E. coli* will be contained in any particular sample and vice versa. The distribution of coliform positive reactions in the range of volumes tested from each water sample is referred to probability tables and the probable number of coliform bacilli in 100 ml of the original water sample is deduced. The standards required vary for different types of water but, in general, an unchlorinated water with more than 5 coliform bacilli per 100 ml would be regarded as unfit for drinking. Chlorinated water would obviously be expected to have no viable coliforms present.

DIFFERENTIAL COLIFORM TESTS Fermentation of lactose in the presumptive coliform count may be due to any one of a number of coliform bacilli and not necessarily to *E. coli*. It is essential, therefore, to determine whether the organism detected in the count is the type known to be parasitic in the gut. The organism is first isolated on solid medium from one of the tubes showing fermentation and differential biochemical tests are performed. If biochemical reactions typical for *E. coli* are obtained and if the presumptive coliform count is in excess of the maximum number permitted, the water is regarded as potentially dangerous and condemned as unfit for drinking.

Books and articles for reference and further study

Soil

ALEXANDER, M. (1977). *Introduction to Soil Microbiology.* 2nd Edn. John Wiley, New York and London, 467 pp.

ALEXANDER, M. (1971). *Microbial Ecology.* John Wiley, New York and London, 511 pp.

BAKER, K. F. and COOK, R. J. (1974). *Biological Control of Plant Pathogens.* W. H. Freeman, San Fransisco, 433 pp.

BERKELEY, R. C. W., LYNCH, J. M., MELLING, J., RUTTER, P. R. and VINCENT, B. (1981). *Microbial Adhesion to Surfaces.* Ellis Horwood, Chichester, 559 pp.

BROCK, T. D. (1966). *Principles of Microbial Ecology.* Prentice-Hall, Englewood Cliffs, New Jersey, 306 pp.

BURGES, A. and RAW, F., eds. (1967). *Soil Biology.* Academic Press, London and New York, 532 pp.

BURNS, R. C. and HARDY, R. W. F. (1975). *Nitrogen Fixation in Bacteria and Higher Plants.* Molecular Biology, Biochemistry and Biophysics; **21**. Springer-Verlag, Berlin, Heidelberg and New York, 189 pp.

CAMPBELL, R. (1977). *Microbial Ecology.* Blackwell Scientific Publishers, Oxford. 148 pp.

GRAY, T. R. G. and WILLIAMS, S. T. (1971). *Soil Microorganisms.* Oliver and Boyd, Edinburgh, 240 pp.

HIGGINS, I. J. and BURNS, R. G. (1975). *The Chemistry and Microbiology of Pollution.* Academic Press, London, 248 pp.

LASKIN, A. I. and LECHEVALIER, H., eds. (1974). *Microbial Ecology.* CRC Press, Cleveland, Ohio, 191 pp.

MCLAREN, A. D. and PETERSON, G. H., eds. (1967). *Soil Biochemistry.* Vol. 1, Edward Arnold, London and Marcel Dekker, New York, 509 pp.

MCLAREN, A. D. and SKUJINS, J., eds. (1971). *Soil Biochemistry* Vol. 2. Marcel Dekker, New York, 527 pp.

MITCHELL, R., ed. (1972). *Water Pollution Microbiology.* Wiley-Interscience, New York and London, 416 pp.

PAUL, E. A. and MCLAREN, A. D., eds. (1975). *Soil Biochemistry* Vol. 3. Marcel Dekker, New York, 334 pp.

POSTGATE, J. R., ed. (1971). *The Chemistry and Biochemistry of Nitrogen Fixation.* Plenum Press, London and New York, 326 pp.

POSTGATE, J. R. (1978). *Nitrogen Fixation.* Studies in Biology No. 92. Edward Arnold, London, 72 pp.

POSTGATE, J. R. (1979). *The Sulphate Reducing Bacteria.* Cambridge University Press, Cambridge, 151 pp.

SCHALLEK, F. (1968). *Soil Animals.* University of Michigan Press, Ann Arbor, 144 pp.

STEWART, W. D. P. and GALLON, J. R. (1980). *Nitrogen Fixation.* Academic Press, London, 451 pp.

SWIFT, M. J., HEAL, O. W. and ANDERSON, J. M. (1979). *Decomposition in Terrestrial Ecosystems.* Blackwell Scientific Publishers, Oxford, 372 pp.

WALKER, N., ed. (1975). *Soil Microbiology. A Critical Review.* Butterworths, London, 262 pp.

Air

GREGORY, P. H. (1973). *The Microbiology of the Atmosphere.* 2nd Edn., Leonard Hill Limited, London, 377 pp.

GREGORY, P. H. and MONTEITH, J. L., eds. (1967). *Airborne Microbes.* 17th Symp. Soc. gen.Microbial. Cambridge Univ. Press, 385 pp.

LEACH, C. M. (1976). An electrostatic theory to explain violent spore liberation by *Drechslera turcica* and other Fungi. *Mycologia* **68**, 63–86.

WILLIAMS, R. E. O., BLOWERS, R., GARROD, L. P. and SHOOTER, R. A. (1966). *Hospital Infection.* 2nd Edn., Lloyd-Luke Limited, London, 386 pp.

Water

GARETH JONES, E. B., ed. (1976). *Recent Advances in Aquatic Mycology.* Elek Science, London, 749 pp.

HUTCHINSON, G. E. (1957, 1967). *A Treatise on Limnology.* Vol. I, *Geography, Physics and Chemistry*, 1015 pp. Vol. II, *Introduction to Lake Biology and the Limnoplankton*, 1115 pp. John Wiley and Sons, Inc., New York.

JACKSON, D. F. ed. (1968). *Algae, Man and the Environment.* Syracuse University Press, Syracuse, 554 pp.

KRISS, A. E., MISHUSTINA, I. E., MITSKEVICH, N. and ZEMTSOVA, E. V., trans Syers, E. (1967). *Microbial Populations of Oceans and Seas.* Edward Arnold, London, 287 pp.

LEWIN, R. A., ed. (1962). *Physiology and Biochemistry of Algae.* Academic Press, New York and London, 929 pp.

Report (1957) *The Bacteriological Examination of Water Supplies*, No. **71**, H.M.S.O., London.

ROUND, F. E. (1965). *The Biology of the Algae.* Edward Arnold, London, 269 pp.

RUTTNER, F., trans. Frey, D. G. and Fry, F. E. J. (1953). *Fundamentals of Limnology.* University of Toronto Press, 295 pp.

SOUTHGATE, B. A. (1950). *Treatment and Disposal of Industrial Waste Waters.* H.M.S.O., London.

Chapter 12

Symbiotic interactions with other organisms: mutualism

Introduction

One of the major habitats for micro-organisms is other living organisms. Indeed, under natural conditions, most larger organisms are inhabited to a greater or lesser extent by micro-organisms. Obviously there must be many different types of interaction between micro-organisms and their living hosts. These interactions are traditionally classified into three major categories: *symbiotic*, *parasitic* and *commensal*. One of the difficulties of this classification is the different meanings given to the term 'symbiosis'. The word was introduced by De Bary in 1879 to cover all examples of associations between dissimilar organisms (including parasitism). Soon afterwards, some biologists restricted its meaning to cover only mutualism, but others continue even today to retain the broad definition originally intended by De Bary. Further problems in classification may arise with those associations which have both parasitic and mutualistic phases; in others, it may be difficult to assess whether 'benefit' or 'harm' is occurring.

Recently, an entirely new approach to the problem of classifying organismic interactions has developed retaining the original De Bary definition of symbiosis so that it covers almost all associations between micro-organisms and their hosts. However, each type of association is described in terms of a range of criteria covering various aspects of how organisms may interact. Some of the criteria used in the classification are as follows. (1) *Durability or lifespan of association*: whether the association is transient, permanent or prolonged. (2) *Type of contact*: e.g. whether a symbiont is extracellular or intracellular. (3) *Degree of morphological integration*: whether the association has morphological characteristics additional to and different from those of the isolated symbionts (e.g. formation of structures such as a legume nodule). (4) *Types of functional dependence*: the degree to which a symbiont is dependent on its partner genetically, repro-

ductively, metabolically, etc. (5) *Degree of specificity* of symbionts for each other. (6) *Degree of necessariness*: whether the association is facultative or obligate. (7) *Nutritional*: whether nutrient movement between symbionts is necrotrophic (from dead host cells) or biotrophic (from living host cells). (8) *Valuational*: whether the interaction can be assessed as antagonistic, neutral or mutualistic.

The particular value of this approach is that it highlights the fact that the same phenomenon (e.g. specificity) may be achieved by similar mechanisms in otherwise different types of interactions. It draws attention away from broad and sometimes vague concepts of 'harm' and 'benefit' and concentrates on more precise aspects of interaction.

For practical purposes, the criteria of the greatest immediate use are the nutritional and valuational.

Nutritional criteria

Almost all symbiotic associations involve a flow of nutrients between the symbionts in one or both directions. Three principal types of nutrient flow can be recognized: saprotrophy, nectrotrophy and biotrophy.

Saprotrophy is synonymous with saprophytism and involves the digestion of previously dead organic material. *Necrotrophy* describes situations in which one organism first kills part or all of the other organism—many of the more primitive kinds of parasite fall into this category. In biotrophy, nutrients move between the living cells of the symbionts, and there is little or no tissue destruction. *Biotrophy* is evolutionarily the most advanced form of nutrition, and is characteristic of nearly all mutualistic associations as well as of many parasites; very many fungi have biotrophic nutrition. It is considered advanced because the donor cells can continue providing nutrients for the recipient over a period of time, whereas in necrotrophy only the nutrients present at one particular time are available.

Valuational criteria: parasitism and mutualism

In general terms, this is one of the most important criteria to human beings in that it attempts to distinguish between harmful and beneficial situations. Problems arise, however, in the precise application of these criteria. Extreme situations of harm and benefit are readily recognized, but it is the intermediate categories which are difficult, especially when 'harm' or 'benefit' become human value judgements and do not involve easily measurable factors. For example, algae are widely assumed to 'benefit' from existence in a lichen (p. 281) but in fact, no experimental evidence showing nutritional, protective or other 'benefit' has ever been obtained.

These considerations reinforce previously outlined difficulties of the precise definition of mutualism and parasitism. Nevertheless, it is possible from a human point of view to distinguish between associations which are obviously pathogenic and those which are not. The latter include some of major ecological and economic significance which will be discussed in this chapter. The precise assessment of some of these interactions on the 'valuational' criterion may not always be clear, but for practical considerations they will be grouped together here as 'mutualistic'.

Organisms may often change in a variety of ways when they enter into 'symbiosis'. It is thus dangerous to predict the behaviour of a microorganism in symbiosis from its behaviour in isolated culture; in most associations, the whole does not equal the sum of the parts. In discussing the various associations below, particular stress will therefore be laid upon the nature of the intact association.

Symbiotic dinoflagellates

A variety of aquatic lower invertebrates contain unicellular symbiotic algae. The hosts belong primarily to the phyla Protozoa, Porifera, Coelenterata, Platyhelminthes (Turbellaria) and Mollusca (Gastropoda and Bivalvia). In marine environments, dinoflagellates (p. 265) are the predominant type of symbiont, becoming particularly important in tropical regions.

They are universally present in reef-building corals, which are estimated to have a total world area of 2×10^8 km^2. Coral tissues are so densely packed with dinoflagellates that as much as half the protein they contain is algal. Giant clams, the most prominent bivalves in the Red Sea, Indian Ocean and West Pacific, also contain a rich dinoflagellate flora in their mantle tissue. Apart from corals,

virtually every other marine coelenterate in the tropics (except for non-reef-building corals) also harbours vast numbers of dinoflagellates. Symbiotic dinoflagellates thus make an important contribution to primary benthic productivity in tropical oceans.

In the great majority of animal hosts, the algae are intracellular and are restricted to particular tissues, e.g. gastrodermis in most coelenterates and mantle tissue in clams. Within the host cells, the symbionts are spherical, thin-walled yellow-brown cells about 10–12 μm in diameter. The flagellae and the theca which gives the characteristic shape to free-living dinoflagellates are lacking, although these features rapidly develop if the symbionts are isolated into pure culture.

Because the algae lack some important morphological characters in symbiosis, reliable taxonomic studies can be made only on cultures. There have been relatively few cultural investigations, but they all suggest that the symbiont of benthic animals belongs to the species *Gymnodinium microadriaticum*. Naturally, doubts have been expressed as to whether the symbiont of many different types of animal in both the tropics and temperate zones belong to the same species, and there is some evidence that different races of this alga may occur. In pelagic hosts, the symbionts have been assigned to various species of the genus *Amphidinium*. In the older literature, symbiotic dinoflagellates were included in 'zooxanthellae'—an umbrella term to cover all symbionts of animals which were yellow-brown in colour.

In almost all host animals except clams, there is no evidence that the symbionts are digested by the host, although they may be ejected under certain conditions of stress. In giant clams the extent of digestion is uncertain, and may involve only senescent algae. Forms of these animals without algae are not found under normal conditions in nature, so the association is ecologically obligate.

Interactions

Most studies of the physiological interaction between symbiont and host have been carried out on sea-anemones and corals, with some on giant clams. The role of the dinoflagellate symbionts is evidently complex, and may be considered under the following main headings: (a) provision of photosynthetically fixed carbon; (b) recycling of waste animal nitrogenous products; and (c) marked stimulation of calcium carbonate deposition by reef-forming corals.

At least half of all the carbon fixed in photo-synthesis by symbiotic dinoflagellates is rapidly passed to their animal hosts. Nevertheless, the importance of the algae in animal nutrition has been questioned because most hosts still retain their conventional holozoic feeding mechanisms. Indeed, corals have the greatest ratio of prey capture surface area to tissue bulk of any group of animals, and the polyps rapidly ingest any food provided for them. However, it must be remembered that corals and many other symbiotic animals live in tropical seas where the supply of animal nutrients (zooplankton) is so poor that they have been described as 'aquatic deserts'. Various methods show that the zooplankton may in fact supply as little as 10 % of the energy requirements of the coral—for although corals exist in nutrient deserts, they are one of the most productive ecosystems known, and this high productivity is clearly due to the symbiotic algae. Corals retain their conventional feeding mechanism through which they obtain essential complex compounds such as vitamins, etc.

Studies with ^{14}C show that the principal organic compounds supplied by symbiotic dinoflagellates to their host are glycerol and alanine. The host animal tissue produces hitherto unidentified 'factors' which stimulate the massive release of glycerol and alanine from their algal symbionts. These compounds rapidly become converted to fats and protein in the animal tissue. The flow of alanine from alga to animal highlights the important role of the symbionts in conserving nitrogen within the system. Algae are able to convert animal nitrogenous waste products such as ammonia back to amino acids which can be utilized by the animal. This recycling of nitrogen may be as important as photosynthesis in maintaining the high productivity of coral reefs. In terms of the nitrogen economy of the coral reef ecosystem, symbiotic dinoflagellates supplement the nitrogen-fixing activities of blue-green algae on the reef flats.

In the light, reef-forming corals deposit calcium carbonate at an average rate ten times faster than in the dark. The stimulatory effect of the algae on deposition is of vital importance in helping reefs to withstand and survive the constantly destructive force of wave action. Although the importance of the algae is unquestioned, the mechanism by which they stimulate deposition remains obscure. The central problem is how CO_2 removal in photosynthesis by algae in the gastrodermis stimulates carbonate deposition by the calicoblastic ectodermis. The situation is rendered complex by such observations as that in the 'stag's horn' coral, the tips of the branching coral have the fewest symbiotic algae, yet show the highest rates of deposition in the light.

The high photosynthetic activity of symbiotic dinoflagellates leads to high oxygen production. Indeed, some ecologists base estimates of the productivity of coral reef ecosystems upon measurement of oxygen production in the light. This massive oxygen release must undoubtedly affect metabolism in the tissue, though this problem has been little studied.

Insect symbionts

With over 700 000 species, insects constitute by far the largest and most dominant group of terrestrial invertebrates. Approximately 10 % of all insect species exist in a symbiotic association with one or frequently more kinds of micro-organisms. There is an overall broad correlation that insects on restricted diets (such as plant sap, wood, blood, stored grain, etc., i.e. the great majority of economically important insect pests, possess symbionts.

The types of micro-organisms in insects include: flagellate protozoans, yeasts, bacteria, actinomycetes, rickettsias and unidentified unculturable forms. There is no clear-cut correlation between the type of insect and type of micro-organism, but there is a general tendency for lower orders to have protozoans and higher orders rickettsias. Microbial symbionts within insects can be broadly divided into intracellular and extracellular. Insect species may or may not contain both kinds of symbionts.

Intracellular symbionts

Many insects possess cells, originally termed 'mycetocytes' by De Bary, which contain unidentified symbionts. The 'mycetocytes' are often aggregated into discrete organs called 'mycetomes'. The terms 'mycetocyte' and 'mycetome' derived from an original belief that the symbionts were yeasts. Because the symbionts still remain unidentified and unculturable, it is better to describe them by the neutral term 'Blochman Body' rather than expressions such as 'bacteroid'. In anatomical location, mycetocytes can be arranged in a series of increasing dissociation from the gut. In some leaf bugs, they occur in some of the cells surrounding the gut; in other insects they may be associated with Malpighian tubules. Where mycetocytes aggregate into mycetomes, they may occur in the body cavity quite distinct from the gut.

Blochman Bodies are membrane bound, of the

same general size as mitochondria but differing from them in lacking internal cristae. Cockroach symbionts have muramic acid, characteristic of bacteria. In aphids, the morphology is suggestive of rickettsias (which also possess muramic acid). In cicadas, two morphologically distinct types of Blochman Body occur in a common mycetome either side of the abdomen. While such polymorphism is suggestive of rickettsias and mycoplasmas, the DNA of Blochman Bodies is shorter and smaller (approx. 2×10^6 Daltons) than that of rickettsias (1–2.5×10^9) or mycoplasmas (6×10^8–1×10^9). Some, but not all Blochman Bodies may be suppressed by antibiotics and sulfa drugs. There have been suggestions that they may constitute a new group of prokaryotes distinct from any of the free-living types.

Direct investigations of the function of Blochman Bodies are difficult because they are unculturable, and it is difficult to remove them from most species. In a few insects (cockroaches, aphids, and body-lice), however, certain types of treatment—especially with antibiotics—produce aposymbionts; in others, mycetomes have been removed by direct surgery. These insects always show substantially less vigour than symbiotic controls. Various supplements have been added to the diets of aposymbionts with the assumption that those compounds which restore vigour must be similar to those supplied by Blochman Bodies. Other types of experiment have compared the fate of ^{14}C and ^{35}S labelled compounds added to the diets of symbiotic and aposymbiotic insects. No clear-cut picture of the role of Blochman Bodies has so far emerged, but experimental evidence suggests one or more of the following functions for them in various insects: vitamin synthesis, sterol synthesis, utilization of nitrogenous animal wastes, and organic sulphur metabolism.

Blochman Bodies are transmitted from generation to generation through the female line, and in all carefully investigated cases, through the egg cytoplasm. The principal problem not yet clearly resolved is how the symbionts move from mycetome to egg. As with most other aspects of Blochman Bodies, the still fragmentary investigations yield no cohesive picture. Electron micrographs of aphids suggest direct infection of eggs by mycetocytes; in cockroaches, implantation of mycetocytes into aposymbiotic females result in no infection of eggs although the mycetocytes flourish; in lice, removal of mycetomes results in failure of the female to reproduce. In some types of insects, Blochman Bodies may reappear again a few generations after

their apparent removal by antibiotic treatment, as if they can develop from dormant particles. Even more obscure are the mechanisms by which, in the next generation, the Blochman Bodies are transferred from the fertilized egg cytoplasm back to the eventual mycetomes during embryogeny. The involvement of special host cells for the movement of symbionts into and out of the egg is indicated in some cases, as is the occurrence of transitional embryonic mycetomes to house the symbionts before they reach their eventual sites (the fat bodies).

Despite lack of definite knowledge about Blochman Bodies it is clear that they are symbionts essential to insect pests of major economic importance.

Extracellular symbionts within insects

These include micro-organisms normally housed in modifications of the gut such as various types of chamber or distensions of specific regions of the alimentary tract, or crypt guts which are either connected or not to the gut lumen.

Normally, symbiotic insects contain a variety of extracellular symbionts. Many different mechanisms exist for ensuring transmission of symbionts to the offspring, but all are essentially directed towards ensuring contact between eggs or newly hatched larvae and microbial suspensions extruded from the gut.

The role of extracellular microbial symbionts has been easier to establish because many can be grown in culture and because it is less difficult to produce aposymbiotic hosts—usually achieved by interfering with transmission of symbionts to offspring. The best studied group of insects are termites (Fig. 12.1). About 80% of termite families are symbiotic and harbour a vast population of flagellates (Hypermastigina + Polymastigina) as well as bacteria and rickettsias. The flagellates are obligately anaerobic and so can be eliminated by exposing the termites to oxygen tensions in excess of one atmosphere. Such de-faunated animals cannot survive longer than a month on the normal termite diet of wood and cellulose, but can be restored to normal vigour by reinfection through feeding on anal droplets of normal termites. The life of de-faunated termites can also be prolonged if they are fed glucose. Evidence therefore suggests that the flagellates are the main agents in wood and cellulose digestion by termites. Termite guts resemble the rumen (p. 279) because they are anaerobic and because the symbionts break down glucose to H_2,

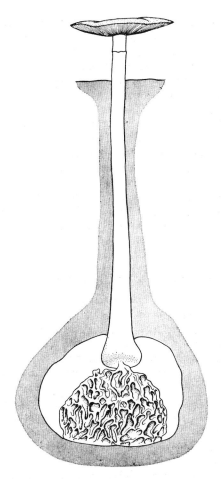

Fig. 12.1 Fungus-insect association. *Termitomyces* sp. ($\times\frac{1}{3}$). Occasionally the fungus in a termite fungus garden (see Fig. 14.2) forms a fruit-body which grows through the soil above the chamber containing the garden until it reaches the surface. The base of its stipe connects with the fungus garden.

CO_2, acetic and other fatty acids. Recently, it has been shown that termites contain organisms able to fix atmospheric nitrogen, though these have not yet been identified (see p. 303; Fig. 14.2).

In other kinds of wood-eating insects, protozoa appear to play a similar role in polysaccharide breakdown. The role of symbionts in insects feeding on other types of diet is much less well investigated. In the beetle *Lasioderma* the symbionts are replaceable by feeding B vitamins and sterols, and it is suggested that yeasts are the key symbionts in providing these essential compounds. The blood-sucking bug *Rhodnius* possesses the actinomycete *Nocardia*, and again it is suggested that its principal role may be in supplying vitamins to the host.

In general, it may be concluded that of all the symbiotic associations of major economic importance, the insect–microbe associations are the least well studied. Undoubtedly, these associations are complex. One example of complexity is the flagellate *Myxotricha paradoxa*, isolated from the gut of the primitive Australian wood-eating termite *Mastotermes*. This symbiotic flagellate is itself a complex series of symbiotic associations. Its surface is covered by what superficially seem to be cilia which can beat in rhythm, but which close investigation shows to be spirochaetes. Further, the spirochaetes are anchored to the surface by bracket-like projections which are bacteria. Within the body of the protozoan are further bacterial symbionts.

The rumen symbiosis

In terrestrial ecosystems, most of the carbon fixed in photosynthesis by primary producers accumulates in cellulose and lignin, so that the ability to digest these compounds must be an essential property of most herbivores. It is therefore paradoxical that vertebrates, which provide the dominant herbivores in most terrestrial ecosystems, do not possess cellulase, xylanase or pectinase. They have been able to become dominant herbivores only by adapting their alimentary tract to house and utilize efficiently micro-organisms which can break down cellulose and lignin. In turn, most of the micro-organisms have become highly specialized to their habitat, so that a true mutualism has been established.

Herbivorous animals are broadly divisible into two main kinds: (a) *ruminants*, in which the main microbial action precedes the action of the animal's own digestive enzymes; and (b) *non-ruminants*, in which the main microbial action occurs after the action of the animal's own enzymes (Fig. 12.2).

In both ruminants and non-ruminants, there are three principal requirements of the way in which the alimentary tract is modified to house microbial action. (1) *The fermentation chamber must be sufficiently large.* In the cow, the rumen is about 100 l in capacity. In a non-ruminant such as a horse, fermentation occurs in the caecum and great colon, whose combined volume can approximate to 120 l (the stomach has a volume of only 10 l). (2) *Microbial fermentation must be anaerobic* so that cellulose is only partially broken down to products (usually fatty acids) which can then pass into the blood stream where they become subject to the aerobic metabolism of the host. If microbial fermentation were aerobic, very little nutritional value would accrue to the host. (3) *Food flow through the animal must be regulated* so that fibrous material can

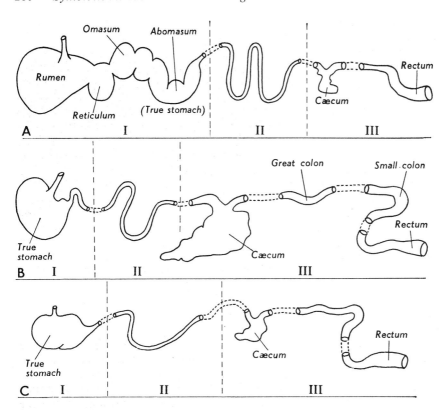

Fig. 12.2 The digestive tracts of different animals showing how ruminants differ from other animals. (**A**) Herbivorous ruminant, e.g., cow. (**B**) Herbivorous non-ruminant, e.g., horse. (**C**) Omnivorous animal, e.g., rat. (**i**) pre-absorptive part of gut. (**ii**) small intestine. (**iii**) large intestine. (After Kon, S. K. (1953) *Voeding*, **14**, 137)

spend a relatively long time (8–12 hr) in the fermentation chamber.

Although there might well be much in common between the microbial activities in ruminants and non-ruminants, the latter have been much less studied. The following account will therefore concentrate upon what is known of the rumen symbiosis, especially as it has similarities to that of the alimentary tract of some symbiotic insects.

The rumen is essentially a large fermentation chamber with a dense culture of various protozoa (10^5–10^6 per ml) and bacteria (10^{10}–10^{11} per ml). The rumen contents are continuously mixed by contractions of the wall at one to two minute intervals. There is a large input of saliva (equivalent to one to three times the volume of the rumen per day) which contains 100–140 mM bicarbonate and 10–50 mM phosphate and is thus an alkaline buffer which can neutralize the fatty acids produced by fermentation so keeping the pH favourable for further microbial action. The saliva is removed in the omasum, the chamber of the alimentary tract after the rumen, and is then recirculated. The dense population of protozoa and bacteria in the rumen includes forms capable of breaking down a variety of different substrates, but there are two main net

effects of their activities. Firstly, the principal products of fermentation are the fatty acids: acetic (47–50%), propionic (18–23%) and butyric (19–19%). These are absorbed and metabolized by the host. Secondly, there is a yield of microbial cells which pass from the rumen and are digested in the abomasum and intestine.

The rumen is highly anaerobic, having an oxidation–reduction potential of -3.35V, i.e. it would be in equilibrium with 10^{-22} M oxygen. The composition of the rumen atmosphere is: 60–70% CO_2, 30–40% CH_4, and traces of H_2 and H_2S. Most rumen micro-organisms are extreme obligate anaerobes, and this has made it difficult to study them, and especially to classify the function of individual species and how they relate to each other. Nevertheless, at least three major functional groups can be recognized. *Firstly*, there are the cellulose digesters which break down cellulose to hexoses (principally glucose), cellobiose, etc. These must act very rapidly since they form only one to five per cent of the total population, although cellulose typically comprises about 30% of the food intake. Some, such as *Bacteroides succinogenes* adhere very closely to the fibrous substrate and indeed extracellular cellulase cannot be detected in most strains in culture. Others, such as *Clostridium locheadii* are less exclusively cellulolytic and can also use starch as well as being proteolytic. *Secondly*, there are the starch and sugar digesters, which are principally responsible for forming the volatile fatty acids. *Thirdly*, there are the methanogenic bacteria which convert hydrogen produced by various processes into methane, generally according to the reaction:

$$CO_2 + 4H_2 \rightarrow CH_4 + 2H_2O$$

The true function of the methanogenic bacteria is not entirely clear, although their probable role is to remove free hydrogen. Experiments have shown that a rise in free hydrogen in the rumen inhibits the activities of bacteria involved in producing the hydrogen (such as some of those involved in formate and pyruvate metabolism). Besides these three main groups, micro-organisms metabolizing a variety of other substrates such as organic acids, xylans, lipids, etc. are also known.

The rumen protozoa are a highly specialized group which consists—apart from a few small flagellates—chiefly of two ciliate groups, holotrichs and entodiniomorphs. Holotrichs convert sugar to starch, although the role of this process in the rumen has not been clarified. Possibly, they function in the regulation of sugar levels, preventing excess growth of sugar-metabolizing bacteria. Entodiniomorphs produce cellulase (but much less than the total amount produced by bacteria) and also metabolize starch to sugar. They also ingest bacteria, sometimes forming semi-stable relationships with them. It has been suggested that they may play a role in increasing or regulating bacterial turnover in the rumen. In general, protozoa may not be essential since ruminants can be reared free of protozoans.

Finally, there is the function of the rumen in producing a continuous crop of micro-organisms to be harvested in the abomasum. This process is likely to be of particular importance in the nitrogen nutrition of the animal. All protein in the food entering the rumen is converted to ammonia. The ammonia is then absorbed and converted to microbial protein. Compared to that of other vertebrates, the pancreatic juice of ruminants is particularly rich in RNAase, an adaptation to the abundant ribosomes in the bacteria to be digested.

Lichens

A lichen consists of an association between a fungus and an alga (Fig. 12.3). In most lichens the bulk of the thallus consists of fungal hyphae, with the algae

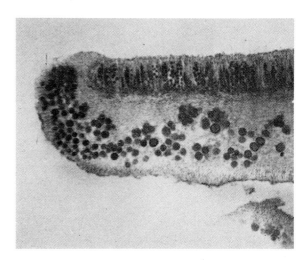

Fig. 12.3 *Physica* sp. (× 100). Section through part of an apothecium of a disco-lichen. Note hymenium of asci on upper surface and spherical algal cells among the fungal hyphae of the hypothecium (basal part) of the fruit-body. (L. E. Hawker)

restricted to a thin layer beneath the surface and comprising only about 5%–10% of the volume of the whole. Above the algal layer is a cortex of compact fungal material in which distinct hyphal structure may or may not be discernible. Within the algal layer, the hyphae are thin-walled and closely appressed to the algal cells. The algal cell wall may be penetrated by peg-like haustoria, but the cell membrane is normally penetrated in only a few lichen species. The true extent of haustorial penetration is not at all clear: there are undoubtedly some lichens in which it never occurs; where haustoria have been reported, no estimate of their frequency or their relationship to the health and condition of the lichen has been given. In healthy lichens, dead or digested algal cells are observed only rarely. Beneath the algal layer the hyphae become much thicker walled, forming a tissue called the medulla. In many lichens, hyphae may form other tissues such as a lower cortex, or rhizines for attachment to the substratum. In only a few species are the algae distributed throughout the thallus and not confined to a distinct layer.

Lichens may reproduce in three ways. Firstly, fragments of thallus may be dispersed by wind, rain, etc. Secondly, since the fungus in many lichens forms long-lived fruit-bodies which disperse abundant fertile ascospores (Fig. 12.3), synthesis of lichens probably occurs through union of germinating fungal spore and algal cells although this has rarely been observed in nature. Thirdly, many lichens have patches of *soredia* on their thallus.

Each soredium consists of an algal cell surrounded by fragments of fungal hyphae, and on dispersal can germinate to form a new plant.

Lichens are a large group, and contain about a quarter of all described fungal species. In almost all cases the fungi are ascomycetes, and only a few basidiomycete lichen fungi are known. With very few exceptions, lichen fungi do not occur free-living, but are clearly related to a number of different free-living groups, both ascohymenial (including apothecia and perithecia producers) and ascolucular. Twenty-seven different genera of algae, both green and blue-green (now classed as bacteria p. 171), have been reported from lichens. The commonest, found in over 70% of species, is *Trebouxia*, a unicellular member of the Chlorococcales, which so far has not been found free-living. All other algae are included in mainly free-living genera, with *Nostoc* and *Trentepohlia* being the commonest after *Trebouxia*. Since taxonomically diverse types of fungus and alga have combined to form lichens, this common and widespread symbiosis presumably evolved on a number of separate occasions.

Formation of a symbiotic association with algae has enabled lichen fungi to colonize a wide range of habitats where conditions are unfavourable for the growth of free-living species. Lichens are resistant to extremes of temperature, and their very slow growth rate (less than one millimetre per year for most mature crusts on boulders) enables them to tolerate habitats where the supply of mineral nutrients is low. They also withstand long periods of desiccation. They not only tolerate frequent cycles of wetting and drying, but attempts at maintaining lichens in laboratory environments indicate that periodic wetting and drying is a positive requirement for healthy growth.

A principal role of the alga is that of maintaining a high level of soluble carbohydrates in the fungal hyphae—this being a major factor in the ability of lichens to withstand environmental extremes. The movement of photosynthate from alga to fungus in lichens is better understood than in any other autotroph/heterotroph association. The simple structure of lichens makes them much easier to study experimentally than associations involving complex organisms such as higher plants. Over forty lichen species involving a variety of algae have been examined, and three characteristics of photosynthate movement are the same in all cases: (1) Movement is substantial, amounting to 50%–70% of all the carbon fixed. (2) Movement out of the alga is predominantly as a simple carbohydrate: glucose from blue-green algae such as *Nostoc* and a sugar

alcohol from green algae (either erythritol as in *Trentepohlia*, ribitol as in *Trebouxia*, or sorbitol as in *Hyalococcus*). (3) The substantial flow of carbohydrate out of the alga ceases rapidly after isolation from the thallus, resulting in a substantial increase in polysaccharide formation within the cell.

A central problem is therefore to understand why existence in symbiosis causes a massive efflux of a particular carbohydrate molecule from the alga. The solution of this problem may be of significance not only for lichens, since it may indicate how other kinds of biotrophic fungi function, such as those involved in many mycorrhizal associations or rust diseases. So far, no evidence has been obtained that the fungus secretes chemical substances which stimulate efflux from the alga. Also, transport out of the alga does not depend upon fungal uptake acting as a sink, since it continues unabated even if fungal uptake is blocked by various kinds of inhibitors. A key feature seems to be that efflux only occurs if there is *physical* contact with the fungus. It has therefore been suggested that efflux is induced by some physical effect of the fungus upon algal membrane properties. For example, one possibility might be that the fungal cell wall carries charges which change the algal membrane potential—known to be an important determinant of the vectorial characteristics of membrane transport.

About 10% of lichens have blue-green symbionts, and all of these so far investigated are very active in nitrogen fixation. Over 90% of the fixed nitrogen passes rapidly to the fungus, probably as ammonia. In these associations, the fungus can apparently stimulate efflux of both ammonia and glucose.

So far we have been concerned only with nutrient movements from alga to fungus. Because lichens have always been regarded as an outstanding example of mutualism, it has been automatically assumed by almost all biologists that the fungus must in some way 'benefit' the alga. Hence, there are two common assumptions that 'the fungus supplies the alga with minerals' and that 'the fungus protects the alga'.

With regard to mineral supply, there is as yet no published experimental evidence that the fungus actively supplies the alga with minerals. Indeed, there is no necessity for substances entering the alga to have first passed into the fungal cytoplasm rather than diffusing from the exterior of the thallus along inert cell walls. In studies of the effects of pollution on lichens, it is widely assumed that lichen algae are particularly sensitive to sulphur dioxide—but nowhere is it assumed that the sulphur dioxide has first to traverse living fungal cells before being released to the alga.

The concept that the fungus in some way 'protects' the alga is again not founded upon particularly solid experimental data and is unconvincing since related free-living algae always occur in lichen habitats, frequently as epiphytes on thalli. Potentially misleading observations suggesting a 'protective' role for the fungus are that certain lichen algae such as *Trebouxia* and *Coccomyxa* grown in isolated culture will bleach irreversibly at light intensities above 2000 lx. It is argued from this that the upper cortex 'protects' the alga from insolation. However *Coccomyxa* bleaches in culture only if grown in an organic medium, showing the danger of translating behaviour in culture to that in the thallus. *Trebouxia* occurs in about 70 % of all lichens, many in exposed habitats where, even on rainy days in summer, light intensities far in excess of 2000 lx occur. Even allowing for light interception by the upper cortex, they can probably tolerate rather greater light intensities in nature than in culture.

Other interactions occur between the symbionts besides the flow of photosynthate and fixed nitrogen. The fungi grown in isolated culture show none of the sometimes elaborate morphology and tissue differentiation of the intact thalli. They are little different from cultures of related free-living fungi except in having exceptionally slow growth and frequently forming rather compact, somewhat elevated colonies on agar. Although the algae only occupy a small volume of the lichen, they are evidently able to evoke a remarkable morphogenetic response from the fungus. This is particularly illustrated by a few cases where it is now known that certain lichen fungi are able to associate separately with different algae and adopt strikingly different morphology with the different symbionts. The lichen *Sticta filix* contains the green alga *Coccomyxa*, but a lichen formed by the same fungus with the blue-green alga *Nostoc* is so different in appearance that it is assigned to a different genus, *Dendriscocaulon*.

Although it is relatively easy to isolate most lichen algae and fungi into pure culture, it has proved very much more difficult to achieve a successful laboratory synthesis to the stage when the original lichen is recognizable. Indeed, despite substantial efforts by a variety of investigators, especially in the first part of this century, no success was achieved until recently and even then in only one or two rather special cases. However, it has become clear from the varying degrees of failure that one of the essential conditions favouring synthesis is an extremely nutrient-poor medium. Lichen symbionts inoculated on to normal nutrient-rich media can grow but show no tendency whatsoever to associate. Two other conditions favouring synthesis are the appropriate substrate (i.e. rock, wood, or soil, not agar) and periodic slow drying and wetting of the cultures.

Mycorrhizas

'Mycorrhiza' is the term applied to a wide range of associations between diverse fungi and the roots, rhizomes or thalli of plants. Probably, most species of higher plants will prove to be mycorrhizal. Four distinct groups of mycorrhizas can be recognized in higher plants: sheathing (or ectotrophic), vesicular–arbuscular, orchidaceous and ericaceous. The last three were formerly described under the collective name of 'endotrophic' but are so different from each other that they merit separate terms.

Sheathing mycorrhizas

These principally occur, especially in temperate regions, in many forest trees—including conifers, beech, oak, birch and their relatives. The feeding roots are completely enclosed in a sheath or mantle of fungal tissue formed of compact hyphal cells (Fig. 12.4a). The mycelium enters the root cortex, but does not normally penetrate cells, remaining between them and forming the 'Hartig net'. The fungus is also directly connected to an extensive mycelium in the soil. The morphology of the host root changes in degree of branching, cell shape and meristematic activity, so that infected roots have an appearance different from that of uninfected ones.

The fungal components are mainly basidiomycetes such as agarics, boleti, etc.—indeed, some of the fruiting bodies commonly found in woodlands are those of mycorrhiza formers. Some fungi are quite specific, but others less so: for example, *Boletus elegans* forms mycorrhizas only with larch, while at the other extreme, *Cenococcum graniforme* has been found associated with 7 genera of conifers and 9 of deciduous trees. A few ascomycetes (*Tuberales*, *Elaphomyces*) may also be involved. Mycorrhizal infection is not universal, and trees on soils rich in mineral nutrients may be quite uninfected. On poor soils, especially if the roots have high carbohydrate levels, infection is normal.

The striking feature about these mycorrhizas is that all nutrients entering the host root must first pass through the fungus. Carefully controlled com-

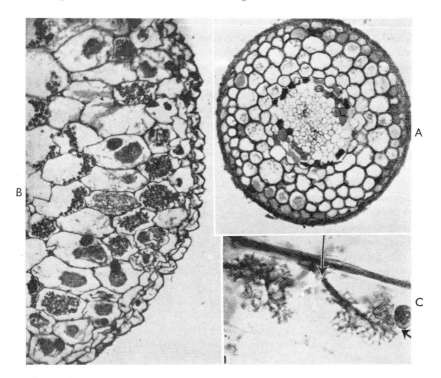

Fig. 12.4
  (**A**) T.S. mycorrhizal root of pine (×100). Note external mantle of fungal hyphae, enlargement and separation of cortical cells with fungal hyphae in intracellular spaces (Harti net) and tannin cells (showing as dark cells) in cortex. (L. E. Hawker)
  (**B**) T.S. outer tissues of mycorrhizal root of *Neottia nidus-avis* (×100) showing coiled peletons of hyphae in some outer cortical cells and partially digested amorphous masses in others. (L. E. Hawker)
  (**C**) A cortical cell of a root of *Allium ursinum* containing a much-branched arbuscle of a vesicular-arbuscular endophyte (×500). Note origin of arbuscle as a branch from an intercellular aseptate hypha. The dark-stained body on the right is the nucleus of a host cell. (Hawker, L. E., Nicholls, V. O., Harrison, R. W., and Ham, A. M. (1957) *Trans. Bri. mycol. Soc.*, **40**, 375)

parisons between the growth of mycorrhizal and uninfected trees clearly showed not only that infected trees had greater vigour, but that the total amount of nitrogen, potassium and phosphorus absorbed is greater. The soil mycelium attached to the mycorrhiza can presumably tap a greater volume of soil than the host root, and this is believed to be the major factor in the improved nutrient uptake of infected trees. Various laboratory experiments confirm the flow of nutrients from fungus to root. They have also shown that there may be a substantial flow of photosynthetically fixed carbon into the fungus. Unlike many other basidiomycetes, mycorrhiza-formers are unable to breakdown cellulose and lignin, so that host photosynthate is probably the main source of organic nutrients. Indeed, it has been calculated for some Swedish forests that the amount of carbon entering mycorrhizal fungi may be equivalent to about 10% of total timber production. In healthy mycorrhizas little breakdown or destruction of host tissues occurs, so that nutrient movements between the symbionts are biotrophic.

Vesicular–Arbuscular mycorrhizas

This is undoubtedly by far the commonest type of mycorrhiza, being found in almost all families of flowering plants as well as in many ferns, liverworts and some gymnosperms (though not pines). It is present in plants of economic importance such as grasses, cereals and legumes (including those also bearing nodules). As with sheathing mycorrhizas, infection is not universal. It is only in recent years that the fungal components have begun to be identified, and so far they all appear to be members of the phycomycete family Endogonaceae.

The fungus does not form an external sheath, though it is always connected to a soil mycelium. The hyphae generally infect unspecialized parenchymatous tissue of the cortex. There is typically extensive intracellular penetration, coils of hyphae often forming in the cells. Branches of the main hyphal system, usually of smaller diameter pass into or are formed in the cells where they produce haustoria which are often dichotomously branched—the 'arbuscules' (Fig. 12.4c). On the main hyphae there are often intercalary or apical swellings—'vesicles'. Thus, the infection gets its name from the abundant occurrence of these vesicles and arbuscules.

Although they are so common and are likely to be of major economic significance, it is only recently that studies have begun on the relationship of fungus to host. It is becoming increasingly clear that there is a broad functional similarity to sheathing mycorrhizas. When grown in a nutrient-poor soil, plants infected with vesicular–arbuscular mycorrhizas show greater vigour and nutrient content than uninfected (although there is little difference in nutrient-rich soil). The beneficial effects of infection are probably due to the large volume of soil explored by the mycelium connected to the mycorrhiza. Vesicular–arbuscular mycorrhiza occur on many trees in the South Temperate and Tropical areas, and they probably play an analagous role in mineral nutrition to that of the sheathing mycorrhizas of the North Temperate forests.

Orchidaceous mycorrhizas

The Orchidaceae are one of the largest and most diverse of plant families, with 500–800 genera and 20000–30000 species. Orchid seeds, which are produced in vast numbers, are very tiny and may contain only 8–100 cells. In almost all cases, unless they become infected with a fungus soon after germination, they do not develop beyond the protocorm stage. The fungus penetrates the cortical cells, forming a coil inside them (Fig. 12.4b). The hyphae later swell, become disorganized and are then digested by the host. The cell may again be infected and the cycle repeated. Digestion both restricts exploitation by the fungus and releases hyphal contents to the host cell. Only cortical cells are affected since other tissues are resistant— probably owing to production of phytoalexin-like inhibitory substances (p. 296). The fungi of orchid mycorrhizas are all basidiomycetes such as *Armillaria mellea*, *Fomes*, and *Corticium*.

Once infected, the protocorm continues development, but in all orchids it spends a shorter or longer period as a saprophyte. Some orchids, such as *Neottia nidus-avis* (Fig. 12.4b), may remain permanently saprophytic, and in others the protocorm may develop beneath the soil for as long as several years.

During the saprophytic phase of growth, the fungus supplies the orchid with carbon compounds. Thus, the direction of carbon flow in orchidaceous mycorrhizas is opposite to that in sheathing mycorrhizas. Another difference is that the fungi of orchidaceous mycorrhizas can break down cellulose and lignin in the soil and this presumably is the source of carbon for the orchid. The relationship with the fungus presumably continues in the case of those orchids which remain saprophytic throughout their life. It is less clear whether the fungus continues to aid in the nutrition of those orchids which are green when mature, or even if there is any reversal of carbon flow from orchid to fungus.

Ericaceous mycorrhizas

The mycorrhizal infection of the Ericales (heather and related plants) involve septate fungi which have never been conclusively identified. Ericaceous mycorrhizas form a more or less homogenous group in which there is an external sheath of mycelium, sometimes almost as well-developed as in sheathing mycorrhizas. Unlike sheathing mycorrhizas there is intracellular penetration, with the fungi sometimes even forming coils within the cells. As in orchidaceous types there may be repeated digestion of the hyphae by the host cell.

Much experimental investigation was carried out in the early part of the century into the physiological function of these mycorrhizas, but the results remain in doubt since it was claimed, for example, that the infecting fungi could fix nitrogen, a view no longer credible.

A complex relationship has recently been demonstrated for *Monotropa*, a saprophyte. This shares a common mycorrhizal fungus with trees like spruce and pine, and it has been demonstrated that sugars move from the tree, into the fungus, and then into *Monotropa*. Such a relationship is also found in some saprophytic orchids, and has been termed 'epiparasitism'.

Legume nodules

Only certain prokaryotes can carry out biological nitrogen fixation, and they may be either symbiotic or non-symbiotic. In terrestrial ecosystems it is believed that symbiotic fixation greatly exceeds non-symbiotic, and the outstanding example of symbiotic fixation is the association between bacteria of the genus *Rhizobium* and roots of Leguminous plants.

The Leguminosae are one of the larger angiosperm families, with 12000–14000 species. About 90% of the subfamilies Papilionatae and Mimosoideae, and about 30% of the remaining family (the Caesalpinoideae, usually considered the most primitive) form nodules. Legumes are most prevalent in the tropics, believed to be their place of origin. About 200 species are of agricultural importance, as various pea and bean crops, forage crops and ornamentals.

The essential characteristic of a *Rhizobium* is that it is a Gram-negative rod which can form a nitrogen-fixing symbiosis with a legume. Certain

Rhizobium strains form effective associations only with certain legume species, and this led to the original classification of *Rhizobia* and legumes into 16 cross-inoculation groups, with the groups of *Rhizobia* being given specific rank, e.g. *Rh. trifolii*, *Rh. lupinii*, etc. More recently, some *Rhizobia* have been found to infect more than one group of legume and the taxonomy has become complex. *Rhizobia* are widespread as free-living bacteria in the soil, although the soil population is greater if legumes are present. However, under natural conditions free-living *Rhizobia* cannot fix atmospheric nitrogen; indeed, they can be induced to fix nitrogen in culture only under very special conditions. It was previously believed that *Rhizobia* could form nodules only on the roots of legumes, but there is a recent report of their occurrence in nodules on the roots of *Trema aspera*, a member of the *Ulmaceae*.

The infection of a root and formation of a nodule occurs in the following way. Free-living *Rhizobia* aggregate around root hairs in response to some unidentified attractant exuded by the root. In many (though not all) cases the root hair curls. The causal agent has again not been identified, though secretion of indole-acetic acid by the bacteria is one widely suggested possibility. In the next stage, the host root secretes extracellular polygalacturonase in response to a bacterial stimulus; this seems to be the stage when host/symbiont specificity begins to operate. The function of the enzyme appears to be to weaken the host cell wall so that it begins to invaginate to form a structure called the infection thread. Bacteria become enclosed in the infection thread which moves through the cortex. The thread has a cellulose cell wall and seems to be entirely of host origin. As the infection thread proceeds through other cells, it normally passes close to their nuclei. Coincident with the passage of the thread, cortical cell division is stimulated to form the bulk of the eventual nodule tissue. The infection thread usually terminates in a polypoid host cell. By some means not entirely clear, the bacteria become deposited in the host cell where they become enclosed in vesicles or host membranes. The bacteria may undergo several divisions, but then swell to about 40 times the volume of the original bacterium to become bacteroids—often with simple branches to give 'Y' or 'X' shapes. The infection thread may branch and spread through other tetraploid cells as they are formed. In the mature nodule, many small groups of bacteroids enclosed in vesicles occur in each infected host cell. A pigment, leghaemoglobin, is also formed by host tissue and colours the nodule pink. Only if this is formed can the nodules fix nitrogen.

There are no major differences between *Rhizobia* and other nitrogen-fixing prokaryotes in the characteristics of the nitrogen fixing enzyme nitrogenase. In all cases, the enzyme contains two protein components, a molybdo-iron protein and an iron protein. When the proteins from different bacteria are recombined they will frequently form an effective nitrogenase, although homologous recombinations will always produce a more effective enzyme than heterologous—indicating that there are expected minor differences between different nitrogenases. The Mo–Fe component of *Rhizobium* will form an effective enzyme when combined with the Fe protein from *Azotobacter vinelandii* or *Bacillus polymyxa*, but not that from *Clostridium pasteurianum* (in general, components from anaerobic bacteria fail to show cross-reactivity with those from aerobic types).

Nitrogenase is very sensitive to oxygen, yet the fixation of nitrogen requires high levels of energy—normally provided much more readily by aerobic than anaerobic metabolism. The legume nodule probably resolves these conflicting features in the following way. It is generally believed (though not yet completely proven) that the function of the leghaemaglobin is to combine with molecular oxygen so that the nitrogenase is not inhibited, but the combined oxygen can become available for oxidative metabolism in the host cytoplasm near the bacteroids. The original source of energy for fixation is host photosynthesis, which also provides the carbon skeletons necessary to combine the ammonia which is the first stable product of fixation. This relationship between photosynthesis and nodule fixation is illustrated by the fact that in soybeans the amount of nitrogen fixation can be increased fivefold by increasing the CO_2 concentration around the leaves (i.e. the rate of photosynthesis is limited by CO_2 concentration).

The legume nodule is thus an excellent example of a truly mutualistic association. The bacteria provide the key enzyme which no eukaryote is able to manufacture. The host plant root provides the optimum environment in which there is an abundant supply of energy coupled with protection of the enzyme against molecular oxygen. Structural organization of the nodule also promotes rapid removal of the products of nitrogen fixation (the accumulation of which inhibits fixation).

Nodules of non-leguminous plants

A number of non-leguminous plants (e.g. *Alnus*,

Ceanothus, Hippophae) also form root nodules capable of nitrogen fixation. The prokaryote symbiont has not been conclusively identified though it is definitely not *Rhizobium*. The most commonly observed organisms in the nodules are believed to be actinomycetes, though the crucial test of successfully re-inoculating hosts with isolates has yet to be carried out. At least 300 species belonging to 14 genera of flowering plants can form these nodules. They have a world-wide distribution, and some of them are particularly common in marginal areas where vegetation is not well established. For exple, *Casuarina* is common throughout the tropics growing on coral sand on coral islands. Such plants may be of particular importance in natural ecosystems, though none are of major agricultural importance.

Non-leguminous root nodules have been much less studied. They do not contain any haemoglobin-like pigments, though in many other aspects the processes of infection and nodule formation resemble that described for legumes.

The leaves of some woody members of the Rubiaceae, e.g. *Pavetta*, produce nodules which contain a nitrogen-fixing species of *Mycobacterium* (p. 171), *M. rubiacearum*.

Books and articles for reference and further study

HALE, M. E. (1974). *The Biology of Lichens.* 2nd Edn., Edward Arnold, London, 181 pp.

HARLEY, J. L. (1969). *The Biology of Mycorrhiza.* 2nd Edn., Hill, London, 329 pp.

JENNINGS, D. H. and LEE, D. L., eds. (1975) *Symbiosis. 29th Symposium Soc. exp. Biol.* Cambridge University Press, Cambridge, 633 pp.

MARKS, G. C. and KOZLOWSKI, T. T., eds. (1973) *Ectomycorrhizae, their Ecology and Physiology.* Academic Press, London and New York, 444 pp.

SANDERS, F. E., MOSSE, BARBARA and TINKER, P. B., eds. (1973) *Endomycorrhizae.* Academic Press, London and New York, 626 pp.

NUTMAN, P. S. and MOSSE, BARBARA, eds. (1963) *Symbiotic Associations. 13th Symposium Soc. gen. Microbiol.* Cambridge University Press, Cambridge, 356 pp.

Chapter 13

Diseases of plants

Introduction

Symptoms of disease in plants result from the attack of pathogenic organisms, including fungi, bacteria and viruses. It has been estimated that plant diseases cause an 18 to 20 per cent reduction in world agricultural yields. Plants also suffer from a number of disorders caused by environmental conditions and not due to any pathogen. Damage is also caused by insect pests and nematodes, which are beyond the scope of this book.

The many micro-organisms parasitic on plants include bacteria, viruses and fungi. Among fungi, the phycomycete *Phytophthora infestans* can produce devastating losses of potato crops, and was the cause of the Irish potato famine in the nineteenth century that led to extensive emigration. Rice crops are often drastically reduced by the blast fungus, *Pyricularia oryzae*, a Deuteromycete. An exceptionally virulent strain of the ascomycete *Ceratocystis ulmi* has been responsible for the decimation of elms in the United Kingdom by Dutch Elm Disease (Fig. 13.1a). Among Basidiomycetes, the rust and smut fungi are entirely composed of pathogens. Some bacteria, mostly non-sporing Gram-negative aerobes, cause plant diseases: fireblight, caused by *Erwinia amylovora*, is a serious disease of pears and related ornamental plants. Although actinomycetes are so common in the soil, few cause disease in plants. *Streptomyces scabies* causes the extremely important scab disease of potato, and other streptomycetes cause scab of mangel and sugar beet, and pox or soft rot of sweet potato. It is now recognized that the Mycoplasmata (p. 168) are important agents of disease, being responsible for many disorders, e.g. the 'yellows' diseases, formerly attributed to viruses. The viruses themselves cause many diseases that reduce crop yields, e.g. barley yellow dwarf virus and the complex of viruses that attack potatoes.

The macroscopic symptoms of plant diseases vary widely depending upon the host and the pathogen and are also greatly affected by external factors such as temperature and daylength. For some diseases (e.g. powdery mildews) the presence of the pathogen is the most noticeable feature to the naked eye. For many others the plant response, which may take the form of altered pigmentation, stunting, localized or general necrosis, wilting, etc. is characteristic. Alteration in the reproductive capacity of host plants as is caused by many smut fungi, is a commonly observed result of infection. Another frequent symptom is the formation of galls, i.e. undifferentiated masses of tissue arising as a result of excessive cell division and enlargement (Fig. 13.1b).

Symptomology is important in the identification of plant viral diseases, the development of symptoms on a range of indicator plants often being diagnostic. A virus disease may be systemic, that is distributed throughout the plant, or localized, producing only local lesions (necrotic or chlorotic spots) on those parts of the plant actually inoculated. The same virus may produce a systemic infection on one host and only local lesions on another. Thus most strains of tobacco mosaic virus are systemic on commercial tobacco (*Nicotiana tabacum*) but cause only local lesions on the related *Nicotiana glutinosa* (Fig. 13.2a and b).

Alterations of pigmentation are amongst the commonest symptoms of systemic virus infection. Such changes may take the form of generalized yellowing such as occurs in sugar beet following infection by beet yellow virus or may assume a pattern of restricted chlorosis usually referred to as mosaic or mottle (Fig. 13.2a and d). Infected plants are often stunted compared with healthy plants, may crop poorly (Fig. 13.2c) and often bear deformed or mis-shapen fruit. A few viruses are deliberately maintained for their effects on foliage and flowers, for example, the 'breaking' of the anthocyanin pigment in tulip petals caused by tulip mosaic virus. Environmental factors greatly affect symptoms of virus infection such that it is possible to observe totally different symptoms in summer and winter. For example, infection of glasshouse tomatoes by tobacco mosaic virus in winter leads mainly to a reduction in the area of the leaf lamina, described as the 'fernleaf' effect. In summer a typical and often severe mosaic pre-

Fig. 13.1 Examples of plant diseases. (**A**) Dutch Elm Disease (caused by *Ceratocystis ulmi*): much of the tree is already dead. (**B**) Club root of cabbage (caused by *Plasmodiophora brassicae*). Plant with large gall on main root. Aerial parts of plant stunted and wilted. ($\times\frac{1}{2}$). (**C**) Leaf spot of cabbage (caused by *Mycosphaerella brassicicola*). The individual spots are small but numerous. ($\times\frac{1}{3}$). (**D**) Common scab of potato (caused by *Streptomyces scabies*) on tuber. ($\times\frac{3}{4}$). (**A**, Miss L. R. Cooke; **B** and **D**, W. C. Moore)

dominates. The provision of favourable growing conditions and in particular high temperatures may enable crop plants to 'grow away' from infection and symptoms may become masked. Thus above 20°C symptoms of crinkle and mosaic on potato become masked although plants still contain virus and can act as a source of infection for other plants.

The relation between host and parasite

Types of parasitism

The dependence of parasites on their hosts varies greatly. The behaviour of a parasite is often classified in two groups on the basis of its mode of nutrition when infecting the host.

Fig. 13.2 Examples of virus diseases of plants. (**A**) Tobacco plant (*Nicotiana tabacum* var. White Burley) systemically infected with tobacco mosaic virus (TMV). (**B**) Leaf of *N. glutinosa* showing necrotic lesions caused by TMV. Left half of leaf inoculated with 10^{-3} mg/litre suspension of TMV; right with 10^{-2} mg/litre suspension. The concentration of virus in a suspension can be estimated by counting the number of lesions caused. (**C**) Leaf roll of potato; left, healthy plant with tubers produced from it; right diseased plant, showing stunting, curled leaves and reduced crop. (**D**) Leaf of White Burley tobacco systemically infected with spotted leaf virus. **B**, photo by W. C. Moore; **A**, **C** and **D** by courtesy of Rothamsted Experimental Station (copyright).

Parasites behaving as *biotrophs* derive their nutrient from living host cells. *Nectrotrophic* parasites, by contrast, derive their nutrient from cells that they have already killed; they differ from saprotrophs (saprophytes) because these cannot kill living host cells.

Formerly, parasites used to be classified on the basis of their ability to grow in axenic culture (p. 69), but now that some strains of rust fungi can be thus grown the distinction between biotrophy and necrotrophy is a more valid criterion.

Most biotrophs cause only minimal damage to the host tissue; frequently their host range is narrow, and often they cannot be grown in axenic culture. The fungi causing rust and powdery mildew diseases are typical biotrophs. The hyphae of bio-

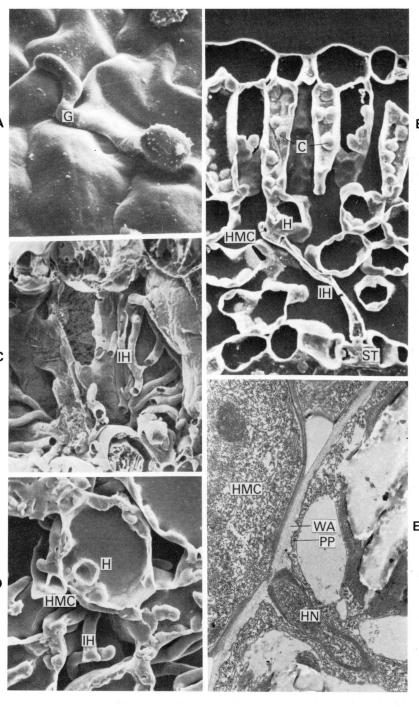

Fig. 13.3 Infection of French bean leaves by the rust fungus *Uromyces phaseoli*.
(**A**) Uredospore of *U. phaseoli* germinating on leaf surface. Critical point dried (×650).
(**B**) Penetration has occurred through a stoma, followed by intercellular growth, formation of a haustorial mother cell, and penetration of a plant cell forming an haustorium. Etched resin-embedded preparation (×600). (From Pring and Richmond (1975) *Trans. Br. mycol. Soc.*, **65**, 291–294, Plate 37, Fig. 1) (**C**) Part of a heavily infected leaf showing intercellular hyphae ramifying throughout the region of the palisade cells. Freeze fracture (×450). (**D**) Infected tissue showing intercellular hyphae, haustorial mother cell and haustorium supported on the haustorial neck. Etched resin-embedded preparation (×850). (**E**) Penetration of a leaf cell. This thin section has just missed the actual penetration pore. The invagination of the host plasmalemma by the invading fungus is shown, and also host wall apposition around the penetration site. Fixed in glutaraldehyde and osmium tetroxide (×10 000). G = germ tube; H = haustorium; HMC = haustorial mother cell; HN = haustorial neck; IH = intercellular hypha; ST = stoma WA = wall apposition; PP = plant plasmalemma; C = chloroplasts. (Photos: R. J. Pring)

trophs within host tissue are often confined to the intercellular regions, although many form special branches (*haustoria*) that penetrate the walls and invaginate the host plasmalemma (Figs. 13.3 and 13.4a). It is assumed, but without final proof, that these haustoria absorb food from the host and pass it to the hyphae. If the pathogen ultimately kills the host, it usually sporulates first.

By contrast, necrotrophs rapidly kill the host cells and then live saprophytically on the dead tissues. They frequently have a wide host range, and can be grown axenically. Typical examples are species of

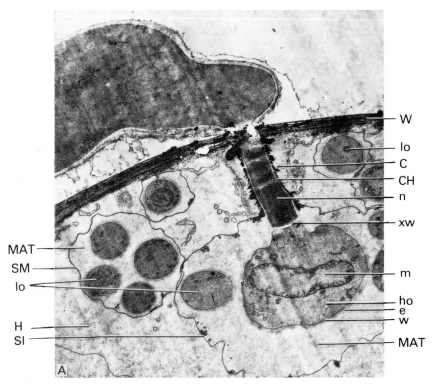

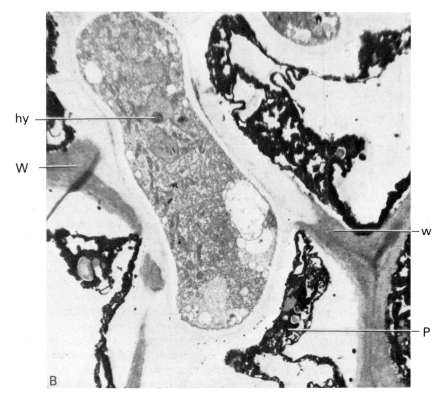

Fig. 13.4 (**A**) Electron micrograph of a cell of *Hordeum vulgare* infected by the biotrophic pathogen *Erysiphe graminis* (KMnO₄ fixation). White spots are tears in the section. The haustorial neck connects the external hypha to the body of the haustorium through the epidermal cell wall. Since this is not a median section through the haustorial septum, the septal pore is not evident. ho=haustorial body, C=collar, CH=channel between haustorial neck and collar, e=fungal ectoplast, H=host cytoplasm, lo=haustorial lobe, m=mitochondrion, MAT=sheath matrix, n=haustorial neck, SI=sheath invaginations, SM=sheath membrane, w=fungal wall, W=host wall, xw=cross wall. (From Bracker (1968) *Phytopathology*, **58**, 12–30, Plate 8) (×6500)

(**B**) Pear fruit tissue infected by the fungus *Monilinia (Sclerotinia) fructigena*, a necrotrophic pathogen. An intercellular hypha (hy) has caused localized wall (w) degradation and invaded a cell, pushing back the dead protoplast (P), its plasmalemma still intact (×3500). (Photo: F. D. Calonge)

Botrytis and *Pythium*, and bacteria that cause soft rots.

Some parasites can switch from the biotrophic to the necrotrophic habit in the course of their life cycle. Thus the cereal smut fungi, although present in the host throughout its development, induce no visible symptoms until the host reaches its reproductive stage, when they attack the floral parts and become necrotrophic.

Spread of pathogens

The methods by which a parasite reaches a new host plant or a new part of the original host are many and varied.

It seems likely that many soil-inhabiting fungi parasitic on plants are capable of growing at least short distances through the soil and thus reach the host roots in the form of active mycelium. Some, such as *Armillaria mellea*, the honey agaric, which is a serious parasite of many woody plants, produce rhizomorphs which can grow through the soil and attack nearby roots with which they come into contact. Others produce resistant sclerotia on the surface of perennating parts of plants. On germination the sclerotia produce hyphae. For example, *Rhizoctonia solani* produces flat black sclerotia on potato tubers, which germinate giving hyphae that spread to and attack the susceptible young shoots of the host.

Many plant pathogenic fungi are seed-borne. They may be present as dormant mycelium within the seed (as with the loose smuts of wheat and other cereals), as spores adhering to the surface (as with bunt or covered smut of wheat, or the species of *Cochliobolus* which cause foot rot of cereals) or as spore-bearing structures (as with celery leaf blight, caused by *Septoria apiicola*, which produces spores in pycnidia), or as sclerotia. Seed-borne fungi are in a position to attack on germination of the seed or to cause 'damping-off' before or soon after seedlings emerge, and a variety of foot-rots or anthracnose diseases in which the plant is weakened even if it survives.

A few virus diseases (e.g. lettuce mosaic) are seed-borne. Although the incidence is usually low, this method of survival can be important in both providing an initial focus of infection within a crop, and in extending the geographical range of the pathogen.

The aerial parts of plants are attacked by infection units (propagules) carried to them by wind, water (including rain splashes), insects or other animals, or by contact with infected plants.

The infection unit of parasitic bacteria is the bacterial cell itself, and raindrops or insects are common dispersal agents. With fungi the infection unit is usually a spore and here the process is three-fold; namely the release of the spore from the parent mycelium or fruit-body (spore discharge), its transport to the host plant (spore dispersal), and the alighting of the spore on the host surface.

Many spores, e.g. those of the powdery mildew fungi, are particularly adapted to wind dispersal because of the low shear forces needed to liberate them. Air currents are also important in the dispersal of ascospores which, in many fungi, are released near to soil level and must thus rely on atmospheric turbulence if they are to be carried large distances. Water ('splash') dispersal, by contrast, is usually limited to a few metres. The surface of these propagules is often adapted in being hydrophilic and sometimes mucilaginous.

Weather is of importance in determining the production of spores or other propagules, their successful transport to the new host and their development up to the moment of initial penetration of the host tissues. Many fungus spores will germinate only when submerged in water. Those of the powdery mildews are an exception in germinating best in humid air. For the majority a thin film of rain or dew on the host surface will suffice, but if this dries up before· penetration takes place the germ tubes may die. Temperature also influences the rate of germination.

The germination of spores or other propagules may also be influenced by secretions from the host plant. Thus germination of conidia of *Botrytis cinerea* is stimulated or inhibited by volatile or diffusible substances given off by various plants; pollen is also stimulatory. Most fungal spores contain sufficient essential nutrients to permit germination, but growth of germ tubes soon ceases in the absence of an external supply of food.

Many plant virus diseases are transmitted entirely by insect vectors under field conditions. The vector does not usually show any ill-effects from the presence of the virus, although some plant viruses are known to multiply within their insect vectors. A few other arthropods also act as vectors, e.g. the eriophyid mite which transmits wheat streak mosaic disease.

Most vectors of plant viruses are sucking insects (aphids, leafhoppers and, to a lesser extent, thrips, whiteflies and mealy bugs). A few viruses especially those which are very stable in sap extracts, e.g. tobacco mosaic and potato X viruses are transmissible by biting insects such as grasshoppers and various beetles. Transmission by biting insects during feeding is probably largely mechanical and differs little in principle from the sap inoculation techniques used for experimental transmission of some plant viruses.

Transmission by sucking insects is more complex and there may be considerable specificity between virus and vector. Some viruses can be transmitted by aphids which have fed for only a short period (minutes) on infected plants. Such viruses appear to be stylet-borne and persist in the vector for only a period of hours. Virus diseases of this group are spread most rapidly by 'restless' aphid species such as *Myzus persicae* which indulge in much exploratory probing with their mouthparts.

By contrast a number of viruses transmitted by aphids and many or all of those transmitted by leafhoppers, whiteflies and thrips remain in the insect throughout its life and are termed *persistent*. Characteristically there is a latent period of hours or even days following acquisition, during which the insect is unable to transmit the virus to healthy plants. This latent period is correlated with the time taken for virus particles to pass down the gut wall, into the bloodstream and thence into the saliva. Some viruses of this group are capable of multiplication within the vector, for example potato leaf roll virus within *Myzus persicae*; and some have been shown to be transmitted to the insect progeny for a number of generations, raising speculation as to whether such viruses are primarily parasites of plants or of insects.

Various arthropods, especially members of the leafhopper group are responsible for transmitting various 'yellows' diseases associated with mycoplasma-like organisms. Generally these organisms show a similar relationship to their vector as is exhibited by the persistent group of viruses, with which they were at first confused.

Non-insect vectors include certain soil-living nematodes, capable of transmitting Arabis mosaic and related viruses within strawberry and some other crops, and the chytrid fungus (p. 195) *Olpidium brassicae*, able to transmit tobacco necrosis virus and bigvein virus to lettuce, the virus entering the plant during infection by the zoospores.

Viruses and mycoplasmata are not the only micro-organisms transmitted by vectors. Insects are important in the transmission of some bacterial diseases, e.g. fireblight of apple and pear. Infection by the causal bacterium *Erwinia amylovora* occurs through nectaries to which it is carried by bees and other pollinating insects. Fungal propagules are also known to be disseminated by insect vectors.

Conidia of the ergot fungus *Claviceps purpurea* and spermatia of certain rusts are produced together with sugary secretions attractive to small insects. Spores of the campion smut fungus *Ustilago violacea* are produced in the anthers where they replace the pollen. These spores are then spread by the insects which normally pollinate the flowers. *Ceratocystis ulmi* the cause of Dutch elm disease is spread by a bark boring beetle.

Vertebrate animals can also act as disease vectors. In the USA the chestnut blight fungus *Endothia parasitica* is thought to be spread by woodland birds, in particular woodpeckers and tree creepers. Man himself is probably the prime vertebrate vector; his agricultural operations are the agencies of both dispersal and inoculation for many pathogens.

Penetration of the host

Many bacteria and some fungi enter the host only through natural openings, such as stomata (Fig. 13.3b) or lenticels, or through wounds. When fungal spores or bacteria alight on the newly cut surface of a vascular bundle they may actually be sucked into a vessel, and there is evidence that some disease organisms, such as the apple canker fungus (*Nectria galligena*), enter passively in this way.

The relative importance of mechanical pressure and of enzymes that degrade and hence weaken the cuticle and cell wall is still controversial. Penetration of the intact cuticle by certain fungi is almost certainly mechanical. Such fungi attach themselves firmly to any hard surface and may form an *appressorium* (pad) at the point of contact. From this, fine penetration hyphae grow out and bore through the cuticle and epidermal cell wall. Artificial membranes of the right degree of hardness are similarly penetrated.

Electron microscopy has also supported the theory that mechanical pressure is involved, because the fibrils of the host wall can often be seen to be distorted. By contrast, in other host–pathogen systems changes in the electron density of the walls near the penetration point suggest enzymic attack.

After the fungus has entered a susceptible host by one means or another, the hyphae may advance through the host tissues. In some diseases the advance of hyphae is limited so that small leaf spots form (Fig. 13.1c); in others, such as rots and leaf blights, progressive lesions result.

For many viral diseases the agent of transmission, the vector, is also the agent of inoculation.

Plant viruses lack any mechanism of active penetration into plants and are totally reliant on vectors and wounds.

Disfunction in the host

Pathogens affect host cells in many different ways which are reflected in the disease symptoms that develop.

1 **Structural effects**, resulting from enzymic action, frequently occur in the cell wall area (Fig. 13.4b), due to the activity of pectolytic or cellulolytic enzymes.

2 **Physiological effects** also occur. Nectrotrophic pathogens bring about drastic increases in the permeability of the membranes of host cells and organelles. This leads to leakage of intracellular host metabolites, and to many interactions between them. Typical of these is the oxidation of host polyphenols, by means of the enzyme polyphenol oxidase, leading to the brown colour of much diseased tissue. In wilt diseases, the pathogen affects the plant's vascular system so that water losses from the leaves by transpiration cannot be made up from the roots.

The action of biotrophic pathogens is more subtle, and can involve changes in respiration and photosynthesis rates, and changes in the transport of host metabolites, with the infected tissue acting as a metabolic 'sink'. In viral and at least some fungal infections, there are drastic changes in the proteins synthesized by the host cells.

TOXIC AGENTS

The toxic compounds which are involved in pathogenicity are chemically diverse but for convenience may be classified as follows:

(i) **Extracellular enzymes**, which attack specific plant structures, e.g. cellulose in the cell wall, pectin in the middle lamella region. There are various biochemical types of enzyme, but a full description is beyond the scope of this chapter.

(ii) **Non-enzymic toxins**, which may be of high or low molecular weight. Examples of compounds of high molecular weight are polypeptides and bacterial polysaccharides. Simpler compounds include fumaric and fusaric acids. Of particular interest are the *host-specific toxins*, which act only on a particular host, or even a particular susceptible cultivar.

Toxins often act at a distance from their site of secretion—thus, the 'silver leaf' toxin secreted by *Chrondrostereum purpureum* is produced by the fungus in the trunk of the affected tree, but the marked silvering symptoms appear in the leaves (which are not invaded by the fungus) as a result of the separation of the epidermis by the toxin from underlying cells.

(iii) **Growth substances**, of which gibberellins secreted by *Gibberella fujikuroi* afford a good example. These compounds are responsible for the excessive elongation of the internodes of rice seedlings infected by the fungus, giving rise to the 'bakanae' disease. Ethylene, often produced during infection, is also in this group.

The action of groups (ii) and (iii) is characterized by modification of host metabolism rather than structural degradation.

Resistance of plants to infection

Plants resist the entry and internal spread of pathogenic micro-organisms in a number of ways.

DISEASE ESCAPE

Plants may escape infection if they pass through the susceptible stage at a time or under conditions when the pathogen is absent. Thus early potatoes in England often escape late blight (caused by *Phytophthora infestans*) since their growth is completed and they are harvested before the period in late July and August when the inoculum has developed and the conditions are warm and humid enough to permit sporulation and consequent spread and development of the disease.

STRUCTURAL FACTORS

Static The presence of a thick cuticle has often been thought to confer resistance to attack by parasitic fungi. Reappraisal of the significance of the cuticle as a barrier to infection now indicates that it is probably of little direct importance. Indirectly, however, it may be involved, e.g. resistance of plum varieties to brown rot (*Monilinia* spp.) has been correlated with a tough skin, but the effect of this is probably to prevent minor injuries which wound parasites need for entry.

Some varieties of strawberry appear to be relatively resistant to *Botrytis* grey mould because the fruits are borne clear of the foliage and thus dry more quickly after rain or dew.

Dynamic The formation of structural barriers. Attempted invasion by a pathogen is often countered by a structural change as the host's response to the presence of the invader, or as a natural reaction to wounding. Thus scab diseases of potato are usually limited to the surface layers of the host by the active formation of a cork barrier in response to attack. The speed of formation of lignin, or of cork or callus over a wound surface, is often important in disease resistance. Varietal resistance to *Fusarium oxysporum* f. *narcissi,* the cause of basal rot of *Narcissus*, is correlated with the degree to which the natural wounds made by the emerging roots are sealed off by cork formation.

CHEMICAL FACTORS

There is now good reason to believe that, even more important in the resistance of potential host plants, are various chemical substances. These may either be already present in healthy plants, or formed in response to attempted invasion.

Static Many plants contain compounds toxic to micro-organisms and apparently in concentration sufficient to inhibit a potential invader. The compounds for which such a role has been postulated include polyphenols and their glycosides, coumarin derivatives, oxazolinones and alkaloids. In addition, the pH of the cell sap may act as an important deterrent to the invasion of organisms with a different pH optimum for growth. Nutritional factors have also been shown to play an important role in resistance. A high nitrogen level often leads to susceptibility. The sugar level of plants has been suggested as an important determinant of their susceptibility to certain disease organisms, some pathogens being favoured by high and others by low levels.

Dynamic Great interest has recently been focused on the significance of chemicals which may or may not normally be present in healthy plant tissue, but substantial amounts of which are formed by the host as a result of attempted invasion.

A great advance in the study of host–parasite relations followed Müller's suggestion that substances not normally present in the host tissue might be formed by the plant in response to the presence of a potential parasite. He termed such compounds *phytoalexins*. The hypothesis gained great impetus when Cruickshank demonstrated a phytoalexin, which he named *pisatin*, in pea pods subjected to attempted invasion by conidia of *Monilinia fructicola*, which is not a parasite of peas. Later, he and his colleagues identified pisatin chemically. Many other phytoalexins have since been identified, and certain resistance factors described earlier by other

workers may perhaps be regarded as phytoalexins though there are divergent views on terminology.

A phytoalexin may now be defined as a *fungus-induced host metabolite, causally controlling fungal growth in a host–parasite interaction*. The phytoalexins appear to be specific to the host which produces them, but non-specific qualitatively to the invader, and in this respect differ from antibodies produced by higher animals in immunological responses: also, chemically they are generally aromatic compounds and are not proteins.

Although phytoalexins can be formed in response to a chemical or physical stimulus, their synthesis by the host may normally be triggered by specific *elicitor* molecules produced by the would-be pathogen or even liberated by host particles.

The success of a pathogen in its attempt to infect a host which is capable of producing a phytoalexin has been envisaged as being determined by two factors: the quantity of phytoalexin which the particular pathogen induces the plant to produce (successful pathogens generally being less stimulating) and the toxicity of the phytoalexin to the given invader (frequently being greater to the non-pathogens).

The study of phytoalexins has not so far provided a complete explanation of host–pathogen specificity, especially that involved in single-gene resistance.

Another 'dynamic' effect involving host metabolism, but not of intact living cells, is the phenomenon of hypersensitivity, which has been studied with particular reference to biotrophic parasites. If the response of the host cell to invasion is early death, with accompanying loss of the semipermeability of the cell membranes, the biotroph can no longer live on the tissue, and as a result its attack is limited. With necrotrophic parasites, somewhat similar events may occur when the semipermeability of host cell membranes is destroyed, either by physical wounding or by the pathogenic activities of the invader.

Control of plant diseases

Intelligent control of a plant disease depends upon a detailed knowledge not only of the life cycle of the parasite, particularly those stages at which it is most vulnerable to interference, but also of its mode of attack and the environmental factors which influence its growth and reproduction.

Cultural control

Much can be achieved by *plant hygiene*, including

the destruction of diseased crops and the remains of harvested ones, and the removal of species capable of acting as alternate hosts.

The use of *clean planting stock* eliminates one source of disease. Virus diseases of many crops, including potatoes, strawberries and apples, are kept in check by such means: potatoes for 'seed' are commonly obtained from Scotland and Ireland, where the comparative absence of aphis vectors, together with careful 'roguing' of infected plants in the crop reduces the number of infected tubers produced. Apple rootstocks and cultivars are now available from stock which has been freed of virus by carefully controlled heat therapy. *Legislation and import controls* are designed to prevent the spread of disease from one country to another. An example of successful legislation is the greatly reduced incidence of the wart disease of potato (caused by *Synchytrium endobioticum*) following the introduction of laws to restrict planting of susceptible cultivars.

Adjustment of cultural conditions may also prevent or lessen the effects of disease. For example, many parasitic fungi are favoured by surface wetness, and wider crop spacing results in more rapid drying of the plants with a consequent decrease in disease incidence. Similarly, many diseases, such as barley mildew, are favoured by a high level of nitrogen nutrition of the host. Water content, H-ion concentration of the soil, and light intensity are other factors which influence the incidence of diseases.

Chemical control

Chemical treatment is often necessary for successful disease control, since changes in cultural practice are frequently insufficient by themselves to keep diseases in check. Agricultural practice results in many similar plants growing in close proximity, thus greatly facilitating the spread of pathogens and increasing the need for control measures.

The chemical control of plant disease usually rests on *selective toxic action*, and discovering chemicals which although toxic to the pathogen are non-toxic, at least at the applied dosage level, to the plant. This requirement is not easily met, and many compounds of high toxicity to bacteria or to fungi have not found general application because they are *phytotoxic*, i.e. toxic to the plant. Apart from a lack of phytotoxicity, a low mammalian toxicity is also necessary, to obviate hazards to the spray operator and to the consumer of food crops.

Biochemical studies on mode of action suggest that many fungicides inhibit a fairly wide range of enzymes in their 'target' organism, and do not owe their selective action to the fact that they inhibit enzymes present only in the pathogen. Their selectivity is due rather to the fact that they are taken up in much greater quantities by the parasite than by the host, which may be protected physically (for example by its cuticle) at the site of application.

A chemical may be used in one of several ways to control plant disease.

AS AN EXTERNAL PROTECTANT

Applied to a plant to prevent its becoming infected by fungus or bacterium. Ideally, a continuous deposit of chemical is applied as a spray or dust. In practice, complete coverage is rarely obtained, and the deposit after the spray dries is eroded by rain and, to a lesser extent, other factors. Nevertheless, the residue of a good protectant fungicide will prevent most infection for several days, after which a further application is often made to protect new growth. The toxic principle present in the residue may reach the fungal spore by diffusion through a layer of water or it may act in the vapour phase (Fig. 13.5), particularly against spores which do not require free water to germinate. Fungicides for this type of application require a comparatively low water solubility or the ability to adsorb strongly to the host surface.

AS AN ERADICANT

Applied to destroy a fungus already established, or becoming established on or in the host plant (or in its immediate environment). For this purpose, the chemical must be capable of killing the fungus without damage to the plant. Few compounds possess these attributes, but organo-mercurials and substituted phenols have often proved useful, while many volatile hydrocarbon soil fungicides and nematicides operate in a similar way but, as they are too phytotoxic to be used in the presence of plant roots, are usually used only in fallow soil. Compounds that prevent sporulation are also valuable in disease control: they resemble eradicants in reducing inoculum.

AS A SYSTEMIC

Capable of entering the living plant and acting against the fungus from within. In order that the plant is not adversely affected, great specificity is essential, together with the ability to be trans-

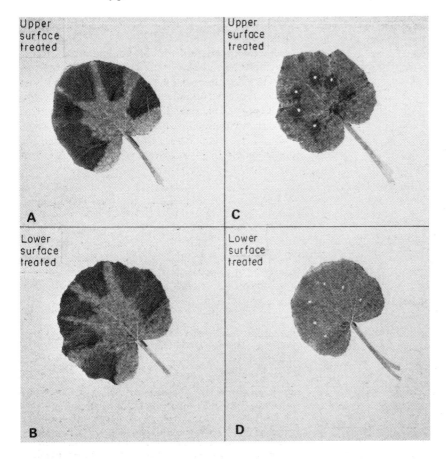

Fig. 13.5 Mechanisms of fungicidal action. Six drops, each of 0·001 ml and containing 5×10^{-9} mole of fungicide, were applied to either upper or lower surfaces of leaves of vegetable marrow. The next day the *upper* surface was inoculated with spores of *Oidium* sp., and then incubated in a ventilated greenhouse for a further 8 days, when the *upper* surface was photographed. Infected areas show light grey in the photograph, healthy areas are darker.

(**A** and **B**) Bupirimate. Activity through the leaf and to the margin indicates systemic activity.

(**C** and **D**) 4,6-Dinitro-2-(1′-propylpentyl)phenol. Absence of trans-laminar effect and a circular zone of inhibition around the deposit zone indicate vapour action. Note white spots indicating tissue necrosis at the site of application. (Photographs: Long Ashton Research Station, by courtesy of D. R. Clifford)

located. This usually occurs in the xylem flow and through the intercellular spaces. The systemic fungicide benomyl (a substituted benzimidazole) and the related compounds carbendazim and thiophanate-methyl, have been extremely successful against a wide spectrum of fungal pathogens, excluding phycomycetes. Other systemic fungicides are also extensively used. In some instances it seems that the applied compound is not itself fungicidal *per se*, but confers resistance to the host.

RESISTANCE TO FUNGICIDES

Is a factor of increasing importance, and strains of pathogens resistant to the systemic fungicides have frequently developed in the field. In practice, systemic fungicides are particularly liable to this because, to avoid effects on the plant, their action is often on a single site in the pathogen. Hence, a change in a single gene can overcome their toxicity. Most protectant fungicides, by contrast, act against a wide range of enzymes, and are thus less prone to this drawback.

Biological control

The desirability of avoiding the use of chemicals is a target that has long been recognized, but efforts to achieve disease control by the encouragement of organisms antagonistic or parasitic to the pathogen have only recently been successful. Thus, the attack of tree stumps by the virulent pathogen *Fomes annosus* can be prevented by the prior inoculation of the stump by the saprophytic basidiomycete *Peniophora gigantea*.

The control of plant parasitic eelworms by predacious soil fungi (p. 320) has had some success and is worth further investigation.

Control of vectors

Many viruses and some bacteria and fungi are carried by insects or by other arthropods and control of the vector, where possible, will also prevent the spread of the disease. This is not easy, since a single aphis may transmit a virus disease to several healthy plants, and it is seldom possible to achieve a hundred per cent kill of insect pests.

Breeding for host resistance

In theory, the ideal way of controlling plant diseases is by the use, or breeding, of resistant cultivars. The cereal rusts, certain wilt diseases, wart of potato, and a number of virus diseases, e.g. some of those attacking potato, have been successfully controlled in this way. Nevertheless, the success of breeding has often been short-lived, particularly with cereal rusts, and late blight disease of potatoes. The breakdown of resistance is due to the remarkably flexible genetic system of the pathogen, which results in the continual production of new races of modified pathogenicity which are rapidly 'selected' by confrontation with the host. Much recent study of the phenomenon has been based on the 'gene-for-gene' hypothesis of Flor, which states that for each gene which conditions reaction of the host to infection there is a corresponding gene in the parasite which controls pathogenicity. Resistance based on a single gene is thus easily overcome by the pathogen.

Tolerant or carrier varieties are also useful as a means of producing crops in the presence of a disease, although their use has the serious disadvantage that they are a danger to susceptible varieties of the same or other species growing near. Thus some cultivars of strawberry (e.g. Cambridge Vigour) show tolerance to levels of virus diseases which cripple others (e.g. Royal Sovereign), and the same type of phenomenon accounts for the tolerance of some hop cultivars to Verticillium wilt.

Some special methods used to control certain plant viruses

Heat treatment is sometimes used to free planting material from various pathogens. It has proved specially useful in freeing clonal material of vegetatively propagated plants from viruses. Plants are grown for periods of up to several months at elevated temperatures just below their own thermal death point. Sometimes this treatment is combined with the technique of meristem tip culture in which apical meristematic tissue is excised and cultured in an appropriate nutrient medium. If precautions are taken to exclude bacterial and fungal contamination a proportion of the meristems will develop into young plantlets some of which will be free of viruses present in the original plant. The success of these methods depends upon the differential sensitivity of virus and host to heat, and the erratic distribution of viruses in higher plants such that embryonic and meristematic tissues are often virus free.

The formulation of suitable control measures for a given disease clearly requires detailed knowledge of many factors, relative to the pathogen, the host plant, the environment, and the choice of crop-protection chemicals. These factors are so varied that each disease needs to be considered on its own. Nevertheless, progress is being made to lessen the losses caused by plant disease which man's 'exploding' population can ill afford.

Books and articles for reference and further study

AGRIOS, G. N. (1969). *Plant Pathology*. Academic Press, New York, 629 pp.

AINSWORTH, G. C. (1981). *Introduction to the history of plant pathology*. Cambridge University Press, Cambridge, 315 pp.

ALBERSHEIM, P. & ANDERSON-PROUTY, A. J. (1975). Carbohydrates, proteins, cell surfaces, and the biochemistry of pathogenesis. *A. Rev. Pl. Physiol.*, **26**, 31–52.

BROWN, W. (1965). Toxins and cell-wall dissolving enzymes in relation to plant disease. *A. Rev. Phytopath.*, **3**, 1–18.

CRUICKSHANK, I. A. M. (1966). Defence mechanisms in plants. *World Rev. Pest Control*, **5**, 161–175.

GARRETT, S. D. (1970). *Pathogenic Root-infecting Fungi*. Cambridge University Press, Cambridge, 294 pp.

HORSFALL, J. G. (1956). *Principles of Fungicidal Action*. Chronica Bot., Waltham, Mass., USA, 279 pp.

HORSFALL, J. G. and COWLING, E. B. (1977–80). *Plant Disease: an advanced treatise*. Vols. I–V. Academic Press, New York and London.

LARGE, E. C. (1940). *The Advance of the Fungi*. Cape, London, 488 pp.

MARSH, R. W. (1977). *Systemic Fungicides*, 2nd Edn., Longman, London, 401 pp.

TARR, S. A. J. (1972). *Principles of Plant Pathology*. Macmillan, London, 644 pp.

TORGESON, D. C., ed. (1968). *Fungicides*. (Vols. I and II). Academic Press, New York and London.

WHEELER, B. E. J. (1969). *Plant Diseases*. John Wiley, London, 374 pp.

WOOD, R. K. S. (1967). *Physiological Plant Pathology*. Blackwell, Oxford, 570 pp.

WOOD, R. K. S., BALLIO, A. and GRANITI, A., eds. (1972). *Phytotoxins in Plant Diseases*. Academic Press, London and New York, 530 pp.

BAWDEN, F. C. (1964). *Plant Viruses and Virus Diseases*. 4th Edn., Cambridge University Press, Cambridge, 361 pp.

ESAU, K. (1968). *Viruses in Plant Hosts*. University of Wisconsin Press, Wisconsin, USA, 225 pp.

GIBBS, A. and HARRISON, B. (1976). *Plant Virology: The Principles*. Arnold, London, 292 pp.

MATTHEWS, R. E. F. (1970). *Plant Virology*. Academic Press, London and New York 778 pp.

SMITH, K. M. (1972). *A Textbook of Plant Viruses*. 3rd Edn., Longman, London, 684 pp.

Chapter 14

Micro-organisms and Animals

Non-pathological relationships between micro-organisms and animals

Vertebrates and non-pathogen micro-organisms

Introduction

The skin and mucous membranes of animals are continuously exposed to contamination with micro-organisms. These contaminants include saprophytes and pathogens. Most organisms fail to establish themselves in any particular site on the body, partly as a result of the self-sterilizing action of many tissues and partly owing to competition between micro-organisms. Certain species, not normally pathogenic, are particularly suited to colonization of particular sites; some are present under almost all conditions, others may be there only temporarily. Thus in man *Escherichia coli* is a normal intestinal parasite whereas *Staphylococcus aureus* is present on the skin of only some individuals and only at certain times.

Organisms regularly present constitute the 'normal flora'. Little is known of the nature of the association between micro-organisms and the healthy host tissue, but it is thought that there exists a dynamic equilibrium rather than mutual indifference between the two. Some organisms of the normal flora may assume a pathogenic role under certain circumstances, e.g. when the host resistance is lowered by injury to tissues, or by diseases that diminish resistance. Pathogenic organisms may or may not be present in the normal flora but most have a special capacity to spread from host to host, establish themselves, and later invade deeper tissues from the site initially colonized.

Normal flora

The term 'normal flora' may be misleading since the environment can greatly influence the types of micro-organisms present. Nevertheless, there are certain bacterial species which are invariably found in particular sites in the healthy animal. Thus it is possible to give general descriptions of normal flora for a given animal species. The normal flora of corresponding sites in different animal hosts varies widely and is profoundly influenced by such factors as body temperature and diet. Most work on this subject has been done on man.

NORMAL FLORA OF THE HUMAN SKIN

The normal flora of the skin and sebacious glands is restricted because of the natural disinfecting secretions of the skin. It consists of a few species of bacteria including Gram-positive cocci (mainly *Staphylococcus albus* and *Sarcina* spp.) and diphtheroids. These are all non-pathogenic but sometimes the pathogenic species *Staphylococcus aureus* is found on the face and hands, particularly in individuals who are nasal carriers. These organisms persist even after intensive washing. Many other organisms, not generally considered to be part of the normal flora, may be present temporarily in numbers largely determined by the standard of personal hygiene, especially in moist crevices of the skin, and surrounding the body orifices, such as the anus. In the fatty and waxy secretions of skin, lipophilic yeasts are common; *Pityrosporum ovalis* occurring most commonly on the scalp and *P. orbiculare* on the glabrous skin. In the secretions of ears and genitalia, saprophytic acid-fast bacilli (e.g. *Mycobacterium smegmatis*) and diphtheroids are common. The tears secreted on the conjunctiva of the eyes contain the enzyme lysozyme, which attacks bacterial cell walls causing lysis, and this keeps the number of organisms in this site at a low level.

NORMAL FLORA OF THE HUMAN RESPIRATORY TRACT

The anterior nares (nostrils) are always heavily colonized, predominantly with *Staph. albus* and diphtheroids, and often with *Staph. aureus*, this being the main carrier site of this important pathogen. The healthy sinuses are sterile. Large numbers of other species of bacteria colonize the normal upper respiratory tract—predominantly mixed viridans streptococci (e.g. *Streptococcus salivarius*) and *Neisseria* spp. (e.g. *N. pharyngis*), and frequently pathogens, such as pneumococci *Haemophilus influenzae*, *Streptococcus pyogenes*, and sometimes meningococci (*N. meningitidis*) and allied organisms.

The trachea, bronchi, and pulmonary tissues are, in the healthy state, almost free of micro-organisms. Air is inhaled through the tortuous nasal passages which act as effective filters. Bacteria adhering to particles of dust are removed. Most organisms that pass this filter are trapped on the mucous surfaces of the nasopharynx and trachea, being swept upwards by ciliated epithelial linings of the tract, and are subsequently removed by expectoration or swallowing. There is thus no normal flora of the mucosa of trachea and bronchi.

NORMAL FLORA OF THE URINO-GENITAL TRACT OF THE HUMAN FEMALE

In the sexually mature human female the vagina is acidic, owing to the production of lactic acid from glycogen in the epithelial linings. Consequently the flora consists of mixed acid-tolerant organisms, including the predominant *Lactobacillus acidophilus* (Doderlein's bacillus), together with coryne-

bacteria and anaerobic streptococci. Mycoplasmas (p. 168) have been isolated on a number of occasions. Infections due to *Candida albicans* —called vaginal thrush—are particularly likely in pregnancy or in patients with diabetes. Before puberty and after menopause the vaginal secretions are alkaline and contain normal skin organisms, e.g. staphylococci, streptococci, coliforms, and diphtheroids. The organs of the urinary tract, other than the lower urethra, are generally free of bacteria, possibly as a result of the flushing action of the urine.

NORMAL FLORA OF THE HUMAN ALIMENTARY TRACT

The presence of particles of food, epithelial debris, and secretions makes the mouth a nutritionally favourable habitat for a great variety of bacteria. The flora is subject to great variation but always includes very large numbers of α- and non-haemolytic streptococci and, in addition, non-pathogenic species of *Neisseria*, anaerobic spirochaetes, *Fusobacterium* (particularly between gums and teeth) and lactobacilli (in the acidic cavities of teeth, especially in the presence of dental caries). The yeast-like fungus *Candida albicans*, the cause of 'oral thrush' and other diseases, is also normally present (Fig. 14.1) but causes disease only in persons, particularly infants, suffering from malnutrition or in adults with debilitating diseases,

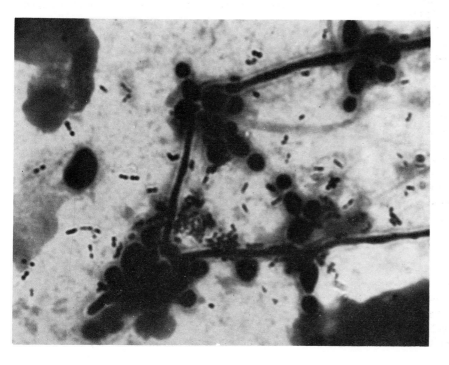

Fig. 14.1 Smear of human sputum showing pneumococci and *Candida albicans* as budding forms and pseudomycelium. (Gram stain; ×000; A. H. Linton)

e.g. diabetes, or in whom the normal bacterial flora is disturbed by the use of antibiotics, steroids and cytotoxic drugs.

The majority of vegetative organisms, other than acid-fast bacilli, originating in food or from the mouth are destroyed by the hydrochloric acid of the gastric juices so that the stomach, when empty, is relatively free of living micro-organisms. In the small intestine the influence of bile secretions becomes gradually apparent; the highly buffered foods revert to an alkaline pH and the numbers of bacteria increase, reaching a maximum in the caecum and colon. The flora includes coliform bacilli (particularly *Escherichia coli*), enterococci, clostridia (mainly *Clostridium perfringens*), species of *Bacillus*, Gram-negative pleomorphic non-sporing anaerobes of the genus *Bacteroides*, saprophytic acid-fast bacilli and many Gram-negative rods of varied types. *Bacteroides* spp. constitute 90% of the total flora in man. Large numbers of organisms die during their passage down the gut and approximately one-third of the dry weight of faeces consists of bacteria.

Diet influences the predominant flora. This is illustrated by the changes in flora in a child as a result of variations in diet during early life. In the breast-fed infant the flora consists predominantly of *Lactobacillus bifidus* which, during the early days of life, may form 90 per cent of the intestinal flora. With change in diet other organisms, characteristic of the adult alimentary canal, gradually become predominant. A high meat diet results in an increase in proteolytic organisms.

Origin of the alimentary flora   The alimentary tract of the foetus is sterile but at birth becomes rapidly and progressively infected by way of the mouth and anus. Immediately after birth, organisms commonly found in the mouth of the new-born are of the same type as those in the vagina of the mother. A few days later other organisms, characteristic of the normal mouth flora of adults, develop and it is thought that these originate from the nasopharynx of the mother. The intestinal flora of infants gradually changes with time as other organisms gain access until a complex mixture of organisms resembling that in the adult is attained.

The importance of the mammalian intestinal flora   Considerable variation in intestinal flora is found among mammalian species. *Escherichia coli*,

for example, is far less frequently found in herbivores than in carnivores. In guinea pigs lactobacilli comprise approximately 80% of the flora, the remainder consisting of organisms from soil and air. Consequently the importance of the intestinal flora must be considered in relation to the host species. Unlike the flora of herbivores, discussed below, the flora of carnivores is probably not essential to life. Animals, including guinea pigs and chicks, have been successfully reared on a balanced and sufficient diet under completely sterile conditions without showing stunted growth. However, such animals are very prone to infection when exposed to normal flora. With chicks, any lack of normal ability to digest their food completely was made good by eating more. In these animals, therefore, life has been shown to be possible in the absence of a gut flora, although there is an indication that the enzymic activities of the bacteria assist in the digestion of the food. However, the administration of broad-spectrum antibiotics over a long period may result in vitamin deficiency, presumably because these vitamins are normally synthesized by intestinal organisms.

In contrast to carnivorous or omnivorous animals, which depend wholly upon the secretion of enzymes into their alimentary canal to digest their food, herbivorous animals rely upon the enzymic activities of micro-organisms to bring about the initial digestion. In these animals the gut flora is essential for the digestion of food and the synthesis of vitamins, and a symbiotic relationship exists between the host and the microbial flora of the gut (see Chapter 12).

SYNTHESIS OF VITAMINS OF THE B COMPLEX   The amount of B vitamins in rumen contents is found to be several times greater than that expected on the basis of the vitamin content of the food ingested. This is the result of synthesis by rumen bacteria but not protozoa. The vitamin concentration in the rumen is not influenced by the amount present in the diet; metabolic control of bacterial synthesis maintains a steady level. This explains the fact that the concentration of these vitamins in ruminant's milk remains steady irrespective of fluctuations in the diet. Although the synthesis of vitamins by bacteria in the alimentary canal is not limited to ruminants, it is probably of particular importance to these animals since for the larger part of the year they graze on dried pastures of poor quality which are deficient in vitamins.

The influence of antibiotics on alimentary flora Considerable changes in faecal flora result from the administration of sulphonamides and broad-spectrum antibiotics. Antibiotics, often in combination, were sometimes prescribed prior to alimentary surgery in order to render the bowel virtually free of living bacteria, with a view to avoiding septic complications. This met with mixed success. In a proportion of patients, the normal bowel flora was superseded by antibiotic-resistant staphylococci, resulting in a staphylococcal enteritis. Overgrowth of other antibiotic-resistant micro-organisms, such as *Candida* spp., also produced complications.

The addition of small amounts of antibiotics to the feed of young poultry, calves and pigs has been shown to increase food conversion and rate of weight-gain. The mechanism of antibiotic growth stimulation is not fully understood but it is thought that the antibiotics check the growth of toxic organisms during the early days of life when the gut of the new-born animal, which is thinner and absorbs nutrients more efficiently than the adult gut, is first colonized by micro-organisms.

The practice of using antibiotics as growth promotors is not without certain disadvantages. Even sub-inhibitary levels select for antibiotic-resistant organisms. These organisms may themselves cause infections (e.g. urinary tract infections with gut organisms) or, if they carry genetic determinants of infective drug resistance (p. 54), they may transfer these factors to enteric pathogens. The antibiotic-resistant pathogen resulting would not yield subsequently to therapy by the corresponding drug. In the hope of reducing this risk, the practice of feeding therapeutically useful antibiotics as growth promotors has been prohibited in the United Kingdom.

Invertebrates and non-pathogenic micro-organisms

The part played by micro-organisms in the preparation and digestion of the food of invertebrates

Micro-organisms may break down complex materials thus rendering them suitable as food for many invertebrates, such as the nematodes and insect larvae inhabiting the soil. It is significant that the adults of many sorts of insect are attracted by odours of fermentation or putrefaction and feed upon the decomposing substances or lay their eggs therein. Termites and the death-watch beetle (*Xestobium*) tend to attack timber already partially rotted by fungi. The life cycle of this beetle may take as long as five years on sound timber and as little as

eighteen months on partially decayed wood.

A more specific association exists between certain wood wasps and species of the basidiomycete genera *Stereum* and *Daedalea*. Fungal oidia contained in tiny pouches at the base of the ovipositor are introduced into wood when the female lays her eggs. The larvae that hatch feed on wood that is partly digested by the advancing mycelium of the implanted fungus. A similar relationship exists between certain flies of the genus *Hylemyia* and bacteria which by rotting plant tissues facilitate penetration by the grubs.

Many species of termites cannot themselves digest the wood that they eat, relying instead on intestinal Protozoa to do this for them. However, there are some termites that lack these Protozoa and

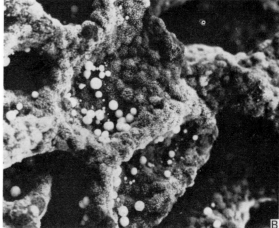

Fig. 14.2 Fungi associated with insects. (**A**) Fungus gardens in subterranean nest of termites in West Africa. (M. F. Madelin) (**B**) The surface of a termite fungus garden, showing surface growth of mycelium and spherical white masses of bromatia (× 8). (M. F. Madelin)

instead cultivate fungus gardens. It is likely that the fungi in these gardens decompose cellulose and also provide vitamins for the termites. The fungus gardens lie in subterranean chambers with the termite nests (Fig. 14.2). They are made of partly digested faecal matter from the termites. Mycelium of a particular fungus, usually a species of *Termitomyces*, permeates the mass and forms spherical groups of bromatia at the surface. The termites feed directly on the material that they have built into the fungus garden. Sometimes the masses of bromatia are eaten too, as they are by the fungus-growing ants (see below). *Termitomyces* species are commonly prevented from fruiting while the garden is actually being tended.

Micro-organisms as food for invertebrates

Bacteria, fungi, Protozoa, and even small algae are used as food by some insects and other invertebrates. For example, aquatic micro-organisms serve as food for mosquito larvae. Algae, as 'primary producers', directly or indirectly provide food for aquatic animals. The larvae of many flies and some beetles feed upon the fruit-bodies of the higher fungi. Other insects have a closer relationship with specific fungi which live in their nests and on which they feed. Ambrosia beetles are wood-boring insects that feed on a yeast-like diet of 'ambrosia', a fungal growth which lines their tunnels. Some ambrosia beetles depend wholly on this for food. They transmit the fungus to suitable host trees and fertilize it with their excreta. The ambrosia fungi are dimorphic. The ambrosial yeast phase depends on the presence of the insects, though the nature of the dependence is unknown. In the absence of beetles most ambrosia fungi revert to their mycelial form, and contaminant fungi swiftly spread. Two sorts of ambrosia fungi have been distinguished: primary ones which are more or less specific for particular species of beetle, and which appear soon after tunnels are excavated; and auxiliary ones which are not specific for particular beetles, and which become conspicuous towards the end of brood rearing.

A somewhat similar relationship exists between certain species of wood-boring and bark-boring beetles and the fungus *Ceratocystis ulmi* (the cause of Dutch elm disease) and the related blue-stain fungi, *Ceratocystis* spp. These fungi are carried from diseased to uninfected trees by the beetles.

Ants of the tribe Attii of the family Myrmicinae cultivate fungi in the so-called 'fungus gardens' made from pieces cut from leaves. The fungi, which have not been fully identified, are induced to produce swollen hyphal tips—bromatia—in spherical clusters on which the larvae are fed.

Pathological relationships between micro-organisms and animals

Vertebrates and pathogenic micro-organisms

Introduction

In contrast to the favourable equilibrium which is established between the normal flora and the host, in disease the host–parasite relationship may be tipped in favour of the parasite, at least initially, whilst in recovery the host usually overcomes the infection and the microbial pathogen is eliminated from the body. Disease-producing parasites either damage the animal tissues directly or disturb the body function. Physical changes in the host result and these are recognized as signs of disease and range from minor tissue damage to death. Infectious diseases of animals may be caused by bacteria (Fig. 14.3 and 14.4), fungi, protozoa, or viruses. A few parasitic bacteria and fungi are able to propagate in nature outside the animal host and assume a pathogenic role when introduced into the body. For example, *Aspergillus fumigatus* is commonly present in soil and on plants or plant products such as grain or straw, but when the spores, in sufficient quantity, are inhaled by birds, or occasionally by other animals including man, the respiratory disease aspergillosis may result. In contrast most pathogens, whilst able to survive for varied periods of time outside the animal host, normally do not multiply away from the animal. The majority of pathogenic bacteria and fungi have been grown under laboratory conditions devised to resemble the host environment. Many grow relatively easily; others only with great difficulty because they require highly nutritious media and

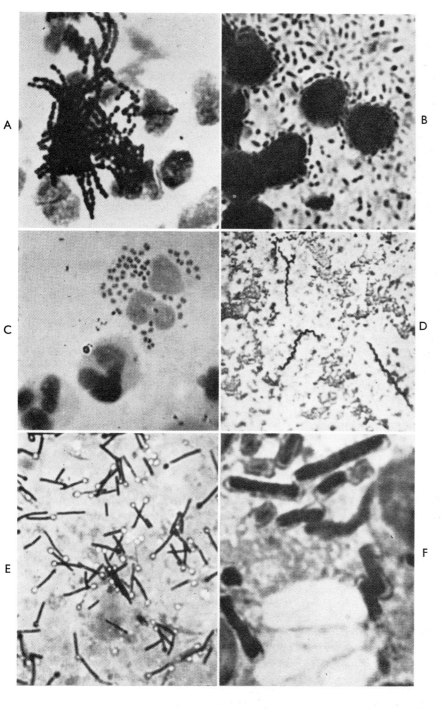

Fig. 14.3 Various bacteria in preparations of pathological material. (**A**) *Streptococcus agalactiae* in pus from mastitis of bovine udder (×2200). Gram-stained. (K. E. Cooper). (**B**) *Diplococcus pneumoniae* in deposit of cerebro-spinal fluid from a human case of meningitis (×2000). Gram-stained. (A. H. Linton). (**C**) *Neisseria gonorrhoeae* in smear of pus from the cervix of a woman with gonorrhoea (×2000). Note that the diplococci are phagocytosed by a polymorph leucocyte. Gram-stained. (A. H. Linton). (**D**) *Treponema pallidum* in a section of liver from a baby with congenital syphilis (×2500). The spirochaetes have been impregnated with silver (Levaditi's stain) to make them visible (Department of Bacteriology, University of Bristol). (**E**) *Clostridium tetani* in a smear from an infected wound (×2200). It is very rare to find so many organisms in an infected site—usually they cannot be detected in direct smears. Gram-stained (K. E. Cooper). (**F**) *Bacillus anthracis* in a smear from spleen of an infected mouse (×2200). Note the capsulated, non-sporing bacilli arranged singly or in pairs. This is characteristic of the appearance of this organism in body tissues and contrasts with the long chains of non-capsulated, usually sporing bacilli, seen in smears from artificial cultures. Methylene blue-stained. (K. E. Cooper)

prolonged incubation. Thus *Mycobacterium johnei*, the causal pathogen of chronic enteritis in cattle and sheep, will grow on egg media containing extracts of other mycobacteria and on primary isolation take upwards of two months to produce visible growth. A few have not so far been grown in artificial culture, e.g. *Treponema pallidum*, the causal pathogen of syphilis.

The survival of a pathogenic micro-organism depends on many interdependent factors. Under natural conditions, if an obligate parasite is to survive, it must be able to infect a susceptible host in which it can propagate. A high degree of infectivity is essential to the survival of those organisms which do not have a resistant phase. Other factors, such as the numbers of susceptible hosts available, also limit the survival of the pathogen. In island populations and other isolated communities, where the number of susceptible hosts is limited, certain diseases, such as measles, may flourish for a time then rapidly die out owing to this limiting factor. An organism which seriously damages its host may not survive, since if the host is killed the organism perishes unless it can first reach another susceptible individual. On the other hand, an organism of low virulence may fail to survive owing to inability to overcome the body defences of the host. Among the most successful of the common pathogens are those which have a high degree of infectivity but relatively low virulence under natural conditions, e.g. the virus of the common cold.

Ecology of microbial pathogens and mechanism of microbial pathogenicity

ESCAPE OF PATHOGENS FROM INFECTED HOST

The successful survival of a parasite depends on its ability to escape from one host and be transmitted to and infect another susceptible animal. Pathogenic micro-organisms usually escape from the living animal in excreta from the alimentary tract, in secretions from mucous membranes of the respiratory tract, in organic debris such as desquamated skin and hair, and in pus from infected wounds. Coughing and sneezing assists the escape of organisms from the respiratory tract (p. 261).

TRANSMISSION OF PATHOGENS BETWEEN HOSTS

This usually occurs between hosts of a single animal species but sometimes infection arises between hosts

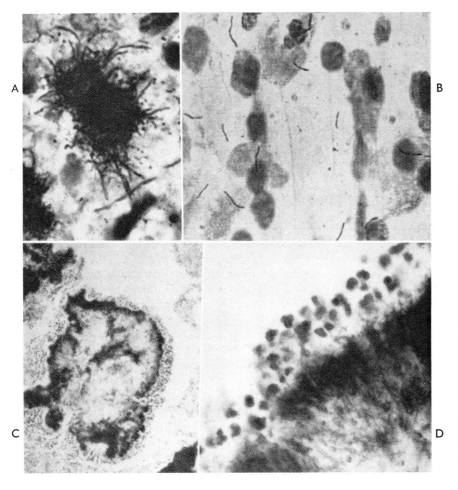

Fig. 14.4 Various actinomycetes in preparations of pathological material. (**A**) *Nocardia* sp. in a smear of pus from an infected mesenteric lymph node of a cat (×2000). Gram-stained (A. H. Linton). (**B**) *Mycobacterium tuberculosis* type *human* in human sputum (×2000). Stained by Ziehl Neelsen's method (K. E. Cooper). (**C**) *Actinomyces bovis* in a section of an actinomycotic lesion of a bovine jaw (×140). Gram-stained (A. H. Linton). (**D**) A portion of the edge of the 'ray fungus' structure shown in (**C**) (×2000). The organism is seen growing out into the surrounding granulation tissue from the central mass of mycelial growth. Gram-stained (A. H. Linton)

of different species. For example, man may be infected by dermatophytic fungi from infected domestic animals, or with *Brucella abortus*, by drinking contaminated raw milk; dogs, on the other hand, may become infected with tubercle bacilli derived from their masters.

Certain diseases are contracted by direct association with an infected host. This may occur by inhalation of droplets from an individual with a respiratory disease (e.g. tuberculosis, common cold virus) or by direct contact with an infected host as in venereal diseases (e.g. syphilis and gonorrhoea in man and vibriosis in cattle).

Contamination of specific sites may result in infection. For instance, infections of the human external ear may be caused by *Aspergillus niger* and *A. fumigatus*; following mastoid operation the resultant cavity, in which debris collects owing to impaired drainage, frequently becomes infected by any of the following: *Candida* spp., *Aspergillus niger*, *A. fumigatus*, *A. flavus*, and *A. terreus*. Bacteria also often infect these sites.

Many diseases are contracted indirectly by ingestion of food contaminated at source from an infected animal or contaminated subsequently by sewage polluted water or by excreta from an infected patient or healthy carrier (e.g. food poisoning, typhoid, tuberculosis, brucellosis, cholera, etc. p. 334). Horses may be indirectly infected with anthrax by feeding on hay grown the previous summer on anthrax-contaminated soil, since the spores survive throughout the winter months. Another method of indirect infection is by penetrating the skin with a contaminated instrument, as often occurs with tetanus, or by contamination of an abrasion, as in anthrax in man.

INVERTEBRATES AS AGENTS OF DISPERSAL OF MICRO-ORGANISMS

Viruses, bacteria, and the spores of many fungi are readily carried by insects, mites or other small invertebrates. While invertebrates, and particularly insects, play a large part in dispersal of micro-organisms picked up by chance, there are also many examples of specific relationships between micro-organism and dispersal agent. The most important aspect of dispersal of micro-organisms by animals (which are referred to as vectors) is the transmission of animal diseases.

Dispersal of viruses by arthropods Arthropods may be both reservoirs and vectors for animal viruses. In the former the virus is passed from one generation of arthropod to the next so that all the descendants of an infected progenitor will harbour the virus. For example, Colorado Tick Fever virus is passed from generation to generation via the egg. Vectors of animal viruses are primarily arthropods, and they may be further classified as mechanical (or casual) vectors and biologic (or essential) vectors.

MECHANICAL VECTORS provide a means of transport for the virus from host to host. The biting and sucking mouth parts of the vector become contaminated with virus, new hosts becoming infected when subsequently bitten or sucked.

The epidemiological paradox created by mechanical vectors in the epizootic transmission of rabbit myxomatosis in Great Britain and Australia, in attempts to control the rabbit populations in the two countries from 1950 to 1954, is of interest. In Great Britain rabbits infected with the virulent myxomatosis virus took to their burrows and died there. After death, the rabbit flea (*Spilopsyllus cuniculi*) left the carcass to find new hosts, thereby facilitating transmission of the virulent virus. On the South Coast, and in France also, *Anopheles atroparvus* played a relatively minor role in the transmission. In Australia the same initial pattern of the disease was seen, the fatally infected animals dying within their burrows. However, the Australian rabbit flea (*Echidnophaga* spp.) is of the 'stick fast' or 'stick tight' variety, which does not leave its host. Flea transmission was thereby largely eliminated. Only those few animals which recovered from the infection or those infected with an avirulent strain of myxomatosis virus, came out of the burrows to feed, and mosquitoes (e.g. *Anopheles annulipes*) there facilitated the transmission of avirulent myxomatosis on an epizootic scale. Since the avirulent virus provided full immunity, subsequent attempts to reintroduce the virulent virus met with complete failure. Carnivorous birds, feeding upon the carcasses of myxomatosis-infected rabbits, may also carry the rabbit flea to new localities and thus assist in the spread of the disease.

BIOLOGIC VECTORS are those in which the virus undergoes a separate replication cycle but does not produce disease in the vector. There is no transmission to subsequent generations by way of the egg. The arthropod-borne (Arbo) viruses are characterized by their dependence upon biologic arthropod vectors for infection of their vertebrate hosts. Yellow Fever virus, for example, is ingested by the mosquito (*Aedes africanus* and *Aedes simpsoni* in Africa; *Haemagogus* spp. in South America) whilst feeding on the infected animal host; the virus replicates in the cells of the gut of the mosquito, passes to the salivary glands via the haemolymph and there replicates a second time and is discharged into the saliva from which infection of the next host occurs. The period required for virus replication within the vector varies with the environmental temperature, being prolonged at reduced temperatures.

The viruses of Dengue Fever, Rift Valley Fever and the various Encephalitides are similarly mosquito-transmitted. Sandfly Fever is transmitted by *Phlebotomus papatasi*, a 'sand-fly' or 'owl-midge' which is a blood-sucker also. Louping-ill and Colorado Tick Fever viruses are transmitted by ticks, which are also blood-sucking. All of these viruses, excepting Dengue Fever and Sandfly Fever viruses, infect other vertebrate animals in addition to man.

Dispersal of bacteria by arthropods Insects and mites are usually externally contaminated with large numbers of bacteria and other micro-organisms picked up from the surroundings. The house fly in particular, owing to the nature of its skin, the hairy pads of the feet and its habit of crawling over and feeding on putrefying organic matter, including excreta, invariably carries a varied and numerous bacterial flora, which may include species, particularly enteric pathogens, pathogenic to man and domestic animals.

One of the best-known examples of pathogens carried by specific insect or other arthropod vectors is that of the rickettsiae (p. 168). Many rickettsiae propagate in the lumen and epithelial linings of the alimentary canal of certain arthropods, such as lice, fleas, and mites, and sometimes invade the salivary glands. Infection of vertebrates follows biting by the arthropod vector or, more usually, the rickettsiae from a crushed arthropod or its faeces gain access through an abrasion of the skin often caused by scratching. Some are non-pathogenic to the arthropod host (as in murine typhus); others are pathogenic both to the vector and the vertebrate host (as in classical typhus).

Rickettsial infections of man may be divided into those which are 'demic', that is, transmitted from man to man, and those which are 'zootic', that is, transmitted from animals to man. The first group

include, *Rickettsia prowazekii*, the cause of classical typhus, which is transmitted by the body louse, *Pediculus humanus corporis*, the infection entering the human host by the bite of the vector or through an abrasion caused by the patient's scratching. The second group may be further divided according to the arthropod vector by which they are transmitted to man. Flea-borne rickettsiae include *R. mooseri*, the cause of murine typhus or shop typhus of Malaya. This organism produces a mild endemic infection in rats and is transmitted from rat to rat by the rat louse *Polyplex speculosus* and from rat to man by the rat flea *Xenopsylla cheopis*. Those transmitted to man by ticks include *R. rickettsii*, the cause of Rocky Mountain Spotted Fever, which is transmitted from rodents, sheep, and dogs to man by ticks such as the wood tick, *Dermacentor andersoni*. Q-fever, an influenza-like disease of man, caused by *R. burneti*, is tick-transmitted between animal hosts such as opossums, bandicoots, dogs, cattle, sheep, and goats but usually infects man by inhalation of contaminated dust in abattoirs, laundries, etc. where infected animals or contaminated clothing are found. Mite-borne rickettsiae include *R. tsutsugamushi*, the cause of scrub typhus which is transmitted from rats to man by mites such as *Trombicula akamuchi* var. *deliensis*. Some are pathogenic only to animals, e.g. *R. ruminatium*, the cause of heart-water disease of sheep, goats and cattle, which is transmitted by the tick *Amblyoma hebraeum*.

In contrast to the rickettsiae very few bacterial pathogens are vector transmitted. A most important example is that of *Yersinai pestis*, which, when flea-borne, causes bubonic plague in man. The animal reservoir of infection is the large grey rat (*Rattus norvegicus*), and the smaller black rat (*Rattus rattus*). *Y. pestis* is pathogenic to the rat and at the terminal stage of the illness the ovoid bacteria are present in the blood in enormous numbers. Fleas common to both rat and man (*Xenopsylla cheopis* and *Ceratophyllus fasciatus*) engorge the organism-laden blood and when the rat dies, leave the corpse and in the absence of the normal host will bite man thereby transmitting the plague bacilli.

Certain spirochaetal diseases are transmitted by insects or ticks. *Borrelia recurrentis*, which causes the European type of relapsing fever, is louse-borne, man usually becoming infected by contamination of scratches with the body fluids of lice in which the spirochaetes occur.

Arthropod vectors of protozoa Protozoa have been

found associated with most insects and many other arthropods. Many of these Protozoa are parasites of other higher animals including man. The life cycles of many are intimately linked with the insect vector and the vertebrate host.

SURVIVAL OUTSIDE THE HOST

The length of time during which pathogenic organisms survive outside their hosts varies enormously from species to species. Some, like the organisms responsible for human venereal diseases, (*Neisseria gonorrhoeae* and *Treponema pallidum*) which are mainly transmitted during coitus, rapidly die outside the body. Many pathogens of the respiratory tract are more hardy but are favoured by fairly rapid transmission of large numbers during the infectious period in cough spray, although infection may persist in contaminated dust and may be inhaled later; this has often been shown to occur in cross-infection in fever hospital wards (p. 337). In general, pathogens of the intestinal tract have much greater resistance and can survive for many days, or even months, outside the body. Spore-bearing pathogens, such as the causal organisms of tetanus, gas-gangrene, and anthrax, survive for years as spores in a fully virulent condition in manured or otherwise contaminated soils.

ROUTES OF INFECTION

Many fully virulent micro-organisms are able to cause disease in the animal body only if they enter by specific routes. *Clostridium tetani*, the cause of tetanus, is frequently found without harmful effects in the intestinal tract, especially of domesticated animals (horses, pigs, etc.), to which it gains access via the mouth. It assumes a pathogenic role only if introduced into a wound, either following surgical manipulation or accidental penetration, in which anaerobic conditions essential for its growth are found. Many alimentary infections, on the other hand, are unable to establish themselves when introduced into abrasions but readily infect via the mouth. The same organism may show varying degrees of virulence when introduced into the same host by different routes; thus anthrax bacilli and staphylococci are more virulent when they enter through skin abrasions than when ingested; the tubercle bacillus is more virulent when it enters by the respiratory than by the alimentary route, probably owing to the differences in the defence mechanisms at local sites.

Closely linked with the portal of entry is the affinity of many organisms for particular tissues, e.g. the pneumococcus for lung tissue, the leprosy bacillus for skin tissue, the meningococcus for the meninges of the brain, leptospirae for the liver and kidney, and *Vibrio cholerae* for the intestine.

SIZE OF THE INFECTING DOSE

It is very rare for a single bacterial cell to give rise to disease. More usually a critical minimum number of cells must be present in the infecting dose before invasion of the host can occur. Very large numbers of spores of certain pathogenic fungi, such as *Aspergillus fumigatus*, are necessary to establish an infection. Even when micro-organisms are able to set up an infection, the rate of onset of symptoms, the severity of the disease, and the mortality rate, are often directly proportional to the size of the initial infecting dose. The more virulent the pathogen, the smaller the minimum dose required to produce infection.

Properties of microbial pathogens

Since Koch's observations in 1878 on the characteristic symptoms and anatomical lesions in animals caused by six morphologically distinct microorganisms, it has been well established that each pathogenic organism produces a specific range of disease. Certain pathogens produce the same symptoms in a variety of animals hosts, e.g. paralysis resulting from the toxin of the tetanus bacillus. Others are able to produce different forms of disease in the same host, e.g. *Streptococcus pyogenes* (Fig. 14.5) in man can cause tonsillitis, scarlet fever, erysipelas, or wound infections. Some organisms are highly pathogenic to certain hosts and non-pathogenic to others, e.g. *Corynebacterium diphtheriae*, which is pathogenic naturally to man and experimentally to the guinea pig but to which the rat is completely resistant. Other pathogens produce disease of varying severity in different animal hosts, e.g. *Bacillus anthracis*, which produces an acute, usually fatal septicaemic disease in cattle, sheep and horses, inflammatory oedema of the pharynx in pigs and dogs, and a malignant pustule at the site of entry through the skin in man.

VIRULENCE

Only those strains of pathogenic species able to

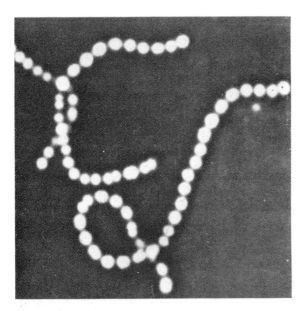

Fig. 14.5 Chains of Lancefield Group A Streptococci isolated from a septic sore throat. Negative staining (×3600). (C. F. Robinow)

overcome the resistance of the host and propagate in the animal body are virulent; those which cannot do this are termed avirulent. There is, however, no clear distinction between the two, since virulence is a relative term dependent upon many factors which can be altered under a variety of environmental influences. The virulence of many bacterial and viral pathogens may be diminished by such procedures as artificial culture, culture at temperatures above the optimum, or passage through animals in which the disease is not normally found. These methods are used to produce attenuated vaccines for artificial immunization (p. 325). Frequently loss of virulence is associated with a change from the 'smooth' colonial form to the 'rough' (p. 89). Selection of smooth virulent colonies from a rough avirulent culture is not readily accomplished, but in certain instances can be achieved simply by animal passage when the few virulent organisms present in the culture will assume a pathogenic role and may be isolated from pathological material in pure culture. Rough avirulent pneumococci can be transformed into a virulent form by growing them in the presence of killed suspensions of whole, smooth peumococci, or simply in the presence of the genetically important transforming substance extracted from smooth pneumococci (p. 50). Virulent toxin-producing strains of the diphtheria bacillus have been isolated from cultures of avirulent non-toxin-producing strains after exposure to particular

bacteriophages, indicating that in this species there is an association between lysogeny with β-phage (p. 131) and virulence; but not all lysogenic strains of the diphtheria bacillus are virulent.

The properties which distinguish a virulent from an avirulent bacterium are twofold, namely, 'invasiveness', i.e. the ability of the organism to establish itself and propagate in the host tissues, and 'toxicity', i.e. the ability of the organisms to destroy or damage tissues or impair their physiological functions.

INVASIVENESS

Apart from those organisms which are introduced into the body through wounds, most micro-organisms initially lodge on the skin or mucous membranes through which they must penetrate in order to invade the underlying tissues. Organisms impinging on the membranes of the nasopharynx are constantly being removed by ciliated epithelial cells or engulfed by wandering phagocytic cells (p. 323) which carry them away to the lymph channels with which the mucous membranes are liberally supplied near the surface. If the organisms escape these defences and multiply, infection may be initiated. Local multiplication of organisms such as *Staphylococcus aureus* (Fig. 14.6) may occur in crypts of the skin, in hair follicles, or sebaceous glands, thereby setting up local damage followed by penetration into the sub-epithelial tissues. Skin secretes bactericidal substances and these must be overcome by micro-organisms before they can multiply. Many pathogens are thus destroyed before penetration is achieved. After penetration

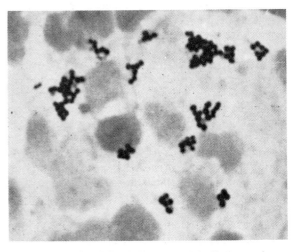

Fig. 14.6 Clusters of *Staphylococcus aureus* in smear of pus from abscess. Gram-stain (×1500). (A. H. Linton)

the subsequent course of the infection varies with the type of organism and the nature of the internal defences of the host. The infection may be confined to local abscess formation, as is common with staphylococci, or the organism may penetrate through subcutaneous tissues to give rise to spreading skin lesions, as in streptococcal skin infections, e.g. erysipelas in man caused by *Streptococcus pyogenes*, or a clinically similar condition in swine caused by *Erysipelothrix rhusiopathiae*. Some pathogens after local multiplication spread *via* the lymph channels to the lymph nodes, which rapidly become enlarged owing to the stimulation of defence mechanisms in these glands. These may check the spread of infection; otherwise organisms continue to spread either in the lymph or the bloodstream. Organisms entering the bloodstream are rapidly circulated to all parts of the body (bacteraemia) and they may set up local foci of infection in various organs from which pus may escape into the bloodstream (pyaemic spread). Progressive multiplication of the organism in the blood stream is termed septicaemia.

TOXICITY

The ability of different pathogens to invade the body varies considerably. Some with very little power to invade are extremely harmful by virtue of their ability to produce highly potent toxins. The tetanus bacillus, for instance, has no invasion powers, but, when introduced into a deep wound, multiplies at the local site and generates a powerful toxin which has a specific affinity for nerve cells and brings about destruction of those parts of the central nervous system which it reaches, resulting in symptoms of paralysis. The diphtheria bacillus does not invade beyond the tissues of the throat and nasopharynx, but its toxin, produced *in situ*, is absorbed and circulated around the body in the bloodstream to produce the various generalized symptoms of diphtheria, in particular, damage to heart muscle tissue. Many strongly invasive organisms, such as streptococci, and in particular the gas-gangrene anaerobic sporing bacilli, produce many potent toxins in gross tissue destruction. The various bacterial toxins are divided into two classes, namely, endotoxins and exotoxins.

Endotoxins are structural components of bacterial cells released from living and dead cells and may be obtained in bacteria-free filtrates of cultures. In the body endotoxin release occurs when the bacteria are

phagocytosed or othersise lysed by natural components of the body fluids (e.g. complement in the presence of antibody, p. 327). Endotoxins are produced mainly by Gram-negative pathogens and non-pathogens (e.g. species of *Salmonella, Escherichia, Proteus, Brucella,* and *Neisseria*). Irrespective of the organisms from which the endotoxin came the symptoms produced in the host are similar and include fever, diarrhoea, local haemorrhage, especially in the visceral tissues, fall of blood pressure and shock. The severity of the effect depends on the dose of endotoxin and the sensitivity of the animal to it. Endotoxins have been chemically extracted from a number of bacterial species and have been found to consist of complex molecules of lipopolysaccharides similar to the O-somatic antigens of Gram-negative bacterial cell walls (p. 331). Serum antibodies produced against these cell wall components, combine with the O-somatic antigens in the whole cell producing antibacterial effects (e.g. bacteriolysis) but the antibodies do not completely neutralize the toxic action of the endotoxins. Endotoxins are more heat stable than exotoxins but do not act at such high dilution. They produce their effects on the host in a variety of ways. The vascular system is particularly susceptible. Endotoxins damage small blood vessels by a hypersensitive type of reaction resulting from the release of pharmacologically active substances such as histamine, leading to shock and even death. The hypersensitivity of the patient arises from past exposure to the same antigens which induced the hypersensitive response (p. 331).

Exotoxins in contrast to endotoxins, are metabolic products of bacteria and are readily separable from the living producer cells. They may be obtained by growing the bacteria in a suitable liquid medium and separating the toxin from the cells by centrifugation or filtration. They are produced mainly by Gram-positive bacteria. Those which have been isolated as pure chemical substances (e.g. tetanus and diphtheria toxins) are simple proteins. They are frequently highly toxic for the host, the lethal dose of pure crystalline tetanus and botulinus toxins for mice being of the order of 0.0000001 mg. They act as enzymes and as such are generally unstable to chemicals (e.g. formalin slowly converts them to toxoid, p. 325), to oxygen (e.g. oxygen-sensitive haemolysins), and to heat. Sensitivity to heat varies widely; diphtheria toxin is sensitive to 65°C, *Clostridium perfringens* α-toxin to 100°C, and staphylococcal enterotoxin and botulinus toxin to higher temperatures. Some exotoxins exhibit specific affinity for tissues; for instance, tetanus bacilli and other bacteria produce neurotoxins which damage cells of the central nervous system, the symptoms of paralysis being produced several days after the toxin is released into the body. Other toxins are rapid in action and are named according to the tissues they damage; haemolysins, as produced by streptococci, lyse red blood cells; leucocidins, e.g. the leucotoxin of *Staphylococcus aureus* (Fig. 14.6), destroy leucocytes, the wandering phagocytes of the blood stream. The necrotoxins exhibit more general tissue destruction, such as the general destruction of tissue by the gas-gangrene group of *Clostridia*. The same toxin may produce a number of effects. For instance, the α-toxin of *Staphylococcus aureus* acts as a haemolysin, a leucocidin, a necrotoxin, and a lethal toxin. In most instances the mechanism of action is unknown. The α-toxin of *Cl. perfringens* acts as a lecithinase, a lecithin-splitting enzyme which can attack lecithin-containing cell membranes such as those of red blood cells. The toxin of diphtheria exerts its primary effect by inhibiting protein synthesis in susceptible tissues. This is brought about by a specific inactivation of the translocating enzyme, aminoacyl transferase II, found only in eukaryotic cells.

Some organisms are able to produce toxins resulting in a tissue rash (e.g. *Streptococcus pyogenes*, which produces an erythrogenic toxin causing the rash of scarlet fever). Specific neutralizing antibodies (antitoxins) can be demonstrated in the bloodstream of an animal following infection with an exotoxin-producing bacterium, or these may be stimulated by artificially immunizing the animal with a toxoid (p. 325).

Other cell products Some bacterial cells produce substances which influence the course of an infection in ways other than by toxic action. For instance, pathogenic staphylococci produce the enzyme coagulase which induces the coagulation of blood plasma, thereby setting up a barrier to defences of the vascular system and assisting the organism to establish itself in the host body. In contrast, streptococci produce the enzymes streptokinase, which dissolves fibrin clots, and hyaluronidase, which breaks down hyaluronic acid (a muco-polysaccharide which holds together the mesodermal structures of subcutaneous tissues), thereby facilitating spread of the organism from the primary focus.

Virus diseases of vertebrate animals

GENERAL CONSIDERATIONS

Unlike many bacterial pathogens viruses do not produce exotoxins and the diseases they cause are the direct result of their replication cycles within the various tissue cells of the animal body. Under experimental conditions however the inoculation of large amounts of inactivated ('killed') virus can produce symptoms of cytotoxicity as does the inoculation of large amounts of other foreign antigen.

The replication of viruses in tissue cells leads to their biological malfunctioning and if large numbers of cells are involved, the whole organ is affected. This may result in death of the animal. Some animal viruses replicate in a limited range of tissue cells, e.g. influenza virus replicates only in cells of the respiratory tract, while others replicate in a wide variety of tissue cells, e.g. smallpox virus in cells of the skin, lungs, and other internal tissues. Some viruses can therefore spread from a primary site of infection to other susceptible tissues by blood-borne dissemination.

CYTOPATHIC EFFECTS

Replication of virus in host cells often results in observable cytopathic effects (CPEs). These include: (1) *hyperplasia*, or cellular proliferation (e.g. herpes simplex virus); (2) *polykarycytosis*, or syncytium formation (e.g. measles virus and ectromelia virus); (3) *inclusion body formation*; inclusions are formed intracytoplasmically (e.g. poxviruses) or intranuclearly (e.g. herpesviruses) but only occasionally at both sites at the same time (e.g. measles, distemper and rinderpest viruses); (4) *malignant transformation*, when the host cells acquire tumour cell-like characters and may give rise to tumours *in vivo* (e.g. papova viruses); (5) *vacuolation*, of either the cytoplasm (e.g. SV (Simian virus) and VA (vacuolating agent) groups of viruses) or the nucleus (e.g. pig pox virus); (6) *latency*, when the viral genome resides in a suppressed or inactive state within the host cell and (7) *necrosis*, which initially exhibits 'rounding-off' (e.g. polioviruses) or a 'tailing off' (e.g. adenoviruses) of the host cells and terminally a shrinking and a contraction in cell volume with heavy nuclear staining (pyknosis), observable in stained preparations (e.g. polioviruses). With some viruses a single CPE is observed in the host cell; with others more than one are found. When necrosis is one of these it is always the terminal cytopathic effect. CPEs may be observed also in monolayer tissue cultures (Fig. 7.14) and have diagnostic value.

Hyperplasia is the characteristic CPE of oncogenic viruses. Cellular proliferation may be so rapid as to give tumour-like masses in the body, e.g. the Shope papilloma and the Shope fibroma of rabbits, and the Rous sarcoma of chickens. Less localized proliferation may give rise to the diffuse swellings seen in rabbit myxomatosis. More recently, a herpesvirus has been observed in association with the proliferating lymphoid cells of Burkitt's lymphoma and its viral DNA has also been detected in the nuclei of lymphoma cells by *in situ* molecular hybridization techniques. This herpesvirus may prove to be the first oncogenic virus with a proven aetiological role in a malignant disease of man.

Inclusion bodies are so characteristic as to be diagnostic; this was so widely recognized and it became the custom at one time to name them after their discoverer (e.g. Negri bodies in the rabid brain). They are acidophilic in staining reaction and most are homogeneous in composition. Inclusions most frequently indicate the original 'factory sites' of virus replication within the cell and are not usually composed of aggregates of virions.

Latency is a cytopathic effect which is receiving considerable attention at present, not least in relation to oncogenesis (i.e. tumour-forming). Some consider that viruses are involved in all oncogenic lesions and the latent oncogenic viruses (retroviridae) constitute the largest single cause of tumours in man. Recent developments indicate that Retroviridae possess the novel enzyme 'reverse transcriptase' (properly, RNA directed DNA polymerase) which promotes the formation of a DNA transcript of the RNA genome of the virus, enhancing its ability for integration within the host cell genome and hence for latency. Further, the ubiquitous nature of these viruses may provide recognizably non-oncogenic viruses, such as measles, respiratory syncytial virus and influenzavirus, with the chance to 'borrow' the enzyme and themselves adopt a state of latency.

Cellular necrosis is usually focal in nature. Pus is never observed at these sites except when there is secondary bacterial infection, e.g. smallpox pocks on the skin. Necrosis in a vital organ or system, e.g. liver, kidney, or CNS, may lead to malfunction. For example, the characteristic jaundice of yellow fever is the result of reduced liver function.

VIRUS INFECTIONS

With a few exceptions the signs and symptoms of virus diseases in animals are insufficient to be diagnostic. However, they may permit a presumptive diagnosis to be made, while awaiting the laboratory identification of the causal virus.

The signs and symptoms presented by a particular virus disease reflect closely the primary and secondary replication sites in the body and the ensuing CPEs. Two examples illustrate this. Poliovirus infects man principally *via* the oral route and is insensitive to the low pH of the stomach. The primary replication sites are the tonsils and the various lymph nodes of the digestive tract. During this stage, virus is found in abundance in the faeces and diarrhoea may be present. Both virulent and occasionally avirulent strains pass via the intestinal wall into the blood stream (viraemia) by which the virus is disseminated throughout the tissues. The viraemia results in an elevated temperature and may produce an invasion of the central nervous system (CNS). Secondary replication sites next become infected including the CNS, lymphatic structures and brown fat and, at this stage of the disease, overt paralysis is observed. Since the virus induces necrosis in its host cells it follows that widespread necrosis of neurons in the CNS, a tissue system with a poor regenerative capacity, results in a permanent paralysis which varies in degree with the extent of the necrosis.

Smallpox virus in man usually infects *via* the respiratory route, although infection *via* the dermal tissues may sometimes occur. The enteric route is much less likely because of the susceptibility of the virus to a low pH, such as would be encountered in the stomach. Replication of the virus may produce transient signs of respiratory disease. From this primary site the virus passes to the regional lymph nodes and thence, via the blood stream, to the liver, spleen, and other tissues where the secondary replication takes place. From these tissues the virus proceeds via the blood stream to the skin and mucous membranes where the typical focal rash is formed. Both viraemias will produce temperature elevations, i.e. a diphasic fever curve, the second being most severe (*ca.* 104–105°F; 40–40.6°C). During the second viraemia the patient suffers a toxaemic illness of about 4 days' duration with fever, systemic symptoms, and prostration, often with accompanying headache, backache, pains in the limbs, and vomiting. The typical focal rash is seen on the third or fourth day, involving the buccal and pharyngeal mucosa and the skin, principally of the face, forearms, and lower legs, although the trunk is involved. The rash, at first macular, soon becomes papular, vesicular, and finally pustular

(8–9 days) when crusts commence to form.

From these two examples it is evident that viruses have a characteristic replication sequence and CPE in the animal body. Similar signs, symptoms, and lesions are, however, not necessarily evidence for common patterns of replication. For example, both smallpox and papilloma (wart) viruses give lesions in the skin, but whereas smallpox virus replicates in internal tissues the papilloma viruses are confined entirely to the skin tissues. Smallpox and chicken-pox (varicella) give very similar skin eruptions which differ in their essential distribution, yet smallpox virus replicates intracytoplasmically while chicken pox virus (herpes zoster) replicates intra-nuclearly. Also lesions in the CNS are en-countered in poliomyelitis, rabies, and the various encephalomyelitides, but each virus has a different pattern of replication in the body.

'SLOW VIRUSES'

The term 'slow infection' was used by Sigurdsson in 1954 to describe certain animal infections, such as Maedi and Rida of sheep and Johne's disease of cattle, each of which have long latent periods (p. 305) extending over months or years, a protracted course followed by an invariably fatal outcome, and strict host specificity. Subsequently, the term 'slow viruses' was used to describe viruses and virus-like agents which produced similarly protracted diseases in their hosts. Since this term is loosely descriptive rather than scientific it almost certainly encom-passes a variety of agents which, at the present time, are incorrectly considered to be 'slow'. These in-clude subacute sclerosing panencephalitis (SSPE), progressive multifocal leukoencephalopathy (PML), lactic dehydrogenase (LDH) viral infection and lymphocytic choriomeningitis (LCM) viral infection. Despite its shortcomings the term 'slow viruses' is likely to find continued use by virologists.

Scrapie of sheep, the related diseases of kuru and Creutzfeldt-Jakob in man and transmissible mink encephalopathy (TME) are the most commonly known diseases produced by 'slow viruses'. Scrapie has been most extensively studied. Because of the unusual stability of the scrapie agent to a wide range of physical and chemical treatments which in-activate conventional viruses (e.g. heat, UV ir-radiation, formaldehyde), and failure to detect conventional virions in infected tissues by electron microscopy, many workers prefer to use the term 'agent' rather than 'virus'. In their respective hosts these agents induce slow, degenerative en-cephalopathies of the central nervous system.

The biochemical nature of the scrapie agent is not clear. Most of the infectivity, as determined by inoculation experiments, is associated with cell membranes but whether this is associated with membrane-bound nucleic acid has yet to be dem-onstrated.

The slow progress which has been made on research with the scrapie agent has been due to the difficulties encountered with the agent in the lab-oratory. The only means for detecting the agent in laboratory animals is by infectivity assays in mice; rabbits and guinea pigs are insusceptible. Assays in tissue culture, by immunology or electron micros-copy are not possible because of the failure of the agent to replicate consistently *in vitro*, to induce immunological responses, or to produce recogniz-able morphological particles. Natural scrapie in sheep may result from infection with more than one strain of the agent at any one time; hence cloning of the agent is a prerequisite for all definitive in-vestigations.

In mice the gene *sinc* (gene for *s*crapie *inc*ubation) controls the basic level of replication of the agent, and this governs the incubation period of the induced disease. In some instances high intra-cerebral doses may require incubation periods of more than half of the $2-2\frac{1}{2}$ year lifespan of the mice, and hence the calculated incubation period of a low dose inoculated by the less efficient intraperitoneal route may paradoxically exceed the natural lifespan of the animal. The overdominance which is ex-pressed by the gene *sinc* upon the replication of some strains of the agent is of considerable import-ance for the elucidation of the molecular events governing replication of the agent. Since such positive control decisively contributes to the 'slow' nature of the disease, a detailed study of gene *sinc* is essential to further progress with scrapie. It is likely that a similar genetic control exists in the related diseases of man and mink.

Invertebrates and pathogenic micro-organisms

A number of viruses, bacteria, fungi, and Protozoa cause diseases of various invertebrates. Owing to the economic importance of the hosts, diseases of insects have been studied, but little is known of diseases of other invertebrates.

Virus infections of invertebrates

INTRODUCTION

Two major types of virus disease of insects are the

inclusion body diseases and non-inclusion diseases. In inclusion body diseases the virus particles are occluded in polyhedral protein crystals (polyhedra), in granules (capsules), or in ovoid inclusion bodies. These constitute the inclusion bodies that are to be seen under the light microscope in the tissues of the diseased insects. Polyhedral inclusion bodies may lie either in the nuclei or in the cytoplasm of the host's cells. In relation to this the corresponding viruses are divided into those that cause nuclear and cytoplasmic polyhedroses respectively. These two sorts of virus differ not only in their sites of multiplication but also in their morphology and in the symptoms they produce. Diseases caused by viruses which become occluded in granules are called granuloses. So far, these are known only in larvae of Lepidoptera. Another group of viruses producing inclusion bodies are the entomopox viruses. These closely resemble the pox viruses of vertebrates and may form distinctive fusiform protein crystals (spindles) associated with the ovoid inclusion bodies. The non-inclusion viruses differ strikingly from the foregoing types in that they lie quite free in the tissues of the hosts. Additionally there are the arboviruses (arthropod-borne viruses) that infect both vertebrates and blood-sucking arthropods and a corresponding group of viruses that multiply in plants and insects.

SURVEY OF DIFFERENT GROUPS OF INSECT VIRUSES

Viruses that cause nuclear polyhedroses Nuclear polyhedroses occur most commonly in larvae of Lepidoptera (the moths and butterflies) but are known also in Hymenoptera and Diptera. The viruses multiply within nuclei in chiefly the skin, blood, fat-body, and tracheae. The virus particles contain DNA and are rod-shaped (*ca.* 20–50 nm diameter, 200–400 nm long), and become enclosed either singly or in bundles within paracrystalline protein which forms microscopically visible polyhedra. These enlarge and rupture the nuclei in which they form, and ultimately the cell bursts. As the disease progresses the tissues of the host disintegrate and liquefy, and the skin, which by now has become very fragile, ruptures. The liberated fluid contains myriads of virus-containing polyhedra. These are insoluble in water and resistant to ordinary bacterial decay. If diseased insects are left to decay in water, the polyhedra eventually form a layer at the bottom of the vessel. In nature, the polyhedra are ingested together with food plants and dissolve in the alkaline juices of the gut of susceptible insect species. Though they are insoluble

in water, polyhedra may dissolve if the pH falls below 5 or rises above 8.5. It appears that their dissolution in the gut and the consequent liberation of occluded virus particles is not enzymic. Analyses of polyhedra reveal traces of silicon (e.g. 0.1–0.3 per cent). Solution of silicates incorporated in the structure of the polyhedra by alkaline agents in the gut may account for dissolution of the polyhedra. From the gut viruses enter susceptible tissues and disease develops usually 10–12 days after ingestion of the virus. In addition to infection via the mouth there is much evidence to suggest that nuclear polyhedrosis viruses may be transmitted through the eggs.

Viruses that cause cytoplasmic polyhedroses Many cytoplasmic polyhedroses are known. Most affect only Lepidoptera. As in nuclear polyhedroses the virus particles are occluded in polyhedral protein crystals, but these do not dissolve completely in weak alkali. A rather sponge-like residue remains which is pitted with the sockets in which the nearly spherical RNA-containing virus particles were formerly located. The individual virus particles are regular twenty-sided (icosahedral) bodies 60–65 nm diameter.

The development of cytoplasmic polyhedroses is different from that of nuclear polyhedroses. The viruses multiply only in cells of the gut, usually the epithelial cells of the mid-gut, from which they spread to fore and hind guts. The skin is not attacked, and as a result the body does not become fragile and burst. The polyhedra appear in the cytoplasm of infected cells. They vary much in size, even within the one insect, but are clearly visible with the light microscope. Though the polyhedra themselves are located in the cytoplasm, recent autoradiographic work has indicated that the viral RNA may in fact be synthesized within the nucleoli of infected cells. From thence it presumably passes into the cytoplasm where the protein coat of the virus particle is formed. The polyhedra in the gut wall frequently show through the dorsal integument of the host as pale yellow or whitish areas. Infected larvae develop more slowly than do healthy ones, and thus are smaller; they have reduced appetites and sometimes disproportionately large heads or long bristles. When the disease is far advanced large numbers of polyhedra are liberated in the lumen of the gut from which they may be regurgitated or voided in faeces. Besides spread in this way there is almost certainly hereditary transmission of virus via the egg.

Viruses that cause granuloses Granuloses have so far been encountered only in larvae, and very occasionally pupae, of Lepidoptera. The diseases are characterized by the appearance in the tissues of millions of granules, so small (*ca.* 300–500 nm diameter) as to be barely visible beneath the light microscope. Each granule is a capsule that invests one or occasionally two rod-shaped virus particles which contain DNA. The granules occur in both nuclei and cytoplasm of the host's cells of which those of the fat body seem to be the main site of virus development. As the disease progresses other tissues are affected, and normally these include the skin. In this event the marked liquefaction of the internal tissues after death, usually 4–20 or so days after infection, is followed by rupture of the fragile body wall, and the symptoms of the granulosis then closely resemble those of nuclear polyhedroses.

Granulosis viruses in general appear to remain infective for up to two years and even may survive in the soil for this time with little deterioration. Percolating water does not readily remove them from the soil. A granulosis virus has been found to remain infective on leaves for up to four months. Spread of granuloses in nature among individuals of the same generation is mainly by ingestion of contaminated food material, but transmission from one generation to the next can be achieved by way of the eggs.

Entomopox viruses The entomopox viruses have a worldwide distribution and infect several orders of insect including Coleoptera, Lepidoptera, Orthoptera and Diptera. The virions have the complex structure of pox viruses (see p. 234 and Fig. 8.11 of *original* text). A lipoprotein envelope surrounds a 'lateral body' composed of amorphous protein and a plate-like core which encloses the nucleic acid of the genome. In some entomopox viruses a rope-like structure, possibly an enzyme-DNA complex, lies in this core. In at least three entomopox viruses the nucleic acid is known to be DNA. The virions are *ca.* 400 nm long and 250 nm wide in Lepidoptera and are smallest (320×230 nm) in Diptera. The majority of the virus particles generated in the host cells are integrated within large proteinaceous spherules which develop in the cytoplasm and eventually after decay of the insect come to lie in the substrate where they infect feeding insects. Some entomopox viruses have large spindle-shaped protein crystals associated with the ovoid inclusion bodies, some have small ones, and some have none. Presence and size of spindles are features of taxonomic significance.

Viruses that cause non-inclusion diseases The number of non-inclusion or free viruses known in insects and other invertebrates has increased rapidly in recent years, and is now rather more than a dozen. The majority of these are located in the cytoplasm of host cells. Only two—densonucleosis of *Galleria mellonella* and the flaccidity virus of *Antheraea eucalypti*—are known to have an affinity with the nucleus. Other free viruses include several that infect bees, for example acute (ABPV) and chronic (CBPV) paralysis viruses and sacbrood virus (SBV): transparency or 'wassersucht' virus of beetles; paralysis virus of crickets; the virus causing 'clear heads' disease of silkworms; lethargy virus in cockchafer; Malaya disease of Indian rhinoceros beetles; a virus disease of *Cirphis unipuncta*, 'iridescent viruses' in *Tipula* (TIV), *Chilo* (CIV), *Sericesthis* (SIV), and mosquitoes (MIV); and hereditary sigma virus of *Drosophila*. Paralysis diseases of crabs and mites are also caused by free viruses. However European foulbrood of bees is now known to be caused by a bacterium.

Non-inclusion viruses are a rather heterogeneous group and vary among other respects in the shape of the virus particles and the sort of nucleic acid contained. The majority are more or less spherical, but CBPV particles are slightly elongated and variable in size, and the Malaya disease virus is rod-shaped in its mature form but spherical at an earlier stage of organization of the particles. Densonucleosis virus, TIV, SIV, and CIV contain DNA, but transparency, *Antheraea* flaccidity, CBPV, ABPV, sacbrood, and European red mite viruses contain RNA. The sort of nucleic acid has not been determined for all.

TIV has received much study. It causes a disease of larvae of the crane fly (*Tipula paludosa*) in which the body fluid becomes brilliantly iridescent owing to the presence of microcrystals that are built up of crystalline arrays of rather large (130 nm diameter) virus particles. The latter in addition to DNA contain lipid and chiefly protein. Though DNA-containing, they lie in the cytoplasm of the host's cells. At a late stage of the disease a quarter of the dry weight of the insect may be virus. This is the highest proportion known for an animal disease. Cross inoculation tests with this distinctive virus have shown it will infect other Diptera as well as Lepidoptera and Coleoptera.

Drosophila flies infected with Sigma virus become permanently paralyzed if exposed to carbon dioxide in doses that merely anaesthetize uninfected flies. This virus is passed on from generation to generation.

Arboviruses Arboviruses are acquired by blood-sucking arthropods (mosquitoes, sand-flies, certain gnats and ticks) when they feed· on vertebrates in whose blood there is a sufficiently high level of virus. The viruses multiply, become established in the salivary gland, and subsequently are transmitted back to susceptible vertebrates, mainly wild animals including rodents, birds and reptiles. Two hundred and four arboviruses had been catalogued up to 1967, and of these about 75 cause disease in man. The nucleic acid in all arboviruses for which information is available is RNA, and the shape of the majority of arboviruses is spherical. Their antigenic relationships allow about three-quarters of known arboviruses to be distributed among a number of antigenic groups, more than half of which include human pathogens. The human pathogens are found mainly in groups A and B.

Typical arboviruses are generally believed to produce no cellular damage in the arthropod host. In susceptible vertebrates the symptoms of infection range from influenza-like ailments to severe disorders of the central nervous system and from mild to severe haemorrhagic disease. Arbovirus diseases in man include St. Louis encephalitis, Western encephalitis, Venezuelan equine encephalitis (that also affects horses), dengue, and yellow fever. Epidemics of arbovirus diseases sometimes cause considerable public concern. Though vaccination has proved effective in controlling yellow fever, it is possible that for arbovirus diseases generally, the best means of control is reduction of populations of the appropriate arthropods. The effective reservoirs of arboviruses are not generally known, but might include ticks, in which life-long infections may occur, and in temperate regions certain vertebrates such as hibernating rodents.

Plant pathogenic viruses in insects There is good evidence for certain plant pathogenic viruses multiplying in their insect vectors, for example potato leaf-roll and sowthistle yellow-vein viruses in aphids, and wound tumour and European wheat striate viruses in leafhoppers. Additionally, the transovarial transmission of rice dwarf and rice stripe viruses from generation to generation of vector repeatedly is powerful indirect evidence. Whether such viruses should be considered primarily plant or insect viruses is arguable.

Some general aspects of the biology of viruses that infect insects Two different viruses may con-currently infect the one insect. It has been found in the laboratory that administration of even heat-inactivated granulosis viruses may enhance the virulence of nuclear polyhedrosis viruses, though no marked effect has been found with the converse arrangement. The feeding to an insect of a virus not known to occur in it can stimulate a latent virus to activity. This can be a complicating factor in experiments on cross infection. Latent viruses are extremely frequent in insects, and a variety of factors ('stressors') which can induce a state of stress in the insect are effective in activating them. Such stressors include unfavourable temperatures or humidities, overcrowding, unsuitable food, feeding with certain chemicals, and inoculation with a foreign virus.

Bacterial infections of invertebrates

Some of the bacteria that cause disease in man and higher animals, and for which arthropods are the vectors, harm their vectors also. For example *Yersinia pestis* (the cause of plague). *Pasteurella tularensis* (the cause of tularemia), and certain species of *Borrelia* (that cause relapsing fevers) and *Salmonella* all cause disease in their arthropod vectors. The same is true of many rickettsiae. For example, the rickettsiae that some arthropods transmit multiply at the expense of the arthropod. While some multiply on the surfaces of host cells (e.g. of the gut epithelium) and cause no disease, others multiply in the cytoplasm of infected cells or even enter the nuclei. Their relationships with arthropods range from obligate commensalism to parasitism accompanied by fatal disease. For example, *Rickettsia prowazekii* (the cause of classical typhus in man) and *R. typhi* will invade cells of the epithelium of the gut of lice and multiply in their cytoplasm. The cells enlarge and are discharged from the gut epithelium, which is left irreparably harmed. The lice soon die. Organs other than the gut may become infected if the rickettsiae are experimentally introduced into the body cavity.

Varieties of *Bacillus thuringiensis* (e.g. vars. *thuringiensis*, *entomicidus*, *sotto*) comprise an important group of microbial insect pathogens that form ellipsoidal endospores accompanied by proteinaceous parasporal bodies. They cause a coupled toxaemia–septicemia in larvae of butterflies and moths. The septicemia results from ingested spores, but successful infection does not occur until after the epithelial cells of the midgut have been damaged by the action of the δ-endotoxin of the parasporal body. In silkworms, signs of poisoning can be detected within minutes of ingestion of parasporal bodies. By 45 minutes, erosion of the gut epithelium is virtually complete, and usually within 90 minutes there is general paralysis. In most other susceptible larvae cessation of feeding is not followed by paralysis but by sluggishness and death in 2 to 3 days. The precise mode of action of this toxin has not been established. The once-favoured possibility that the toxin disturbed cation regulation in insect tissues has now been rejected. The δ-endotoxin is produced abundantly when the bacterium sporulates but is synthesized in small amounts during the phase of vegetative growth also. The bacteria also produce a water-soluble exotoxin, the β-exotoxin, in the culture medium. It is toxic to fly larvae, but it is possible to eliminate it from preparations of *B. thuringiensis* by strain selection or by purification. Preparations of *B. thuringiensis* are commercially produced for use as microbial insecticides and contain a mixture of bacterial spores and toxic crystals. They are formulated for application to crops as sprays or dusts. They are harmless to higher animals including man, and even to beneficial insects such as honey bees, predators and parasites. The cessation of feeding and slow death of caterpillars after treatment with these preparations are consequences less spectacular than those which may follow use of chemical insecticides. Nevertheless, these preparations provide effective control which may be applied to crops even close to their harvest period when chemical insecticides are no longer allowed. It has recently been found that the effectiveness of *B. thuringiensis* in laboratory and field against the spruce budworm, a very important forest pest in North America, can be enhanced by adding chitinase to the preparation, but the mechanism involved has not been elucidated.

A second important group of bacterial pathogens contains *Bacillus popilliae* and *B. lentimorbus*, the causes in white grubs of types A and B milky diseases respectively. The large-scale use in North America of Type A milky disease for the control of the Japanese Beetle, *Popillia japonica*, represents one of the major successes for artificial biological control of insect pests by microbial pathogens. The pathogens affect only certain closely related beetles of the family Scarabaeidae. The bacterium gains access to the host's blood in which it multiplies and produces a marked turbidity, whence the name of the disease. In an area in which the disease becomes established, spores of the pathogen come to be present in the soil in enormous numbers. The pathogen cannot continue its development in dead insects, and requires special media and conditions to be cultured artificially. Commercial preparations of these bacilli for use as microbial insecticides are prepared from the bodies of specially infected Japanese beetle larvae.

Bacillus cereus, a widely distributed, common, soil saprophyte to which *B. thuringiensis* is closely related, is known sometimes to cause disease in divers insects. It multiplies in the gut and produces much phospholipase which either kills the host directly or facilitates the invasion of the body cavity by bacterial cells. Other insect-invading bacilli include *Bacillus larvae*, the cause of American foulbrood of bees. Obligate anaerobes are also known to cause diseases in insects and include *Clostridium brevifaciens* and *Cl. malacosomae*. Clostridia are probably commoner insect pathogens than was formerly thought. Their detection requires the use of special culture media.

A number of pseudomonads and members of the Enterobacteriaceae cause lethal septicemias in a range of insect hosts when small doses (e.g. 10^4 cells) of them are injected into the body cavity, but do not readily infect when ingested unless injury helps them penetrate the gut wall. Such 'potential pathogens' are common causes of low and sporadic mortality in laboratory populations of a wide range of insects but rarely cause epizootics in either the laboratory or the field.

Among rickettsiae with a distinct affinity solely for arthropods *Rickettsia popilliae* causes the fatal 'blue disease' of the Japanese beetle, *Popillia japonica*. The name of the disease refers to the colour that results from the scattering and reflection of light by the rickettsiae themselves in the fat-body cells of the host. *Rickettsia melolonthae* attacks cockchafer grubs and has been introduced into other insects. It usually proves fatal to the grubs.

Protozoan infections of invertebrates

Many Protozoa live in association with invertebrates in relationships that range from commensalism to parasitism. About 1200 of the approximately 15000 known Protozoa are associated with insects. Some of these are vertebrate pathogens and the insect serves as a vector, but there are Protozoa that multiply within the vector and harm it. There are trypanosomes that spend part of their life cycle in the gut or salivary glands of insects and the rest in higher plants or vertebrates. Infection of mosquitoes with species of *Plasmodium*, including the malarial pathogen, may shorten their lives and prejudice their survival of unfavourable conditions. Other protozoa are serious and more or less specific pathogens of insects, helminths, molluscs, and other invertebrates. Such parasites are to be found in all four subphyla of the Protozoa.

Among the Sarcomastigophora there are both flagellates and sarcodines that cause insect diseases of which one of the best-known is amoebic or 'spring' disease of the honey bee. The amoeba (*Malpighamoeba mellifica*) develops as an extracellular parasite in the Malpighian tubes of the bee and causes a malfunctioning of the tubes that lowers the resistance of the bees to other infections and is fatal for actively flying bees. A similar amoebic disease affects grasshoppers. Among ciliates pathogenic for insects there are species of *Tetrahymena* that besides parasitizing and killing insects such as mosquitoes and chironomids, can also live freely. Many of the Sporozoa are associated with insects vectors to which they do little harm but others are strict insect pathogens. Among these are many gregarines. Because schizogony (p. 201) leads to the formation of a new generation of feeding individuals that need much host tissue for their nutrition it is only among what are termed schizogregarines that serious insect pathogens are found. For the same reason, entomophilous eugregarines, being incapable of schizogony, are nearly all more or less harmless commensals that live in the lumen of the gut. Little is known of the host specificity of schizogregarines. Sometimes they cause severe outbreaks of disease among insects in the field. Coccidians, another group of Sporozoa, are mainly parasites of invertebrates and some have insect vectors, but a few species are true parasites of insects that include stored products pests, aquatic insects, and fleas. Another group, the haplosporidians, invade such freshwater invertebrates as Copepoda and Cladocera, but also some insects, including mosquitoes, black flies, ants, and the bark beetle *Ips typographus*. Infection ranges from chronic to fatal. The haplosporidian *Minchinia nelsonii* is the probable cause of extensive mortality among oysters in the Delaware and Chesapeake Bay regions of the United States, while *Haplosporidium tumefacientis* causes a disease in the Californian sea mussel. Among the Cnidospora most microsporidians are parasites of insects and some are well known as causes of loss of silkworms and honey bees.

Fungal infections of invertebrates

Many aquatic Phycomycetes have been recorded as growing in or on aquatic invertebrates which include flatworms, eelworms, rotifers, water fleas, copepods, crustacea, mites, and molluscs. Many of these records undoubtedly relate to parasitic infections, but in few cases has the pathology been investigated. Most of what is known about fungal infection of invertebrates relates to entomogenous fungi. These are fungi that grow upon insects, living or dead. Many infect live insects and cause diseases ranging from fatal to mild cutaneous infections. Other entomogenous fungi characteristically grow on or in living insects in relationships of commensalism or even mutualism (p. 275). Yet others are relatively unspecialized wound parasites, secondary invaders, or saprophytes.

Most of the fungi that produce fatal infections spread vegetatively through much or all of the host's body as hyphae or free cells or both. Such endoparasitic fungi probably infect by way of the integument in most instances. Three sizeable specialized groups of endoparasites merit special mention. *Coelomomyces* is a genus of flagellate aquatic fungi (order Blastocladiales) of which there are around fifty known species, almost all parasitic upon mosquitoes; almost all of the remainder attack certain Diptera whose larvae live in water or moist habitats. The genus has a virtually worldwide distribution. Dramatic discoveries since 1974 have revealed that *C. psorophorae* is obligately heteroecious, having a second phase within the copepod *Cyclops vernalis*. There is evidence that *C. punctatus* also has an alternate stage in this same copepod. Prior to this discovery, obligate heteroecism in fungi was known only in the Rusts (Uredinales, p. 198). The discovery is likely to expedite studies on the life cycles of other species of *Coelomomyces* and speed the use of these fungi for the biological control of mosquitoes and other susceptible insect pests. In the Entomophthorales (class Zygomycetes, p. 196), there are many species that are specialized for parasitism of insects in subaerial environments. In most the parasite spreads to new hosts by means of spores that are forcibly discharged into the air. Thick-walled perennating spores are also formed. A strange amoeboid vegetative phase of the fungus in the host's haemolymph occurs in at least some species. The third important group are the numerous species of the ascomycetous genus *Cordyceps*. Many of these parasitize insects that eventually die in litter or in the soil. A somewhat club-shaped fructification that rises into the air after the death of the insect provides for the dispersal of the parasite's ascospores. Similar endoparasitism is displayed by a variety of other species widely distributed among the major groups of fungi. Among the parasitic Deuteromycetes, species of *Beauveria* and *Metarrhizium* are particularly common and widespread.

Ectoparasitism that generates mild non-fatal cutaneous disease is particularly well developed in the order Laboulbeniales (class Ascomycetes). Each fungal thallus comprises a minute cellular axis that is attached to the surface of the host's integument and bears antheridia or perithecia or both. Probably these fungi are nourished by way of haustorial penetrations of the host's integument, though only in relatively few species have these actually been observed.

There are some fungi that appear to be strictly superficial commensals. These live attached to the integument or gut lining of insects and certain other arthropods. This habit is characteristic of the class of Trichomycetes, the members of which produce anchored filaments of limited growth. They probably are nourished by materials dissolved in the aqueous phase that surrounds them.

Attempts to use entomogenous fungi to control insect pests of crop plants have not, on the whole, been successful. The introduction of the parasite into an area is seldom sufficient alone to achieve control, since it is usually present already but does not reach epidemic proportions unless conditions are just right. Reinforcement of the natural inoculum of the pathogen in the early stages of annually recurring epizootics, so that the latter develop more swiftly, is one approach currently receiving attention.

Predacious fungi

A number of fungi, including some members of the Hyphomycetes (p. 199) and Zygomycetes (Zoopagales, p. 196) actually trap nematodes, soil amoebae, and sometimes rotifers, and subsequently invade and devour them. The hyphomycetous predacious fungi (species of *Arthrobotrys*, *Dactylella*, etc.) trap their prey by complex loops of hyphae (Fig. 14.7); by hyphal rings, the component cells of which inflate, thus constricting the aperture when an eelworm attempts to pass through (Fig. 14.7); or by hyphal branches with terminal sticky knobs to which the eelworm adheres. Once the prey is trapped the hyphae penetrate the skin and ramify through the body. The members of the Zoopagales shed sticky conidia which adhere to the prey, usually an amoeba or other small soil animal, and put out germ tubes which penetrate into the body where a

small thallus is formed. The water mould *Zoophagus insidians* captures rotifers by traps of the 'lethal lollipop' type. When the peg-like trap is triggered by a browsing rotifer it rapidly secretes a glue which holds the prey by its cilia. Thereafter an extension of the trap grows into the rotifer and digests it.

Vertebrate host defences against microbial attack

Introduction

The success or otherwise of a virulent parasite in establishing itself in a host is not solely a property of the parasite, but is greatly influenced by the resistance of the host. This may be characteristic of the host species or may be peculiar to the individual.

INNATE RESISTANCE

It is well known that certain diseases are limited to particular species of animals, as with gonorrhoea and diphtheria found naturally only in man. Reasons for host specificity are not usually known but the following are examples where some information is available. Anthrax, a pathogen of most animals, is not found in birds, probably owing to their higher body temperature; it is well known that this organism when grown at a temperature of 42°C becomes avirulent. Differences in susceptibility of various animal species to brucellosis is due in part to the presence of the carbohydrate erythritol, a growth-promoting substance for *Brucella* spp., found in significant concentrations in the tissues where the organisms normally localize in susceptible animals, but not in the corresponding tissues of non-susceptible ones.

Not only does susceptibility vary between species, but different breeds or races within a species may show variation in resistance to the same pathogen. It has been shown that breeds of pigs not possessing the receptor sites in their gut to which pathogenic *E. coli* attach, are thereby insusceptible to infection by these strains.

INDIVIDUAL RESISTANCE

The individual animal, in addition to its innate resistance, has many forms of defence against the pathogenic organisms; these may be considered under two headings: general host resistance and specific immune responses or acquired immunity.

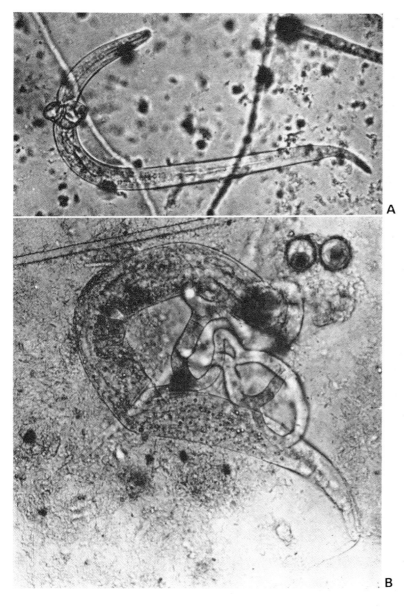

Fig. 14.7 Predacious fungi. (**A**) *Dactylaria gracilis* (× 500). An eelworm is shown caught in a constricting hyphal ring. (C. L. Duddington, by permission, *New Scientist*). (**B**) *Arthrobotrys robusta* (× 500). An eelworm captured in a network of adhesive hyphae. (C. L. Duddington)

General host resistance

HEALTH

Animals in good health are better able to resist infections than those in poor health. The health of an individual can be influenced by many environmental factors.

Factors affecting the primary defences of the body Before an organism can reach the body tissues, the epithelial coverings, e.g. skin and mucous membranes, must be penetrated (p. 309). In

a healthy animal, where these are intact, many pathogenic micro-organisms are destroyed or removed before they can invade the internal tissues. Factors damaging these natural barriers deprive the animal of this protection.

Diet Adequate food is essential for good health, and there is no doubt that variations in diet are associated with alterations in resistance to infection. Epidemic diseases frequently follow times of food deficiency such as in war or famine.

The influence of diet on resistance is a very

complex problem interconnected with many other factors, and few experimental data are available. Many workers claim that deficiency of vitamins A and C, and of proteins, particularly during the period of rapid body growth, decreases resistance to infection. It is thought that this is due to a lowering of efficiency in the production of phagocytes and antibodies (see below p. 323). Indirectly a deficiency of one nutrient in the normal diet may lead to unusual habits which predispose to disease. In certain areas of the cattle ranges of South Africa, a deficiency of minerals in the soil leads to a deficiency in the grass on which the cattle feed. The latter then chew bones of animals that have died on the open veldt. The carcasses from which these bones come are frequently contaminated with *Clostridium botulinum* type C which produces a potent neurotoxin in the decomposing carcass. Cattle chewing the bones ingest the preformed toxin and contract the disease 'Lamziekte' which is a form of toxic food poisoning analogous to botulism in man (p. 335).

Fatigue Fatigued animals are more readily infected than others. Where animals are transported long distances, without proper facilities for rest and under badly ventilated conditions, the resistance of the animal is lowered and organisms normally carried in their respiratory tracts assume a pathogenic role, infect the lungs, and produce pneumonia. This appears to be the main predisposing factor in the so-called 'shipping fever' of animals. This clinical condition possibly involves a number of infectious agents which frequently include a myxovirus of parainfluenza group 3 and *Pasteurella multocida*.

Climatic variations Sudden changes of weather, particularly increased humidity and chilling after a warm, dry spell, are known to influence the resistance of animals, particularly to respiratory infections. Prolonged exposure to cold has been shown to depress antibody formation (p. 324). On the other hand the incidence of infection by the common cold virus has been shown under carefully controlled experimental conditions not to be associated with chilling.

Occupational hazards Exposure of individuals to unusual hazards, such as close contact with infectious agents or damage to normal defences, increases the risk of infection. Particles of silica or other material regularly inhaled by stone-masons, miners, and others cause irritation of the lungs which predisposes to pneumonic diseases, tuberculosis in particular. The effects of increased contact with infectious agents may be illustrated by a number of examples; brucellosis in veterinary surgeons due to contact with infected cattle; woolsorter's disease due to inhalation of anthrax spores by workers sorting untreated imported wool; erysipeloid (in contrast to erysipelas caused by streptococci) in butchers and abattoir workers as a result of infection with *Erysipelas rhusiopathiae* through abrasions caused by contaminated pig bones; Q-fever by inhalation of dust (contaminated with rickettsias) by workers in close contact with infected cattle or their products.

Irradiation The effect of radiation on the resistance of the animal is dependent on the degree of exposure. Small doses of X-rays have been shown by some workers to increase resistance, probably owing to a stimulation of the immunity-producing centres of the body. On the other hand, there is no doubt that large doses of ionizing radiations, whether X-rays or atomic radiations, decrease the resistance of the host by damaging the lymphoid tissue so that it can no longer produce antibody in response to infection (p. 324). Radiation may also damage bone marrow tissue, thereby impairing the production of red blood cells and leucocytes in the adult mammal.

The presence of other parasites Damage caused to the host by one organism may increase the risk of infection to a second, as in black disease of sheep, in which infection of the liver with *Clostridium oedematiens*, a blood-borne infection from the gut, occurs if liver flukes have previously caused necrosis of the liver, thereby producing suitable anaerobic conditions for the rapid propagation of the clostridia.

AGE AND SEX

Susceptibility to infection varies considerably with the age of the animal. New-born animals are frequently very susceptible to organisms, such as those of the normal gut flora (p. 301), which are not usually pathogenic to older animals. Infection often occurs during the critical period whilst the animal is acquiring its gut flora and before it has built up a resistance against these organisms. Thus young children succumb to diarrhoea by specific strains of *Escherichia coli*; young lambs frequently become infected with *Clostridium perfringens* from the soil and develop lamb dysentery. New-born animals may, however, be more resistant to certain infections than are slightly older ones if they have received a temporary immunity from the mother (p. 325), Once this temporary immunity is lost, young animals are frequently more susceptible to certain infections than are mature ones. For instance, babies in their first year of life are usually immune to measles because of passive immunity transferred from the mother but in the next few years are highly susceptible. With some diseases, mature animals are actually more susceptible than the young. *Brucella abortus* is unable to establish itself in calves but readily invades the uterus and udder in the cow and the testicles in the bull. Growth promoting substances for the microbe (p. 320) present only in the adult animal, are essential for the establishment of infection by this pathogen.

CELLULAR DEFENCES

When bacteria pass through the natural barriers of the primary defences of the body they are treated in the same way as inert particles by the cellular defences. Specialized cells, known as phagocytes, rapidly engulf the organisms and, unless killed by the toxins of the bacteria, begin to digest and destroy them. These phagocytic cells may be fixed in the liver, spleen, and lymph glands and include macrophages or histiocytes; other phagocytes may be circulating in the blood and consist mainly of large mononuclear leucocytes and polymorphonuclear leucocytes, the so-called white cells of the blood.

Many virulent organisms kill phagocytes, resulting in the formation of pus, and continue to multiply. The subsequent course of the infection then depends on whether the animal is stimulated to produce sufficient phagocytes to overcome the rapidly multiplying bacteria. In this the cellular response is aided by immunity and any artificial assistance that may be provided by specific chemotherapeutic treatment (p. 333). Frequently virulent organisms are able to resist phagocytosis by means of special properties of their capsule which give them some degree of protection.

Specific immunity

It is widely known that many diseases of childhood rarely occur in adult life, the early occurrence of the disease establishing a lasting immunity in the individual. This acquired immunity is highly specific and arises after exposure of the host to the pathogen. The basic function of the immune system is to detect and eliminate from the body any substance which is recognized as *foreign* and this includes microbial pathogens. In this function, the host employs a wide variety of cells and cell products, each interacting with one another in the removal of this material. Usually this is successfully accomplished and the foreign substance is neutralized, destroyed or removed; occasionally the host response to certain types of foreign material may be exacerbated and lead to harmful results as in the immunologically mediated diseases.

ANTIGENS

Substances which initiate immune responses are termed antigens. Usually they consist of protein, polysaccharide or lipid or combinations of these. The protein is concerned mainly with the antigenic stimulus which provokes the immune response whilst polysaccharides, especially terminal groups of sugar side chains on the molecule (the determinant groups), are concerned with the specificity of the response. Antigens have a large molecular weight (they are too large to be disposed of by the host's normal physiological functions), they can be soluble or particulate and must be foreign to the host. Not all stimulate a response to the same extent; some are good whilst others are poor antigens. Micro-organisms and many of their products, serum and tissue transplants are good antigens; gelatin, although a protein is relatively poor. The microbial cell is a complex of numerous antigens and the immune mechanism of the body may respond to each one which is exposed on the cell surface or which diffuses away from the cell.

HOST IMMUNE RESPONSES

An antigen may enter the animal body and gain access to the tissues during the course of a natural infection or it may be introduced artificially. The host immune mechanisms respond to the antigenic stimulus by either or both of two principle ways:

1 by the formation of antibodies (immunoglobulins—Ig's) which are released into the blood and other body fluids. They combine with the type of antigens which stimulated their formation thereby impairing or neutralizing their pathological function.

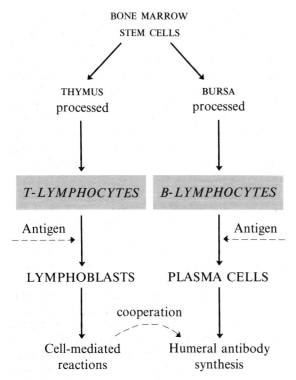

BONE MARROW
STEM CELLS

THYMUS
processed

BURSA
processed

T-LYMPHOCYTES   **B-LYMPHOCYTES**

Antigen   Antigen

LYMPHOBLASTS   PLASMA CELLS

cooperation

Cell-mediated   Humeral antibody
reactions   synthesis

Fig. 14.8 Processing of bone marrow cells by thymus and gut associated lymphoid tissue to produce T- and B-lymphocytes respectively. Proliferation and transformation of cells to lymphoblasts and plasma cells occurs under antigenic stimulus.

2 by the stimulation of cell-mediated immunity in which certain whole cells become sensitized to and react with the stimulating antigen. This produces various results. For instance, foreign transplants are rejected by sensitized cells which migrate to the site: in others the sensitized cells play a role in delayed hypersensitivity.

Both types of immune response are associated with lymphocytes derived from primitive bone marrow cells (Fig. 14.8). According to how these primitive cells are processed two populations of lymphocytes are formed. The B-lymphocytes are processed by gut associated lymphoid tissue. In chicks this tissue is located in the Bursa of Fabricius sited near the anus and hence are termed B-lymphocytes; in other species the lymphoid tissue is located in the tonsil, appendix and Peyer's patches. Under antigenic stimulus B-lymphocytes transform into plasma cells which synthesize specific antibody; this is released unbound into the body fluids.

The T-lymphocytes are derived from primitive bone marrow cells processed by the thymus gland. Under antigenic stimulus these cells are trans-

formed into lymphoblasts which possess specific antigen-receptor sites on their plasma membranes. Since the receptor sites are cell-bound the whole cell is involved in combining with the antigen—hence this type of immunity is cell mediated.

ANTIBODIES

Antibodies are secreted into the body fluids, and into serum in particular, in response to an antigenic stimulus. They constitute a group of immunoglobulins which can be separated by their distinctive mobilities under the influence of an electric field. Four major immunoglobulins are recognized, IgG, IgA, IgE and IgM. The first three have molecular weights around 150000; IgM is much larger, having a molecular weight of around 900000. More than one type of immunoglobin may be formed under the same antigenic stimulus. The dominant component in the serum at any time depends upon a number of factors. For instance, IgM is chiefly formed in young animals and some lower animals, such as elasmobranchs. (More primitive vertebrates and invertebrates do not possess the capacity to produce antibody). A primary antigenic stimulus results in IgM formation whilst subsequent stimulation by the same antigens produces IgG. Only IgG can pass from mother to foetus by the placental circulation and thus only IgG is found in the serum of the new-born. Particulate antigens tend to stimulate IgM production. IgM is usually associated with particular cells, e.g. mast cells, and is involved in certain types of hypersensitive reactions (p. 331). IgA is secreted at mucous surfaces, e.g. mouth, alimentary tract, and constitutes an important local defence.

Since antigens are highly specific, the antibody produced in response to them is equally specific, irrespective of the class of immunoglobulin produced. This forms the basis of specific immunity to disease in the host and also the specificity of serological reactions in the laboratory (p. 326).

NATURAL ACTIVE IMMUNITY

The production of antibody by the animal host is known as active immunity. The titre of antibody increases during the course of an infection and remains high for months or years, depending on the identity of the original infection, but eventually begins to fall. When, however, the animal, which is specifically immune, is exposed to another attack by the same pathogen, antibody production occurs rapidly and only rarely does the infection establish itself a second or subsequent time. This sequence of events may be compared with antibody production following a second or subsequent 'booster' injection as practised in immunization (p. 326). Frequently a subclinical infection, which does not produce symptoms of disease, stimulates antibody production and may confer a lasting immunity on the individual. This may explain why some animals never succumb to a particular disease to which others of the same host species fall ready victims.

NATURAL PASSIVE IMMUNITY

During early foetal life antibodies are not formed under an antigenic stimulus but later the foetus is able to produce antibodies and full immunological competence is reached soon after birth. Antibody to specific pathogens can, however, often be demonstrated in the blood serum of newly born animals. These antibodies have not been produced by the young animal itself but by its mother in response to antigenic stimuli to which she has been exposed. The antibodies are passed to the young either via the placenta whilst the animal is *in utero*, as in man, rabbits, guinea pigs and rats, or in the first milk (colostrum) which is ingested by the animal immediately after birth, as in cattle, horses, sheep, goats, and pigs. Passive immunity persists in the young for a relatively short time (up to several months) but is nevertheless of great value in giving protection against infections that are likely to be met prior to the animal becoming fully immunologically competent. If calves are deprived of their mother's colostrum, they frequently succumb to pathogenic *Escherichia coli* present in their environment which produces a lethal septicaemia, associated with profuse scouring and emaciation; calves receiving colostrum usually exhibit only scouring and survive. The colostrum must contain specific antibodies to the types of *E. coli* encountered if protection is to be achieved. By natural exposure to particular serotypes of *E. coli* before the calf is born, the cow is stimulated to produce specific antibodies which are later passed on to the calf in the colostrum.

ARTIFICIAL ACTIVE IMMUNITY

The methods by which artificial active immunity may be achieved simulates that resulting from natural exposure to infection. Antibody production can be induced artificially by administering relevant antigens to the patient in the form of vaccines of dead or living micro-organisms or their products. It is extremely important that vaccines be prepared from organisms antigenically similar to those to which the individual is exposed naturally so that antibodies corresponding to the virulent pathogens may be formed. Only occasionally are fully virulent organisms used as vaccines, because of the danger of initiating the disease in susceptible individuals. Immunization against avian infectious laryngotracheitis utilizes the fully virulent virus, but this is introduced at an unusual site on the mucous membrane of the Bursa of Fabricius, a pouch of the cloaca, and care must be taken to ensure that the virus does not gain access to the respiratory system. A harmless inflammatory reaction results but immunity against the virulent respiratory disease is achieved.

Living vaccines are usually prepared with organisms of lowered virulence. These are known as attenuated strains and may be obtained by growing organisms under artificial conditions in the laboratory (e.g. the Bacillus of Calmette and Guerin—BCG vaccine for tuberculosis), or attenuated mutants may be selected such as the S19 strain of *Brucella abortus* and the non-paralytic strain of poliovirus oral vaccine (Sabin) used for protection against poliomyelitis. Occasionally, related strains, such as vaccinia virus for vaccination against smallpox, or strains adapted by passage through an animal host (e.g. goat-adapted rabies virus) or egg-adapted viral and rickettsial vaccines grown in chick embryos or their membranes, may be used.

Immunity to certain infections can be achieved by the use of killed bacterial and inactivated viral vaccines. These include bacterial vaccines against whooping cough, the enteric fevers (TAB, a vaccine against typhoid and paratyphoid A and B fevers), cholera, plague, and leptospirosis and viral vaccines against poliomyelitis (Salk vaccine, in contrast to the Sabin vaccine already mentioned), foot-and-mouth, swine fever, influenza, and measles. In many diseases in which major damage to the host is the result of bacterial exotoxins, protection can be achieved by stimulating the production of antitoxins by the host. Toxins are not used for this purpose on account of their toxicity but this can be reduced either by ageing or by treatment with formalin, the toxin being converted to toxoids. Toxoids are antigenic but not toxic and can be safely used to produce antitoxin, e.g. against the toxin of tetanus, diphtheria, and staphylococci.

Animals may be protected against many anaerobic infections caused by clostridia (e.g. lamb dysentery, black disease of sheep, blackleg of cattle) by the use of bacterins which are (formalin-killed,

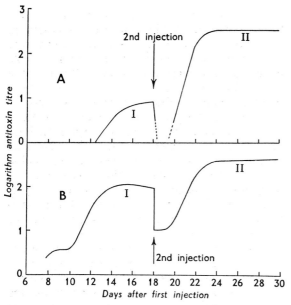

Fig. 14.9 The primary and secondary antitoxic responses of two rabbits to staphylococcal toxoid. Curves I (A and B) show the slowly appearing primary response following injection at time 0, by A, intravenous and B, subcutaneous routes respectively. Curves II (A and B) show the rapid secondary response in both animals to intravenous injection made eighteen days after the primary injection. The fall in antibody titre immediately following a second injection of a large amount of the corresponding antigen is known as the *negative phase*. This is thought to be due to removal of circulating antibody by combination with the antigen. (Modified from Burnet, F. M. and Fenner, F. (1949) *The Production of Antibodies*, Macmillan, London)

whole cultures which include both organism and toxoid).

When a non-living vaccine is used there is no continuous antigenic stimulus as in an infection and usually two or more injections are required to produce an adequate level of immunity. The first injection sensitizes the animal and gives a low titre of antibody which can be detected in the bloodstream from 8–12 days, dependent upon the site of the injection (Fig. 14.9). This is termed the 'primary' response. A second injection given several weeks later then results in a more rapid and marked response, as indicated by a rising antibody titre which can be demonstrated from about 48 hours onwards. This is termed the 'secondary' response. The titre once established, persists for varying periods of time but usually begins to fall after several months. A 'booster' dose given at infrequent intervals stimulates a rapid response, comparable to the secondary response of the initial immunization, and restores a significant titre and corresponding immunity.

ARTIFICIAL PASSIVE IMMUNITY

Several weeks are needed to establish a satisfactory immunity by active immunization and, to be fully effective, inoculation must be carried out before the animal is exposed to infection. Circumstances arise when it is necessary to give immediate protection, as when a susceptible animal is known to have been in contact with infection or is showing symptoms of the disease. It is then possible to protect by injecting antiserum containing the specific antibody against the infection. This is prepared by immunizing a suitable animal (usually the horse) and separating the serum from blood aseptically taken. Thus a dog which is likely to be exposed to the distemper virus could be given a prophylactic dose (i.e. treatment to ward off infection) of antiserum containing neutralizing antibodies against the virus. Similarly when diphtheria was prevalent in Great Britain infected children were treated with diphtheria antitoxin to neutralize the lethal effects produced by the toxin of the diphtheria bacillus. Just as passive immunity naturally transferred from mother to offspring is of a relatively short duration, so artificial passive immunity gives protection for only about three weeks. Administration of this foreign serum to the patient presents certain risks. The serum proteins from a foreign host are themselves antigenic and produce antibodies upon injection. Patients who have previously received injections of the same foreign serum may develop a hypersensitivity to it and produce an anaphylactic reaction upon receiving a later injection (p. 332). For this reason antiseral therapy is not practised as frequently as it used to be, more reliance being placed on combined toxoid immunization and antibiotic therapy as soon as symptoms arise.

In sheep, a combination of active artificial and passive natural immunity is used to protect young lambs against lamb dysentery within the first few days of life. The pregnant ewe is actively immunized with a formalin-killed whole culture of *Cl. perfringens*. The resultant high titre of antibody is transferred to the new-born lamb in the colostrum of the ewe's milk, thereby providing adequate protection. Typical values of antitoxin titres in the sera of the ewe and lamb, together with the titre in the colostrum over a period of time, are shown in Fig. 14.10.

ANTIGEN–ANTIBODY REACTIONS

When an antigen is mixed with antiserum containing its homologous antibody, physico-chemical

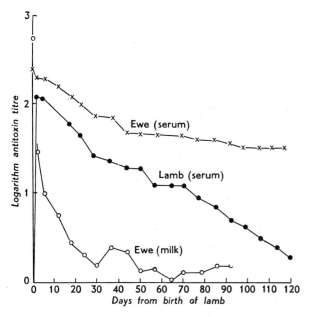

Fig. 14.10 Passive transfer of antitoxin in colostrum. The concentration of antitoxin in the blood of a ewe, in her colostrum and milk, and in the blood of her lamb from birth onwards. The ewe was immunized during pregnancy. (After Mason, J. H., Dalling, T. A., and Gordon, W. S. (1930) *J. Path. Bact.*, **33**, 783)

forces induce the antigen and antibody to combine together. Each antibody molecule possesses two combining sites of equal specificity, thus one molecule can bridge two molecules of antigen. Since antigens are multivalent the result is the formation of a lattice structure which is often macroscopically visible as in the agglutination of red blood cells or bacteria.

The detection of an antigen–antibody reaction is only possible if demonstrable reactions take place. These include agglutination, precipitation, complement fixation, killing and lysis of whole cells, phagocytosis, and neutralization of the pharmacological activity of the antigen. Antibodies used to be named according to the type of reaction they produced, such as agglutinin, precipitin, lysin, and opsonin. It is now appreciated that the same antibody may produce a number of effects depending on the environmental circumstances. An agglutinating antibody, for instance, may render cells susceptible to phagocytosis and killing or lysis in the presence of complement (p. 328), in addition to agglutinating the particulate antigen. On the other hand, some antibodies may be deficient in certain properties, as when an agglutinating antibody fixes complement but poorly.

Agglutination reaction This reaction occurs between homologous antibody and particulate antigen in the presence of electrolyte. When a suspension of cells is mixed with homologous antiserum a visible clumping or agglutination occurs. By use of a standard suspension of cells and a series of dilutions of antiserum the antibody can be titrated. This test is widely used for blood grouping and the identification of bacteria and diagnosis of their infections. Agglutinins, in addition to agglutinating bacteria, may also immobilize motile ones, an observation used in the Treponema Immobilization Test for the diagnosis of syphilis. Agglutination of virus particles occurs in the presence of homologous antibody, clumps of particles being observed under the electron microscope.

Soluble antigens may be absorbed on to carrier particles, e.g. bacterial polysaccharide on to red blood cells (sometimes treated with tanning agents, e.g. tannic acid or bis-diazobenzidine, to aid the adsorption) and these cells agglutinate when mixed with antiserum against the bacterial polysaccharide.

Precipitin reactions Soluble antigens, such as serum and bacterial polysaccharides, when mixed with homologous antibody in suitable proportions, precipitate in the presence of electrolyte. The reaction may be used as a ring test, in which a ring of precipitate forms at the fluid interphase when antigen is layered onto an antiserum; as a flocculation test when floccules of antigen–antibody complex are formed in a mixture of both reagents; as a gel diffusion test (often called the Ouchterlony test), in which both reagents diffuse towards each other in an agar or gelatine gel (Fig. 14.11).

In gel diffusion, lines of precipitate, corresponding to each antigen–antibody reaction, are formed in the gel at points where optimal levels of reactants meet. This technique is sometimes superimposed on a primary separation of immuno-globulins by immuno-electrophoresis. Antigens are caused to move through an agar gel under the influence of an electric current by which separation is accomplished. Antiserum, containing antibody to the various antigens, is later diffused from a source at right angles to the prior movement of the antigens and lines of precipitate form at the points reached by the various antigens (Fig. 14.11). This technique is used for the identification of viruses.

Lysis by complement Homologous antibody will

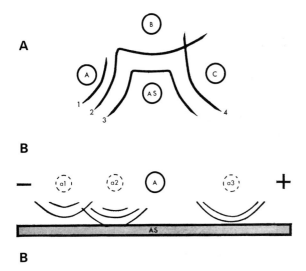

A

B

B

Fig. 14.11 Precipitation reactions in agar gels. (**A**) Lines of precipitation laid down in an agar gel when three antigen complexes (A, B, and C) are set up in an *immunodiffusion* test against an antiserum (AS). 1. Antigen common to complex A only. 2. Antigen common to complexes A and B (hence joined). 3. Antigen common to complexes A, B, and C. 4. Antigen common to complex C only (hence crossing line 2). (**B**) Immunoelectrophoresis. Antigen A, placed in an agar cup, has been subjected to an electric field which separates fractions a1; a2; and a3. Each of these fractions includes a number of antigens which migrate in the electric field at the same rate. Subsequent normal diffusion in the agar gel against a composite antiserum (AS) reveals the constituent antigens.

combine with living bacteria and usually bring about their agglutination, but the organisms still remain viable and fully pathogenic. However, a component of normal serum, known as complement (see below), will combine with bacterial cells treated with homologous antibody bringing about their death and, in some species, their lysis. Both death and lysis frequently occur with the Gram-negative bacteria, as with *Vibrio cholerae*, but no lysis occurs with Gram-positive bacteria. This is thought to be due to the greater thickness of cell walls in Gram-positive bacteria. Complement combines with most antigen–antibody systems with certain exceptions, e.g. toxin–antitoxin complexes.

Red blood cells are also lysed in the presence of both homologous antibody and complement; this has been used as a model system for the elucidation of the mechanisms of complement lysis and is now extensively used in the laboratory for detecting the presence of free complement *in vitro*.

Electron microscope studies have shown that complement lysis of red cells is the result of a large

number of holes being formed in the cell membrane (80–100 nm diameter) through which the haemoglobin leaks out. The initial addition of homologous antibody to red cells does not cause lysis; only the final addition of complement produces the holes. Quantitative studies have shown that one molecule of IgM (p. 324) is sufficient for the formation of one hole, whereas 3000–6000 molecules of IgG are required per cell in order to form one hole. The greater efficiency of IgM is thought to be due to its larger molecular size covering a greater number of denaturation sites on the cell membrane.

Complement is a non-specific component of all animal sera and is not increased by any immunization procedures. The serum level of complement varies from animal to animal, but guinea pigs fed on green food produce a sufficiently high level for their serum to be employed as the laboratory source of complement. Complement is not a single substance but a mixture of at least nine substances (C_1, C_2, C_3, etc.) of globulin or mucoprotein composition. Certain components are heat liable and heating an immune serum at 56°C for 30 minutes 'inactivates' complement; activity can be restored by the addition of fresh serum. When complement is added to an antigen–antibody mixture the components are adsorbed in a defined order (i.e. C_1, C_4, C_2, $C_{3,5,6,7,8,9}$) at a site adjacent to the antigen–antibody junction.

The complement lysis of red cells or bacterial cells, which can be observed to take place *in vitro* also take place in the tissues of the immune animal. For instance, complement participates in haemolytic anaemia of the new-born, bringing about the haemolysis of red blood cells in the presence of specific antibody. A further possible role of complement is in assisting polymorphs to lyse ingested bacterial cells.

Complement fixation The lysis of red blood cells by complement is the basis of a widely used serological test known as the complement fixation test. Many antigen–antibody reactions, involving viral or soluble antigens, are not macroscopically visible, but it is possible to demonstrate the reaction by superimposing the visible lysis of red blood cells in the presence of complement.

To detect specific antibody in a patient's serum, a standard volume of the neat or diluted serum (heated at 56°C for 30 minutes to destroy its natural complement) is mixed with a standard amount of the specific antigen. To this is added a minimal quantity, based on titration, of fresh normal serum

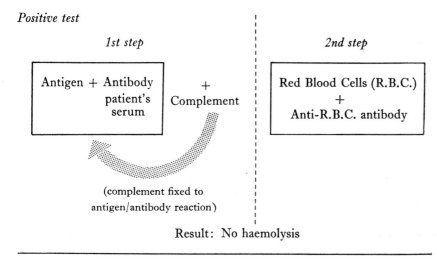

Positive test

1st step

Antigen + Antibody patient's serum	+ Complement

(complement fixed to
antigen/antibody reaction)

2nd step

Red Blood Cells (R.B.C.) + Anti-R.B.C. antibody

Result: No haemolysis

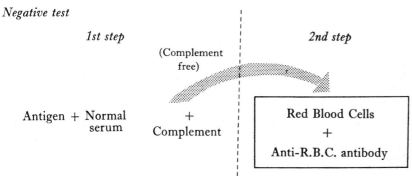

Negative test

1st step

Antigen + Normal
serum

+
Complement

(Complement
free)

2nd step

Red Blood Cells + Anti-R.B.C. antibody

Fig. 14.12 The complement fixation test. (See text for explanation)

(usually guinea pig) as a source of complement. This combines with the antigen–antibody complex if the patient's serum contains antibody. To demonstrate that this reaction has occurred, sensitized sheep red blood cells (i.e. red blood cells treated with homologous antiserum) are added. These cells will only lyse in the presence of free complement and hence act as an indicator of its presence or absence. If antibody is present in the patient's serum, it will combine with the antigen and fix complement—the consequence will be that the red blood cells are not lysed. On the contrary, if no antibody is present, the complement will be free and available to lyse the sensitized red blood cells. These reactions are illustrated in Fig. 14.12.

Phagocytosis Foreign particles found in the body are phagocytosed by macrophages and monocytes in the tissues and polymorphonuclear leucocytes in the peripheral blood of the host. Unlike inert particles, virulent bacteria are removed only after prior modification, usually by antibody. In the presence

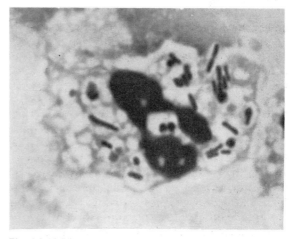

Fig. 14.13 Phagocytosis of *Escherichia coli* in a polymorphonuclear leucocyte in a centrifuged specimen of urine from a case of cystitis.

of complement, the antibody-treated bacteria are killed, after which they are removed as inert particles (Fig. 14.13). Viruses, after being neutralized by homologous antibody, are similarly removed.

Neutralization Antibody is able to neutralize the toxic or infective power of a micro-organism. Neutralizing antibodies constitute the basis of animal immunity against virus diseases. These antibodies are stimulated by the protein components of the virus nucleoprotein and the neutralization of virus infectivity entails the firm adsorption of the homologous antibody molecules to the surface proteins. This reaction occurs *in vivo* and *in vitro*. Subsequent penetration of an animal cell by the neutralized virus particle results in its total digestion by phagolysosomal enzymes rather than its establishment within the cytoplasm as an infectious particle.

The assay of the virus neutralizing antibodies contained in a serum is carried out in the laboratory by means of a neutralization test. Standard amounts of the virus are mixed with a range of dilutions of the serum, allowing neutralization to proceed to finality. Aliquots of each mixture are inoculated into either animal tissues or tissue cultures to determine the greatest serum dilution which is able to neutralize the virus infectivity.

Virus neutralizing antibodies persist for a long time in the animal, thereby providing prolonged immunity. For example, the immunity against yellow fever and German measles tends to be life-long. The apparently short-lived immunity to certain viruses (e.g. influenza, common cold viruses) is not due to a failure to produce a sufficient immune response but to the large number of serotypes, or to their genetic variation, consequently infection with one does not provide immunity against another.

Bacterial toxins are antigens and the antibody (antitoxin) they stimulate combines with the toxin to neutralize its toxic effects on the animal host. Endotoxins are poor antigens and their toxicity is usually incompletely neutralized by antibody stimulated by them. Exotoxins can be assayed by determining the quantity of the more stable homologous antibody required to neutralize their lethal effects or local skin reactions in animals. Other tests, such as the specific neutralization of haemolytic and lecithinase activity of particular exotoxins can be determined *in vitro*. Exotoxins are not destroyed by antitoxins but may be recovered from toxin–antitoxin complexes, and because of this, complexes of these reagents are unstable as immunizing agents.

Haemagglutination inhibition Certain viruses or products of their intracellular replication (haemagglutinins) are able to agglutinate red blood cells of various animal species *in vitro*. This reaction is termed haemagglutination (HA) and, since under controlled conditions the degree of agglutination is proportional to the quantity of virus, the HA test can be used as an assay method for the virus. The different degree of agglutination are readily detected by allowing the agglutinated red cells to settle out in round-bottomed tubes or in the cups of perspex plates, giving different settling patterns. The sensitivity of the HA test is not great; with influenza virus one haemagglutinating unit (i.e. the quantity of virus which will agglutinate 50% of the red cells in the tube) may be 10^6 virus particles. Further, since both infectious and non-infectious particles haemagglutinate, the test is a measure of the total virus in a preparation, not of infectivity.

Following infection with a haemagglutinating virus, antibodies are formed which inhibit the HA reaction, and these are termed haemagglutination-inhibiting antibodies. These may be assayed in the laboratory by means of the haemagglutination-inhibition (HAI) test. Standard amounts of virus are mixed with a range of dilutions of the serum, and, after a suitable incubation period, standard amounts of a one per cent red cell suspension are added to each mixture. The settling patterns of the red cells in the tubes or cups determine the HAI endpoint of the serum. Serum from a convalescent patient gives a fourfold or greater rise in titre compared with serum taken from the same patient during the acute stage of the illness.

SPECIFICITY OF ANTIGEN–ANTIBODY REACTIONS

Antigen–antibody reactions are highly specific, making it possible to recognize differences between closely related proteins and other antigens of plant or animal origin and even between tissues within the same host. This specificity depends primarily on the nature of the chemical groupings of opposite charge on molecules attracting each other. In addition, the spatial arrangement of the molecules, dependent upon the *ortho-*, *meta-*, or *para-* positions of chemical groups and stereoisometric spatial arrangements (*laevo-*, *dextro-*, or *meso-*), imprint their influence upon the shape of the molecules and enable the antigen and antibody to combine together in close proximity, thereby allowing short-range intermolecular forces to operate. These include van der Waal's forces, electrostatic forces, and hydrogen bonding. The union of an antigen with its antibody is usually very firm but the avidity (i.e. the strength of the attracting forces between antigen and antibody) varies from system to system and in some cases the union is reversible. No profound changes occur in either molecule as a result of combination, as is seen in antitoxin–toxin reactions from which the unchanged toxin can be fairly readily released.

The intact bacterial cell possesses multiple antigens on the cell wall surface in addition to those of flagella, fimbriae, and capsules. Many antigens are common to different species of the same genus and occasionally different genera, as for instance, antigens common to the rickettsiae (p. 308) of typhus and typhus-like fevers and certain strains of *Proteus*, where the sharing of antigens forms the basis of the Weil-Felix diagnostic test for these diseases. Antigens which are specific for individual species or strains are those most useful in distinguishing between micro-organisms.

Our present knowledge of the chemical structure is most complete for polysaccharide antigens. In some organisms the nature of the chemical group conferring specificity has been determined; the determinant group is often a relatively small part of the whole molecule. The antigenic specificity of the polysaccharide capsule of Lancefield Group A streptococci is dependent on the presence of *N*-acetyl glucosamine protruding as a side chain from a backbone of rhamnose in the polymer; in Group C, specificity is attributed to a side chain of *N*-acetyl galactosamine.

Similar work has been done on the chemical basis of serotyping in *Salmonella*. The specificity has been shown to be due to the presence of specific carbohydrates on the polysaccharide molecules of the cell wall structure. The terminal dideoxyhexoses (of which only five have been identified) determine the specificity of the antigenic determinants. Each *Salmonella* species examined has been shown to possess only one of these sugars which is characteristic of the serological group. For instance, paratose is characteristic of O-antigen 2 in species falling in Group A, abequose of antigen 4 in Group B, tyvelose of antigen 9 in Group D and colitose of antigen 35 in Group O. The specific sugar always occupies the terminal position but the antigen specificity may be modified by the way the sugar is linked to the molecule or by the sugar immediately next to it on the side chain.

THE ROLE OF INTERFERONS IN IMMUNITY

In contrast to other microbial infections, antibody immunity (p. 324) is not now considered to be a major factor in recovery from virus infections; whilst formed against virus antigens, antibodies are ineffective intracellularly and do not, therefore, inhibit virus multiplication. There is increasing evidence that interferon is one of the major factors in recovery. Interferons are soluble proteins, with a molecular weight of about 30000 and their production in tissue cells is induced by many viruses and some non-viral agents. Inactivated or avirulent viruses are usually more active inducers than infective virulent viruses since the latter rapidly inhibit cellular protein and RNA synthesis, both of which are required for interferon production. Interferons exhibit species specificity and inhibit the multiplication of a variety of viruses only in the same cells or cells from the same animal species.

There is little uptake of interferon by treated cells and it has no effect on the virus extracellularly. It does not interfere with virus adsorption, penetration or assembly of progeny virions. Interferon may act at the plasma membrane of or within the infected cell to derepress the host cistron coding for inhibitor (antiviral) polypeptide. The polypeptide is made in the cell via DNA-dependent RNA synthesis, and interferes with the transcription of the infecting viral genome by preventing polysome formation. The synthesis of early viral enzymes, and thereby viral nucleic acid and viral structural proteins, is drammatically inhibited, with a considerable reduction in the output of progeny virions. The infected cell still dies, but the net result is a failure of the infecting virus to establish itself in the tissues of the host with the usual rapidity, thus allowing more time for the build-up of immunological defences. Alternatively, interferon produced by one infecting virus may make superinfection by a second impossible, though there are a number of examples of dual virus infection where this inhibition is not manifested, e.g. vaccinia and herpesvirus, measles and poliovirus.

Attempts to utilize interferon itself for the prevention and treatment of virus infections have not proved fruitful, even though studies on its mode of action have yielded considerable information on virus-cell interactions at the molecular level. The present hope for interferon therapy lies in the development of harmless interferon inducers, especially synthetic RNA analogues. So far the trials with the most promising analogue, poly 1:C (a complex of 2 homopolymers of polyriboinosinic acid and polyribocyticylic acid), have given indefinite results in man, and it may well prove that natural RNA, from whatever source, is still more efficient inducer than the synthetic analogues. The problem associated with the administration of natural (viral?) RNA is that of possible genetic 'engineering' of the patient, with unknown long-term effects.

HYPERSENSITIVITY

Contact with an infectious disease usually results in an increased resistance by the host to subsequent exposure to the same infectious micro-organism. Occasionally the converse occurs, the animal becoming abnormally sensitive to the infectious agent or other environmental antigens. Many types of hypersensitivity are encountered which may be divided according to whether the host response to contact with the sensitizing agent is immediate or delayed. In the former the host reaction is the result of antigen–antibody reactions; in the latter the sensitivity is mediated by host cells specifically sensitized to the antigen.

Immediate hypersensitivity Immediate hypersensitivity has been classified into three types diagrammatically represented in Fig. 14.14. Type 1 occurs in patients previously sensitized to a par-

ticular foreign antigen. The antibody produced is thought to be absorbed to the surface of tissue cells. When a second injection of the same antigen is given, the interaction of antigen–antibody at cell surfaces results in tissue damage and the release of pharmacologically active substances such as histamine. These substances cause the smooth muscle of particular organs to contract which results in profound shock. The classical example is *anaphylaxis* (meaning 'against protection') which may follow a second injection of a foreign serum, as in serum therapy (p. 326) or by exposure to antibiotics. In small animals anaphylactic shock may result in death; in man and large domesticated animals symptoms of shock include bronchial spasm, pulmonary oedema (outpouring of fluids into the tissues), and urticaria (skin rashes). When the antigen is introduced locally, as in hay fever and asthma, local symptoms of excessive secretion of the mucous membranes or bronchial spasm occur although the patient may be generally sensitive.

Type 2 occurs when circulating antibody reacts with an antigenic component of tissue cells bringing about their damage. This may occur in acute

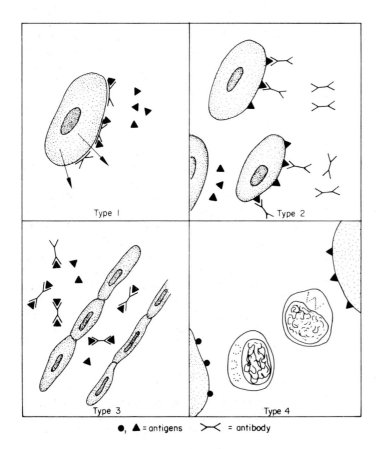

●, ▲ = antigens   >< = antibody

Fig. 14.14 Highly diagrammatic illustration of the four types of allergic reaction which may be deleterious to the tissues and harmful to the host. Type 1. Free antigen reacting with antibody absorbed to the cell surface results in the release of histamine and other pharmacologically active substances from the cell. Type 2. Antibody reacting either with cell surface or with antigens or haptens which becomes attached to cell surfaces: complement plays a major destructive role. Type 3. Antigen and antibody reacting in antigen excess forming complexes which are toxic to cells. Type 4. Specifically modified mononuclear cells (indicated by intracellular shadow of antigens) reacting with antigen deposited at a local site. (After Gell, P. G. H. and Coombs, R. R. A., 1963)

glomerular nephritis, a disease of the kidneys, and rheumatic fever. Both these conditions follow infection with specific types of streptococci. Antibodies produced against the streptococci cross-react with the basal membrane of kidney substance and heart muscle respectively. Since antibodies begin to appear in the serum after about ten days, the symptoms of these diseases do not arise until sometime after the streptococcal infection has cleared up.

Type 3 is demonstrated by the toxic symptoms which follow an antigen–antibody reaction in the blood in the presence of excess antigen, as in serum sickness, which frequently follows injection of large volumes of serum into certain hosts. Antibodies produced against the foreign serum are demonstrated in the patient's serum after the normal period of about 10 days. These react with excess antigen (the foreign serum) still circulating in the blood and form toxic complexes which produce symptoms of serum sickness.

Delayed hypersensitivity Delayed hypersensitivity (Fig. 14.14, type 4) occurs in patients with chronic infections, such as tuberculosis, and in virus infections. In these diseases cell-mediated immunity plays a greater role than antibody immunity. In tuberculosis, exposure to the antigens of tubercle bacilli results in lymphocytes and other mononuclear cells becoming sensitized to products of the bacterial cells (i.e. tuberculin), with which they react specifically when these substances are introduced locally either by natural infection or by injection. The area becomes red and swollen owing to the infiltration of cells into the local site and the reaction reaches a peak after 48–72 hours; hence the term delayed hypersensitivity. This reaction is used as a clinical test to determine if a patient has been previously, or is at present, infected. No tissue reaction occurs in a patient who has had no previous exposure to the tubercle bacillus. In animals the test is known as the Tuberculin test and in man, the Heaf test. A Heaf-negative patient can be converted to a Heaf-positive patient by BCG vaccination (p. 325).

Cell-mediated immunity provides another mechanism of host resistance. It can be passively transferred by living lymphoid cells but not by serum and, as with passive antibody immunity, passive cell-mediated immunity is of relatively short duration. Not all cell-mediated immune responses are advantageous to the host. Exposure to foreign tissues induces this type of immune response and homografts (i.e. tissue transplants from another host of the same species not genetically identical) are rejected after about two weeks by infiltration of specifically sensitized lymphocytes.

Chemotherapy as an aid in defence

Specific immunity helps an animal overcome an infectious disease either by neutralizing bacterial toxins or by inactivating the parasites so that they can be removed by the phagocytic cells. Chemotherapeutic agents, by their bacteriostatic or bactericidal actions, similarly aid the body defences to deal with the invading organisms. Chemotherapeutic agents are selectively toxic for specific parasites with little or no toxicity for the host and are extremely valuable in controlling infections already established in the body. In particular, the antibiotics are highly selective inhibitors, each one exhibiting a defined spectrum of activity against micro-organisms owing to their specialized modes of action (p. 105). The clinically useful antibiotics can be grouped according to their range of antibacterial activity (Table 14.1). Some are active against Gram-positive bacteria and pathogenic Gram-negative cocci but relatively inactive against Gram-negative bacilli, others are active to a greater or lesser degree against most species of bacteria, the larger viruses and rickettsiae, and are referred to as broad spectrum antibiotics whilst a third group is particularly active against the Gram-negative bacilli. A few, e.g. streptomycin, are active against the tubercle bacillus.

Certain bacteria, such as *Streptococcus pyogenes*, pneumococci, and meningococci, are consistently sensitive to penicillin. Other bacteria, normally susceptible to a particular drug, frequently produce resistant variants by a number of possible mechanisms; with these organisms the choice of antibiotic for treatment of an infection must be based on prior tests of their sensitivity *in vitro*.

The use of chemotherapy during the early stages of an infection prevents production of antibody immunity since the stimulating micro-organisms will be killed and removed from the body. Chemotherapy may not therefore always be immediately desirable since it may prevent development of lasting immunity.

Prevention and control of infectious disease

Most microbial diseases of animals and man arise from an exogenous source. One important method of disease control is achieved, therefore, by treating

Table 14.1 Some of the anti-microbial agents used in clinical practice

	Narrow spectrum[1]	Broad spectrum[2]
Anti-bacterial agents	Benzyl penicillin Phenoxymethyl penicillin Methicillin Flucloxacillin	Ampicillin Amoxycillin Carbenicillin
		Sulphonamides
	Cephalosporins Erythromycin Lincomycin Clindamycin	Trimethoprim Tetracyclines[3] Chloramphenicol
		Streptomycin
	Vancomycin	Kanamycin Neomycin
	Fucidin	
		Gentamicin
	Bacitracin	Amikacin
Anti-tuberculous agents	Streptomycin Para-aminosalicylic acid Isoniazid Rifampicin Ethambutol	
Anti-fungal agents	Nystatin Griseofulvin Amphotericin B 5-fluorocytosine Miconazole Clotrimazole	
Anti-viral agents[4]	Methisazone Idoxuridine Cytosine arabinoside Adenine arabinoside Amantadine	

[1] Usually Gram-positive bacteria and Gram-negative cocci.
[2] Gram-positive and Gram-negative bacteria.
[3] Tetracycline is also the drug of choice for many Mycoplasma, Rickettsial and Chlamydial infections.
[4] Drugs have very limited uses—often best given prophylactically.

or removing the source of infection to prevent pathogenic micro-organisms reaching susceptible hosts.

WATER SUPPLIES AS A SOURCE OF INFECTION

Some pathogenic organisms are water-borne and when present in drinking water may infect susceptible hosts. Water-borne pathogenic organisms usually gain access to the water from animal and human excreta and sewage. Outbreaks of diseases, such as typhoid fever and cholera in man, were at one time widespread but the control of water supplies and improved hygiene in food handling have virtually eradicated these diseases from the technically more advanced countries. Water may also be polluted by poisons of non-microbial origin from factory wastes. Constant control is therefore necessary to ensure that drinking water conforms to both chemical and bacteriological standards of purity. Drinking water may be artificially purified (p. 272) and chlorinated.

Chlorination of drinking water Neither natural purification nor filtration (p. 272) renders water absolutely safe for drinking; this may be achieved by the use of chlorine. Chlorination destroys all water-borne pathogenic bacteria and also pathogenic intestinal protozoa, such as *Entamoeba histolytica*, the cause of amoebic dysentery. When chlorine is added to water some combines with reduced organic material in the water and is no longer effective. This 'chlorine demand' varies considerably with different waters according to the amount of organic matter present and consequently the amount of chlorine that must be added to give a safe residual amount of free chlorine (0.2 to 0.4 part per million) also varies. The small amount of free chlorine necessary to render water safe is harmless and usually cannot be detected, but waters with an abnormally large 'chlorine demand' may have an unpleasant flavour after chlorination, due to the presence of such chlorine compounds as chlorophenols.

FOOD AS A SOURCE OF INFECTION AND DISEASE

Many diseases may be contracted by the ingestion of contaminated food. The food may be infected at source, e.g. meat, milk or eggs from infected animals or birds (many of the natural infections of these animals being common to man) or vegetables grown in contaminated soil or irrigated by polluted water (e.g. watercress beds). Foods may also become contaminated with pathogenic organisms during processing or preparation for eating. Pathogens may be introduced by flies or by the handling of foods by human carriers of food poisoning organisms (see below). Risk is high with foods prepared for consumption the following day, the organisms multiplying if the food is kept at a temperature conducive to growth.

Food poisoning Diseases contracted by the ingestion of food are designated 'food poisoning'.

Outbreaks of these are usually explosive in character and may be related to ingestion of a specific meal or common food source. Poisons may be derived from plant or animal sources, from chemicals added inadvertently or as preservatives at too high concentrations, or to the presence of harmful micro-organisms or toxins produced by them.

Bacteria causing food poisoning In Great Britain the term 'bacterial food poisoning' is legally restricted to poisoning resulting from the ingestion of food containing certain bacteria (or their toxins) which are not included among other specific infectious diseases notifiable to the Minister of Health. The types of foods involved in food poisoning are those favouring growth of the causative organisms, so that large numbers of bacteria or their products are present in the food at the time it is ingested. Outbreaks of food poisoning frequently involve a large proportion of the people who have eaten a particular food. Two groups of agents are recognized: pathogenic bacteria which can infect the alimentary canal and produce symptoms of gastro-enteritis, and bacteria (or their toxins) which do not infect the gut but produce various toxic symptoms when ingested with food.

INFECTIVE FOOD POISONING In this form of food poisoning the infective agent is able to establish itself in the gut, multiply rapidly and produce symptoms of disease over a number of days, often for about a week or longer. This form of poisoning is mainly limited to species of *Salmonella*. When infection is established, endotoxins (p. 311), resulting from the breakdown of bacterial cells, cause irritation of the gut linings resulting in symptoms of acute gastro-enteritis.

The *Salmonella* group includes many hundreds of types of serologically related organisms which are found in almost every species of animal, including birds and reptiles. In Great Britain meat, milk, synthetic cream, and eggs (particularly duck eggs) are the foods most commonly contaminated with salmonellae. Meats from cattle, pigs and poultry have been found to be a frequent source of infection, particularly when made up into processed foods such as brawns, sausages or pies, which may be incompletely cooked or subsequently contaminated during handling. Ideally *Salmonella* should be absent from food but if adequately cooked, these pathogens are destroyed and there is no risk of food poisoning, since there are no residual toxins to produce food poisoning symptoms.

Symptoms of salmonella food poisoning vary in severity but include acute gastro-enteritis accompanied by severe headaches followed by nausea, vomiting, abdominal pain, and diarrhoea. Fever frequently occurs. Symptoms usually arise between 12 and 72 hours after the meal has been consumed, by which time the organisms are established in the gut. In favourable cases the symptoms subside within a week, but in a small percentage of patients organisms invade the tissues from the gut and may cause death. Even when symptoms disappear, the patient may continue to excrete the pathogens and thereby remain a potential source of infection for susceptible individuals and contamination of food. If this continues for a long period after recovery, the patients are termed *carriers*. It is most important to ensure that such people are not employed in food handling.

BACTERIAL CELLS OR THEIR PRODUCTS AS CAUSES OF FOOD POISONING In this form of food poisoning no infection of the gut occurs but bacteria or exotoxins of bacterial origin produce symptoms of food poisoning within 6 hours when ingested in foods. Since these toxins are absorbed by the gut wall they are called enterotoxins. They are preformed in the food by growth of the pathogenic organism and although the vegetative cells may be killed by heating the food prior to eating, many of the toxins, being moderately resistant to heat, remain and cause symptoms of food poisoning when ingested. The most important bacterial toxic food poisonings are caused by the toxins of specific strains of *Staphylococcus aureus* and *Clostridium botulinum*, the latter frequently causing a fatal form of food poisoning known as botulism. Strains of *Clostridium perfringens* also cause toxic food poisoning; certain non-pathogenic bacteria, such as *Proteus* sp., not generally considered to be toxigenic, produce acute gastro-enteritis when consumed in sufficiently large numbers.

POISONOUS FUNGI These are considered in Chapter 15 (p. 340).

Control of food supplies The increase in urban populations during the present century and improvements in methods of food preservation have led to large-scale transport of basic foods from the producer to the consumer areas. This has inevitably increased the risk of infection of many people from a common food source. This risk can be considerably reduced by suitable precautions.

Methods employed for improving the keeping quality of a food (i.e. avoiding spoilage) are often adequate to render a food safe for eating. The handling of vegetables, eggs (p. 350), meat (p. 345) and milk (p. 347) are considered in Chapter 15.

ANIMAL AND HUMAN RESERVOIRS OF INFECTION

The spread of infection from man to man, from animals to man, and to a lesser extent from man to animals, constitutes some of the more important pathways of cross-infection. The term zoonoses has been coined to include diseases (protozoan, fungal, bacterial, and viral) which are naturally transmitted between vertebrate animals and man; more than 100 zoonoses are recognized. Prevention, control and eradication of sources of infection are the concern of many countries.

Eradication of infected animals Eradication programmes obviously cannot be practised in the control of human diseases although under certain circumstances examination and treatment is compulsory. Eradication is, however, proving a most successful approach to the control of a number of diseases of domesticated animals. Great Britain, separated from the Continent by a sea barrier, has already achieved eradication of a number of serious diseases including tuberculosis and foot-and-mouth disease in domestic animals, and a programme is already in hand to eliminate brucellosis. Initially a defined area is cleared of infection and designated disease-free, and this area is enlarged year by year until the whole country is free of infection. Reliable tests to detect infected animals are essential to success. For instance, in order to maintain the national herd free of tuberculosis, farm animals are regularly tested by the tuberculin test (p. 333) and positive reactors immediately destroyed. The occurrence of tuberculosis in badgers has necessitated their eradication in certain regions to avoid infection of dairy herds.

Anthrax, an acute, fatal disease of cattle, is controlled in Britain by the enforcement of special methods of disposal of anthrax-infected carcases. *Bacillus anthracis* readily produces highly resistant spores when exposed to air. Hence, if death from anthrax is suspected, the carcase must not be opened for post-mortem examination; this avoids exposing the organism to the air and contamination of the surrounding ground or pasture by spores which would thereby be produced. Instead the cause of death can be confirmed by microscopical examination of blood smears taken from a peripheral vein and, if anthrax is diagnosed, the carcase is either incinerated on the spot or buried in quicklime well below ground level.

In foot-and-mouth disease stringent measures of eradication are adopted in every outbreak both in Britain and N. America. The cost of compensating farmers for loss of livestock, even in such a serious outbreak in Britain as that of 1967–8, is considerably less than the widespread economic loss in milk yields and value of carcase meat which would result if the disease were allowed to become endemic in the animal population.

Isolation of infected animals The spread of infection may be prevented by the segregation of infected animals from susceptible ones. Quarantine measures are widely practised today in the control of animal and human diseases but are now based on knowledge of the length of time the patient remains infectious. Importation into Britain of many animal pets is controlled by quarantine measures at the ports to avoid introducing infections not normally present in the country, such as rabies in dogs and psittacosis in birds. Many countries require veterinary certification that dogs are free of leptospiral infection before they are allowed in. Immigrants are medically examined at ports of entry to ensure that they are free of infection. Patients infected with smallpox or similar highly infectious diseases are treated in isolation hospitals to restrict spread of infection.

Interference with the normal cycle of infection Many viral and rickettsial infections are vector-transmitted (p. 308). The most successful control of these infections requires destruction of the transmitting vector. For instance, the general application of mosquito control measures has virtually stamped out epidemic yellow fever in man, the disease now existing only as an enzootic infection of jungle monkeys with man as an incidental host.

NATURAL AND APPLIED CONTROL OF INSECT PESTS BY VIRUS DISEASES

Spontaneous virus diseases are commonly responsible for severe outbreaks of disease in insect populations in the field and their recurrence each season may contribute to pest control. Though outbreaks may result from acquisition of new infections it is possible that latent infections that become activated by stress are of more importance.

There are several examples of the successful use by man of viruses to control insect pests, chiefly those which feed openly on the host plant. A number of circumstances serve to heighten the chance of success in any given instance. The virus used should be highly virulent and should kill the insect quickly enough to prevent it doing much damage before it dies. Because nuclear polyhedroses and most granuloses result in liquefaction of the body contents and disintegration of the skin, they offer the best chance of further liberation and spread of the virus. It is advantageous if the insect to be controlled is one which is gregarious and feeds openly on the foliage. As in all infectious diseases, a high host population density favours spread of the disease. Inclusion body viruses offer the advantage that the inclusion body protects the virus and so allows its longer storage and extends its retention of infectivity once distributed in the environment. Virus preparations have been applied as dusts or sprays. Administration of a number of insect viruses to man and other mammals have revealed no harmful effects, but in view of the host specificity known among viruses that infect mammals it would be unwise to generalize from the limited data available. The very existence of arboviruses argues for caution. Although some insect viruses have been successfully produced in tissue culture, current technology virtually requires that viruses for control purposes must be produced by rearing and infecting host populations in the laboratory.

Control of air-borne cross-infection The majority of pathogens are obligate parasites and consequently most are spread from infected animals and man to susceptible hosts. With few exceptions air-borne animal infections are therefore limited to indoor atmospheres (p. 261).

Respiratory infections arise by inhalation of pathogenic organisms in droplets of moisture from infected hosts or contaminated air-borne dust.

Cross-infection by 'droplet' infection can be restricted by avoiding overcrowding. This may be achieved by the adequate spacing of beds in hospital wards, of desks in schoolrooms, and of stalls in stables. Recent work has shown that subdividing large hospital wards into smaller units is more effective in reducing cross-infection than wide spacing of beds. It has long been held that the common cold virus is transmitted from patient to patient by droplet infection but prolonged personal contact, such as occurs in the home or school, is more important than the casual contact. It is often difficult to separate a purely air-borne route from that of personal contact and invariably both pathways of infection have to be controlled at the same time.

The provision of adequate natural or mechanical ventilation with filtered air, reduces air contamination. The fact that ventilation is better in summer may account, in part at least, for less respiratory disease occurring at this time of the year. In many-storied buildings with mechanical ventilation requiring sealed windows, movement of contaminated air can occur from one floor to another by way of lift shafts or laundry chutes and these present a real hazard, especially in hospitals where frequently a build-up of infection in one ward may be passed to another on a different floor. In operating theatres positive air-pressure within the theatres excludes air contamination from adjacent rooms and corridors in addition to removing air contaminated from movement of personnel.

Both droplet and dust-borne air contamination are partly controlled in special environments, such as operating theatres and wards of infected patients, by the wearing of sterile protective clothing and face masks by the staff. Face masks must be frequently changed and although they filter only the larger salivary droplets this is considered adequate to prevent direct contamination of wounds.

Dust derived from animal and human sources is invariably laden with micro-organisms. This comes from hair, desquamated skin and fragments of textiles resulting from friction between layers of clothing or activities such as bed making. This is of particular importance when the bedding has been used by an infected patient. Contaminated dust settles on to horizontal surfaces, but is readily redispersed into the air by air currents, dry sweeping, dusting, and similar activities. By the use of water or, more effectively, oils and other mixtures which cause individual particles to adhere together in large aggregates, the raising of dust can be greatly reduced. Vacuum cleaning is a far more efficient way of removing dust without contamination of the air than other means, but precautions must be taken to prevent fine particles of dust from being blown out of the machine. These preventive measures alone, however, seldom provide complete control.

Organisms may be destroyed in air by the use of ultra-violet radiation, and aerosols or sprays of disinfectants. Ultra-violet radiation of wavelength

254 nm, whilst not being the most bactericidal part of the spectrum (p. 99), is readily available from mercury vapour type lamps and is used to reduce the count of air-borne bacteria and viruses within buildings. Since ultra-violet radiations produce an irritant effect on the conjunctiva of the eye the radiation source must be shielded from occupants in the room. This correspondingly reduces its field of effectiveness. Ultra-violet radiations have a limited range of action which varies as the square root of the distance from source; it is far less effective against bacterial and fungal spores than against vegetative cells; organisms must be directly exposed to the rays and screening them from direct radiation or protection by thin layers of protein materials reduces the 'kill'. Ultra-violet radiations are more effective against organisms suspended in droplets than against those attached to dust particles. In contrast, chemical disinfectants are most effective at low humidity.

Chemical disinfectants have the advantage of penetrating to all parts of a room. To be efficient they must make contact with the air-borne organisms and, for this reason, those in a gaseous or vapour state are more effective than others but less volatile substances are sometimes effective if sufficiently finely dispersed. They are used either as sprays or in a volatile form dispersed as aerosols. Disinfectant sprays consist of droplets larger than 150 nm in diameter and result in wet mists, the drops rapidly settle out, lay the dust and disinfect horizontal surfaces. Aerosols, of droplets size 5 to 150 nm in diameter, produce dry mists or fogs, which remain suspended in the atmosphere for long periods of time, travel for considerable distances from the source and, by virtue of their large surface area/volume ratio, rapidly build up high local vapour concentrations of volatile substances. All effective air disinfectants act at comparatively low levels of concentration when dispersed in the atmosphere, a feature of great practical importance since these levels are well below those toxic for man and animals. They can therefore be used in occupied buildings. For instance, one gram of sodium hypochlorite dispersed in 40 000 litres of air has been found to be bactericidal in a few minutes. The lethal effects are thought to be due to the vapour derived from the aerosol particle producing a high lethal concentration localized around the micro-organisms. The particle therefore acts merely as a mobile source of vapour, aiding its persistence and distribution. Humidity usually plays an important role, providing residual moisture on the organismal surface in which the germicide dissolves.

Not all disinfectants are suitable for air disinfection. The most useful include chlorine, sodium hypochlorite, hypochlorous and lactic acid, propylene, and triethylene glycols and phenols of low volatility such as hexyl resorcinol and resorcinol. Many of these have been used successfully for the sterilization of air in public buildings and for the safeguarding of young poultry against pullorum disease in brooder houses. Formalin vapour, due to its toxicity, cannot be used in occupied buildings but is a most effective means of sterilizing the air of unoccupied animal and poultry houses. Reductions greater than 99 % in the coliform counts have been achieved by use of this reagent.

Prevention of cross-infection in hospitals Cross-infection by any route is an important problem in hospitals where seriously ill and highly susceptible patients are often at risk from particularly virulent antibiotic-resistant organisms derived from patients already infected by these. Such strains have been selected over many years by the extensive use of antibiotics in hospitals. In addition to the preventive measures already considered, numerous precautions are now taken in hospitals to prevent infection by direct and indirect contact. Indirect contact, the transfer of pathogens from an infected to an uninfected patient *via* hospital equipment, surgical instruments, or the hands of patients or staff, constitutes the greatest problem in hospital cross-infection. Many of these infections can be avoided by very careful aseptic and antiseptic techniques in surgical and nursing procedures, and in recent years many hospitals have introduced improvements such as central sterile supply departments to ensure that instruments and dressings are correctly disinfected or sterilized. In addition, improvements are constantly being introduced in disinfectant measures such as the pre-operative disinfection of skin, disinfectant soaps and creams for doctors' and nurses' hands, disinfectant creams and sprays to prevent nasal colonization by staphylococci, and the treatment of carriers. Many of these improvements have been made in the past two decades but still more are needed.

Books and articles for reference and further study

AINSWORTH, G. H. and AUSTWICK, P. K. C. (1973). *Fungal Diseases of Animals*. Commonwealth Agricultural Bureau. (Review Series No. 6), 2nd Edn., 216 pp.

AJL, S. J., CIEGLER, A., KADIS, S., MONTIE, T. C. and WEINBAUM, G. (1970–71). *Microbial Toxins*. Vols. I–V, Academic Press, New York and London, 517 pp., 412 pp., 548 pp., 473 pp. and 507 pp.

ANDREWS, C. H. (1967). *The Natural History of Viruses*,

Weidenfeld and Nicolson, London, 149–161 pp.

BARRON, C. L. (1977). *The Nematode Destroying Fungi*, Canadian Biological Publications, Guelph, 140 pp.

BATRA, S. W. T. and BATRA, L. R. (1967). The fungus gardens of insects. *Scient. Am.*, **217**, 112–120.

BELLANTI, J. A. (1971). *Immunology*. W. B. Saunders Co., Philadelphia, London and Toronto, 584 pp.

BRUNER, D. W. and GILLISPIE, J. W. (1973). *Hagan's Infectious Diseases of Domestic Animals*. 6th Edn., Bailliere, Tindall and Cox, London, 1385 pp.

BUCHNER, P. (1965). *Endosymbiosis of Animals with Plant Micro-organisms*. Revised English Edn. (trans. B. Mueller). Interscience, New York, London and Sydney.

BURNET, F. M. (1960). *Principles of Animal Virology*. 2nd Edn., Academic Press, New York and London, pp. 390–392.

BURNET, F. M. and WHITE, D. O. (1972). *Natural History of Infectious Disease*. 4th Edn., Cambridge University Press, Cambridge, 278 pp.

BUXTON and FRASER, G. (1977). *Animal Microbiology*, 2 vols. Blackwell Scientific Publications, Oxford, London, Edinburgh and Melbourne, 830 pp.

COHEN, D. (1966). Epidemiology of virus diseases. In *Basic Medical Virology*, **202**, edited by J. E. Prior, Williams and Wilkins Co., Baltimore.

DICKINSON, A. G. (1976). Scrapie in sheep and goats. In *Slow Virus Diseases of Animals and Man*, edited by R. H. Kimberlin, Elsevier, Amsterdam, pp. 209–241.

DUDDINGTON, C. L. (1968). Predacious fungi. In *The Fungi*, edited by G. C. Ainsworth and A. S. Sussman, 329, Academic Press, New York and London.

EKLUND, E. M. (1966). Arbovirus. In *Basic Medical Virology*, edited by J. E. Prior, Williams and Wilkins Co., Baltimore.

FENNER, F. (1968). *The Biology of Animal Viruses*, **2**, Academic Press, New York and London, pp. 765–769.

GELL, P. G. H., COOMBS, R. R. A. and LACHMANN, P. J. (1975). *Clinical Aspects of Immunology*. Blackwell Scientific Publications, Oxford, 1754 pp.

GRAHAM, K. (1967). Fungal-insect mutualism in trees and timber. *A. Rev. Ent.,* **12**, 105–126.

HARVEY, W. C. and HILL, H. (1967). *Milk : Production and Control*. 4th Edn., Lewis and Co Ltd., London, 711 pp.

HEIMPEL, A. M. (1967). A critical view of *Bacillus thuringiensis* var. *thuringiensis* Berliner and other crystalliferous bacteria. *A. Rev. Ent.*, **12**, 287–322.

HOBBS, B. C. (1974). *Food Poisoning and Food Hygiene*. 3rd Edn., Edward Arnold, London, 308 pp.

HORTON-SMITH, C., (ed.) (1957). *Biological Aspects of the Transmission of Disease*. Oliver and Boyd, Edinburgh.

HOWIE, J. W. and O'HEA, A. J., (eds.) (1955). *Mechanisms of Microbial Pathogenicity, 5th Symp. Soc. gen. Microbiol*. Cambridge University Press, Cambridge, 333 pp.

INGOLD, C. T. (1953). *Dispersal in Fungi*. Clarendon Press, Oxford.

LEADER, R. W. and HURVITZ, A. I. (1972). Interspecies patterns of slow virus diseases. *Ann. Rev. Med.*, **23**, 191–200.

LOVELL, R. (1958). *The Aetiology of Infective Diseases*. Angus and Robertson, Sydney, 136 pp.

LUDERITZ, O., STAUB, A. M. and WESTPHAL, O. (1966). Immunochemistry of O and R antigens of salmonella and related Enterobacteriaceae. *Bact. Rev.*, **30**, 192–255.

MADELIN, M. F. (1966). Fungal parasites of insects. *A. Rev. Ent.*, **11**, 423–448.

MADELIN, M. F. (1968). Entomogenous Fungi. In *The Fungi*, Vol. 3, edited by G. C. Ainsworth and A. S. Sussman, Academic Press, New York and London, pp. 227.

MARAMOROSCH, K. and KURSTAK, E. (eds.) (1971). *Comparative Virology*, Academic Press, New York and London, 584 pp.

NUTMAN, P. S. and MOSSE, R. (eds.) (1963). *Symbiotic Associations. 13th Symp. Soc. gen. Microbiol*. Cambridge University Press, Cambridge, 356 pp.

RAMSBOTTOM, J. (1945). *Poisonous Fungi*. King Penguin Books, London.

ROITT, I. M. (1972). *Essential Immunology*. Blackwell Scientific Publications. 220 pp.

SMITH, H. and TAYLOR, JOAN (1964). *Microbial Behaviour, 'in vivo' and 'in vitro'. 14th Symp. Soc. gen. Microbiol*. Cambridge University Press, Cambridge, 296 pp.

SMITH, H. (1968). Biochemical challenge of microbial pathogenicity. *Bact. Rev.*, **32**, 164–184.

SMITH, H. and PEARCE, J. H. (1972). *Microbial Pathogenicity in Man and Animals. 22nd Symp. Soc. gen. Microbiol*. Cambridge University Press, Cambridge, 451 pp.

SMITH, K. M. (1974). *Plant Viruses*. 5th Edn., Chapman and Hall, London, 211 pp.

STEINHAUS, E. A. (ed.) (1963). *Insect Pathology*, Vols. 1 and 2, Academic Press, New York.

WILLIAMS, R. E. O., BLOWERS, R., GARROD, L. P. and SHORTER, R. A. (1966). *Hospital Infection : Causes and Prevention*. 2nd Edn., Lloyd Luke Ltd., London, 386 pp.

WILSON, G. S. and MILES, A. A. (1975). *Topley and Wilson's Principles of Bacteriology Virology and Immunity*. 2 vols. 6th Edn., Edward Arnold, London, 2693 pp.

WILLIAMS, R. E. O. and SPICER, C. C. (eds.) (1957). *Microbiol Ecology. 7th Symp. Soc. gen. Microbiol*. Cambridge University Press, Cambridge, 388 pp.

Chapter 15

Microbiology of food and beverages

Introduction

Food microbiologists are concerned with the level of spoilage organisms in the raw materials, the standard of hygiene during processing, the efficiency of preservation and storage methods as well as the incidence of organisms responsible for food poisoning. Particular attention will be paid in this chapter to the changes in the natural microflora by methods of preparation and preservation. The public health aspects are discussed elsewhere (p. 334) and will be considered here only in relation to incorrect processing methods.

The natural microflora

The natural microflora of a food or beverage consists of two components, associated with the raw material, or acquired during processing, some of which survive preservation and storage. The latter can be subdivided into harmless organisms, producing either desirable or undesirable flavour changes in the food, and pathogens forming dangerous enterotoxins (p. 335).

Micro-organisms spoiling the aroma, taste, colour, or texture of a product may be moulds, yeasts, or bacteria. The precise flora that is present at any stage will depend on the nutrient status of a food, its temperature, pH, water content, etc., as well as on the nature of the organisms themselves. Even foods without very marked characteristics carry an association of bacteria, e.g. fresh meat and fish have *Pseudomonas* and *Achromobacter* species that give way to a micrococci/lactobacilli association when the flesh is cured. Often the natural flora is succeeded by a factory flora, as in cheese and cider making. Many foods and beverages are stable only because of the early development of particular bacteria which reduce the pH by producing lactic acid and thus inhibit the development of food pathogens. Again, foods are often rendered unpalatable by spoilage organisms before pathogens have had time to develop in sufficient numbers. Sequential changes in the microflora of a number of foods are detailed later in this chapter.

Food pathogens include salmonellae (p. 335), *Clostridium botulinum* (p. 335), *Cl. perfringens* (p. 335), and some of the staphylococci (p. 335) and streptococci. Generally, food poisoning bacteria grow optimally at $37°C$ and very little at refrigeration temperatures. All non-sporing pathogens are killed by pasteurization of the food but preformed enterotoxins are not destroyed at such temperatures. Where the raw material contains pathogenic organisms, there is a danger of re-infection of the pasteurized product. Staphylococci are more salt-tolerant than many spoilage organisms and this can give them an advantage in some preserved foods. Faecal streptococci (*Strep. faecalis* and other group D streptococci) and coliforms, including *Escherichia coli*, are common in foodstuffs; but in this situation are not reliable indicators of faecal contamination, though they may be so in water supplies (Chapter 11). These organisms grow readily on food residues in badly maintained processing lines. Thus their numbers in foods are generally considered to give an indication of the effectiveness of factory hygiene. They are normally absent from liquids pasteurized and processed in a closed system. Massive numbers (10^6 to 10^7 per g) in other foods would be regarded as a potential hazard to health, since in such circumstances they have been suspected of causing food poisoning.

Poisonous fungi

Fungi can cause poisoning if eaten intentionally but mistakenly, as for example when poisonous mushrooms are confused with edible ones, or when consumed unwittingly, as when certain moulds or mould products are present in human foodstuffs or in feedstuffs for animals.

Poisonous mushrooms are only a minority, but unfortunately cannot be recognized as harmful by any simple test since they contain different toxins. Some mushrooms cause gastro-intestinal upsets, some cause disorientation, feverishness and changes in pulse rate, some interfere with the central nervous system, others harm the liver and kidneys. Often the victim recovers, but sometimes the harm is sufficient to kill.

The species to which most deaths from mushroom poisoning are attributed in Europe is the 'death cap', *Amanita phalloides*, while in North America most are attributed to *Amanita virosa*. The genus *Amanita* contains a number of highly poisonous species. The poisonous constituents of *A. phalloides* have received much study but there is still uncertainty about the toxins which actually cause the poisoning when the mushroom is eaten. The fungus contains two classes of poisonous cyclopeptides, the amatoxins which are octapeptides and the phallotoxins which are heptapeptides. The phallotoxins are quicker acting than the amatoxins, but less toxic. Both primarily attack the liver, but while phalloidin (a phallotoxin) binds to the plasmalemma of liver cells and leads to the accumulation of actin filaments in the cytoplasm, amatoxin causes the nucleoli to disintegrate. It is a potent inhibitor of DNA-dependent RNA polymerase in the nucleoplasm. The gastro-intestinal distress caused by *A. phalloides* has been attributed to the amatoxins. There is, however, doubt whether phalloidin is naturally a cause of poisoning since it does not kill if administered orally. Indeed it has recently been suggested that the cyclopeptides may merely be components of complex high molecular weight toxins named myriamanins which also contain a polysaccharide.

The high proportion of fatalities resulting from ingestion of *Amanita phalloides* is partly due to the fact that the first symptoms of poisoning, namely vomiting and diarrhoea, do not show till 10 to 24 hours after the delicious-tasting toadstool has been eaten, by which time the victim's liver is being damaged. If the victim survives the sickness and diarrhoea, there may be a transient remission before the liver damage leads to death. The administration of thioctic acid, followed by treatments to avert hypoglycaemia, is a recently developed therapy which appears to be very effective.

The toxic components of fungi are often mixtures. Thus the 'fly agaric' *Amanita muscaria*, contains ibotenic acid which has insecticidal properties; muscarine, which has certain structural similarities with acetylcholine which also occurs in this fungus; and unidentified indole bases. There is evidence even that the alkaloid content of *A. muscaria* from different parts of the world differs. Rather curiously, ibotenic acid and a related insecticidal substance, tricholomic acid from another fly-killing mushroom, *Tricholoma muscarium*, both have great tastiness as well as synergistic taste

action with nucleotide seasonings, so are of interest for their possible use in food formulations.

The consumption of certain poisonous fungi specifically for their hallucinogenic effects is a part of rituals performed by some tribes of Central America, by whom the mushrooms are collectively known as 'teonanacátl'. These 'sacred mushrooms' include particular species of *Conocybe*, *Panaeolus*, *Psilocybe* and *Stropharia* and induce predominantly visual hallucinations. Two Central American species of puffball, *Lycoperdon mixtecorum* and the unpleasant smelling *L. marginatum*, by contrast produce purely auditory hallucinations. The psychotomimetics in teonanacátl are the 4-hydroxyindole derivatives psilocybine and psilocin.

The unwitting consumption of fungi also can lead to hallucinations, as occasionally happens when bread made from rye infected with the ascomycete *Claviceps purpurea* (ergot) is eaten. More often the effect of ingestion of this fungus is gangrene of the extremities of the body owing to constriction of the vascular supply. Ergot contains a number of potent alkaloids of pharmacological importance including ergine, ergonovine and ergotamine. Other fungi which attack growing crops also are known to cause 'mycotoxicoses'; for example certain *Fusarium* species on grain crops can cause illness in man, and *Fusarium graminearum* on barley causes disease in horses and pigs. Similarly *Stachybotrys atra*—best known as a troublesome cellulolytic mould—may grow on straw which thereby becomes poisonous to horses. *Pithomyces chartarum* growing on pasture grasses produces sporidesmin which causes liver damage in sheep which become jaundiced and thereby sensitized to light so that 'facial eczema' develops. Another disease of sheep involving liver damage is believed to result from ingestion of *Phomopsis leptostromiformis* which grows on lupin fodder. The growth of fungi on grain and seed crops after they have been harvested can also be injurious to man and animals. *Fusarium*, *Penicillium* and *Aspergillus* species on cereals, groundnuts, cotton seed, etc., are known sometimes to produce toxins which include luteoskyrin, islanditoxin, rubratoxin, aflatoxin, sterigmatocystin, citrinin, and ochratoxin. Some of these not only cause liver or kidney damage, but also in experimental animals cause hepatic carcinoma. There is evidence that in those areas of the world in which there is higher than average incidence of hepatic carcinoma the population is exposed to aflatoxins (produced by *Aspergillus flavus* and *A. parasiticus*). Moulding of food and feedstuffs thus raises many questions concerning hazards to public health, even including the communication of toxins to man by way of meat from animals fed on mouldy grain.

Methods of food preservation

Any product, whether solid or liquid, can be sterilized if heated long enough and/or treated with a suitable concentration of a germicide. Usually the treatment with substrates of neutral pH is so drastic that accompanying changes in the flavour, texture, and colour render the product unacceptable. Hence, for this type of product, techniques are designed to inhibit multiplication of the spoilage flora prior to consumption of the food.

Chemical inhibition

Lowering the pH is an obvious method of controlling micro-organisms by chemical means. Lactic acid is preferred in many products, not only for its lack of distinctive flavour, but also because of its very flat neutralization curve, the pH range 3.5–5.5 being of particular interest in food-preservation. This allows considerable pH change without unpleasant increase in acid taste. Both additions of the acid and encouragement of growth of lactic acid bacteria are used in practice. Growth of food pathogens is rare below pH 5.0; lactic acid bacteria grow readily down to pH 3.8 but only very slowly down to *ca.* pH 3.0. Many moulds and yeasts will continue to grow at pH 2.3.

Organisms can also be inhibited specifically by the addition of small quantities of a chemical preservative. The ideal properties of an antimicrobial food preservative were formulated at the Fourth International Symposium on Food Microbiology (Molin, 1964) and are summarized below:

1 It is preferable that the preservative should kill rather than inhibit the micro-organisms. Having killed them it should itself decompose into innocuous products.
2 A bacteriostatic preservative would be equally satisfactory if it is destroyed only during final cooking, but if it is to be used in conjunction with thermal control processes it would need to have adequate heat resistance.
3 The range of specificity should correspond with the range of micro-organisms able to develop in the food, i.e. it must inhibit both food poisoning and spoilage organisms.
4 Any preservative used as a supplement to thermal processing should give a similar protection against *Clostridium botulinum* as that given by the standard thermal processing alone, i.e. a reduction in viable spores by a factor of 10^{12}.
5 A preservative should not be inactivated or removed by chemical reaction with the food, by some specific inhibitor in the food, or by products of microbial metabolism.
6 The preservative should not stimulate the development of resistant strains, and should be avoided totally if the same substance is also used therapeutically or as an additive to animal feeds.
7 There should be a chemical method for analyzing the effective portion of the preservative.

No known antimicrobial food preservative has all these properties and there is a dearth of suitable compounds active in the pH range 5.0 to 7.0 which is important especially for wet foodstuffs of high nutritive value, such as fish, meat, and milk.

Food preservatives can be divided into two groups according to their mode of action. In the first, which includes acids, esters, and phenols, the compound is absorbed by the solid components of the bacterial cell and, if of high lipoid solubility, is concentrated on the cell membrane and on various cell structures. The metabolic effects of salicylic acid, for example, are related to its influence on ATP formation in the mitochondria (p. 181). The action of such preservatives becomes increasingly less effective with an increase in the lipid and solid content of the food to which it is added. The antimicrobial effect of the second group, quinones and nitrofurans, depends primarily on their ability to penetrate cell membranes. Quinones are able to penetrate any type of cell, whereas nitrofurans have a selective action on bacteria which they can penetrate, but are unable to penetrate yeasts. Once inside the cell, excessive amounts of co-enzyme are required to produce the corresponding hydroquinones and aminofuran compounds, ultimately disrupting electron transport in the cell. The role of preservatives in the prevention of spore germination is also important. No known preservative will prevent germination of *Bacillus cereus* spores, but at minimum inhibitory concentrations nisin, subtilin, diethyl pyrocarbonate, and sodium nitrite prevent growth immediately after germination. Spores that shed their spore wall are prevented from elongating into vegetative cells by sodium benzoate, whereas tylosin, sodium sorbate, sodium metabisulphite, and sodium chloride allow some increase in length but prevent cross-wall formation. At greater concentrations all preservatives prevent any development after germination. Nisin is used to prevent gas formation by clostridia in cheese, but attempts to use it in other foodstuffs have not been entirely successful; part of its activity is lost during heating or curing and on storage it is eventually lost completely so that any bacterial spores present can then develop.

Acid foodstuffs and beverages are more easily conserved chemically. Sulphurous acid (sulphur dioxide) controls the acid-tolerant bacteria of ciders and wines, but moulds and fermenting yeasts, common to such environments, are highly resistant. Sorbic acid is used in cheese wrapping to inhibit surface mould growth. It has some activity against yeasts and is permitted in wines, together with sulphur dioxide, which is antibacterial and an antioxidant. Benzoic acid, used as a preservative in soft drinks, is effective against many yeasts, but *Saccharomyces bailii* can metabolize it to some extent in the presence of sugar. These compounds, whose undissociated molecule is antimicrobial, are more effective at lower pH levels. All substances suggested for use as antimicrobial food preservatives must be submitted to a statutory two-year evaluation programme, that includes feeding to rats and dogs, before being permitted in foods and beverages.

Chemical inhibition of microbial growth can also be obtained with gases. Thus, with the original Boehi process non-sterile apple juice was impregnated with eight atmospheres pressure of carbon dioxide and stored at ambient temperature. The process had to be modified, since contaminating lactic acid bacteria (acid-tolerant and micro-aerophilic) grew and spoiled the product. The juice is now concentrated to a third or fourth of its original volume, saturated with the same percentage of carbon dioxide (0.8 per cent wt) and held at 2°C.

Smoking meat, fish, cheese, etc. (normally in conjunction with salting), is an ancient method of food preservation brought up to date. Essentially it is a vapour absorption process, the solid particles playing only a minor role. Recent work has shown that the active components of the vapour include volatile fatty acids which have both a bactericidal and a residual mild bacteriostatic effect. Usually micrococci and staphylococci are inhibited, leaving a flora consisting mainly of lactic acid bacteria.

Dehydration and the use of concentrated solutions

A wide range of foods is preserved by dehydration or by the use of concentrated solutions of salt or sugar. The lower the moisture content of a product, the less liable it is to support microbial growth. A similar effect is obtained the higher the osmotic pressure of a food or beverage, whether this is due to added salt or sugar. Hence the terms xerophile, halophile, and osmophile are used to describe organisms likely to be found in extremes of such environments.

Moisture requirements of a micro-organism for survival or growth are expressed as water activity (a_w), i.e. equal to one-hundredth part of the corresponding relative humidity. Water activity can also be related to osmotic pressure and absolute temperature (for details see Scott, 1957).

Each species of micro-organism has its own characteristic optimum and range of a_w values. Bacteria normally exist only between 0.995 to 0.990 a_w, although staphylococci can exist down to 0.86 a_w and a few halophiles down to as low as *ca.* 0.75 (saturated sodium chloride solution). Yeasts can withstand drier conditions than bacteria; 'osmophilic' species such as *S. rouxii* can exist at a_w values as low as those tolerated by most moulds. However, some moulds can exist at lower a_w values than any other micro-organisms, even as low as 0.62. Reducing the available water below the optimum serves merely to increase the lag phase and decrease the rate of growth; at a_w values 0.65 to 0.62 mould spoilage would be unlikely to become serious in less than one-and-a-half to two years.

It is possible to compare the susceptibility of all types of foods and beverages to microbial growth by using the concept of available water. Thus the a_w in frozen foods is given by the ratio of the vapour pressure of ice to that of water at the temperature under consideration. At $-5°C$, $-10°C$ and $-15°C$ the corresponding a_w values are 0.9526, 0.9074, and 0.8642 respectively, so that the inability of bacteria to grow on frozen food below *ca.* $-5°C$ could be due to the limitations of unsuitable a_w. The ability of specific moulds to grow below this temperature is due to their tolerance both of low temperature and low a_w values. Some specific effects are also due to the toxic effect of ions; for an osmophilic yeast these are in the order of toxicity $K^+ < Na^+ < Mg^{++} < Ca^{++} Li^+; < Cl^- < SO_4^{--}$. It must not be forgotten that enzyme reactions can still continue even in a dry substrate at 0.35 a_w, albeit very slowly indeed and irrespective of whether water is a reactant (hydrolytic enzymes) or not (oxidizing enzymes).

Temperature

HIGH TEMPERATURE

In practice the terms 'sterilization' and 'pasteurization' are used to differentiate different levels of heat treatment. The second usually implies that only some of the spoilage organisms or food pathogens are destroyed. With both processes it is first necessary to know the amount of heat required to inactivate an organism or, if spore-forming, its spores. This will vary with the chemical nature of the food, e.g., its pH, water activity (a_w), and presence of inhibitors. Secondly, heat penetration studies must be carried out on the food in its actual container during the heating process. The rate of penetration is modified by the volume and shape of the container and the viscosity of the contents, presence of solid particles, etc. Thirdly, since the death/time curve approaches zero only at infinity, it is never possible to kill every organism in every container in every batch produced. Hence, some statistical limit must be agreed initially, e.g. a reduction in the number of viable spores of *Cl. botulinum* by a factor of 10^{12} in the canning of certain foods, such as wet foods at neutral pH.

Cases of botulism have been reported in the past, mainly in the USA, following the consumption of vegetables, mushrooms, etc. canned or bottled in the home. The heat treatment given was sufficient to destroy the spoilage organisms that would normally keep *Cl. botulinum* in check, but left spores of the latter undamaged. If the subsequent storage temperatures were sufficiently high the spores grew and produced toxin without altering the physical appearance of the food. Hence, low acid foods must not be heat preserved in the home as heavy duty domestic pressure cookers are no longer available. With acid foods and beverages a relatively mild heat treatment will give satisfactory results, since *Cl. botulinum* will not develop at low pH.

The amount of heat required to inactivate bacterial spores (units of lethal heat) is usually calculated in terms of F values, or the number of minutes at 121°C. This assumes instantaneous heating and cooling but, as this is impossible in practice, the equivalent heating effects occurring during the heating and cooling processes must also be calculated from standard sets of tables or graphs. The values would be inconveniently small if used for calculating heat requirements in pasteurization, which is carried out below 100°C. Instead, Pasteurization Units (PU) are used; for pickles the reference temperature is usually 82°C, while 60°C has been proposed for beer. It is often difficult to determine the effectiveness of sterilization or pasteurization, except by incubating large numbers of samples. With clear liquids, membrane filtration of the contents of a specified number of containers is a routine measure, while chemical determination of the remaining amount of the enzyme phosphatase is a suitable test for milk and beer. For milk there is also the Short Methylene Blue Test carried out 24 hours after heating, to determine the amount of post-pasteurization infection (p. 348).

Normally canned goods other than acid or cured products are processed to at least the *Cl. botulinum* spore standard but difficulties are sometimes experienced with other heat-resistant organisms, e.g. the spores of thermophilic bacteria. The problem is greatest with very viscous products of neutral pH, like canned rice pudding, whose flavour and appearance would be spoiled by the excessive heating needed to kill extreme thermophiles. Whether the contents of the can remain sound will depend on the minimum growth temperature of the organisms remaining after heating and the temperature at which cans are stored. Canners now have rigid standards for the numbers of thermophiles they will accept in sugar, starch, and other additives.

Staphylococcal food poisoning from consumption of commercially canned vegetables has now been eliminated following recognition that the source of the infection was the operatives handling the newly processed cans. The cans are sterile when leaving the cooker, but 2 to 3 per cent of cans with good commercial seams will leak temporarily during cooling. Hence, infected liquid on the outside of the seam could be drawn into the can while it was cooling. The following measures have now been adopted in modern canneries to overcome this problem: (1) rigid control of cooling water chlorination, (2) no mechanical damage to the can seam while it is in motion, (3) keeping the wet runways between cooler and dryer to a minimum (surface count on the runway must be < 500 organisms/4 sq. in.), (4) keeping any subsequent runways dry, (5) rigid sterilization programme for the equipment, (6) no manual handling of wet cans. Typhoid outbreaks, such as the one in Aberdeen in 1964, associated with imported canned meats have usually originated from the use of non-chlorinated river water, contaminated with sewage, for cooling the cans after heating.

Considerable developments are now taking place in sterile canning. The product is flash-heated in a heat exchanger to the required temperature, then filled into sterilized cans that are sealed and held for the required time before cooling. Many viscous products suffer much less damage with the improved heat penetration. Basically the same process is used for the UHT process for milk (138°C for a few seconds).

LOW TEMPERATURE

Food and beverages may be chilled to between 0 and 5°C to delay spoilage while awaiting sale or consumption, or to −18°C for extended storage. Only deep frozen foods will be considered in this section. Three main groups of organisms are important in this respect—food poisoning bacteria, psychrophiles, and moulds. Psychrophiles are defined as organisms that grow well on solid media at 0°C, forming colonies visible to the naked eye in one week. It is important to note that this is not their optimum temperature. Most psychrophilic bacteria are strains of the genus *Pseudomonas*, with some from the genera *Flavobacterium*, *Achromobacter*, *Alcaligenes*, *Escherichia*, and *Enterobacter*. They are very rare amongst Gram-positive bacteria. Psychrophilic yeasts usually belong to the genera *Candida* and *Rhodotorula*. Vegetative forms of moulds are more sensitive to cold than are spores. Spores of certain species of the genera *Monilia*, *Chaetostylum*, *Cladosporium*, *Aspergillus*, *Fusarium*, *Mucor*, *Thamnidium*, and *Botrytis*, are particularly resistant. Some of these are able to grow slowly at low levels of water activity and prevention of growth therefore requires the addition of CO_2 or exclusion of air from the package.

It is difficult to determine accurately the minimum temperatures at which organisms will grow. Generally it is accepted that (1) the lowest temperature for bacterial growth is −12°C, (2) growth of *Staph. aureus* and *Salmonella* spp. has not been reported at −4°C (Enterotoxin production by staphylococci is unknown below −18°C), (3) *Cl. botulinum* (types A, B, C, and D) do not grow or produce toxin below −10°C, and (4) *Cl. botulinum* type E can grow and produce toxin at −3.3°C after prolonged incubation, but there is no recorded case of botulism arising from this type of activity at low temperatures.

Frozen foods are not sterile and most spoilage is due to improper handling prior to freezing. Freezing is not generally bactericidal unless it is carried out slowly and the produce subsequently stored at a comparatively high temperature (0 to −10°C). Fluctuating storage temperatures within this range also reduce the bacterial load; food pathogens are found to die more rapidly between these limits (maximal at −2°C) than they do below −17°C. Hence the commercial freezing temperatures cannot be guaranteed to destroy pathogens, particularly with modern freezing techniques using rapid freezing at −35°C followed by storage at −18°C. Hence prevention of food poisoning from frozen

foods depends on the use of good quality foodstuffs, on maintaining impeccable hygiene standards during processing and on the avoidance of contamination and multiplication of micro-organisms in the thawed product. All vegetables and some meat and fish products are heat-treated before freezing. The cooling period between the two processes should be as short as possible to avoid bacterial growth.

After a frozen food is thawed it does not then spoil any more rapidly than it would have done had it not been frozen unless the cell walls have been damaged excessively by poor freezing techniques. However, microbial control must be exercised during thawing; this is done either in cold rooms or by using dielectric heating, the latter being too rapid for bacterial growth to occur. Similarly, food being de-frosted reaches the minimum temperature for growth of the psychrophilic microflora before that of the toxigenic staphylococci. Danger would arise in products containing fairly high concentrations of salt, since staphylococci may then grow faster than less salt-tolerant spoilage bacteria. No frozen food should contain salmonellae, irrespective of whether it will be cooked before being eaten or not. This is perhaps an ideal situation, practicable only in pasteurized products and frozen vegetables. Faecal indicators and enterococci rarely grow below 5°C but their presence in substantial numbers in a food thawed above this temperature could falsely indicate insanitary processing conditions.

MISCELLANEOUS METHODS

Clear liquids can be clarified and then filtered free of all micro-organisms (p. 101). Many millions of gallons of wines, beers, ciders, clear fruit juices, soft drinks, sugar syrups, etc. are sterilized annually in this manner.

In contrast to this well-tried method, efforts have been made over the past twenty years to use atomic or ionizing radiation for food preservation. Unfortunately off-odours and taints tend to form at the dose levels needed to sterilize foods.

Specific foods and beverages

Only very brief mention can be made of some of the microbiological problems involved in the preparation of particular foods and beverages.

MEAT

The microflora of freshly dressed meat from healthy

animals is derived only slightly from the flesh (mainly the lymph nodes) but mainly from dirt on the animal, its faeces, the personnel and instruments in the abattoir, and the air flora of the chill rooms.

Animals must not be fatigued or distressed before slaughter, otherwise rigor mortis sets in early and the meat putrefies rapidly. The high muscle glycogen content of a rested animal ensures the continued presence of some lactic acid, which is unfavourable to spoilage bacteria in the meat. Efficient stunning suspends the heart action. Otherwise contaminants from the knife or from the cut area circulate around the body during the bleeding process. Strains of *Serratia*, *Achromobacter* and *Pseudomonas*, which are resistant to the bactericidal action of fresh mammalian blood, are particularly important in this respect. Immobilization of pigs with carbon dioxide prior to bleeding is helpful for both these reasons. During evisceration it is essential to remove the intestines from the carcass cutting area immediately; this reduces the amount of contamination by bacteria from the gut.

When the carcasses are hanging in the chill room (2° to 3°C for 2 to 3 days) before cutting up, the microflora can include bacteria, moulds and yeasts (Ayres, 1955).

Most of the bacteria are mesophiles and tend to die out during chilled storage; only a few are associated with any specific defect or spoilage of stored meat. Moulds and yeasts may then form between one and ten per cent of the flora. Growth on the uncut surfaces which are covered with a layer of fat and connective tissue is very limited.

During chilled storage any of the following changes may take place, depending on the original microbial 'load':

1 Psychrophilic bacteria grow readily on moist surfaces of the meat, this producing individual colonies that coalesce, especially if there is condensation, forming a slime and off-odours. *Pseudomonas* sp. are virtually the sole slime formers when the meat is stored at 10°C, while at 15°C approximately equal amounts of *Micrococcus* and *Pseudomonas* occur.
2 Putrefactive spoilage can occur in the deep tissue of large thick pieces.
3 The mould *Cladosporium herbarum* can grow as black spots on chilled meat during transhipment. Carcasses of home-killed meat, held for long periods in chilled rooms formerly became infected with *Cladosporium*, *Rhizopus*, *Mucor*, and *Thamnidium* species, but these moulds rarely develop with the shorter storage periods and damper conditions now in vogue.

4 Rancidity can be caused by lipolytic bacteria and yeasts.

After a chilling period the carcasses are transported, cut and re-chilled. Each cut re-distributes the organisms and adds to the total number of bacteria, so that minced meat, which is cut most of all, has the highest count. The microfloras of chilled carcasses and cut meat are mainly aerobic surface contaminants with very few anaerobic spore-formers. Good quality meat has an initial load of *ca* 10^3–10^4 bacteria/cm^2 and should not become slimy before 14 days at 0°C. When slime does occur the count is usually 3 to 10×10^7 bacteria/cm^2, including mainly *Pseudomonas* and *Achromobacter* species.

When meat has lost its freshness it becomes offensive and is not likely to be eaten. The consumer, however, may be unable to judge whether apparently fresh meat has come from a diseased animal and he is protected against this risk by qualified inspectors who examine the animal both before and after death, and only meat fit for human consumption is released. Premises where meat is sold or prepared are regularly inspected. The prevention of cross-infection from meat is the alternative to the complete eradication of certain diseases from the domesticated animal population within a country (p. 336).

BACON

Lack of space permits mention of only one cured meat product, bacon. The meat is injected with brine containing sodium chloride, potassium nitrate, sodium nitrite, polyphosphate, etc., followed by steeping in a similar solution for several days. The brine tanks contain large numbers of bacteria (10^9/ml), mainly salt-tolerant micrococci and some lactobacilli, part of whose function is the reduction of nitrate to nitrite. The sliced bacon is vacuum-packed in oxygen-impermeable plastic wrapping; the exclusion of air prevents the attractive pink colour of nitrosomyoglobin from fading. The bacterial flora of micrococci (10^5 to 10^6/g) is fairly stable at low temperatures, but it changes rapidly at higher storage temperatures. Sliced bacon can support the growth of *Staphylococcus aureus* if the storage temperature is 30°C and the numbers of competing organisms are low. Packs of sliced bacon should therefore be kept cool during sale and in the home. The same is also true of cooked or smoked hams.

FISH

The flesh of healthy fish is largely free of organisms,

but their external surfaces carry an appreciable bacterial flora (skin 10^2–10^7/cm^2, gill tissue and intestines 10^3–10^8/g). The aerobic bacteria of marine fish are largely psychrophiles, species of *Pseudomonas*, *Achromobacter*, coryneforms, *Flavobacterium*, micrococci, *Vibrio*, and possibly *Alcaligenes*. The percentage composition of the flora differs in fish from different parts of the world, with larger numbers of mesophiles on fish from warmer seas. The intestines usually contain some strictly anaerobic bacteria, similar to those found in deposits on the sea bottom. In certain parts of the world there has been a progressive increase in the incidence of botulism caused by fish contaminated with *Cl. botulinum* Type E, an organism able to grow at low temperatures. Intestines of fish caught in sewage polluted areas commonly contain typical coliform and food poisoning bacteria.

Fish caught by deep-sea trawlers is eviscerated and packed in ice, that caught inshore may be so treated after landing. Evisceration contaminates the flesh with intestinal organisms and a low storage temperature is essential to prevent these from multiplying with consequent spoilage of the fish. As on meat, *Pseudomonas* species are the most active spoilage organisms at 0°C.

Once the fish reaches the market cooling usually ceases, leading to an increase in the flora already present. Further contamination is likely from fish boxes, filleting knives and boards, etc. The nature and amount of the microflora developing depends upon the temperature during transit and sale.

Many fish are brined and/or smoked, with numerous permutations of treatment on both unsplit and split fish. The overall effect of brining is to reduce the Gram-negative bacteria and to increase Gram-positive types such as micrococci and coryneforms. Cold smoking, while reducing the total bacterial load considerably, has little effect on the composition of the flora. Moulds never occur on fresh fish but they are a problem on smoked fish, the spores being derived from the sawdust used in smoke production. In contrast, hot-smoked fish (i.e. processed at temperatures of 65–75°C for 30 minutes or longer) are usually sterile unless *Cl. botulinum* was present originally and the salt concentration is below 3 per cent.

MILK

Milk drawn aseptically from the udder shows a predominance of staphylococci (the great majority coagulase-negative) and diphtheroids (mainly heat sensitive corynebacteria), both groups being part of the cow's normal skin flora. Some udders harbour mastitis streptococci but, from 1962 onwards, coagulase-positive staphylococci have become the commonest cause of mastitis. However, thermoduric organisms are uniformly absent from milk collected aseptically. Under usual milking conditions further contamination occurs from milking utensils, the dust of the dairy, the udder of the animals, and from the milking operatives. Of these the most important source is the milking utensils, which, if unsterilized, become coated with large numbers of bacteria and contribute enormous numbers of organisms to the milk.

Not unnaturally, single farm samples show wide variations both qualitatively and quantitatively. Counts of thermoduric bacteria (micrococci, corynebacteria, aerobic spore-forming bacilli, and thermophilic strains of *Strept. faecalis*) above 10^2–10^3/ml are indicative of heavy infection from contaminated equipment. Bulk samples, particularly from non-refrigerated tanks, carry heavy contamination and give evidence of bacterial growth, particularly of *Strep. lactis* and *Strep. kefir*. Holding the raw milk unchilled (10 to 20°C) allows an active growth of the two last-named species, coagulase-negative staphylococci and Gram-negative rods (*Alcaligenes viscolactis* and fluorescent and non-fluorescent pseudomonads). Finally, the flora is likely to be dominated by *Strep. lactis*, resulting in souring owing to fermentation of lactose. Lower temperatures favour the growth of Gram-negative rods which turn the milk alkaline and eventually putrid by their proteolytic activity.

Under laboratory conditions, milk pasteurized by being held at 63°C for three minutes (see below) succumbs to deterioration by *B. cereus* (including *B. mycoides*). In commercial practice, *B. cereus* also overgrows the slower growing coryneforms, but is itself overtaken by Gram-negative rods derived from recontamination. These in turn are outgrown by the milk-souring organisms, *Strep. lactis* and *Strep. cremoris*. *B. cereus* causes not only rapid sweet curdling but also 'bitty' cream and 'ropiness'.

Milk supplies Consumer milk is usually transported by bulk-collection services; it is therefore most essential that adequate measures of control are observed since contaminated milk from one source may be mixed with a large volume of clean milk. The first requirement is good animal husbandry and dairy technique to produce a clean product of high quality. As an additional safeguard most milks are heat-treated to kill pathogenic bacteria which may be present and at the same time to reduce the number of contaminants thereby improving keeping quality.

METHODS OF HEAT TREATMENT OF MILK   Heat treatment is widely practised for the preservation of milk. Three methods are currently used.

Pasteurization: This method of partial destruction of the microbial population by heat was first introduced by Pasteur to kill contaminating organisms which interfered with the fermentation processes in the manufacture of wine. Its application to milk was first used in Denmark to safeguard pigs against infection from bovine pathogens, but its widest industrial application today is in treating milk for human consumption. By holding the milk for a defined time at a standard temperature, e.g. 15 seconds at 161°F (71.7°C) in the 'High Temperature Short Time' process, most non-sporing organisms and all non-sporing pathogens are killed. This renders the milk safe for drinking and extends its keeping quality.

Boiling: Greater numbers of micro-organisms are killed when milk is held at a temperature not less than 100°C for a period, which ensures that it will comply with the turbidity test (see below), and the bottles are sealed immediately afterwards. This process imparts a caramelized flavour to the milk and homogenization occludes a visible cream line. Milk subjected to this process is often called 'Sterilized' but total sterility is not achieved and it will not keep indefinitely at normal temperatures.

Ultra heat treated: By this method the milk is exposed to a temperature of not less than 270°F (132.2°C) for at least one second and is then filled aseptically into sterile containers. Usually this treatment renders it sterile and therefore gives it excellent keeping qualities. There is no cream line, the cream being dispersed throughout the milk as in homogenized milk. From 10 to 20 per cent of vitamins are destroyed and a slight flavour is imparted, but this becomes less upon storage. Because of the excellent keeping quality of the milk it is likely to become very popular since less frequent deliveries to the consumer will be possible and export to other countries is facilitated.

STATUTORY STANDARDS FOR MILK   In Great Britain milk supplies are regularly tested to ensure that they conform to a reasonable degree of cleanliness or

have been exposed to the correct heat treatment for the particular designation specified. The one official test to which an 'untreated' milk must comply is the *Methylene Blue Test*. A standard solution of methylene blue is added to a sample of the milk on the day after it left the farm and the sample is then incubated at 37°C. A satisfactory milk will not reduce the dye to the colourless leuco-form within 30 minutes. Methylene blue is a redox potential indicator and is reduced by the microbial activity in badly contaminated samples.

Pasteurized milk is required to pass two official tests prescribed by the Ministry of Health. A Methylene Blue Test similar to the one described for untreated milk is performed on the milk 24 hours after it has been pasteurized. This determines the bacterial cleanliness of the milk at the time when it would normally reach the consumer and measures contamination which may arise from improperly cleaned bottles subsequent to processing. The other test, the *Phosphatase Test*, is a quantitative method of detecting the amount of the natural enzyme phosphatase normally present in milk. In pasteurized milk, most of this enzyme is destroyed and only a minimal amount is detectable. This test checks that the milk has been adequately pasteurized.

'Sterilized' milk must conform to the *Turbidity Test*. Proteins are denatured by boiling, but if the milk has received inadequate heating, soluble proteins remain and may be detected by the turbidity test. The milk is first saturated with ammonium sulphate and filtered; a tube of the clear filtrate is then plunged into boiling water and kept in it for 5 minutes. The formation of a turbidity, due to the heat denaturation of proteins, indicates that the milk had been insufficiently heated. Ultra Heated Milk is tested by a *Colony Count*. A standard loopful of the milk is cultured in 5 ml of Yeastrel milk agar at 37°C for 48 hours. Not more than 10 colonies are permitted if the milk is to pass the test.

BUTTER

Butter is made from separated cream, which is usually soured, since the butter keeps better if this is done. The cream is soured by organisms naturally present in the milk, or the milk may first be pasteurized and specific starters then added. Organisms concerned in the making of butter include *Streptococcus lactis* and *Strep. cremoris*, which ferment lactose and sour the cream, and two other capsulated streptococcal-like organisms, *Leuconostoc citrovorum* and *L. dextranicum*. The latter organisms attack citric acid, a by-product of lactose fermentation, to produce diacetyl, which imparts a buttery aroma to the cream. The cream is soured at around 20°C, low enough to prevent growth of thermophilic spoilage organisms which have survived pasteurization and yet sufficiently high to allow growth of the desirable streptococci.

In butter-making the soured cream is churned. This causes the fat globules to aggregate and form butter, leaving the buttermilk, which can be drained off.

Bacteria are restricted to the moisture droplets throughout the butter. Washing of butter to replace these droplets of buttermilk by water deprives bacteria of nutrients and limits their growth, thus improving the keeping quality of the butter. 'Working' of butter to break the moisture into smaller droplets also improves the keeping quality by limiting the available nutrients within a droplet. This does not apply to moulds, which can penetrate the surrounding wall of fat and move to other droplets.

CHEESE

The raw milk for Cheddar cheese is heated to 68°C for 15 seconds without delay in order to destroy most vegetative organisms and to ensure that any strains of *Staph. aureus* which may be present do not have time to produce enterotoxins. The pasteurized milk is not sterile and will contain thermoduric organisms (p. 347). A starter is required to produce lactic acid *in situ* in the curd; this can either be *Strep. lactis*, *Strep. cremoris*, or a mixture of these, with or without *Leuconostoc cremoris*, etc. Great care must be taken not to build up specific bacteriophages in cheese rooms, particularly where a single strain of bacterium is always employed. Antibiotics used to control mastitis cause problems in cheese making and penicillinase-producing strains of coagulase-negative staphylococci have been used both to remove the antibiotic and because their lipolytic action can contribute to the cheese flavour.

The starter rapidly gives a low redox potential, produces protein degradation products and a pH of 5.0, all conditions essential to the development of the lactobacilli and unfavourable to other bacteria; clostridia are inhibited by the antibiotic nisin produced by some strains of streptococci. If the streptococci fail, so also do the lactobacilli and the cheese develops such faults as taints and gas produced by coliforms and clostridia, and faulty colour produced by a variety of organisms. Toxigenic staphylococci would also find the conditions favourable.

Renneting with calf rennet follows immediately after the starter which also produces the optimum pH for casein decomposition and clotting. The coagulum is cut into small pieces to release the whey, the degree of syneresis being dependent on the temperature and the rate of acid production by the bacteria. Finally the curd is consolidated into blocks under light pressure until they become plastic, when they are broken into pieces, sprinkled with salt and pressed into moulds. The shaped blocks are first dipped in molten paraffin wax to exclude air and thus to prevent external growth of mould and weight loss, and then ripened at 13°C for quick process cheese or 5.5°C for the old slow-ripening type. During cheese ripening the numbers of lipolytic organisms are low; the lactic streptococci used as starters are soon inhibited while the numbers of lactobacilli increase progressively during the later stages of ripening of all cheeses, especially the hard varieties. The main flora consists of homofermentative types (*L. casei* and *L. plantarum*); *L. brevis* and *L. lactis* occur much less frequently. The total number of lactobacilli and streptococci, both alive and dead, is extremely large and probably accounts for as much as 0.1 per cent of the weight of the cheese.

FERMENTED EASTERN FOODS AND BEVERAGES

Fermentations of bland starch (cereal products) and protein (particularly fish) with moulds, yeasts, and bacteria have been practised for centuries in the Orient to add flavour, improve nutrient quality, or prevent spoilage. Some of these fermented foods and beverages are an important source of minerals and vitamins in the diet of large populations. Table 15.1 lists a few of the many preparations known. Space does not permit detailed descriptions of these nor of methods of preparation.

Fermented drinks, prepared from millet or other local cereals, are also an important source of vitamins in the diet of many African populations.

LACTIC PICKLES

While vinegar is commonly used in the Western world in the preparation of pickles, many vegetables are preserved by fermentation with lactic acid bacteria. Thus in the preparation of Sauerkraut, the solid, round-headed cabbages are washed to remove the undesirable Gram-negative bacteria on the outer leaves (*Pseudomonas*, *Flavobacterium*, and *Achromobacter* spp.) and then sliced and mixed with brine. Sugar, mainly sucrose, nutrients, and mineral

Table 15.1 Some fermented foods. (After Hesseltine and Wang, 1967)

Product	Raw material	Micro-organism used	Country of origin
Katsuobushi	Bonito (fish)	*Aspergillus glaucus*	Japan
Nam-pla (fish sauce)	Fish	Halophilic bacteria	Thailand, S.E. Asia
Shoyu	Wheat and soybean	*Aspergillus oryza*, yeast and *Lactobacillus* spp.	Japan
Natto	Soybean	*Bacillus subtilis*	Japan
Tempeh	Soybean	*Rhizopus oligosporus*	Indonesia
Sufu	Soybean curd	*Actinomucor elegans*	China
Hamanatto	Soybean	*Aspergillus oryzae*	Japan (called Tao-Si in Phillipines and Tu-Su in China)
Ontjom	Peanut press cake	*Neurospora sitophila*	East Indies
Sake	Rice	*Lactobacillus* spp. and *Saccharomyces sake* (*S. cerevisiae*)	Japan

salts are leached into the liquid which is then fermented by a succession of three bacterial species, derived from the innermost leaves. The heterofermentative *Leuconostoc mesenteroides* and *Lactobacillus brevis* produce lactic and acetic acids, carbon dioxide, alcohol, mannitol, and dextran while the homofermentative *Lactobacillus plantarum* produces lactic acid only. The optimum temperatures and salt concentrations are 13–18°C and 1.8–2.25 per cent salt. At the upper ends of these limits, the homofermentative species *Streptococcus faecalis* and *Pediococcus cerevisiae* tend to replace the less salt-tolerant *Leuconostoc mesenteroides*. At 3 per cent salt the product is tough and can develop a pink coloration with the growth of a carotenoid-producing yeasts. At too low a salt concentration the product is unpleasantly soft due to the growth of bacteria capable of digesting pectin and cellulose. Pure culture inoculations have never produced the succession of species that gives the best flavoured product. Olives and small cucumbers are also brined and preserved by the action of lactic acid bacteria.

COCOA AND COFFEE

Fermentation is also used to prepare certain products for processing, as with cocoa and coffee beans, where removal of unwanted outer layers is facilitated by the action of micro-organisms.

EGGS

The contents of some 90 per cent of freshly laid hens' eggs are sterile; their shells and the interior of the remainder contain Gram-positive bacteria of ovarian origin. However, the predominant species inside rotten eggs are Gram-negative rods and are extragenital in origin, derived from the nesting material and not the faeces. The predominant species are *Alcaligenes faecalis*, *Aeromonas liquefaciens*, *Proteus vulgaris*, non-proteolytic strains of *Cloaca* sp., *Citrobacter* sp., and *Pseudomonas fluorescens*. These organisms possess combinations of proteolytic and lecithinase activity, pigment and hydrogen sulphide production. The progress of bacterial rotting follows a definite pattern. First there is contamination and penetration of the shell, followed by growth in the cell membranes. The extent of growth is controlled by the antimicrobial action of the conalbumin in the egg albumen. Eventually the yolk makes contact with the shell membrane and the organisms are able to utilize the glucose of the albumen and infect the complete contents of the egg.

Salmonellae can proliferate in eggs without producing any visible symptoms other than a faint turbidity in the albumen, *Salmonella typhimurium*, *S. anatum*, and *S. enteritidis* are frequently encountered even in eggs from clinically normal birds and the importance of duck eggs as a source of human salmonellosis is well established. Humans can become infected by salmonellae not only from fresh whole eggs but also from frozen, liquid, or spray-dried products made from lower grade eggs. Because of the lack of obvious spoilage symptoms, salmella infections are not apparent and the contents of such eggs are added to the main bulk for processing. This is in addition to any contamination from the hands of the process workers or from the containers into which the eggs are poured.

The control of infection from egg sources relies on reduction of the incidence of infection in the flock, reduction of contamination of the shells by faeces, special attention to the sites where eggs are laid, the sale of only clean, unwashed eggs, and the education of kitchen staff to encourage adequate cooking of food containing eggs or its products.

BREAD

In the conventional breadmaking process, a dough of flour, water, salt, etc. is allowed to ferment with a culture of the yeast *S. cerevisiae*. The dough is worked, allowed to rise, then reworked, placed into tins, allowed to rise for a further 50 minutes (final proof) and baked. This is now being replaced by the high-energy system (Chorleywood Bread Process) in which the dough is worked mechanically. The amount of yeast is increased by 50% and the addition of fat and a fast-acting oxidizing agent are essential parts of the process. Production time is decreased by some 60%, much less space at controlled humidity is required in the bakery, the loaf has a lower staling rate, and the yield of bread is increased 4–5%.

Both compressed yeast and bakery products are subject to contamination. Bacteria contaminating compressed yeasts are mainly lactobacilli, normally considered harmless; only heavy infections of the slime-forming *Leuconostoc mesenteroides* are liable to affect the baking properties of the yeast. Nowadays contamination with wild yeasts is rare, but before rigid sanitation programmes were instituted in yeast factories the presence of such aerobic yeasts as *Candida krusei*, *C. mycoderma*, *C. tropicalis*, *Trichosporon cutaneum*, *T. candida*, and *Rhodotorula mucilaginosa* was not uncommon. The widespread use of moisture-impervious wrappings has made control of moulds such as *Aspergillus* and *Penicillium* species and *Oidium (Oospora) lactis* of special importance, since they can form unsightly blotches on the surface of the yeast block beneath the wrapping. Acid calcium phosphate and acetic acid are commonly used as preservatives in bakery products to prevent the development of bacterial ropiness. The addition of propionates in bread, sorbic acid and its salts in flour, and both these in baked products, are now permitted for the prevention of mould growth. Reducing the available water content (a_w) to as low a level as possible without spoiling texture and appearance helps to delay mould growth but in the absence of fungicides cannot prevent it. Strict cleanliness within the bakery is absolutely essential to ensure that any waste flour is quickly removed, otherwise it becomes mouldy and spores are discharged into the atmosphere.

BEER AND LAGER

Beer is made by fermenting a hop-flavoured extract of barley malt with a top-fermentation strain of *S. cerevisiae*. The bottom-fermenting yeast, *S. carlsbergensis*, is required for lager. Most beers in England are made by the first process, whereas lager predominates in the rest of the world.

Barley is converted into malt by allowing the soaked grain to germinate under controlled conditions of temperature and humidity, leading to the formation of α- and β-amylases and the breakdown of the protein hordein to amino acids. The process is stopped by raising the temperature and removing the rootlets and plumules. The exact treatment depends on the type of malt required. Whereas in England the protein is fully converted into amino acids, lager malts still contain appreciable quantities of protein. Again, stout malts are dried at 105°C to give a dark-coloured extract; malts from distilleries and vinegar breweries receive no final kilning and still contain limit-dextrinase.

An extract or wort is now prepared from the ground malt, the exact conditions of extraction again being varied according to the type of beer. In England the starch is hydrolyzed by the action of the amylases to glucose, maltose, malto-triose and -tetrose, and oligosaccharides; glucose and fructose are also present in the extract. Worts from lager also need extensive proteolytic action during mashing. Sometimes flaked barley or maize, or gelatinized unmalted cereal grits are added to supplement the starch supply. When extraction is complete the wort is boiled under pressure with hops, the tannins from which co-precipitate with any remaining proteins. A proportion of the humulone and lupulone complexes are converted into their corresponding isomers during boiling, ultimately, giving bitterness to the beer and conferring some resistance to Gram-positive bacteria. The cooled and filtered wort is 'pitched' with yeast and allowed to ferment. Distillery and vinegar brewers' worts are not boiled (so that further saccharification due to the enzyme limit-dextrinase can take place during fermentation), neither are they hopped. Bacterial infection is often restricted by inoculating part of the wort with *Lactobacillus delbrueckii* and, when acidified sufficiently, boiling this before returning it to the remainder of the wort ready for yeasting; sulphuric acid may be used as an alternative method of reducing the pH. Again, some power-alcohol distilleries sterilize the wort chemically with 1000 ppm ammonium fluoride and use a fluoride-adapted yeast.

Beer wort is yeasted with between 2 and 6 g moist yeast/litre, depending on its original specific gravity. After 12–18 hours at 15°C a rapid fermentation ensues, the temperature is raised by 3 to 7°C and the pH falls from 5.2 to 4.1. The thick yeast head that collects after $2\frac{1}{2}$ days is skimmed off, fermentation slackens and the beer, still containing some sugar, is ready for clarification and natural conditioning in tanks before sale. Some modern breweries use a non-flocculent yeast that ferments very rapidly; the fermentation is terminated at the desired sugar content by centrifugation. Traditionally, part of the yeast crop (often a balanced collection of strains) was collected and used for the next fermentation. Nowadays pure cultures are being used increasingly and are replaced after 12 to 20 brews. Lager has always been made with a pure culture, the crop being collected from the bottom of the tank after the main fermentation. The pitching temperature, 5 to 9°C, is lower than with beer; fermentation at 12°C is more protracted (7 to 14 days), and storage at 2°C may take 6–40 weeks before the lager is considered sufficiently conditioned and stabilized for sale. Distillery worts of high specific gravity are pitched at 21°C with a heavy inoculum of a yeast specially bred for alcohol tolerance; fermentation is rapid, the temperature rising to 25–30°C during the process. Either a fresh culture is used for each fermentation or the previous crop is acid-washed and re-used.

CIDER AND WINE

In the traditional cider-making countries, the raw material consists of special varieties of apples, high in tannin and low in acidity, supplemented with dessert and culinary apples graded as unsuitable for the fresh fruit market; other countries use apples unsuitable for the fresh fruit market as their sole source of raw material. The fruit carries both an internal and an external microflora, mainly *Candida pulcherrima*, *Kloeckera apiculata*, and *Torulopsis famata*, together with the 'black yeast' *Aureobasidium pullulans* and occasionally rhodotorulae, *Candida*, and *Torulopsis* spp. *Saccharomyces* spp. and bacteria are rare in sound fruit but mould-damaged specimens carry large populations, not only of mould spores but also of acetic acid bacteria and fermenting yeasts. Both the moulds and acetic bacteria produce sulphite-binding compounds that are of great significance at later stages of processing. Hence sound, preferably washed fruit, is essential. The apples are milled to a pulp and pressed hydraulically. If the equipment is washed very

infrequently, it soon supports a dense population of *Saccharomyces* species, the yeasts mainly responsible for the subsequent fermentation. The freshly pressed juice is treated with sulphur dioxide, the actual amount being varied according to the pH. Below pH 3.3 100 ppm suffices, while 150 ppm is desirable between 3.3 and 3.8; most factories process a mixture of apples whose juice pH is generally 3.5–3.6. It is vitally important that the concentration of sulphite-binding compounds be kept low in the juice, otherwise no free sulphur dioxide remains in the solution and this is the only effective moiety (sometimes referred to as molecular SO_2). After treatment only *Saccharomyces* spp. remain, so that in effect sulphiting selects a pure culture from the original microflora. However, with higher standards of cleanliness in the press rooms, the juices are virtually free of fermenting yeasts. Hence, the addition of pure cultures of bottom-fermenting wine yeasts, following sulphiting, is becoming almost universal.

A second fermentation of malic acid to lactic acid and carbon dioxide, due to lactic acid bacteria, nearly always occurs during storage of the cider. The organisms that can bring about this change include both homo- and heterofermentative lactic rods and cocci, including the organisms responsible for producing polysaccharide slime or ropiness. The bacterium found most frequently is *Lactobacillus collinoides*. If citric acid is present, as in perry but not cider, diacetyl, lactic, and acetic acids are also formed and the flavour is thereby spoiled. The source of these bacteria is still in some doubt since they appear to be absent from many sulphited juices. Ciders stored in contact with air, especially those held in small wooden containers, soon show surface growths of *Acetobacter xylinum* and the film yeasts *Pichia membranaefaciens* and *Candida mycoderma*. The problem is now rare in large factories where juice sulphiting and very large storage tanks are normal practice.

Basically, wines are made by a similar process but many other fermentation treatments are possible, due to the much wider range of climatic conditions in which grapes are grown. Thus grapes can be 'raisinified' by being left on the vine or by being dried in the sun; wines made from their high gravity juices are fermented with osmophilic yeasts such as *Saccharomyces rouxii*. Sauterne is made from grapes naturally infected with the mould *Botrytis cinerea* that dehydrates the grapes, metabolizes some of the acid and produces glycerin and the mild antibiotic *botryticin*. Consequently the subsequent fermentation of the very sugary juice ceases prematurely, leaving a soft, sweet, luscious-flavoured wine. Climate also has an effect on the type of yeasts found on the grapes. The greater the ambient temperature, the greater the concentration of alcohol-tolerant *Saccharomyces* yeasts and sporing apiculate yeasts (*Hanseniaspora valbyensis*) and the fewer there are of their non-sporing counterparts (e.g. *Kloeckera magna* and *K. apiculata*). Finally a number of *Saccharomyces* species have been isolated from both cider and wine. These are *S. acidifaciens, S. carlsbergensis, S. cerevisiae, S. delbrueckii, S. elegans, S. florentinus, S. fructuum, S. oviformis, S. rosei, S. rouxii,* and *S. uvarum.* In addition, there are other *Saccharomyces* spp. apparently characteristic of each beverage. Again *Brettanomyces* spp. have been isolated twice from cider and it was only in the 1960s that they have been found in wine, in the Bordeaux area and in dry wines made within a 40-mile radius of Cape Town. There is one wine for which no parallel exists in cider-making, namely, sherry. The yeasts responsible for the characteristic fermentation of alcohols and organic acids, sometimes called *S. beticus, S. cheresiensis,* etc. develop as a veil on the surface of the dry wine when it is exposed to air. Unlike *C. mycoderma* and other spoilage film yeasts, the sherry yeasts form acetaldehyde and other characteristic flavours from the alcohol; the remaining process is a complicated blending system (the solera) also used for the sun-baked wine Madeira.

Books and articles for reference and further study

AYRES, J. C. (1955). Microbial implications in the handling, slaughtering and dressing of meat animals. *Adv. Fd Res.* **6**, 109–161.

BEECH, F. W. and CARR, J. G. (1977). *Cider and Perry.* In *Economic Microbiology.* Rose, A. H., ed. Academic Press, London and New York. Chapter 3. Vol. I. 139–313.

BEECH, F. W. and DAVENPORT, R. R. (1970). The role of yeasts in cidermaking. In *The Yeasts.* Rose, A. H. and Harrison, J. S., eds. Academic Press, New York and London. Chapter 3, Volume 3 (in press).

BUTLER, W. H. (1975). Mycotoxins. In Smith, J. E. and Berry, D. R., eds. *The Filamentous Fungi, 1, Industrial Mycology,* Ch. 16, 320–329. Edward Arnold, London.

CARR, J. G. (1968). *Biological Principles in Fermentation.* Heinemann Educational Books Ltd., London, 97 pp.

DAVIS, J. G. (1963). The lactobacilli. II. Applied aspects. *Progress in Industrial Microbiology,* **4**, 95–136.

EMBODEN, W. (1972). *Narcotic plants.* Studio Vista, London, 168 pp.

FRAZER, W. C. (1967). *Food microbiology.* 2nd Edn., McGraw-Hill, London, 512 pp.

HERSON, A. C. and HULLAND, E. D. (1964). *Canned Foods: an Introduction to their Microbiology.* Chemical Publishing Co., New York, 291 pp.

LEITCH, J. M., ed. (1965). *Food Science and Technology.* Gordon and Breach Science Publishers, New York, Volumes 2 and 4.

LITTEN, W. (1975). The most poisonous mushrooms. *Scient. Am.,* **232**, 90–101.

LUTHI, H. (1959). Micro-organisms in non-citrus juices. *Adv. Fd. Res.,* **9**, 221–284.

MOLIN, N., ed. (1964). *Microbial Inhibitors in Food.* Almqvist and Wiksell, Stockholm, 402 pp.

PIRIE, N. W. (1969). *Food Resources Conventional and Novel.* Penguin Books Ltd., Harmondsworth, Middlesex.

SCOTT, W. J. (1957). Water relationships of food spoilage micro-organisms. *Adv. Fd Res.,* 7, 84–128.

SHEWAN, J. M. and HOBBS, G. (1967). The bacteriology of fish spoilage and preservation. *Progress in Industrial Microbiology,* 6, 169–208.

TANNER, F. W., ed. (1944). *Microbiology of Food.* 2nd Edn., Garrard Press, Illinois, 1196 pp.

WIELAND, T. (1968). Poisonous principles of mushrooms of the genus *Amanita. Science,* **159**, 946–952.

Chapter 16

Micro-organisms and industry

Spoilage of materials by micro-organisms

Introduction

Almost all natural organic materials are degradable to some extent when in suitable environments, and economically important losses occur with such diverse substances as petroleum products, paints, adhesives, rubbers and even some plastics. Economically the most important industrial materials, other than foodstuffs, affected by micro-organisms are cellulose and wood products, including wood itself, wood pulp and paper, and textiles made from natural fibres such as cotton, flax, and jute. These materials are attacked by fungi, and to a lesser extent by bacteria, causing loss in strength and discoloration.

Conditions affecting the rate of decay are those governing the growth of the organisms, e.g. pH, temperature, nutrient availability, oxygen supply, and moisture (Chapter 4). Apart from the latter these are generally not limiting and anyway often cannot be controlled under conditions of use. The balance of these factors determines the particular species concerned, for example, bacteria are more common than fungi when the oxygen concentration is low. The substrate must be moist or in conditions of high humidity for decay to occur.

The commonest way of preventing decay is to keep the material dry. When this is not possible (for example, under tropical conditions, when use involves contact with moist soil or water) then chemical preservation methods must be used. There is a wide variety of commercial products available and these are routinely used to impregnate materials likely to be used under moist conditions. Preservatives are ideally toxic to micro-organisms but not to animals, and should be almost insoluble in water so that they are not leached out. Heavy metal salts and phenolics are commonly used, though the choice of a particular biocide depends on the organisms concerned, the level of mammalian toxicity that can be tolerated and on compatibility with other constituents of the substance being protected. No biocide gives complete protection, at best they reduce or delay decay in favourable environments.

Degradation of cellulose products

Importance and method of attack

Cellulose for industrial use is obtained from cell walls of higher plants either directly (cotton fibres are approximately 95% cellulose) or by removing other cell wall components, such as lignin, from wood as in the production of some paper pulp. Cellulose decay may cause great economic loss where large quantities are handled, but losses may also be important in small quantities of such expensive items as high grade paper, or valuable books and documents.

Cellulose is a polymer of β 1–4 D-glucopyranose with a molecular weight of approximately 1.0–1.3×10^6, normally forming unbranched chains up to 400 nm long. In cell walls the chains are grouped into crystalline structures termed micelles. One molecule may be incorporated, along its length, into several micelles forming an interlocking system. The micelles are themselves grouped into microfibrils.

The action of the enzyme cellulase (β 1–4 glucan, 4 glucano-hydrolase), is not fully understood, but two stages seem to be involved; first the crystalline micelle is broken down, and then the amorphous cellulose is hydrolysed (either endwise or in a random manner, depending apparently on the organism), to cellobiose. Cellulase is an induced enzyme and is extracellular, though in bacteria it may be closely bound to the cell walls or to the substrate. The cellobiose is further hydrolysed to glucose by cellobiase (β-glucosidase).

Degradation of cellulose may at first cause little loss in strength, though the decay is visible microscopically as eroded areas on the fibres. It may also be detected at this stage by an increase, as compared with unaffected fibres, in the swelling in alkali solution and this probably corresponds to the breakdown of the crystalline structure. As decay proceeds the cellulose chains are broken up and the strength of the material is rapidly lost.

Organisms associated with decay

Two fungi, *Chaetomium globosum* and *Stachybotrys atra*, are of particular importance in cellulose decay; they have a world-wide distribution, though the latter is more common in north temperate regions. Both occur on rotting vegetation in the soil. *Myrothecium verrucaria* and *Memnoniella echinata* are less common but are often used in laboratory tests for decay resistance. *Aspergillus, Penicillium, Cladosporium, Fusarium, Alternaria*, and *Trichoderma* have less cellulolytic ability but are often isolated from cellulose substrates. Many other fungi, not themselves capable of utilizing cellulose, may degrade dyes, sizes, and other finishers in the fabric or paper. Though they cause no damage to cellulose they may produce stains or otherwise affect the quality of the product; such fungi as *Mucor, Aspergillus, Penicillium*, and *Fusarium* are often isolated from such damaged material.

Bacteria are less important than fungi in cellulose breakdown. They may, however, become of particular interest under anaerobic conditions and they may cause oxidation instead of hydrolysis of cellulose. *Cellvibrio, Cellulomonas, Cytophaga*, and a large number of Actinomycetes are aerobes, and *Clostridium* spp. such as *Cl. cellulosolvens* are anaerobes commonly found on cellulose substrates.

A specialized microflora is associated with wood pulp and paper production where large quantities of cellulose suspension are handled. Many fungi are present in bulk storage vats; over one hundred species were isolated within a year on one survey. The most important genera are *Alternaria, Aspergillus, Cladosporium, Penicillium*, and a basidiomycete, *Polyporus*. They presumably cause a loss of cellulose but more important they produce, in association with bacteria such as *Flavobacterium, Pseudomonas, Desulfovibrio*, and *Clostridium*, a slime which is often coloured and foul-smelling. The slime affects the quality of the paper and increases the cost of machine maintenance. Bacteria may also cause corrosion of machines and pulp-storage tanks.

Synthetic fibres and wool

Synthetic fibres based on cellulose, e.g., rayon, have now been superseded by nylon and polyester fibres such as 'Terylene'. Rayon (cellulose acetate) is more resistant than natural fibres and modern synthetics are usually not degraded. The dyes and finishers may still be attacked.

Wool is largely protein and is quite resistant to decay. In a raw state when contaminated with dirt and oil there may be a fairly high microbial population. However, once the wool is cleaned and manufactured, and provided dyes and finishers are suitably chosen, only some Actinomycetes cause problems.

Degradation of wood

Timber is one of the world's main natural resources and is in great demand for constructional use, furniture, fuel, packaging, and increasingly for pulp production for paper, fibreboard, and fibrous packaging materials. In view of the very large quantities involved even a small percentage of decay during growth, conversion, storage or use represents a large economic loss. Decay of sawn timber by one fungus alone—'dry rot' (*Merulius lacrimans*, Fig. 16.1)—causes a loss in Britain of several million pounds sterling per year, and there are numerous less important rots also attacking timber in use.

Wood is a much more complex substrate than cellulose fibres for the walls of the xylem also contain considerable quantities of lignins, mannans, xylans, etc. The complete breakdown of such a complex mixture is not usually accomplished by a single organism; different species of bacteria and fungi may occur simultaneously or in succession and the relationship between them may be competition, synergism or mutualism (Chapter 12). Different organisms can cause degradation of particular chemical components of the timber, which results in a characteristic appearance of the rotten area.

Types of rot, their occurrence and appearance

BASIDIOMYCETE ROTS

These are caused by Holobasidiomycetes and include most of the important decays of standing and converted timber. There are two main types, 'brown' and 'white' rot. Brown rot occurs when the fungus utilizes the cellulose and leaves the brown lignin, causing a rapid loss in strength. White rots are produced by fungi which attack the lignin and some of the cellulose and by fungi using only cellulose, as in brown rots, but bleaching the

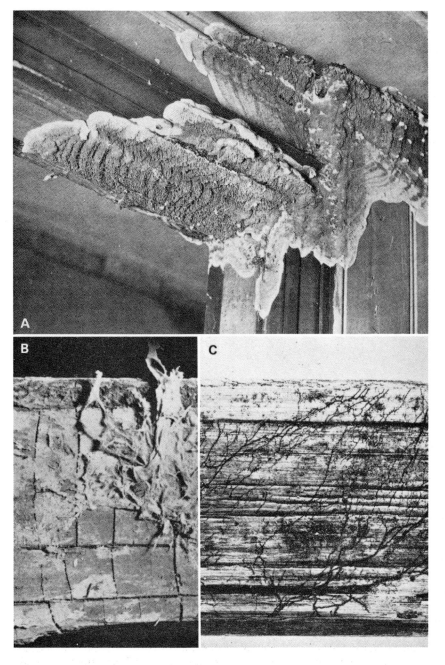

Fig. 16.1 (**A**) *Merulius lacrimans* (about two-thirds natural size): a large fruit-body of the dry rot fungus on a door lintel. (**B**) *Merulius lacrimans* (about natural size): a board showing the typical brown cubical rot of and mycelial strands of the fungus. (**C**) *Coniophora cerebella* (about natural size): a board showing typical longitudinal cracking and black rhizomorphs of the 'cellar fungus'. (From Cartwright and Findlay (1958), by courtesy H.M.S.O.)

lignin. Most white rots produce a less rapid loss in strength than do the brown rots.

Further terms may be used to describe the appearance of the wood or the distribution of the fungus. Thus *Merulius lacrimans* produces a 'brown cubical rot'; the wood is brown and is split into cubical blocks (Fig. 16.1b). Similarly *Fomes annosus*, a very important parasite of standing timber, produces a 'white pocket rot'; the rot is confined to small spindle-shaped cavities in the wood (Fig. 16.2a).

The gross appearance of the wood may also be altered by 'zone lines'. These are clearly defined black lines within the timber caused by masses of dark mycelium or by deposits of resins and gums produced by the tree to contain the rot (Fig. 16.2b).

The hyphae of hard-rot fungi are found within the lumena of the wood cells (Fig. 16.2c). They

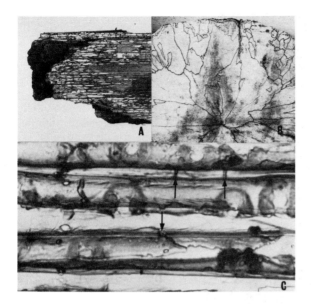

Fig. 16.2 (A) (about half natural size). White pocket rot in oak (caused by *Hymenochaete rubiginosa*) : note the well-defined areas of white rotten wood. **(B)** (about one tenth natural size). Zone lines on the transverse face of a log. **(C)** (×about 600). Microscopic appearance of wood attacked by a brown rot fungus; note the hyphae in the cell lumena and the bore holes (arrowed).

occasionally pass through the cell walls, usually directly across the wall. A hyphal tip comes into contact with the wall and dissolves a narrow hole through which it passes into the next lumen. This bore hole is often enlarged by subsequent enzyme action. In white rots the walls become noticeably thinner, but this does not occur with brown rots until an advanced stage of decay is reached.

Decay has a great effect on the properties of wood, but for many purposes, such as rough packaging and low grade constructional use, a small amount is tolerable. The following effects are produced: (1) The density is reduced. (2) The shrinkage of rotten wood on drying is greater than that of sound timber: a slight amount of decay, which may not otherwise be important, causes distortion of planks. (3) The chemical constitution of the wood is altered. (4) The colour may be changed, though when this becomes obvious the decay is usually so far advanced that the wood is useless anyway. (5) The most important effect is a decrease in strength. There are different types of strength, such as tensile, bending, impact, or crush strength, and the amount of each type lost will depend on the type and degree of rotting.

The decay of converted timber in buildings is particularly important for economic reasons, and in

Britain *Merulius lacrimans*, the dry-rot fungus (Fig. 16.1a and b), is of outstanding importance. It requires damp conditions to become established, but thereafter can spread for long distances by mycelial strands which can pass through brickwork. Sufficient water is produced by oxidation of the glucose from the degraded cellulose to enable the fungus to attack relatively dry timber when humidity is high due to poor ventilation. A less important, but still quite common, fungus causing rot in building timber is the cellar fungus (*Coniophora cerebella*), which is restricted to damp wood, in which it causes a brown rot with mainly longitudinal cracks (Fig. 16.1c). Characteristic thin dark rhizomorphs are formed over the surface of the wood. Other fungi occasionally causing damage include *Phellinus megaloporus* and *Poria xantha* which is common in greenhouses. *Poria incrassata* is particularly common in North America where it is more important than *M. lacrimans*.

Numerous other fungi are found as saprophytes on rotting wood under natural conditions where they are important in the carbon and nitrogen cycles of forests.

ASCOMYCETE ROTS

These are also called soft rots and the main fungi are Pyrenomycetes and Fungi Imperfecti (Deuteromycotina). Soft rot is of interest for two reasons: firstly because it occurs under very wet conditions where rots by Basidiomycotina are not common and secondly soft rot fungi are not controlled by many preservatives routinely used on timber, though copper chromate and copper chromium arsenate are effective. The rot is usually confined to specialized environments such as timbers in constant contact with fresh or salt water and has received particular attention because of the damage caused in softwood slats used in cooling towers of power stations. It is possible that soft rot is quite widespread but usually masked by the faster rotting and more conspicuous Basidiomycotina.

Soft rot is so called because of the soft spongy texture of the outer layers of the attacked wood, usually without any discoloration. The rot is characterized, however, by its microscopic appearance (Fig. 16.3). The hyphae penetrate the wood through the cell lumena and through rays but then enter the secondary cell walls of the wood cells and branch at right angles, forming cylindrical cavities with sharply pointed ends running the length of the cell and frequently following the orientation of the cellulose microfibrils (cf. hard rots where the hyphae

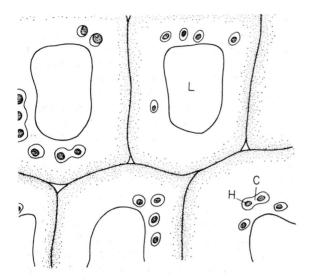

Fig. 16.3 (× about 2,000). Diagram of a transverse section of wood attacked by soft rot fungi; note the cavities (C) containing hyphae (H) in the secondary cell wall, and the absence of the hyphae in the lumen (L) of the tracheids.

occur mainly in the cell lumen). The cylindrical cavities, enlarged by enzyme action, may fuse with each other in advanced stages of decay forming irregular-shaped cavities in the secondary cell wall.

The soft rot causes a loss of all types of strength in the surface layers of the wood. In large bulks of timber, for example marine piling, this may not be particularly important unless it allows excessive abrasion; in smaller sizes of timber, such as plywood hulls of small boats and the slats of water-cooling towers, the destruction may be virtually complete.

Many genera have been implicated in soft rot, but the most common include: *Chaetomium*, *Xylaria*, *Leptosphaeria*, and *Ophiostoma* among the Ascomycotina, and *Phoma*, *Bispora*, and *Phialophora* in the Fungi Imperfecti. There is no notable difference between the rots produced by the different species; nothing equivalent to the distinction between brown and white rots of the Basidiomycotina described above.

BACTERIAL ROTS

The rots caused by bacteria are thought to be relatively unimportant. Decay which occurs under anaerobic conditions such as are found in wood deeply submerged in water or waterlogged soil, is slow and appears microscopically as minute eroded areas in the cell walls of the wood. Bacterial decay may precede fungal attack in the wood. The bac-

teria destroy the pit membranes and may reduce the C/N ratio, making fungal invasion more rapid.

STAINING OF TIMBER BY MICRO-ORGANISMS

As with paper and textiles various fungi may cause stains without seriously affecting strength. The economic loss may, however, be considerable because the wood is unsuitable for decorative work. Common fungi such as *Penicillium*, *Aspergillus*, and *Trichoderma* are frequently isolated, but the most serious staining is caused by *Ceratocystis* spp. in sawn softwoods. These fungi live on the contents of the ray parenchyma cells, do not attack cellulose or lignin to any great extent, and produce blue or greenish discolorations.

Staining may be particularly associated with damage by wood- and cambium-boring beetles. The flight holes may furnish entry sites for naturally occurring fungi, the beetles may passively carry the spores on their bodies or even actively inoculate the brood galleries with fungi on which the larvae feed (p. 304).

Prevention of timber decay

Timbers vary greatly in their natural resistance to decay, depending on their structure (close-grained being generally more durable than less dense types) and their natural content of fungicides such as phenols and tannins. Timbers in common use that are resistant to microbial decay include oak (*Quercus* spp.), red cedar (*Thuja plicata*), walnut (*Juglans* spp.), and many imported tropical hardwoods such as teak (*Tectona grandis*) and greenheart (*Ocotea rodiaei*). Wood from near the centre of the stem, the heartwood, is often more durable than the sapwood.

The most practical way of preventing decay is to store newly cut planks in well-ventilated stacks to dry (i.e. to 'season'). Modern techniques of kiln drying timber under carefully controlled conditions of heat and humidity are better but more expensive. In use, the timber is best preserved by keeping it dry. Fungi will not normally grow when the moisture content is less than 20 % which is barely obtainable under covered outside storage in this country. If the wood cannot be kept dry then chemical preservatives must be used. The commonest one is creosote, a product of the coal-tar industry, which is effective against most fungi and is cheap. It has the disadvantages that it is exceedingly unpleasant to work with, wood treated with it cannot be painted, it smells, and is leached comparatively rapidly by water. A wide range of copper salts are now frequently used where the defects of creosote cannot be tolerated. Impregnation with preservatives is usually done under pressure in large tanks to achieve adequate penetration of the wood.

Degradation of petroleum

Decay depends on the presence of water and minerals but these are not usually limiting since crude oil is contaminated with brine from the oil beds and water is invariably present in storage tanks from condensation or for technical reasons concerned with its transport by sea. Crude oil in transport and storage may be attacked by *Actinomyces* sp., *Mycobacterium* sp., *Pseudomonas aeruginosa*, and *Desulfovibrio* sp. These are originally introduced from soil contamination, but become established as a resident population in tanks and pipelines. However, the damage to crude oil is not usually severe, but includes a decrease in viscosity, loss of oil by oxidation and changes in the relative amounts of the aliphatic and aromatic fractions of the oil. The degradation of crude oil by yeasts and bacteria has recently become an important beneficial effect in connection with marine pollution. The more volatile fractions evaporate and the destruction of the residual tar-like mass is slow. It may take place under alternating aerobic or anaerobic conditions as the oil sinks with the accumulated weight of micro-organisms and then rises again as such gases as methane accumulate in the mass. The final result is a hard pebble-like piece of unutilizable tar residues. The spreading of detergents is sometimes thought to be necessary to break up large patches of oil but can be ultimately harmful as it may kill the bacteria and yeasts, as well as much other marine life, thus actually slowing the destruction of the oil.

Degradation of petroleum products such as gasoline, kerosene, paraffin and lubricating oil is more serious, though again it occurs only in the presence of water. There is no appreciable loss of the substance except that the octane content of gasolene may be reduced. The by-products of the attack may, however, produce deleterious effects such as the hydrogen sulphide content rising to unacceptable levels because of the action of sulphate-reducers. Methane may also be produced and combinations of gases have been blamed for bursting and spontaneous explosions in storage tanks. Attack on aircraft fuel has become important and is particularly associated with the fungus *Cladosporium resinae*, though other fungi and bacteria (e.g. *Pseudomonas*) are also common. The growth of the micro-organisms causes blocking of fuel lines. The organic acids which are produced damage fuel tanks and pipes. The fungus is exceedingly difficult to eradicate as it can live in empty tanks on condensation water and with kerosene vapour as its sole carbon source. The problem is so serious, especially in aircraft operating in the tropics, that frequent inspection of the tanks is needed, though this may be difficult with the intricate systems of modern aircraft. Biocides are used, either continuously as a fuel additive such as ethylene glycol monomethyl ether (EGME) or as an intermittent cleaning agent during servicing. The fuel tanks are also lined with protective coatings of butyl rubber or epoxy resins. Large supersonic aircraft, such as Concorde, have rather different problems for they use the fuel as a heat transfer system to cool the wing edges. *Cladosporium* is not a serious problem, partly because the heating kills it but also because EGME is used anyway as a de-icing fluid in the fuel.

Hydraulic fluids, both oil and glycol based, are frequently attacked, causing hydraulic failures as the pipes are blocked by microbial growth or corroded. Such fluids usually contain biocides. Lubricating oils and antifreeze solutions are not normally degraded because they are heated during use of the engine, but problems may arise with machinery during storage. Cutting emulsions on lathes and other machines are seriously degraded since the emulsion is contaminated with dirt (containing nutrients) during use. Bacterial growth causes corrosion to equipment and to the surfaces being worked: furthermore unpleasant smells can be produced and there may be a risk of dermatitis in workers handling contaminated emulsions.

Microbiology of other materials

A large number of other industrial products are degraded by micro-organisms. Damage to glass is important only in optical instruments where precision-ground surfaces occur. It is particularly common in the tropics where high temperatures and humidity make it a serious problem. Fungi, of which *Aspergillus* is the most common, grow on lens coatings, adhesives, and dust particles. Their metabolic products etch the glass, or in extreme cases the mass of fungal hyphae obscures the lens. Prevention of damage depends on keeping the equipment as dry and dust-free as possible, and having tight fitting seals around the lenses. This will not, however, prevent damage under the worst conditions. The use of fungicidal vapour introduced into the air spaces of the equipment may be effective, but the best method of protection is to keep the apparatus dry.

Paints may be discoloured by fungi growing in the wood beneath, or fungi may become established in dust on the paint surface and subsequently attack the paint itself causing it to peel from the surface of the wood. High humidity and moisture on the surface are usually required. Species of *Pullularia*, *Cladosporium*, and *Phoma* have been most commonly reported. Bacteria, such as *Pseudomonas* and *Flavobacterium* may also cause degradation of paints in use. In addition anaerobic bacteria may attack the paint in storage, before application. Water-based emulsions are more liable to this attack than are oil-based paints, and deterioration may include fermentation and changes in viscosity and colour. A wide variety of antimicrobial additives are routinely used by paint manufacturers to combat this damage.

The deterioration of stone surfaces in cities with polluted air is well known and is attributable to the chemical action of such substances as sulphuric acid derived from sulphur impurities in fuels. The problem may be increased by growth of bacteria such as *Thiobacillus thio-oxidans* which oxidize sulphur compounds, thus increasing the amount of acid.

Natural rubber is subject to attack by bacteria, though synthetics like neoprene are more resistant. The hydrocarbon content may be lowered by many genera including *Actinomyces*, *Serratia*, and *Pseudomonas* and the fungi *Penicillium* and *Aspergillus* have been implicated in the spoilage of raw latex, though it is possible that they were living on impurities. Sulphur oxidizers such as *Thiobacillus thio-oxidans* may attack sulphur in vulcanized rubber, causing the production of sulphuric acid which destroys textile reinforcing and metal joints in hoses.

The damage done to plastics depends on their exact chemical nature, but in general they are resistant to decay, particularly those of very high molecular weight. The hard urea–formaldehyde plastics are not degraded, but polyvinyl chloride (PVC) may be decayed under favourable conditions. Much more common than the destruction of the plastic itself is the decay of fillers and plasticisers. Thus dibutyl phthalate (used in PVC and accounting for up to a third of the finished product) may be decayed and the plastic then becomes very brittle.

The attack on rubber and plastics may be serious in electrical insulation of underground cables and in the complex circuits of electrical equipment in the tropics. In the latter case spores may be carried in by mites and by their growth, initially on the dead mites or dust but later on the insulation, the hyphae may cause short-circuits. Miniaturized electronics and printed circuits often make it impossible to get at the fault and circuit boards or even the whole equipment must be discarded. Under humid conditions fungi may also grow on lacquers, resins and other protective coatings on coils and variable resistances (Fig. 16.4).

Fig. 16.4 Fungi growing on the lacquer coating of a variable resistance coil that had been stored under humid conditions. Note the short-circuiting to the chassis by fungal hyphae (arrow).

This emphasizes that it is the whole manufactured article which must resist decay: one degradable part may negate all the careful testing and quality control on the major components. A common source of failure in equipment is, or used to be, the adhesives. Old adhesives, based on plant and animal gums and proteins, are very easily degraded. Modern epoxy resins and other 'unnatural' chemicals, are resistant to, though not immune from, deterioration. Because of the large number of formulations now available failure of adhesives usually occurs only when they are used outside their manufacturer's specification. One common use which is however still a problem is in adhesives for decorative wall coverings, especially the vinyl 'wallpapers' which are waterproof and therefore tend to trap moisture behind them if the damproofing of the wall and the ventilation are not perfect. The adhesives are used in very small quantities and it is therefore difficult to get sufficient biocide incorporated in them to be effective over a period of years.

The problem of microbial corrosion of metals has already been briefly mentioned in connection with aircraft fuel. Micro-organisms aggravate normal chemical corrosion processes by causing both cathodic and anodic depolarization of the metal surface. Microbial colonies may set up oxygen depletion gradients from their colony margins to the centre and so hasten corrosion and cause a pitting of the surface. Corrosion is particularly serious in anaerobic environments rich in organic matter where the main organism concerned is *Desulfovibrio*. Metal pipes buried in soil are particularly susceptible and are protected by various plastic and resin coatings and by surrounding the pipes by gravel, rather than

soil, which improves aeration and contains no organic matter.

Conclusions

It is desirable, in the long run, that micro-organisms should develop enzyme systems capable of degrading modern synthetic substances such as plastics, detergents, and pesticides (Chapter 11) so that man has some convenient way of disposing of his increasing quantities of rubbish. The reverse approach of designing bio-degradable substances has been adopted in the short term and such detergents are at present in use. Micro-organisms are, however, very adaptable and there are signs that even such resistant substances as polythene can eventually be destroyed by a combination of chemical weathering and microbial attack.

The industrial uses of micro-organisms

Introduction

Micro-organisms were used in industrial processes even before their existence was known. The production of fermented beverages and vinegar, and the leavening of bread are all traditional processes which have come down to us from time immemorial (Chapter 15). The discovery of micro-organisms with their multiplicity of highly specific biochemical activities, has stimulated a steady growth of industrial fermentation processes. Perhaps the most famous of all industrial fermentations is that of acetone–butanol production by *Clostridium acetobutylicum*. During the First World War, acetone, sorely needed for the manufacture of cordite, was produced by this fermentation.

Between the wars, the industrial use of micro-organisms was extended, but it was not until the Second World War that the fermentation industries developed on the massive scale that we know today. The major stimulus was the need to produce antibiotics. Just prior to the war, some of the larger American companies had developed microbial fermentations to produce fumaric acid and this was one of the first examples of the use of neutral conditions. Hitherto, most industrial fermentations had been conducted under acid conditions where sterility was not highly critical. With the advent of neutral fermentations, there were new bio-

engineering problems to be solved in order to maintain pure cultures on a large scale. The neutral conditions and the special sterile precautions required for the production of fumaric acid were exactly those required for the mass cultivation of antibiotic-producing micro-organisms in the postwar period.

Although the industrial processes using micro-organisms are too numerous to mention, many have declined and some have disappeared altogether. With the enormous amounts of crude oil being distilled for petroleum, there are many useful hydrocarbons in other fractions unsuitable for fuels. These form the starting materials for such organic solvents as acetone and butanol which can be produced more efficiently from these sources. With our increasing consciousness of diminishing supplies of fossil fuels and their inevitable rise in price it is probable that we shall return to using such organisms as *Clostridium acetobutylicum* in an attempt to conserve supplies. Certain processes that have always been independent of petrochemicals are discussed below.

The development of fermentation processes to their present degree of efficiency was a lengthy and costly procedure. It is, therefore, not surprising that commercial firms are reluctant to disclose their manufacturing methods. Indeed, some of the methods are such closely guarded secrets that they are

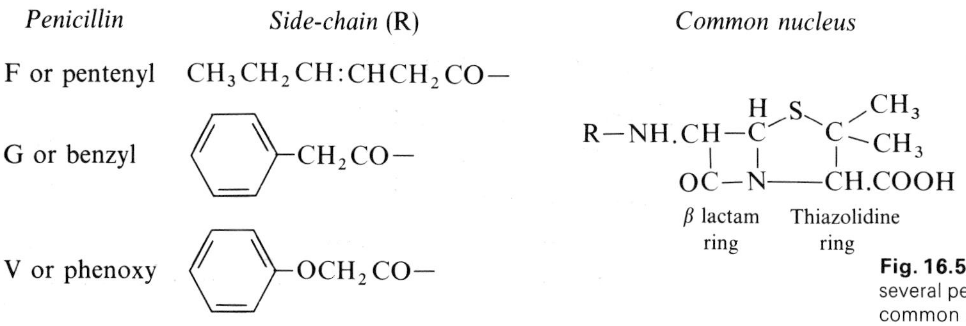

Penicillin	Side-chain (R)
F or pentenyl	$CH_3CH_2CH:CHCH_2CO-$

G or benzil

$-CH_2CO-$

V or phenoxy

$-OCH_2CO-$

Common nucleus

β lactam   Thiazolidine
ring   ring

Fig. 16.5 The structure of several penicillins showing the common nucleus.

not even patented. Thus, while the broad outlines of industrial fermentations are well known, the details often remain undisclosed.

Substances produced as a major end-product of microbial metabolism

Antibiotics

THE PENICILLINS

Fleming's original and dramatic observation of the effect of the mould *Penicillium notatum* on pathogenic bacteria was not pursued until stimulated by the pressures of war in the 1940s. This led to the founding of the antibotics industry and subsequent saving of many lives.

When *Penicillium notatum* was originally grown as a surface organism on liquid media, levels of only 10 units/ml of penicillin were produced (an international unit of benzyl penicillin is 0.5988 μg or 1670 i.u./mg). With improved media this was increased to 200 units/ml. The introduction of the more slowly metabolized lactose, instead of glucose, together with corn steep liquor as a growth medium increased yields even further. This was again improved by the use of *Penicillium chrysogenum* followed by the selection of many mutants produced artificially with the aid of such mutagens as ultra-violet light.

Nowadays, this antibiotic is produced under submerged aerated conditions in stainless steel vats of 20000 gallons capacity with the control of such variables as pH and glucose concentration. Glucose has supplanted lactose because it is now added in controlled doses giving the same slow carbohydrate metabolism. Amounts in excess of 7000 units/ml of penicillin are attained under modern fermentation conditions.

Many other organisms are known to produce penicillins, including *Aspergillus*, *Trichophyton*, *Epidermophyton*, and *Cephalosporium* species. It

was also discovered that penicillin made in Britain differed from the American product. Investigations showed that these two penicillins had the same basic molecular structure with different substituent groups. The American preparation is benzyl penicillin or penicillin G, while the British equivalent at that time was pentenyl penicillin or penicillin F. The various molecular structures are illustrated in Fig. 16.5.

Fig. 16.5 also shows penicillin V which with penicillin G are now only two members of this group of compounds manufactured by large-scale fermentation.

During the course of chemical analyses on fermentation liquors, it was found that there was often more penicillin present than could be accounted for by the measurable antimicrobial activity. This led to the discovery of 6-amino-penicillanic acid (6APA) which has an NH_2 group on the lactam ring (see Fig. 16.5) and hence no side chain. Although 6APA is biologically inactive, side chains can be substituted by acylation either chemically or enzymically. A wide range of organisms that contain penicillin acylases including *Escherichia coli* and *Alcaligenes faecalis* is now known. These will link specific side chains to 6APA producing additions to the penicillin family with novel properties. One such semi-synthetic penicillin, Cloxacillin, is illustrated in Fig. 16.6. It is stable to the action of β-lactamase (penicillinase) and to acid, the latter property allowing administration by mouth.

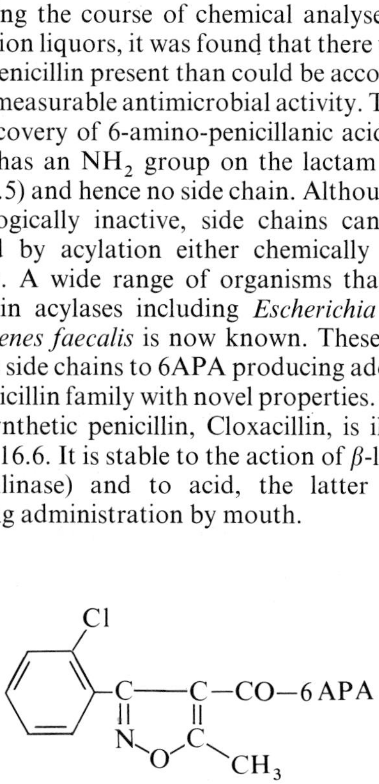

Fig. 16.6 Cloxacillin: sodium-3-ortho-chlorophenyl 5-methyl-4-isoxazyl-penicillin monohydrate.

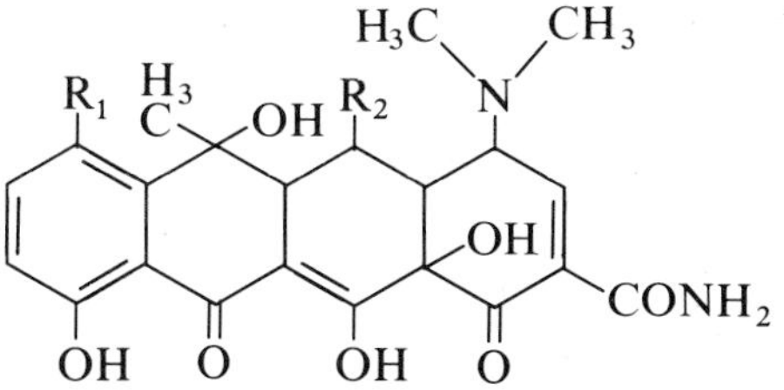

	R$_1$	R$_2$
tetracycline	H	H
chlortetracycline (Aureomycin)	Cl	H
bromotetracycline	Br	H
oxytetracycline (Terramycin)	H	OH

Fig. 16.7 The general structure of tetracyclines and variations.

THE TETRACYCLINES

These form another family of antibiotics with the general formula shown in Fig. 16.7. Oxytetracycline was first obtained from *Streptomyces rimosus* and chlortetracycline from *Streptomyces aureofaciens*. The latter is normally used for commercial production and may be grown on a variety of media including corn steep liquor, ground nut or soy bean meal. For the production of chlortetracycline, lard oil is also added while arachis oil is used in the manufacture of oxytetracycline. Sucrose or starch are the most suitable carbohydrate sources for this organism. Like many antibiotics, the tetracyclines are toxic to the organisms that produce them, and since the toxicity is thought to be due to sequestration of trace metals a plentiful supply of calcium or magnesium ions is added to overcome this effect. These organisms are favoured by a neutral pH and a temperature of 28–33°C. Aeration is essential.

OTHER ANTIBIOTICS

Space does not allow the description of other antibiotics in detail, but Fig. 16.8 shows the sources and structures of two of the more important ones.

It should be remembered that although all the antibiotics discussed here are derived from fungi or

Name	*Source*	*Structure*
Griseofulvin	*Penicillium patulum*	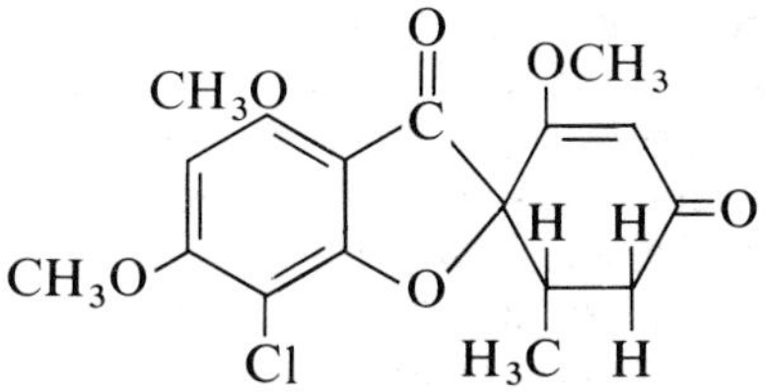
Streptomycin	*Streptomyces griseus*	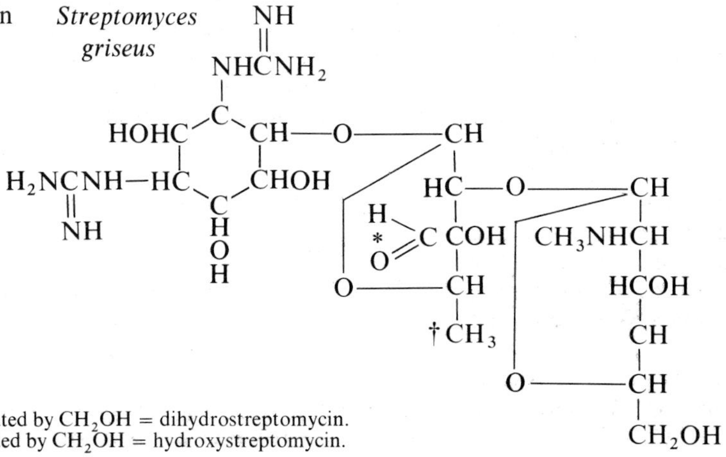

*CHO substituted by CH$_2$OH = dihydrostreptomycin.
†CH$_3$ substituted by CH$_2$OH = hydroxystreptomycin.

Fig. 16.8 The source and structure of some antibiotics.

actinomycetes, others such as subtilin and bacitracin are produced by bacteria. The usefulness of the latter is limited by their greater toxicity.

DEXTRAN

Dextran is a polysaccharide composed of glucose molecules and may be found in slimy masses wherever sucrose is used. Sugar refineries and ham-curing factories are prone to trouble from dextran-forming bacteria. Dextran can also block pipelines, make floors slippery, render food uneatable and beverages undrinkable. It has, however, an important use in blood transfusion where it is employed as a blood volume expander. Its application in this way is possible because it is relatively inert, causing neither pyrogenic (i.e. raised body temperatures) nor allergic reactions (p. 332) and will remain sufficiently long in the circulation to allow time for protein renewal in the blood plasma. It is stable to autoclaving and can be stored for long periods at room temperature without deterioration. One particular advantage is that it can be prepared with a known mean molecular weight. In Britain, for instance, a mean molecular weight of 110 000 is favoured, whereas in the USA 75 000 is the preferred molecular size.

There is only one bacterial strain used in Britain for dextran production, namely, the heterofermentative lactic coccus *Leuconostoc mesenteroides*. This is normally maintained in sucrose agar and is gradually worked up through intermediate volumes to a 1000 gallon (4500 litre) scale for commercial production. The medium used is sucrose plus yeast extract supplemented with ammonium and potassium phosphates. Incubation temperature is 23°C: aeration is unnecessary since the organism is micro-aerophilic. Additions of alkali are made to keep the pH above 5.0 because below this point dextran production is reduced.

The overall reaction for dextran production is as follows:

$$n \text{ sucrose} \xrightarrow{\text{dextran/sucrase}} (\text{glucose}) n + n \text{ fructose}$$

The fructose is subsequently used as an energy source by the bacteria, while the glucose units are joined together into chains mainly with 1–6 linkages but sometimes 1–4 and occasionally 1–3. Branched chains vary in amount according to the particular strain of bacterium used. Hence *L. mesenteroides* is suitable for commercial production because it gives 95 per cent straight-chain molecules, these being the most acceptable for clinical use.

Dextran is removed from the culture fluid by acetone precipitation, and this so-called 'native dextran' is then ready for the final stage of degrading to the correct molecular size. The methods available for this molecular sizing operation are as follows:

1 Acid hydrolysis. Dextran is heated with acid, and the average molecular size is checked by viscosity measurement. This method has the advantage that it can be performed without separating the dextran from the fermentation liquor.

2 Heating under pressure. Degradation may be brought about by heating the dextran under pressure at 160°C in the presence of sodium sulphite to prevent oxidation and calcium carbonate, to keep the pH at 7.

3 Enzymic degradation. Various micro-organisms contain dextranases, the one most frequently used being *Cellvibrio fulva*.

4 Mechanical breakdown. High frequency waves emitted by a piezo-electric quartz crystal when stimulated electrically will break down high molecular chains and give a product of considerable homogeneity. As with method (1), viscosity is the main criterion used for estimating molecular size.

LACTIC ACID

This is one of several acids produced microbiologically on a commercial scale. Others include citric, fumaric, itaconic, and gluconic acids, some of these being produced from moulds which require high aeration rates in the submerged conditions employed. Lactic acid, in contrast, is produced by micro-aerophilic bacteria belonging to the genus *Lactobacillus*. This acid is an odourless, colourless liquid having an acid flavour. It is used in a variety of ways, finding applications in the drinks industry as a flavouring, in the preservation of food as an adjunct or substitute for vinegar, and in the form of calcium lactate as a convenient means of getting calcium into the body. Other uses are in the plastics and leather industries. A number of grades are produced varying in purity and concentration (22–85%) according to the purpose for which they are required. Broadly, the grades may be classified as pharmaceutical, edible, and technical. Irrespective of grades, however, all types are made by the same process, and it is the rigour of the purification processes that decides the final grade.

The most common organisms used in lactic acid production are the homofermentative lactobacilli. In these, as with yeasts, the Embden–Meyerhof–Parnas pathways is a main energy-gaining mechanism. The bacterial pathway, however, differs from that of the yeasts because there is no pyruvic decarboxylase present, therefore no loss of CO_2 and no subsequent acceptance of hydrogen by acetaldehyde. Instead, pyruvic acid performs this role in these bacteria and, by accepting hydrogen, is reduced to lactic acid as shown below.

$$CH_3.CO.COOH + 2\,NADH \rightarrow$$

Pyruvic acid $\quad CH_3.CH.OH.COOH + 2\,NAD^+$

Lactic acid

This type of lactobacillus produces a single end-product, namely, racemic lactic acid. Recent work has shown, however, that by varying temperature and type of medium, the relative proportions of the two optical isomers can be varied.

In the commercial production of lactic acid, a variety of substrates may be used, such as hydrolysed starch with barley, whey or molasses supplemented with nitrogenous material. The purer the starting material the less post-fermentation processes are required to produce pure lactic acid. The starting concentration of sugar is about 15%, and this is reduced to 0.1% in 4–6 days. Calcium carbonate is added to the mix to prevent a fall-off in productivity caused by a drop in pH. The organisms most favoured for lactic acid production are the high temperature group of homofermenters that includes *Lactobacillus delbrueckii*. This allows a working temperature of 50°C which has the advantage of suppressing possible contaminants with lower optimum growth temperatures. To maintain this unusually high temperature in vats that may be of 135 000 litres capacity, heating coils are fitted and the medium is stirred to ensure uniform temperature and composition. Aeration is not applied, since this is an anaerobic process. As with all large scale batch fermentations, it is necessary to build up the inoculum by passing it through a series of tanks of increasing volume until finally the size is sufficient to initiate a rapid fermentation in the large vats used for commercial production.

When fermentation is complete, the yield is about 85% of the fermentable hexose. The liquor is then heated to 80°C to kill the organisms and coagulate the protein. The end-product, present as the calcium salt, is acidified with sulphuric acid which frees the lactic acid and precipitates the calcium as insoluble sulphate. Heavy metals, at one time removed by treatment with sodium sulphide, are now taken out with ion-exchange resins. Lactic acid is decolorized with powdered charcoal and is extracted from other constituents by the use of isopropyl ether. The acid is then extracted from the ether by using water.

Both of these processes can be operated continuously in counter-current towers. For the very pure grades, such as those required in the manufacture of plastics, a lactic acid ester is prepared, the impurities distilled off, and the lactic acid and alcohol recovered.

AMINO ACIDS

These substances are used for various clinical purposes, such as postoperative therapy, as supplements to low protein diets, and as food flavourings. A common ingredient of certain dehydrated foods such as soups is monosodium glutamate. Amino acids can be prepared by chemical synthesis, but this has the disadvantage of producing both optical isomers. In contrast, biosynthesis produces the L-isomer which is the form found in most biological systems. Sometimes, a combination of chemical synthesis and biosynthesis is employed as in the production of lysine from diaminopimelic acid. The production of the diaminopimelic acid is purely chemical, while the final step to lysine is biological.

The main amino acids prepared by fermentation are glutamic acid and lysine (Table 16.1). In the preparation of the former, the carbohydrate (usually glucose or sucrose) is present at a concentration of 10–20%. Urea or NH_4^+ salts provide the nitrogen; magnesium, manganese, and iron, the metallic constituents, and such substances as pelargonic acid are added as stimulators. The pH is maintained between 7 and 8 by the addition of urea or ammonia gas. The concentration of biotin is most important because it controls the yield.

The organism commonly used in the manufacture of glutamic acid is a *Corynebacterium* sp., but strains of the genera *Brevibacterium*, *Micrococcus*, and *Microbacterium* have also been used. Removal of glutamic acid from the growth medium is easy because it is insoluble in acid conditions. High-yielding media are treated as follows: the organisms are removed, the filtrate concentrated and acidified to pH 3.2, and the glutamic acid, thus precipitated, is recovered by filtration. Lower yielding media are treated with a suitable ion-exchange resin, and this method not only separates but also purifies the glutamic acid.

Lysine preparation is similar to that of glutamic acid, the usual organism being a mutant of *Micrococcus glutamicus*. Biotin is less critical than for the production of glutamic acid, whereas the concentrations of homoserine, threonine, and methionine have a profound effect upon yield. Lysine is more difficult to separate from the growth medium

Table 16.1 Organisms and methods for amino acid production

Amino acid	Method	Organisms
Aspartic acid	Produced by transamination from oxalacetate or fumaric acid and ammonia	*E. coli* *S. marcescens* *Bact. succinum* *Bacillus* spp.
L-homoserine	This organism produces L-lysine with L-homoserine. Varying the concentration of biotin changes the ratio of the yields of glutamate, lysine and homoserine	*Micrococcus glutamicus*
L-threonine	Made by a strain which has a growth requirement for diaminopimelic acid or one which also has a requirement for methionine. Mannitol or sorbitol instead of glucose increase yield	*E. coli*
L-methionine	This organism is cultured in a medium containing methyl mercapto-α-hydroxy butyric acid	*Pseudomonas* sp.
L-valine	Strains (a), (b), (c) and (d) all accumulate L-valine in the medium. Strain (e) is an auxotrophic mutant requiring L-leucine or L-isoleucine. Biotin favours L-valine production	(a) *Aerobacter cloacae* (b) *Paracolabactrum coliforme* (c) *E. coli* (d) *Enterobacter aerogenes* (e) *Micrococcus glutamicus*
L-ornithine	This organism accumulates ornithine and requires either L-arginine or L-citrulline as a precursor	*Micrococcus glutamicus*
DL-alanine	This is a transamination reaction involving pyruvic and glutamic acids and a racemization	*Corynebacterium gelatinosum*
L-phenylalanine	Strains (a) and (b) require tyrosine as a precursor. Yield is affected by biotin and may be stimulated by the presence of shikimic, phenyl pyruvic, and D-quinic acids	(a) *Micrococcus glutamicus* (b) *E. coli*
L-tyrosine	Phenylalanine is required to accumulate L-tyrosine. Shikimic, *p*-hydroxyphenyl pyruvic, and D-quinic acids or tryptophan stimulate L-tyrosine production	*Micrococcus glutamicus*
L-tryptophan	Strains (a), (b) and (c) will convert indole in submerged culture to L-tryptophan. Strains (d) and (e) will convert 3 indole pyruvic acid to L-tryptophan by reductive deamination. Strain (f) produces L-tryptophan from anthranilic acid	(a) *Claviceps purpurea* (b) *Rhizopus oryzae* (c) *Ustilago avenae* (d) *Micrococcus* sp. (e) *Serratia* sp. (f) *Hansenula anomala*

than glutamic acid, but ion-exchange resins offer the best means of extraction. In the final stages, it is usually obtained as lysine monohydrochloride and the pure amino acid produced by recrystallization.

Many other amino acids can be made in smaller quantities as summarized in Table 16.1.

The use of micro-organisms to reshape molecules

Organisms used for this purpose have the ability to utilize and change substituent groups without altering the basic molecular structure. Those used with steroids fall into this category.

Steroids are compounds of extreme importance in the economy of the animal body. Steroid hormones are secreted by the adrenal cortex and are responsible for the regulation of carbohydrate and mineral metabolism. Other well-known examples of steroid-secreting organs are the ovaries and testes that regulate the animal's sexual activities. Re-

cently, steroid compounds have been administered for therapeutic reasons. Steroids are administered as anti-inflammatory agents, menstrual cycle regulators, in contraceptive pills and as immune reaction suppressors after organ transplants.

Steroids are complicated molecules with the basic structure as illustrated in Fig. 16.9.

To synthesize a steroid from suitable starting material takes about 32 steps and even to move an

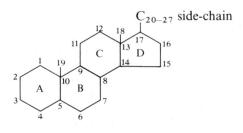

Fig. 16.9 The basic steroid structure.

oxygen from C_{11} to C_{12} may take up to 12 steps. These lengthy chemical syntheses are not practicable for producing steroids on a commercial scale and new methods were therefore sought. The first microbiological steroid transformation was made in 1937 by Mamoli and Vercellone, and since that time a whole range of steroids have been changed in chemical structure by an equally wide range of micro-organisms. The advantage of the microbiological method is that highly specific changes in the steroid molecule can be effected in a single step. The known range of chemical changes includes hydroxylation, dehydrogenations, etc. The organisms concerned include many moulds, particularly species of Zygomycetes (p. 196) and Deuteromycetes (p. 199) and Actinomycetes and a few species of yeasts and bacteria.

Although the first microbial transformation of a steroid was made as long ago as 1937, one factor that discouraged such processes on a large industrial scale was the relative insolubility of these complex compounds. One of the factors that stimulated industrial-scale steroid transformations was the discovery that oxygen at C_{11} position imparted important anti-inflammatory properties. This gave rise to a steady demand for such compounds as cortisone and stimulated new ways of manufacturing this and related compounds on a large scale. Thus, one of the early discoveries was the transformation of compound S (11 deoxycortisol) to cortisol and smaller amounts of cortisone by the mould *Cunninghamella blakesleeana*.

Such transformations as these are now commonplace, usually being performed on a 135 000 litre scale. A suitable growth medium such as glucose, molasses, or corn steep liquor, with yeast extract added as a nitrogen source are provided for the transforming organism. The large-scale tanks are agitated and aerated, the organisms being allowed to reach maximum growth. Towards the end of the growth phase, the appropriate precursor is added in a non-toxic hydrophilic organic solvent such as ethanol, acetone, or propylene glycol. The chemical change can be followed chromatographically, and when complete the newly formed steroid is extracted with organic solvents. Although, owing to the insolubility of steroids, the total weight per batch of culture is small, losses are also small and a 95% recovery is quite common.

There seems little doubt that the transformation of steroid compounds by micro-organisms is a process that will not easily be supplanted by chemical methods. The ease with which the substituent groupings of this complicated molecule are changed

in a way that cannot be achieved by chemical means is a factor that appeals to manufacturers. In spite of the large range of transforming organisms known, little has been done to select high yielding mutants as has been successfully accomplished in the antibiotics industry.

Mixed culture fermentations

There are a number of industrial fermentation processes that use mixed cultures. These include the production of fermented beverages (p. 349), vinegar-making, and ensilage of grass for cattle feed.

The manufacture of vinegar

This is prepared as the result of two microbiological processes, namely: alcoholic fermentation and acetification mediated respectively by yeast and by bacteria of the genus *Acetobacter*. It is interesting to note that different countries have a preference for certain substrates. In Britain, malt wort is commonly used, whereas in France and other continental countries grape must is the usual starting material, whilst in the USA apple juice is the preferred substrate. Since alcoholic fermentation is discussed elsewhere (p. 351) it is the second phase that will be described in some detail.

All *Acetobacter* species are aerobic and usually found in the wild state growing as a pellicle at the air/liquid interface of alcoholic liquors. This is utilized in the old-fashioned French system called the Orleans process. In this, acetic acid bacteria are allowed to grow on the surface of wine in partly filled casks. When the wine has been slowly converted to vinegar, part is drawn off at the base of the cask, the original volume being re-established by replenishing with new wine. Although this is a slow process, it produces vinegar of the highest quality.

Perhaps the most widely used method for vinegar production is the so-called quick or trickling process. The apparatus consists of a large wooden tower packed with beechwood shavings, birch twigs or other non-compressible material A sparge at the top of the tower distributes alcoholic liquor over the packing, which provides a support for the acetic acid bacteria. The liquor gradually acetifies as it trickles down the tower to a reservoir at the base from which it is recycled until the process is complete. Acetification being exothermic, the heat produced causes the tower to act as a chimney drawing cold air in at the base and discharging warm at the top. A series of vents around the base of the tower can be opened or closed to regulate air flow, while condensers at the exit trap all volatiles entrained in the warm air. This method, although slow, is simple in operation and requires the minimum of attention and mechanical maintenance.

The third method of vinegar manufacture is by cultivating acetic acid bacteria submerged in a suitable substrate. Constant and intense aeration is required to keep the organisms in a rapid phase of acetification and provision must be made for efficient dissipation of heat. This is usually effected by the presence of cooling coils in the acetifier that are supplied with coolant by way of a solenoid valve actuated by the temperature of the vinegar. The shape and size of vessels used for submerged acetification may vary, but the commonest is a tall narrow vat which will accommodate froth that accumulates on the surface of the liquor. These vessels may be of wood or stainless steel. The submerged system is not only used as a batch culture but also continuously when a rotating stirrer may also be incorporated to induce more thorough mixing and aeration. Such systems have the advantage of producing large quantities of vinegar in a relatively short time in a minimum of vat space. They do require, however, complex control mechanisms and suffer from the disadvantage that if the air supply is interrupted, even momentarily, the culture may take several days to recover.

Dilute acetic acid produced by synthetic means is sold as a substitute for vinegar. In Britain, at least, the condiment may not be called vinegar unless it has been produced by one of the microbiological processes here described. Furthermore, it is not recognized as vinegar unless it contains a minimum of 4 % of volatile acid.

Ensilage

The process of ensilage is an alternative to the drying of green crops for winter feeding of cattle and sheep. The resulting silage is stored in a succulent form. The process involves the controlled fermentation of green crops such as grass, maize and occasionally sugar-beet tops or kale, and has the advantage of being largely independent of weather conditions. Processing can thus be done at the optimum time, the most nutritious product being achieved with younger, rather than mature crops. Success depends on adequate temperature control and exclusion of oxygen. The green crop is packed and compressed in containers (silos), in pits in the ground or in stacks above-ground enclosed in polythene. Molasses is sometimes added to promote fermentation and increase palatability. As the fermentation proceeds the temperature rises and the product becomes more acid. The acidity prevents the growth of many putrefactive organisms and, as the oxygen is used up, moulds and strictly aerobic bacteria are also inhibited. Only facultative and strict anaerobes continue to grow. At first, species of *Escherichia* and *Enterobacter* predominate, but as the pH falls these are replaced by lactobacilli and streptococci, which ferment the plant material to produce lactic, butyric, propionic, and acetic acids and esters, giving flavours much relished by cattle. Active fermentation continues for three or four weeks, after which the temperature falls and a fairly stable product results.

Fermentation for the production of micro-organisms

Bakers' yeast

All the processes so far considered have been concerned with the production of an end-product from a substrate. The conversion of part of the substrate to biomass is incidental in all these processes, but once the optimum amount of end-product has been achieved the organisms are merely a waste product. In the preparation of bakers' yeast, the roles are reversed, for it is the aim of the producer to achieve the maximum quantity of micro-organisms from the substrate.

Pasteur in 1876 first showed that aerating a yeast culture would improve growth, and it is this principle that is used today for producing panary yeast. The fermentation is carried out as a batch culture in vats of 50 000–200 000 litres capacity. The yeast species used is *Saccharomyces cerevisiae* and this is grown in a medium consisting mainly of beet or cane molasses. This is often deficient in biotin or pantothenic acid, and it may be necessary to supplement the quantities naturally present. Other prefermentation treatments of the molasses involve its dilution, treatment with sulphuric acid and heating to remove sludge protein. Inocula are built up initially from single cells until a sufficient quantity has been prepared to inoculate the vats used for commercial production. Various yeasts have been bred to fulfil the requirements of modern baking, and breeding programmes are in progress to produce yeasts with even more desirable characteristics.

In the large-scale production of yeasts, various technical factors must be considered. The normal sequence of events in alcoholic fermentation is a rapid multiplication of yeast cells with a consequent depletion of oxygen. This halts cell multiplication and the yeasts switch to their anaerobic method of energy-gaining with the consequent production of ethanol and carbon dioxide. It is the aim of the yeast manufacturer to keep the cells in their aerobic multiplication phase, thus producing maximum biomass with minimum alcohol. About 12% of oxygen is utilized in aeration so that about 40 times the absolute requirement must be supplied which is a total of 330000 cu. ft. (9900 cu. m.) of air per ton of commercial yeast. The high rate of aeration presents problems, for example, the amount of energy produced per unit of sugar metabolized aerobically is about 1384 Kcal compared with 177 Kcal for the same quantity consumed anaerobically. Some of this energy is lost as heat and it is necessary to maintain steady conditions and a working temperature of 30°C.

A further complication of rich medium combined with a high rate of aeration is that the yeasts may grow so rapidly that they outstrip the available oxygen. In such a situation, oxygen starvation will occur and instead of the yeast producing $CO_2 + H_2O$, as they do under aerobic conditions, they begin to produce CO_2 and ethanol, which is undesirable. To limit yeast growth and thus avoid oxygen starvation, the nutrients are controlled. At first, they are kept at a low level and as the cells increase exponentially they are added to keep pace with yeast growth. Thus, at any one time there is only a small excess of nutrients over the number of yeasts available to use them. Another variable requiring control is pH. Since nitrogen is often added in the form of $(NH_4)_2SO_4$, the removal of the ammonium ions tends to create acid conditions which are adjusted by periodical additions of alkali to keep the pH in the range 4.0–5.5.

At the end of yeast growth, when the nutrients are exhausted, aeration is continued for 30–60 minutes to 'ripen' the yeast. It has been found that this part of the process stabilizes the yeast and imparts better keeping qualities. After removal from the vats, the yeast is centrifuged and washed several times then chilled to 2–4°C. Water is removed by vacuum filter dehydrators and the yeast is packaged. Packing material can be of importance in the preservation of bakers' yeast, and contamination can only be prevented by strict controls throughout the process (p. 351).

Protein from methyl alcohol

It has been known for some time that certain microorganisms can utilize various hydrocarbons as an energy source providing water and nitrogenous nutrients are also present. This has been carried a stage further by growing the micro-organism *Methophilus methylatrophus* on methyl alcohol derived from oil or natural gas. By the addition of water, ammonia and mineral salts it is possible to produce large quantities of protein from this substrate. Such proteins are used to produce a milk substitute for calves, thus releasing more cow's milk for human consumption. This process is operated by a large chemical concern and it is expected that about 75000 tonnes of the product will be made in 1979. Although this process can only have a finite life it does seem less wasteful than 'burning off' the natural gas as seems to be practised in various Middle Eastern oil producing countries.

Formerly the use of cheap oil to produce fuel and petrochemicals caused such unfavourable economic competition with the commercial production of microbial products that many of the processes have been abandoned. In a very short space of time the insatiable world demand for fuel to feed the internal combustion engine has reversed this trend and thoughts are returning to the use of micro-organisms to produce some of the basic substances that modern technology demands. Indeed, certain countries, notably Brazil, are using ethanol produced from the fermentation of plant material as a substitute fuel. All the processes mentioned in this chapter plus many others now being considered come under the heading of biotechnology.

One area of rapid development in biotechnology has come about by the use of genetic engineering. These techniques have been used to introduce genomes into micro-organisms to induce them to synthesize substances that they would not otherwise produce. Considerable progress has been made in these techniques in the U.S.A., West Germany and Japan. Already substances like interferon and insulin that are either too complex or too costly to synthesize chemically are being produced by suitably manipulated micro-organisms. These techniques offer limitless possibilities for the production of a whole range of organic substances not hitherto produced by micro-organisms, which should help to alleviate many of the shortages that will be caused by the disappearance of our oil supplies.

Books and articles for reference and further study

Spoilage

CARTWRIGHT, K. ST. G. and FINDLAY, W. P. K. (1958). *Decay of Timber and its Prevention*. 2nd Edn., H.M.S.O., London.

CHATER, K. W. A. and SOMERVILLE, H. J. (eds) (1978). *The oil industry and microbial ecosystems*. Heydon, London, 250 pp.

DAVIS, J. B. (1967). *Petroleum Microbiology*. Elsevier, New York, 605 pp.

GILBERT, R. J. and LOVELOCK, D. W., (eds.) (1975). *Microbial Aspects of the Deterioration of Materials*. Soc. Appl. Bact. Technical Series No. 9. Academic Press, London, New York and San Francisco, 261 pp.

LEVY, J. (1965). The soft rot fungi: their mode of action and significance in the degradation of wood. *Adv. Bot. Res., 2*, 323–357.

LIESE, W., (ed.) (1975). *Biological Transformation of Wood by Micro-organisms*. Springer-Verlag, Berlin, Heidelberg and New York, 203 pp.

MILLER, J. D. A., (ed.) (1971). *Microbial Aspects of Metallurgy*. Medical and Technical Publ. Co. Aylesbury, England.

OXLEY, T. A., ALLSOPP, D. and BECKER, G. (eds) (1980). *Biodeterioration*. Pitman, London, 375 pp.

SKINNER, C. E. (1970). Microbial problems in the paint industry. *Paint, Oil and Colour Journal*, January 1970. 177–180.

TURNER, J. N. (1967). *The Microbiology of Fabricated Materials*. J. & A. Churchill, London.

Uses

CARR, J. G. (1968). *Biological Principles in Fermentation*. Heinemann Educational, London, 97 pp.

IIZUKA, H. and NAITO, A. (1967). *Microbial Transformation of Steroids and Alkaloids*. University of Tokyo Press and University Park Press, State College, Pennsylvania.

Process Biochemistry (1967). Volumes 1 et seq.

RAINBOW, C. and ROSE, A. H. (1963). *Biochemistry of Industrial Micro-organisms*. Academic Press, London and New York, 708 pp.

ROSE, A. H. (1961). *Industrial Microbiology*, Butterworths, London, 286 pp.

Index

Principal references are in **bold** type, those to illustrations are in *italics*.